SPATIAL ORGANIZATION

SPATIAL

RONALD ABLER
The Pennsylvania State University

JOHN S. ADAMS
University of Minnesota

PETER GOULD
The Pennsylvania State University

ORGANIZATION

The

Geographer's View

of the World

PRENTICE-HALL, INC., Englewood Cliffs, New Jersey

SPATIAL ORGANIZATION:
The Geographer's View of the World
by Ronald Abler, John S. Adams, and Peter Gould

PRENTICE-HALL INTERNATIONAL, INC., London
PRENTICE-HALL OF AUSTRALIA PTY. LTD., Sydney
PRENTICE-HALL OF CANADA LTD., Toronto
PRENTICE-HALL OF INDIA PRIVATE LIMITED, New Delhi
PRENTICE-HALL OF JAPAN, INC., Tokyo

To the generation before us

CONTENTS

PART **II** *Measurement, Relationship, and Classification*

PART III *Location and*
Spatial Interaction

CHAPTER 7

THE BASES FOR
SPATIAL INTERACTION   193

CHAPTER 8

MOVEMENT AND
TRANSPORT SYSTEMS   236

CHAPTER **9**

LOCATING HUMAN ACTIVITIES 298

CHAPTER **10**

LOCATION AND
THE USE OF LAND 340

PART **IV** *Spatial Diffusion Processes*

CHAPTER **11**

SPATIAL DIFFUSION:
MESHING SPACE AND TIME   389

PART **V** *Spatial Organization and the Decision Process*

CHAPTER **12**

INDIVIDUAL SPATIAL DECISIONS
IN A NORMATIVE FRAMEWORK   455

CHAPTER **13**

INDIVIDUAL SPATIAL DECISIONS
IN A DESCRIPTIVE FRAMEWORK   491

PREFACE

Spatial Organization is an outgrowth of our attempts to introduce students to the way geographers think about the world. Fortunately or unfortunately, depending upon your viewpoint, geographers do not share a single, unified way of thinking about man's spatial experience. Thus, our essay on the basic principles of geography reflects, as any similar exposition must, the philosophical convictions the three of us share.

Underlying our approach to the analysis of human activities in terrestrial space is our belief that human geography is a social and behavioral science. We think the principles which govern human spatial behavior are generally applicable all over the world. Obviously, some elements of human spatial organization are attributable to unique factors, but we feel that what is *common* in the ways people perceive and organize space is more important. Thus, in presenting the principles of geography, we chose to concentrate on the circularly causal relationship between spatial structure and spatial pro-

cess. People generate spatial processes in order to satisfy their needs and desires, and these processes create spatial structures which in turn influence and modify geographical processes.

Understanding the structural and processual consequences of human spatial behavior has now become an absolutely necessary condition of mankind's future welfare. No serious problem exists the solution of which does not require a comprehensive knowledge of the ways man perceives, values, and uses space and places. The ability to explain and predict human spatial behavior and to modify human spatial organization is now quite literally vital. We shall not survive as a species if we cannot predict and modify the world in this fashion.

Spatial Organization is designed to increase the reader's understanding of spatial structures and processes and the ways geographers think about them. Following the Socratic dictum that the unexamined life is not worth living, we feel that the unexamined geography is not worth learning or practicing. For that reason, Part I is devoted to a description of science and geography as ways of thinking about the world. Understanding a particular world view implies some knowledge of the origins of that thought system. The necessary foundation is provided for science in Chapters 1 and 2, and for geography in Chapter 3.

All sciences share a fundamental set of goals and strategies. These are briefly described in Part I, but a more intensive look at the fundamentals of the geographic viewpoint appears in Part II. Chapters 4 through 6 are devoted to measurement, the recognition of relationships between and among phenomena in the world, and the way in which scientists group things into classes. Accurate measurement, the identification of relationships, and classification are all means of molding the world into simpler and more readily understandable forms than exist in the complex stuff of unstructured reality. We can think of these three procedures as filters through which we force our immediate experience. They allow us to ask more penetrating spatial questions and devise more satisfying answers.

Part III goes directly to the heart of all geographical thought—the nature of the circularly causal relationship between spatial structure and process. All spatial processes have a common origin in the areal differentiation of the earth and in the spatial interaction that differences from place to place evoke. Chapter 7 explains why things move through space, whereas Chapter 8 describes the structures of the movement systems which spatial interaction brings into existence. These two chapters provide an understanding of the kind of terrestrial space in which men live today. With this background, we consider, in Chapters 9 and 10, spatial and locational behavior from two basic points of view: first, that of choosing locations for human activities, and second, that of choosing the best activities for given locations or places. Choosing places for things and things for places are the linked processes which produce the geographical organization we see in the landscape around us and on the maps we make of larger systems.

Part IV summarizes the current state of our knowledge of geographical diffusion processes at local, national, and international scales. Space and time are the two most fundamental dimensions of human existence. Geographers have always been more or less historical, and historical geographers have given us some of the best accounts of how spatial structures developed because they devoted attention to both time and space as explanatory variables. Recently, geographers have made important advances toward better theoretical understanding of the ways things and ideas move through space and time simultaneously.

In Part V we focus on individual locational decisions, emphasizing the utility of models—the abstractions and simplifications of reality which we derive from our theories and which make the world intelligible. The spatial abstractions which geography produces and uses are the true kernels of the discipline, because our ability to explain, predict, and thus understand the world can only be as good as our theories and models. Chapter 12 describes *normative* models of human spatial behavior, those which make some assumptions about the way people *could* or *should* make locational decisions. Chapter 13 discusses the value of models which describe the ways decisions are *actually* made. When we use *descriptive* models, we accept the disadvantages of greater complexity and decreased elegance in return for the advantages of greater congruity with reality.

The chapters on spatial decision-making indicate how limited our useful knowledge of human spatial behavior still is, and the point is emphasized in the final chapters of the book. Chapter 14 considers what we know and do not know about locating facilities and activities in order to maximize social benefits and minimize social costs. Chapter 15 provides some indications of the kinds of locational and spatial problems geographers will have to solve in the future if humanity is to thrive, or even survive. Given the size and intensity of the world geographical system which man has now created, problems must be anticipated and solved in advance if disaster is to be avoided.

Throughout the book, our purpose is to present the *strategies* geographers can use to solve the problems that distributions of people and their activities create. *Spatial Organization* is about problem-solving tactics, but we have dealt very little, and in most cases not at all, with the actual intricacies of research methods. Our concern with techniques of spatial analysis is confined to the rather general ways they can be used to solve problems. We must all develop a sharply honed sensitivity to the ends to which techniques of spatial analysis may profitably and validly be devoted. Thus, our focus has been on the utility and goals of geographical inquiry.

Books like this are actually written by the hundreds of people whose ideas we have selected, modified, and brought together at this particular point in space and time. Certainly we are greatly indebted to those who produced the important advances in theoretical human geography of the last two decades. They have made us look with new eyes at old and persistent problems. More specifically, we extend special thanks to Allan Rodgers. He built the exciting, stimulating department at Penn State which brought the three of us together. Moreover, his genial scepticism that we would ever finish the book provided just the goad we needed when the going was rough and spirits flagged. Tony Williams participated in the design and teaching of the course which produced *Spatial Organization*, and in many ways helped us decide the form the book would take. Finally, our students have been persistent and (usually) constructive critics and sources of stimulation. It is to them we owe the greatest debt of all.

RONALD ABLER
JOHN S. ADAMS
PETER GOULD

P.S. The title we chose originally was *The Intelligent Student's Guide to Geography*, but most of the mirthless souls who reviewed the manuscript thought it was a bit much.

TABLE OF
EQUIVALENT MEASURES

Almost all mankind now lives in a single world system, and because most of the world's people have found the Metric System more useful than alternative measurement systems, all measurements in this book are in metric form. The table below will facilitate conversion to the United States Customary System when such conversions are necessary. For a discussion of metrification, see Lord Ritchie-Calder, "Conversion to the Metric System," *Scientific American* (July, 1970), 17–25.

UNITS OF LENGTH

1 millimeter = 0.039 37 inch	1 inch = 25.40 millimeters
1 centimeter = 0.393 7 inch	= 2.54 centimeters
1 meter = 39.369 6 inches	1 foot = 30.48 centimeters
= 3.280 8 feet	= 0.304 8 meter
= 1.093 6 yards	1 yard = 91.44 centimeters
1 kilometer = 3,280.8 feet	= 0.914 4 meter
= 1,093.6 yards	1 mile = 1,609 meters
= 0.621 4 mile	= 1.609 kilometers

UNITS OF AREA

1 square centimeter	= 0.155 square inch
	= 0.001 076 square foot
1 square meter	= 1,550.003 square inches
	= 10.763 square feet
	= 1.195 99 square yards
1 hectare	= 107,639.1 square feet
	= 11,959.90 square yards
	= 2.471 acres
	= 0.003 861 square mile

1 square inch	= 6.451 6 square centimeters
	= 0.006 451 6 square meter
1 square foot	= 929.030 square centimeters
	= 0.092 9 square meter
1 square yard	= 8,361.274 square centimeters
	= 0.836 1 square meter
1 acre	= 4,046.856 square meters
	= 0.404 69 hectare
1 square mile	= 2,589,988.110 square meters
	= 258.998 811 hectares

UNITS OF VOLUME

1 milliliter	= 0.061 023 cubic inch
1 liter	= 61.023 74 cubic inches
	= 0.035 315 cubic foot
	= 0.001 308 cubic yard
1 cubic meter	= 35.314 67 cubic feet
	= 1.307 951 cubic yards

1 cubic inch	= 16.387 064 milliliters
	= 0.016 387 liter
1 cubic foot	= 28.316 847 liters
	= 0.037 037 cubic yard
1 cubic yard	= 764.554 858 liters
	= 0.764 555 cubic meter

UNITS OF MASS

1 milligram	= 0.000 035 3 ounce
1 gram	= 0.035 3 ounce
1 kilogram	= 35.274 ounces
	= 2.2046 pounds
1 metric ton	= 2,204.6 pounds

1 ounce	= 0.028 35 milligram
	= 28.349 5 grams
1 pound	= 453.592 4 grams
	= 0.453 6 kilogram
1 ton	= 907.184 7 kilograms
	= 0.907 2 metric ton

UNITS OF LIQUID MEASURE

1 milliliter	= 0.033 8 ounce
	= 0.002 pint
1 liter	= 33.814 ounces
	= 2.113 38 pints
	= 1.056 688 quarts
	= 0.264 172 gallon
1 cubic meter	= 264.172 gallons

1 liquid ounce	= 29.573 5 milliliters
	= 0.029 6 liter
1 pint	= 473.176 5 milliliters
	= 0.473 18 liter
4/5 quart	= 0.757 082 liter
1 quart	= 0.946 353 liter
1 gallon	= 3.785 liters

UNITS OF DRY MEASURE

1 liter	= 1.816 16 dry pints
	= 0.908 08 dry quart
	= 0.113 5 peck
	= 0.028 378 bushel
1 dekaliter	= 18.161 66 dry pints
	= 9.080 8 dry quarts
	= 1.135 1 pecks
	= 0.283 775 bushel
1 cubic meter	= 113.510 4 pecks
	= 28.377 59 bushels

1 dry pint	= 0.550 6 liter
	= 0.055 1 dekaliter
1 dry quart	= 1.101 2 liters
	= 0.110 1 dekaliter
1 peck	= 8.809 8 liters
	= 0.880 98 dekaliter
1 bushel	= 35.239 07 liters
	= 3.523 907 dekaliters
	= 0.035 239 cubic meter

ABOUT THE COVER

We believe the map on the cover is the oldest one in existence. It comes from central Italy, and shows a neolithic village around the year 5000 B.C. The pattern of stilt houses, enclosed fields, animal pens, and connecting pathways indicates that man has been aware of rather abstract and sophisticated aspects of spatial organization for a long time. The roots of geography go back a long way.

The map is incised upon a huge rock overlooking the valley site of the village. Obviously someone sat upon the hillside thinking about the spatial structure and pattern of the settlement lying below him. Whoever he was, we offer our respectful greetings over seven millenia of time.

The map was called to our attention by Professor Forrest Pitts and is adapted from Piero Maria Lugli, *Storia e Cultura della Città Italiana* (Bari: Editori Laterza, 1967). A photograph of the map may be found in *Capo di Ponte Centro dell'arte Rupestre Camuna* (Luglio: Studi Camuni, 1968).

PART I

Order, Science, and Geography

CHAPTER 1

THE ORIGINS OF SCIENCE

We make a map of our experience patterns, an inner model of the outer world, and we use this to organize our lives.

—Gyorgy Kepes

INTRODUCTION

A certain proportion of those engaged in any occupation must always be somewhat more than practitioners. A practitioner does not devote the majority of his time to questioning the nature of his activity. He concentrates on solving daily problems in accordance with principles which are usually well defined. The general medical practitioner, for example, removes an appendix according to the method he learned in medical school or according to a more recently developed technique described in a journal. He does not ordinarily concern himself with designing new and better ways to remove the organ. Similarly, a geographer called upon to map a distribution of hospitals normally uses established cartographic conventions.

Those who devote their time to thinking about how things should be done, rather than simply doing them, are methodologists. Methodologists are often inhabitants of academic com-

munities. Removed from the pressure of having to solve immediate problems, they can afford to devote much of their time to developing and experimenting with new techniques and reflecting upon what practitioners do. People who call themselves cartographers, for example, are solving the practical and important problem of making the maps we require to communicate spatial information. Improvements in cartographic theory and practice, however, originate in the academic community or in the research branches of mapping organizations. To be sure, innovators engage in the production of maps, but they also question the techniques, methods, and concepts upon which their substantive work is based. Such questions often lead to attempts at improvement.

Some people take yet another step away from immediate problems and devote themselves to thinking about the ways people think about what they are doing. There are always a number of these perverse individuals in any profession; we refer to them as theoreticians, often with a hint of exasperation. When thought processes get as far removed from everyday problems as they often appear to be in theoretical discussion, there is a tendency to ridicule such nebulous carryings on. Yet the insights of theoreticians are as essential to the ongoing success of a professional group as is the daily work of the practitioner or of the methodologist.

At the furthest remove from immediate problems are those individuals who are primarily concerned with thinking about how we think generally, and who worry about the ends to which we should devote our thinking. These are philosophers. They are concerned with problems which are at the same time the most abstract and the most basic of those which man encounters. Philosophers devote themselves to solving abstract problems of a very general nature for all the sciences, arts, and humanities, whereas theoreticians are interested in solving relatively abstract problems for specific disciplines. Philosophers provide the foundations upon which we build conceptual frameworks for our activities. As a mark of profound respect and gratitude for this service, we usually assign full-time philosophers offices in the oldest buildings on college campuses and pay them low salaries.

Problem solving is the central focus of all disciplines (Fig. 1–1). But in any discipline, some fraction of the membership should engage in the kinds of activity represented by each segment of the continuum from complete immersion in practical problem solving at the base of the pyramid to engagement in very abstract contemplation at the top. You may object that our scheme places theoreticians and philosophers in the highest positions—but you should note that in their exalted positions they are supported by those below them and bear the least weight themselves. Theoreticians and philosophers are necessary to the completeness of the intellectual pyramid. Just as your sense of symmetry would be displeased if the pyramid were incomplete, no profession could exist for an appreciable length of time if it were composed solely of any one of its segments. Ideally, a judicious and balanced proportion of the components should be maintained. Disciplines which find their manpower heavily concentrated in one segment simultaneously find themselves in trouble. Geography has had problems of this kind in the past and it does yet to some extent. We are still seriously short of well trained practitioners. Geography has been a closed community of research-oriented university scholars for too long.

We begin our exposition of geography with a discussion of some topics usually considered to be the provinces of philosophers and theoreticians. The value of abstract perspectives on the work one performs as a practitioner is undeniable. Our entire university system and especially our graduate and professional schools are living monuments to our commitment to a theoretical approach to our activities. We feel it especially important that the neophyte geographers to whom we are addressing our thoughts understand the conceptual underpinnings not only of our science, but of all kinds of intellectual activity.

EVENTS AND ORDER

Events and Experience

A convenient starting point for our prelude to geography is the broad class of phenomena we will call *events*. Events are things which happen; they are conditions, processes, or objects which exist for one or more human

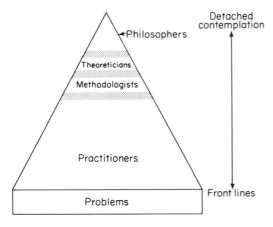

FIGURE 1–1

The structure of a science.

beings. Normally, we use the word "event" to describe something which exists for only a short period of time. This usage reflects man's particular temporal bias. Since human lifetimes are very short, the things which seem important to us are those things which change rapidly, which have an even more restricted temporal existence than we do. On a less egocentric time scale, such as that of geological time, our lives—and indeed entire eras—are very minor events. A mountain range is as much an event as a stage play or a dinner, although in normal discourse we confine the term to things like the play or the dinner simply because our time perspective is so very limited. But using the term "event" to designate anything that has existed, does exist, or might exist in the future will be helpful. By doing so we create a single class which embraces all possible occurrences of any sort and enables us to talk about what we do to them.

We can first divide events into two categories: external and internal. To do this we must be able to separate ourselves from our environment. Somehow we are all conscious of ourselves as discrete entities, distinct from our surroundings. Exactly how we arrive at this distinction is not clear. We do know that the ability to do so is learned. Infants and small children have difficulty distinguishing themselves from events which occur around them. In a mature person, an inability to separate oneself from the environment is a symptom of some forms of mental illness. Accurate delimitation of the boundary or interface between self and non-self is not always pursued. Users of psychedelic drugs describe one of the positive effects of the drugs as the way they enable those who take them to "get outside themselves" and become one with external objects such as flowers or lights.

Aside from instances such as these, most of us feel we have no difficulty separating ourselves from what surrounds us. As individuals we are coincident with our bodies, and mankind as a collective entity sees itself as a distinct component of a man-environment complex. We talk about "man *and* environment." And sometimes we talk of two kinds of science; the social and behavioral sciences which deal with man and his behavior, and the physical sciences which

focus their attention on man's physical habitat.

Because we feel we can make sharp distinctions between ourselves and the environment we usually have no difficulty separating internal from external events. Emotion, thought, desire, and physical pain are internal. Things which take place outside our bodies are external. Our perception of the locations of events determines whether we classify them as internal or external. Recent investigations into human physiology and psychology have demonstrated that such locational perceptions are not always accurate. What we perceive to be external events sometimes have no existence in the external world. Upon occasion we manufacture events. We have all had the experience of perceiving distant water on a highway on a sunny day. This mirage is produced when the central nervous system commits an error. It leads a person to experience an external event where subsequent experience—driving over the "wet" spot—will prove to his satisfaction that the water could not have existed. We commit similar errors when we see optical illusions. The two heavy lines in Fig. 1–2 are perfectly straight, although our immediate experience tells us they are curved. It is necessary to contrive another experience—that of laying a straight edge along them—to convince ourselves that they are indeed straight.

Mirages and optical illusions indicate how important it is for us to make a distinction between events and our experiences of them. The disagreement over what actually happened which often arises between witnesses of the same event is additional evidence that experience and event are not always identical. This

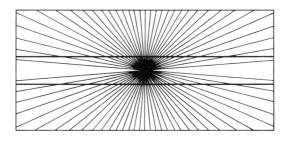

FIGURE 1–2
Hering version of Zöllner illusion.

distinction between event and experience will prove useful not only with regard to external events but also with regard to internal events. *Event* is a convenient notion under which all possible occurrences—external or internal—may be considered. But our *experiences* of external and internal events are more important to us in structuring our world than the events which provoke those experiences.

Experiences are our perceptions of events. Whether an event appears to us to have occurred externally or internally, it is our knowledge of it—the fact that it impinges upon our existence—which is important. Awareness of an event is an experience, and each of us passes his existence in an *experience continuum*. The experience continuum comprises our perceptions of events in the world and within ourselves. As conscious and self-conscious beings, we must concern ourselves with events in our environment and events within us as well as with the relationships between the two.

Usually external and internal events are correlated with our experiences of them in a very satisfactory manner. But ambiguities arise in distinguishing external from internal events. Because we do not experience some internal events, we are sometimes unsure what proportion of an experience is attributable to internal events and what proportion is attributable to external events. In the cases of mirages and optical illusions, the correlation between external event and internal experience breaks down. Recent experiments have demonstrated that the nervous energy output of the eye consists of a series of electrical pulses similar to telegraph signals. Our brains transform these digital signals into the scenes we see and experience by processes of which we are not aware—that is, by processes which are internal events but not experiences.

Events which we locate in the external world are heavily overlaid by internal, unexperienced events. Some events are wholly internal. A stomach ache may be attributable to an external cause, such as an overdose of pizza, but it is an internal experience for the individual who has it. Similarly, most intellectual, emotional, and theological experiences are primarily internal events. Immediate stimuli may be external, but whether a person enjoys a style of music or a brand of bourbon depends on internal events—the attitudes that a person has developed and thus the reception external stimuli meet. Thus, when people describe experiences we classify as aesthetic, emotional, or religious, they always focus on internal experiences. Ultimately, all experience is internal in that we experience within ourselves events which may occur either in the environment or internally.

Our perceptual capabilities reside within our bodies, and thus we live in physiological prisons. But these are open prisons; because of the sensory equipment we possess, we cannot avoid experience. Over millenia, well-developed sensory equipment has had survival value for the higher mammals and by now has developed to the point that we are physically and psychologically incapable of shutting off the experience continua through which we move. We exist as receptors in a world rich with events. What is more, because of his complex sensory and data processing physiology, man's internal experiences are far richer than those of any other creature. Whether internal or external, we can no more ignore the wealth of experience which impinges upon us than most of us can ignore a domestic squabble in the next apartment.

The Necessity to Order Experience

Just as we cannot escape the experience stream which engulfs us, we seemingly cannot refrain from manipulating those experiences into patterns which are meaningful to us. Why we must impose order on our experience continua is not wholly clear. But the existence in all cultures of comprehensive ordering systems which explain experience is evidence that the need is universal. Creation myths tell us how the world was formed and how we came to exist within it. Religious systems tell us how we should order our lives and what our relationships to other human beings should be. They also usually tell us why we exist and what we may expect to experience upon cessation of our earthly experience. Science and its precursors tell a person how to manipulate his environment and interact with it in such a manner as to produce benefits for himself. Order—satisfactory answers to the questions people ask about their experience—is a fundamental requirement of human welfare. The need for order must be fulfilled, even if the order must be created where none can be discovered.

If we wish to explain why this is so we can again cite human evolution. The necessity to order experience is an acquired characteristic, developed over the last four million years. Being able to understand the experience continuum to the point that future experience can be predicted has always had survival value. But in developing the capacity to order his experience successfully, man has played an evolutionary trick on himself. Recent investigations indicate that the brain does not exist to *permit* us to order the experiences which assault us, but that the organ is so specialized in the human being that it cannot exist *unless* it has externally originating experiences to order. Infants raised in the absence of normal stimulation, for example, suffer irreversible retardation by the time they are three years old.

We have long known that people placed in situations wherein experience is unpredictable suffer stress. A very common example is the slight unease felt by many people at social gatherings where they are among strangers. Since the people one will meet at such a gathering are unknown, the kinds of experiences they will initiate are slightly unpredictable. Similarly, one is vaguely aware that his behavior represents an unknown quantity to everyone else. Thus, on occasions when strangers mingle socially we are usually on our best behavior. We want our behavior to be correct—that is, predictable. Most people become slightly apprehensive when one person's behavior varies from that normally observed upon such occasions. The rituals observed when meeting strangers—exchanging introductions, shaking hands, inquiring about mutual acquaintances, commenting upon the weather, offering cigarettes, etc.—are devices designed to produce order for a brief period. After performing the required social rituals, individuals can decide whether they wish to become further acquainted (in other words, whether they wish to enable

themselves to make better predictions about each other's behavior) or to terminate the interaction.

The slight unease felt at social gatherings or when guests are expected but overdue is a familiar example of the stress which unpredictable experiences engender. All of us constantly experience the anxiety arising out of the unpredictability of human affairs, whether it is a homely case of worrying about whether we will be on time for an appointment or the deeper fears aroused by such tremendously disorganizing events as the assassination of President Kennedy in 1963. In crises like this, the stress of uncertainty over what will happen next can rise to almost unbearable levels.

As human beings we have a deeply rooted desire to pass our existence in controlled situations. The evidence that this need is at least partially physiological in nature comes from recent experiments in sensory deprivation. In such investigations, subjects have been cut off from experience streams to the maximum extent possible. Contrary to expectations, the experience of a complete absence of stimuli has been anything but pleasant. In one experiment subjects were dressed in wet suits connected to breathing apparatus and immersed, blindfolded, in large tanks of water in completely silent rooms. Most subjects had thought that the experience would be pleasant, that with all external stimuli eliminated they would be able to think very clearly and concentrate very intensely on some problem or question. The opposite turned out to be the case. All the subjects reported almost immediate loss of spatial and temporal orientation. Moreover, their mental faculties, rather than being enhanced, rapidly became useless. Concentration was impossible. Subjects' minds wandered uncontrollably from random thought to random thought. To a man, the subjects were terrified and refused to repeat the non-experience at any price. Even those who have experienced absolute silence in specially constructed echoless chambers while retaining visual and tactile stimuli report the sensation very unnerving. The enervating effect of sensory deprivation was known long before these modern experiments, as evidenced by the use of solitary confinement as a punishment for recalcitrant prisoners.

The inability of the brain to function normally in the absence of external events demonstrates that the organ relies upon the ordering processes it performs for its own well being. Like maize, which has been selectively bred by man to the point that it cannot reseed without human assistance, our brains have specialized in ordering events for so long that they cannot function without stimulation. Because the ability to order events has conferred great survival value in the past, we have developed an ordering organ which emits distress signals if it cannot perform its accustomed functions. What we recognize as a primarily psychological need for ordered experience has deep physiological-biological bases.

Human beings can tolerate neither disorder nor the absence of an experience stream to order. Experiences which batter us in wholly mystifying patterns are as detrimental to our well being as is complete isolation from the external world. Either situation distresses us greatly. Given the needs human beings have for order, let us examine some of the conceptual frameworks we erect as ordering contexts for the experiences we have.

Primitive Order:
Individuals in Experience Continua

We are the nodes of our experience continua. As a whole, mankind is anthropocentric, and as individuals we are necessarily egocentric. The continuum which surrounds each of us has very obvious facets—time and space—which are the basic contexts of human existence. The fact that the sciences of man have only recently begun to study people's perceptions and attitudes toward time and space testifies to their importance; until recently, we believed that they were absolutes with regard to all human behavior and thus did not require examination. We now know that such is not the case. Scholars such as John Calhoun, Edward T. Hall, Robert Sommer, and Robert Ardrey, among others, are investigating time and space as determinants of experience.

Because we cannot control our movement through time and because it is divided into formal

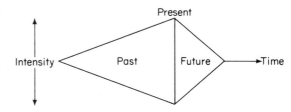

FIGURE 1-3
Intensity of experience in the temporal context.

units like days, lunar periods, and years, we are more aware of time than of space. Our consciousness of existing in time produces three regions along the temporal dimension: the past, the present, and the future. Temporally, we can think of ourselves existing at the common base of a double cone (Fig. 1-3). At each moment we occupy a point in time. At that point, experience is very intense and immediate. Intensity and immediacy diminish as experience moves further into the past or as the events we probably will experience move into an increasingly distant future. If we wish to represent the number and variety of events we experience, we would have to restructure our diagram to represent the present as the vertex of two cones (Fig. 1-4). Present experience ranges are very narrow, since we can attend to but one experience at a time. Immediately, experiences move into our past, and the further into the past we go the richer the experience range. Whether intensity or variety is considered, we see ourselves as occupants of a fleeting present and possessors of a sequence of past experiences and expectations of further events to attend to in the future.

In both Figs. 1-3 and 1-4 we have sketched the future as less important than the past. Many of us do tend to live more in the present

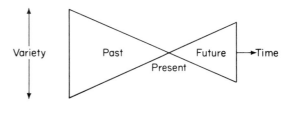

FIGURE 1-4
Variety of experience in the temporal context.

and the past than in the future, although our preoccupation with past experience is not as total as that which prevails in some cultures. Very few people attend largely to future events, although the number who do so is now increasing rapidly. Some of our present intensely unpleasant experiences (pollution and urban social problems, for example) are the products of men's predilection for not worrying about the future. Present experience has convinced us that if we want to avoid similar problems hereafter, we must take pains to order our futures to a greater degree. Both the growth of planning as a profession and the establishment of a chapter of the World Future Society in Washington, D.C., in 1966 are indicative of the increasing attention now being paid to future events.

Spatially, we also occupy a nodal (central) position, although we have no widely used categorization of space analogous to the division of time into past, present, and future. Individuals and groups have spatial ranges of various sizes, but these have not been formalized in any generally accepted way. A number of investigators have begun this work with animals, some of whom have territories which they will defend as their own and ranges over which they travel. All of us have territories and ranges and crudely formed conceptions thereof. For example, we distinguish among short trips around town, long trips out of state, and major journeys to another continent. Perhaps our insensitivity to space is related to the fact that movement in space is voluntary whereas movement in time is wholly involuntary. Our inability to control time makes us more sensitive to it. In any case, the space continuum of existence is far less accurately regionalized than the time dimension with its clear past, present, and future.

One facet of the human spatial continuum which has been well documented by geographers is the effect of distance on interaction. Space is not unidimensional, and we cannot represent our experiences as trailing off along a single axis. Rather, we must represent the individual as the node of a spatial interaction field. Frequency of experience is high close to ego, but drops off with distance from him (Fig. 1–5). Variety and intensity of experience increase with distance (Fig. 1–6). The further we are from

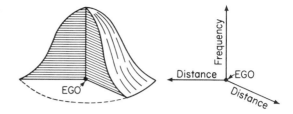

FIGURE 1–5

Frequency of experience in the spatial context.

home, the more unusual and stimulating the events we encounter. Because new places are so exciting, we can often observe new landscapes with a keener eye than we can direct to our usual habitats, whose contents become so familiar over time that we almost stop seeing them.

The distributions of spatial experience represented by Figs. 1–5 and 1–6 are, of course, idealized, for none of us lives in a wholly circular world. Our experience in space is usually highly channeled along selected major axes which we traverse much more frequently than others. Still, the basic relationships are clear: whether we consider time or space, man's mind is the focal point of his existence in these two contexts.

Given current ideas about space and time, it is probably not wise to separate the two except for expository purposes. Because time is unidimensional, space is a part of our temporal continuum and vice versa. We can experience only one place at a time and moving from place to place consumes time. All of us are nodes in personal space and time fields of experience, and as we move through time and space, we carry our nodality with us. As Pierre Teilhard

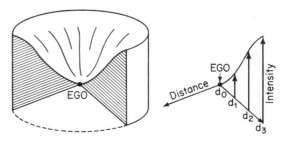

FIGURE 1–6

Intensity and variety of experience in the spatial context.

de Chardin has observed, man is the *center of construction* of his universe and his nodality is wholly personal, although occasionally this personal nodality may coincide with an objective nodality.

It is tiresome and even humbling for the observer to be thus fettered, to be obliged to carry with him everywhere the center of the landscape he is crossing. But what happens when chance directs his steps to a point of vantage (a cross-roads, or intersecting valleys) from which, not only his vision, but things themselves radiate? In that event the subjective viewpoint coincides with the way things are distributed objectively, and perception reaches its apogee. The landscape lights up and yields its secrets.

Time and space are obvious and immediate aspects of human existence. Naturally, existence in these continua has given rise to informal and formal ordering systems. Calendars and clocks provide frameworks for ordering experiences in time. Maps—mental and formal—provide a context for ordering events in space. Time and space are the fundamental contexts of all experience. Experience must be located in time and space before we can begin to process it further. The fundamental nature of time and space is indicated by our lack of any concepts to express the absence of this context for experience. We can say that something is "nowhere" in time or space, but to do so is to say, in effect, that the event does not exist. Locating an event in the spatio-temporal continuum is our first step in ordering our experience of it.

Just as time and space are contexts in which the locations of events may be plotted, we may postulate additional contexts which may be used as frameworks to locate and order the events we experience. We shall discuss the many distinct contexts of an event later. For the moment, let us turn our attention to the nature of order and a discussion of the questions we ask about all our experiences without regard to context.

The Nature of Ordered Experience

Ordered experiences are those which arouse no questions in our minds. If we have questions about an experience, it is, by definition, not completely ordered. We do not bother to ask questions about many experiences, of course, and this is one way of imposing order upon them. We decide they are not worth bothering with since we think they are irrelevant. We dismiss by far the greatest number of experiences we have in this way. We simply assign them to a category labeled "unimportant" and concern ourselves with them no further. We have no questions about them because we choose not to ask any.

But the most important way of producing order is to attend to an event and satisfactorily answer the questions the experience evokes. This method—raising and answering questions about experiences—is what science, theology, art, and the humanities are all about. All are systematic bodies of thought which attempt to answer questions about our existence and its experiential consequences. The number and kinds of different questions which we ask about experience are virtually infinite, but all possible questions can be subsumed under six basic questions. For the remainder of this section we shall concern ourselves with *what* questions, *when* questions, *where* questions, *how* questions, *why* questions, and *who* questions. Satisfactory answers to these six questions constitute order.

What questions may deal with semantics (word definitions), essence, occurrence, or combinations thereof. We ask the majority of our semantic *what* questions when we are children and are acquainting ourselves with the furniture of our ecosystem. "What" may be interpreted "what is the word which designates this thing?" Most often, though, there is more to a *what* question than semantics. After some vocabulary has been acquired, *what* questions are usually *how* questions concerned with the uses to which things can be put. Chemists, metallurgists, and physicists, among others, raise *what* questions of essence. They are concerned with the composition of materials, for example, rare metals or alloys. *What* questions of occurrence are asked concerning either past or future events. We inquire as to what sort of events happened in the past and what we may expect in the future, and what the probabilities are that experiences we think we might have will actually occur. Most *what* questions we ask in adult life are those of essence and occurrence.

When and *where* questions arise in both the past and future contexts, each of which requires somewhat different treatment. If they are asked about past events, they are basic queries which must be answered accurately if productive discourse is to continue. Our immediate questions about a defined experience—whether they are implicit or explicit—attempt to fix the location of the event in time and space. On an accident report form, for example, the first two substantive questions ask for the temporal and spatial coordinates of the incident. Upon being informed that a serious earthquake has occurred in South America, our immediate reply is "Oh really? When did it happen?" In fact, such questions are so common that people relaying information like this usually provide the answers so that we do not have to ask. One would normally say, "There was a plane crash in Chicago this morning." Almost instinctively, we provide the spatial and temporal coordinates of past or ongoing events. If asked about potential future experiences, *when* and *where* questions are locative and related to *what* questions about the future. Since we will all spend the rest of our lives in the future, most of us are concerned not only with what is going to happen, but with exactly when and where future experiences will affect us.

How questions may be questions about the origin of experience, about processes, about utility, or about policy. *How* questions about experience seek information about the processes which have produced the experience in question. Answers to *how* questions are descriptions of series of events which together determine that the experience which evokes the *how* question will occur. These events may be from the past or we may ask about the processes which will produce future events. When asked about processes themselves, *how* questions treat sequences of events as a single event and are focused on the relationships between events rather than on the events themselves. These questions are raised because those who ask them want to know how processes themselves come to be. We can contrast the two questions "How did

World War I start?" and "How do wars start?" to illustrate this point. The first is concerned with a single event, whereas the second seeks further explanation of the process which produced many other wars in addition to World War I. The way we answer a question about process like "How do wars start?" is to describe yet another process which embraces the processes by which wars start. You may be confused by this at the moment, but we will clarify this relationship later on when we discuss law and theory.

We noted before that many *what* questions are *how* questions in disguise. Very often when we say "What's that?" our real interest is in the function the phenomenon serves, in the utility it has within a system. We see it within the context of a larger series of events where it is functionally related to the events adjacent to it. A child, observing his father using an unfamiliar tool, is satisfied upon being told that it is a wrench and that one uses it to turn nuts and bolts. He would be much less satisfied if told that it was a "machined piece of drop-forged steel."

Another way we often use *how* questions in daily discourse is to inquire of matters of policy. In these instances, we wish to obtain some information about the way we should or should not structure some process over which we have control. In all the ways we use them, *how* questions are related to process and connected sequences of events. *How* questions are the most important questions we ask because they are the questions which lead to explanation.

Why questions can be divided into two categories: true *why* questions and *how* questions in disguise. True *why* questions are the most difficult to answer. Fortunately, we ask them rarely. Most *why* questions are *how* questions in disguise. For example, if a traveler's car breaks down one may ask "why did he arrive two hours late?" This is really a *how* question which seeks to identify a specific event in the sequence leading up to the lateness which has more explanatory power than any of the others. Any number of mischances could cause a traveler to be late. In this particular case a faulty auto part was responsible rather than a snowstorm or a late start. When asked about individual behavior, *why* questions are queries about human motivation. Here again, we seek a single event in some sequence which has explanatory power. When we ask why a person commits a crime, we are interested in some aspect of his personal situation and psychology which produced the incident. A statement that an individual stole a loaf of bread because he was hungry usually satisfies our curiosity. We relate how the theft came to occur and identify one event in the sequence as causal. "Why did he steal the loaf?" is the same as "How did he come to steal the loaf?"

True *why* questions, whether asked about human or physical phenomena, reduce ultimately to theology or metaphysics. "Why (not how) did X happen?" cannot be completely answered by any event chain, no matter how long. Ultimately such questions exhaust our knowledge. At that point two answers are possible. We may admit that we do not know and speculate about the events needed to complete the chain to our satisfaction. The alternative is to argue that God designed the world that way. The generally unsatisfactory nature of both these replies need not concern us greatly. Few except cosmologists and theologians are concerned with the ultimate *why's* of our universe.

Who questions may be either disguised *what* questions or hidden *how* questions. In the former case, we are concerned again with questions of identity. We may ask "Who is that?" and the answer will be either the name of the individual in question or some description of his function, or both. On the other hand, when we ask "Who did that?" or "Who will do it?" we are raising what are really *how* questions. Our interest is not in identification as such, but in the designation of causal elements in event chains. *Who* questions then, are special cases of *what* and *how* questions in which people rather than nonhuman phenomena are involved.

These six questions and their combinations and permutations exhaust our curiosity about our experience. Answering them produces a feeling of psychological well being. We apply this basic set of questions to past, present, and potential future experience. On balance, answers to *what, when, where,* and *how* questions about

the future are the most important to us. If we can predict potential events accurately, we can manipulate those events and our own behavior so as to experience only those we deem most beneficial. That prediction is a major goal should not, of course, blind us to the fact that we also require such explanations about past experiences (postdiction) and ongoing events which impinge upon us.

Satisfactory answers to all possible questions about all possible events would produce a state of complete order. That this state does not exist in reality, and cannot exist given the nature of human thought, does not negate its value as an ideal model.

How We Begin to Order Experience: A Scheme

Henry Margenau, the Yale University physicist and philosopher, has devised a conceptual scheme for representing the stages by which we begin to impose order on experience. This conceptual framework is the most fruitful of those which we have examined. A system similar to Margenau's was developed independently by P. van Duijn. The scheme we present here is an extension of the two.

We have seen that our experience has boundaries, that we are prisoners of our own senses. Events become experiences only when they attract our sensory attention. Our senses are the effective limit of our penetration of the external and internal world. We constantly devise instruments like electron microscopes to penetrate observational barriers and extend our range of experience. But the additions to the experience continuum which result must still be channeled back through the media of our senses. When ordering events in the world and even those internal to ourselves, we find ourselves against a sensory frontier.

We can place all events beyond our sensory frontier (Fig. 1–7). Some events cross the sensory frontier and become part of our experience. Margenau calls this frontier the P-plane, where P represents the plane of primary experience or perception. Many events are unperceived and do not become elements of our

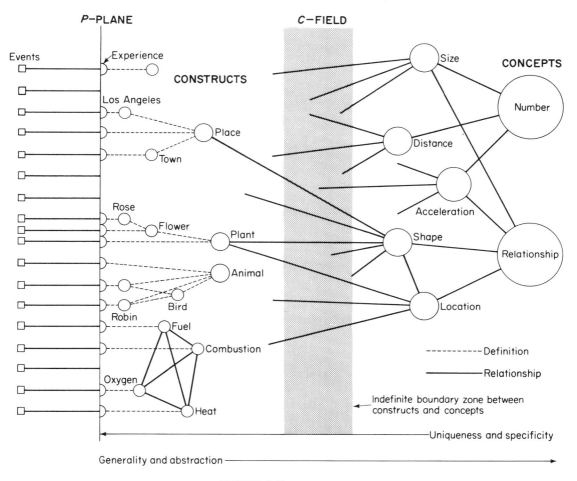

FIGURE 1–7

The P-plane and the C-field.

experience continua. They do not cross our P-plane although they may cross some other people's P-planes. This P-plane of perceived experience is our starting point. We need not concern ourselves with events which are not experienced. Indeed, it is self-contradictory to talk about them.

Related to the immediate experiences which lie on the P-plane are *constructs*. Constructs are ideas about experience which impose preliminary order upon them. Indeed, constructs are often necessary for an event to become an experience. Much of our early education con-

sists of people acquainting us with constructs which make events intelligible experiences. We cannot place a plant in the category of flowers, for example, unless we have previously established that construct in our minds. Similarly, we cannot decide whether a given flower is a tulip or a rose unless we have a fairly good idea of what each looks like. The construct *rose* is more specific than the construct *flower*, which is in turn more specific than the construct *plant*. Each experience must be relegated to an existing construct or we must generate a new construct for it if it is to be meaningful to us.

Constructs are products of and answers to our semantic *what* questions.

The field of simple constructs is quite wide. The closer to the P-plane a construct lies, the more specific it is. Thus *rose* is closer to the P-plane than *flower* or *plant* (Fig. 1–7). Making constructs is essentially a taxonomic (classificatory) process. We assign our experiences to categories of greater or lesser generality. The dashed lines connecting experiences on the P-plane to constructs in the C-field are definitions. They are rules of correspondence between experiences and the constructs we build for them. Constructs are the first step in ordering experience. Experience requires us to build constructs so that we can make sense out of it. The constructs we establish have empirical content. One can give an example of a rose by producing the physical entity for inspection.

Concepts do not have empirical content. They are abstract ideas generalized from innumerable experiences. The number of abstract concepts which exists is very large. Ideas like distance, shape, velocity, and quantity, are a few examples. For our purposes all concepts may be reduced to two megaconcepts, *number* and *relationship*. Both can be applied universally. Any experience or construct may have number, and any may be related to other experiences and constructs. We build our arsenal of concepts out of combinations of the two megaconcepts.

In our schematic representation of our thought processes, lines interconnecting constructs with each other and with concepts will represent relationships. Dashed lines indicate definitional relationships. Solid lines are used to represent any other kinds of relationships between two constructs which might exist. One example might be cause and effect relationships. Combustion, for example, is causally related to fuel, heat, and oxygen. Another example might be a relation of location. Flowers are found in Los Angeles, and if we were interested in constructing networks of relationships concerning the locations of flowers, we could legitimately link the constructs *Los Angeles* and *flower*. In order to avoid a messy diagram, we shall adopt no specific representation for the concept *number*. The capacity to have number is implicit in almost all regions of the C-field and at most locations on the P-plane. At the right side of the diagram we have placed two megaconcepts to represent *number* and the idea of *relationship*. Intermediate between the two megaconcepts and the substantive constructs of the C-field we have placed a few familiar concepts which are derived from *number* and *relationship*.

Concepts are extremely important because they provide us with a means of manipulating constructs. They enable us to devise answers to the questions we ask about our experience. Number provides one framework within which some *what* questions of composition (how many?) can be raised and answered. Relationship provides a means of linking constructs to one another in useful ways. Number and relationship are used together to answer *when* and *where* questions. *How* questions are answered primarily by constructing networks of interrelationships among constructs, but number proves very useful in answering these questions also. Number enables us to describe constructs and relationships accurately by measuring them.

The P-plane and C-field model of our thought processes is a useful structure for representing the way we organize the world. It provides a conceptual framework within which we can move directly from events and immediate experience to classes of experiences and relationships among them. As these relationships become more and more complex—as we begin to construct intricate networks in the C-field—we move toward the development of systems of thought. Thought systems incorporate the very elementary experiences and constructs which lie on or close to the P-plane as well as our most abstract concepts located far into the C-field.

We have a considerable choice concerning the kinds of relationship networks we can construct in the C-field. Two different network structures may connect identical constructs, but they may connect them to each other in different combinations or with different connectors. Different kinds of order will produce distinct structures of connectivity in the C-field.

MODES OF ORDER

Four major systems of order are now in use. Theology, the earliest intellectual ordering system, is one. Aesthetic and emotional order is a second. Common sense—a much valued commodity—is the third form. Science, the most recent and the most powerful of our ordering systems, is the fourth thought system we shall consider. We are forced to discuss these four as if they were separate and separable. In some cases they are, but in many instances they are not. Whether we consider the experiences to which they are applied or the systems themselves, overlap and interpenetration is the rule. Nonetheless, it will be simpler to take them one at a time and discuss them separately.

Theological Order

Theology provides explanations of experience which are intrinsically as valid as those produced by any other system. Order is the absence of questions about our experience and an absence of anxiety over potential experiences. It is clear that theology has at times been very successful in answering questions and reducing anxiety. Moreover, theological ordering systems occupy a primary position in any chronology of the evolution of explanation. All cultures have theological systems, and in all of them theology appeared earlier than any other formal system of thought. Cultural universals like theology and creation myths are very rare. They excite our interest primarily because they are universal. So much human behavior is non-universal that the discovery of such a behavior pattern indicates that we have identified something very basic to the human ethos.

Theological order is non-empirical. The existence of the relationships theology builds between constructs to explain events cannot be tested by inspection or examination. One either believes that such relationships exist or one does not. If one does not, theological order does not exist. Theology is more authoritarian than other systems. In answer to questions about experience, one will be told that a certain set of relationships in the C-field exists. A current example

is the assertion by some prelates of the Roman Catholic Church that the relationship of coincidence between the constructs *member of the Roman Catholic Church* and *user of oral contraceptives* is one which cannot exist. The two elements of the C-field are held to be mutually exclusive and thus nonconnectible. Obviously a large number of lay Catholics and not a few clerics disagree. This debate is instructive since it illustrates the nature of theological order, especially as dispensed by the highly formalized Western religions. No empirical test will ever answer the question of whether or not a Catholic may legitimately use oral contraceptives. Such questions are answered by citing dogma and its implications. The ultimate source of dogma is a Supreme Being whose precepts are held to be overriding.

Thus, to represent theological order fully we would have to add to our previous diagram an additional concept—a very large one—in the rightmost position in the C-field to represent the Supreme Being or greater force. On the assumption that most readers of this book have been inculturated with the idea of the single God of Judaism, Christianity, or Islam, we present the monotheistic case. Were we representing the theological system of the ancient Greeks, we would have a large number of concepts in the right hand position to represent all the gods and goddesses. Similarly, a distinct network of relationships in the C-field would indicate the separate terrestrial domain of each of the various gods. Theological networks interconnecting constructs in the C-field represent admonitions, commands, and proscriptions. If the concept of the Supreme Being or greater force is part of one's general C-field, the connecting relationships tend to follow logically. If such a concept is not part of an individual's C-field, theological explanation does not exist.

Because theology holds that the ordering connectors between constructs in the C-field exist at the wishes of a Supreme Being or his representatives, theology tends to be more concerned with answers than questions. A certain stock of actual and potential answers to questions exists. Questions must be answered in the same way they have always been answered; innovation is usually discouraged. If new questions arise, new answers must not contradict answers given to earlier questions. The primacy of answers rather than questions is one distinguishing characteristic of theological order. We might note here that in this respect theology is similar to law, another social ordering system.

An advantage of theological order is the wide range of experience to which it can be applied. Any construct or concept can be reached by the network of relationships emanating from the Supreme Being. The most mundane and the most divine events of life are equally explicable. Questions as diverse as an inquiry as to why our universe exists or as to why the grass grows can be treated as manifestations of God's will. At the moment, we in the Western world limit our use of theological explanation considerably. We tend to reserve theological explanation for our very ultimate *why* questions. Indeed, theology seems to be the only system which is able to provide reasonably satisfactory answers for such questions. Cosmologists are becoming increasingly successful at telling us how our universe came to be. But explaining why it did is still a very sticky matter.

Our own limited use of theology should not blind us to its importance in times past and its role as a seedbed for other formal ordering systems. Until a few centuries ago theology provided more comprehensive order for members of Western society than most of us can today imagine. We in the West and a few other areas of the world no longer rely on theology as heavily as we once did because we have generated an alternative which serves our particular ends more satisfactorily. Yet in times of great stress and for certain other purposes, we still invoke theology as an ordering system.

Aesthetic and Emotional Order

Aesthetic and emotional order can be more individualistic than the other systems because it is often informal. In some instances, however, aesthetic value systems can become highly formalized. An example would be the dominance of a period by a particular style of art or music.

But often the structure we apply to experiences which are primarily aesthetic and emotional in nature, and the aesthetic and emotional networks we erect among events, are not systematized to any significant degree. No organization tells us what kinds of music we should prefer, or how we should feel about certain experiences. Critics, marriage counselors, and psychiatrists do exist to provide advice on such matters. But to the limited extent that anarchy reigns in the realm of ordering, it reaches its fullest expression in aesthetics and emotions.

The individualistic nature of this kind of order is indicated by the frequency with which we use the expressions "I like" and "I feel." Each person is at the center of his own continuum and builds his own somewhat distinctive network of artistic and emotional links in the C-field. As in theology, answers are usually more important than questions. Once most people form aesthetic and emotional judgments, they change them very little. When we use aesthetic and emotional principles, we usually rely on past answers rather than trying new experiences; we resist the blandishments of competitors and continue to drink the same brand of beer year after year.

Despite the limited anarchy which reigns in this kind of network building, aesthetic structures are not wholly dissimilar. Fads in clothing and musical preferences are evidence that tastes are supra-individual. The fact that you would become nauseous at a serious suggestion that dear old Fido be converted into a pot of stew indicates how deeply some of our emotional attitudes are rooted in our culture. Supposedly, each of us builds a very distinct network of aesthetic and emotional relationships in the C-field. In fact, many of the links are common to large groups of people, even though a number of links are unique in every individual. Despite our inability to agree on the relative superiority of brands of beer, most of us think beer is desirable.

The nominal realm of aesthetic ordering embraces those experiences and constructs we would describe as artistic and personal in nature. We expect preferences for paintings, music, literature, and landscapes to be different in different people. Nominally, we expect emotional judgments to be confined to interpersonal and social relationships and to other areas where gut feelings are valid bases for ordering our activities. It is permissible in our society, for example, to avoid interaction beyond the exchange of greetings with a person down the block whom we dislike heartily. It is not acceptable behavior in another realm of order (common sense) to refuse to interact with one's superior in a corporation because you dislike him. In situations like this, emotional feelings are overridden by other interconnections in the C-field.

The actual use of aesthetic structures is greatly at variance with nominal use. They play far more important roles in realms where they should not be applied than we are prone to admit. Overwhelmed as we are by the explosive growth of science in the last century, we tend to overlook the grip which our personalized emotional and aesthetic systems have on our behavior. Many times when we think we are ordering and manipulating a situation in accordance with rational principles, our emotions are playing a much stronger role than we realize. For examples, one has only to review the scientific justifications given for the persecution and annihilation of Jews in Europe in World War II or the rationally couched arguments deployed by segregationists in the United States today. Our most difficult social and interpersonal problems arise when the differences between emotional connections and scientific connections in the C-field become blurred. Yet even in the realm of science itself, aesthetic and emotional considerations are very important. Some locational problems in geography, for example, can be solved with iterative (trial and error) procedures, yet we continue to search for elegant solutions, and our desire for elegance is based largely on aesthetic preferences.

When we consider aesthetic and emotional networks in the C-field, we must face the fact that they are far more pervasive than is immediately apparent. Parts of the aesthetic structure—perhaps the major fraction of it—are invisible to us. Because it is nominally restricted to ordering certain kinds of events we overlook the importance of this mode in other realms where it is not supposed to be applied. The informal nature of emotional order makes it difficult to identify, yet it exists and must be taken into account. Indeed, it sometimes seems that emotion is the most important ordering mode extant despite our claims to the contrary.

Common Sense

Common sense, the third ordering system, is much sought by theologian, artist, and scientist alike. To some extent it is emotional. What we consider to be common sense depends upon our prejudices. But we generally distinguish it as a separate system of ordering. Indeed, we often contrast common sense to emotion. Whether or not we can separate them, we perceive a difference and insist on the distinction.

In any society, certain forms of behavior are commonsensical and others are not. The roots of commonsense behavior patterns lie in common and repeated experience. Commonsense structures in the C-field are not the sorts of things one can learn to build by going to classes in common sense. Common sense comes from living in a world of experience and adapting to it. Commonsense behavior patterns are learned only by practice or by listening to one's elders discuss their experiences and trying to draw analogies between what happened previously and experiences we are likely to encounter in the future.

Commonsense ordering is transmitted through the medium of folk wisdom. Proverbs distill the collected experience of a culture into essential kernels of truth which can be transmitted to others. That people quote them in times of uncertainty, sorrow, and joy indicates their usefulness in answering questions and allaying stress. Proverbs and homilies can preserve and pass on this kind of order, but the most important means of constructing commonsense structures in the C-field is the actual experience of living itself.

Two drawbacks of the commonsense mode are that it tends to be useful only within a specific culture and that it does not possess very effective change mechanisms. Folk wisdom is well adapted to folk society, but it may be a negative asset in nonfolk situations. A rolling stone, for example, may have gathered no moss in an agrarian era, but people who change jobs and residential locations frequently today do gather more financial moss than those who stay put. Our society pays a premium to those willing to adopt a form of slow-motion nomadism. Like any system which does not contain provisions for changing itself readily, commonsense ordering can become a drawback in a changing world. Yet corporate and military maxims such as "don't make waves," and "it isn't what you know, but who you know that counts" indicate that folk wisdom can adapt to a limited extent.

The major drawback of common sense is that it cannot be transmitted efficiently. Experience is indeed a good teacher—to paraphrase another adage of folk wisdom—but it is also a notoriously slow teacher. When we wish to inculcate certain behavior patterns and structures of thought quickly, experience will not do. Despite such limitations, there is no denying that a lack of common sense has been the bane of existence for some exceedingly brilliant as well as some exceedingly dense people. The ability to absorb common sense is not distributed in proportion to the intelligence quotient. Anyone who does not possess a certain minimum quantity of common sense usually finds the experience continuum of life rather painful.

Scientific Order

Like theological, aesthetic, and commonsense systems, science attempts to answer questions we ask about our experience of the events which impinge upon us. Like theology, but unlike emotive and commonsense ordering, science is now highly formalized and institutionalized. And unlike the other systems, science incorporates formal subsystems for producing change and for verifying the existence of the relationships it asserts to be parts of the C-field.

The institutionalization of science has progressed to the point that critics can write books with titles such as *Science Is a Sacred Cow.* Theology was traditionally the most institu-

tionalized of the ordering systems, but science has now surpassed the major Western churches in the degree to which it has become a self-conscious, organized activity. If possessing a methodological literature is the criterion of institutionalization and formalism, science outdistances theology. The importance of institutionalization and self-consciousness is difficult to overrate. A very large share of the success which science has had in constructing useful C-field structures in the last several decades must be attributed to the institutionalization it has achieved. Organization and success are circularly causal, and science is now a great communications network whereby scientists keep in touch with each other and build upon one another's work.

Scientists not only are aware that they are building structures in the C-field and thereby ordering the world, but many are professionally engaged in doing so on a full-time basis. Over 90 percent of the scientists who ever lived are now alive. The fact that so many people are devoted to ordering events and experience as their life's work contributes greatly to the success of the enterprise. A small number of playwrights, artists, musicians, and entertainers of a less socially approved kind are professionally engaged in providing aesthetic and emotional experiences. A few professional commonsense mongers provide guidance through the advice-to-the-lovelorn columns of newspapers. But no other ordering system can come close to matching the manpower science commands. The resources—human and financial—devoted to science are explanations as well as indices of its success.

Whatever ordering mode we use, it often happens that relationships we believe to exist among constructs and concepts in the C-field do not exist by some objective criterion. That is, they form parts of our individual or collective C-fields but have no counterparts in reality. On the other hand, one may fall into the error of asserting that a relationship between C's does not exist when in fact one does. These mistakes occur in science just as they do in other ordering systems. But science weeds mistakes out more rapidly than do the other systems. Moreover, science produces fewer hypothesized but nonexistent relationships. Science has some rather stringent verification procedures. Good scientists will not accept the assertion of a relationship without some experiences which verify its existence to their satisfaction. Nor will they reject an accepted C-field interconnection until an alternative is available which can replace the one being rejected and do more besides.

A further facet of institutionalization which adds strength to science is that the answers it produces are replicable. There are no varieties of science. One man's geography is the same as another's. Because answers to questions in science are accepted only after testing, far more agreement can exist about answers than is possible in other ordering modes. Any scientist can replicate the series of experiences which led another to assert that two C's are interconnected in some particular way in the C-field. He can thus convince himself that the relationship exists or find some fault in the original experience sequence which will enable him to modify or refute the assertion.

The verification and replication abilities of science endow it with considerable adaptive capacity. Much more than the other systems, science can change its internal structure. In science, the questions asked about experience are most important. Answers change, and current answers to the questions science asks are accepted provisionally until better answers can be devised. This is in sharp contrast to theological, common sense, and aesthetic ordering, where the given set of answers tends to be more important than the questions. Science is adaptable whereas the others are often rigid. Its verification and replication procedures enable science to remain viable in the rapidly changing world it creates. Far more than the others, it is a supra-individual, empirical ordering system.

THE USE OF ORDERING SYSTEMS

Since you are perceptive, you will have noted by now that we are going to come down hard on the side of geography as science. You may think we have discussed theological, aesthetic and emotional, and commonsense systems in order to provide some straw men to further that devious purpose. Perish the thought! We are convinced that science is an intellectual activity which has its origins in the same sources as the other ordering systems. We are even more firmly convinced that there exists a very fundamental unity among all four ordering modes. All are trying to answer the questions we raise about our existence in the world and its implications. Our purpose in discussing the other modes was to demonstrate their common origins and to show that each mode has its own particular realm of experience in which it is superior. No mode is inherently superior to any other. Only when the realm of experience we wish to order and the purpose of the order we wish to produce have been specified can we make value judgments about the relative superiority of the modes.

For the most part the systems are complementary. We divide our experience into categories, each of which is dominated by one of the four modes. The boundaries of the regions into which these four modes divide the C-field shift considerably at times, and sometimes border disputes arise. But for the most part, we feel that we have a reasonably accurate map of our experience regions. The fact that one mode may dominate an experience region does not mean that none of the other modes is relevant to the constructs which that region contains. A number of relationships are incident on every construct, in the same way that our houses are connected to gas, electricity, water, and telephone networks.

The construct *marriage ceremony*, for example, is part of a network of theological constructs and relationships in American culture. But the fact that theological relationships are primary does not mean that no others exist. Marriage ceremonies are constructs which sociologists and anthropologists would claim as elements of their scientific systems. Similarly, enough proverbs exist concerning marriage to provide abundant evidence that commonsense relationships are relevant (marry in haste, repent at your leisure). It is even conceivable that emotional relationships might somehow be involved. Although one network may be dominant because the construct is by its nature primarily a theological, emotional, commonsensical, or scientific phenomenon, less important relationships cannot be ignored.

We can review briefly the categories of experience to which we apply the four modes. Theology attempts to answer our ultimate *why* questions and to develop the implications of its answers in the realm of daily life. We cite theology's answers especially in matters of ethics or morality. Aesthetic systems dominate the realms of artistic experience and our consumptive activities such as eating and listening to music. Emotional considerations are dominant in the area of interpersonal relationships. They are also dominant in many other areas of activity when we think they are furthest from our minds. The realm of emotional order is probably the most poorly delimited of the four. Common sense is a system applied to our secular social and economic activities. Science is invoked to explain a very wide array of events in the external world, and increasingly, experiences which arise within us. Over the last century, science has become particularly aggressive and has made numerous conquests of adjacent territory.

Up to this point, we have been discussing ordering modes as if each one had life and purposes of its own. This is a common convention in expositions of this sort, and is of no great consequence so long as it is clearly understood that all ordering systems are human products and essentially passive. This point is worth emphasizing because we are now concerned with why people invoke the ordering systems they do to explain various experiences. We are also interested in the related question of why science is so much on the ascendancy in our era. The answers to both questions lie in the fact that given a certain purpose, people invoke the ordering system which best serves that purpose. It follows that no discussion of preferences for ordering systems can ignore the human goals to which each of the ordering systems is best suited.

We all have certain short run and long term goals in mind. Ordering systems are means to these ends. Just as we do not use a sledge hammer to repair a wristwatch, we should not choose an inappropriate system to explain an experience at hand. Before we can deal with the question of what kind of order geographers should choose to invoke, we must decide what sorts of questions and answers geographers should be asking and answering.

It is our conviction that the kinds of questions most often asked by geographers are the *how* questions which science handles with such great facility. Geographers ask one basic question. In its disguised *why* question form it is: "*Why are spatial distributions structured the way they are?*" This question may easily be translated into a *how* question about the sequence of events necessary to produce a spatial distribution. Naturally, forms of this basic question may be asked about a past, an existing, or a potential future distribution. Thus the question embraces past, present, and future geographies. Our goal as geographers is to be able to replicate, actually or conceptually, the series of events which have led or will lead to certain distributions of events in space. Alternatively, we are often curious about the influence of spatial structure on processes or sequences of events. In either case, our focus is on the way spatial structures are produced and the effects such structures have on our experience. The *how* questions and the *why* questions which seek to identify a particular causal event are the substance of geographical thought when they are applied to spatial structure.

We want to know about the processes which produce spatial structures because we wish to manipulate them. Whether we focus on distributions of physical or human phenomena, our emphasis is on the manipulation of these distributions and the processes which produce them so that they become more amenable to our purposes. Man has always manipulated his social and physical environment to some degree. In the future we will design our own ecosystems to an extent which is incomprehensible now. Manipulating events to produce maximum benefits for maximum numbers of people will become an increasingly important activity in the future. Understanding and manipulating space and spatial distributions will be the geographer's contribution to human welfare.

Because it asks and answers *how* questions, it is precisely in the realm of manipulating physical and social events that science has had its most spectacular success. The other modes do have some value in the management of experience. Praying for rain, either by gathering in churches or performing dances, will eventually result in rain. More direct and more efficient courses of action are possible, however, such as seeding clouds with certain chemicals. Similarly, one may hope that a slow student will learn faster or one may follow the commonsense suggestion that tells us that a caning would provide him with a suitable negative incentive to do his lessons. A more rational evaluation by psychologists or doctors might reveal that the student is mentally incapable of absorbing what we wish him to learn, or that providing him with eyeglasses would enable him to see well enough to do his work. Since we wish to understand our experience so as to mold future experience closer to our hearts' desires, science is the ordering system we must adopt. The payoffs of using it, in terms of the acquisition of power to manage ourselves and our environments, make it the only viable choice. Given our goals, the choice of theological, aesthetic and emotional, or commonsense systems would be self-defeating.

Science is but one of the several ways of explaining our experience. It developed out of the same desires and needs as did the other systems we have discussed. We conceive of geography as a science because we are convinced that practicing geography scientifically will be more productive and satisfying than practicing geography as a subsystem of other modes. But this does not mean that we hold the other systems to be useless. We are all more or less religious. We all love our wives and children. Two of us appreciate Bach and the third has no musical taste whatsoever. And we all could care less whether science ever discovers what it is that makes Chateau Lafite Rothschild, 1961,

the ecstatic spiritual thing it is, so long as we are permitted to experience it occasionally. To explain the latter by talking about enzymes, polyphenols, and the physiology of taste seems to us as barbarous as to refuse to accept the conclusions of a colleague solely because he wears loud ties. Each particular ordering system is useful in its own domain. To try to order one's whole existence on scientific principles would be ludicrous.

But so far as our professional activities as geographers go, we see no alternative to the scientific viewpoint. The nation and the world in which we live are faced with serious social and physical problems, at least three of which—the Cold War, pollution of the environment, and overpopulation—threaten our very existence. The spatial dimensions of problems like these are tremendously important and poorly explored. If these problems are to be solved so that life for our children and grandchildren will be worth living, geographers will have to accelerate their output of useful knowledge about human spatial behavior and its consequences. Like it or not, geography and the other social sciences have pressing social and moral responsibilities. We see no alternative to practicing geography as a science if we hope to meet these obligations.

Suggestions for Further Reading

GLACKEN, CLARENCE J. *Traces on the Rhodian Shore.* Berkeley and Los Angeles: University of California Press, 1967.

HALL, EDWARD T. *The Hidden Dimension.* New York: Random House, Inc., n.d.

_____. *The Silent Language.* New York: Fawcett World Library, 1959.

LOWENTHAL, DAVID. "Geography, Experience and Imagination: Towards a Geographical Epistemology," *Annals of the Association of American Geographers,* LI (1961), 241–60.

MARGENAU, HENRY. *Open Vistas.* New Haven: Yale University Press, 1961.

PLATT, JOHN R. "The Fifth Need of Man," *Horizon,* (July, 1959). Reprinted in Platt's *The Excitement of Science.* Boston: Houghton-Mifflin, 1962.

Works Cited or Mentioned

ARDREY, ROBERT. *The Territorial Imperative.* New York: Dell Publishing Co., 1968.

CALHOUN, JOHN C. "Population Density and Social Pathology," *Scientific American* (February, 1962), 139–48.

SOMMER, ROBERT. *Personal Space.* Englewood Cliffs, N. J.: Prentice-Hall, Inc., 1969.

STANDEN, ANTHONY. *Science Is a Sacred Cow.* New York: E. P. Dutton & Co., Inc., 1950.

TEILHARD DE CHARDIN, PIERRE. *The Phenomenon of Man.* New York: Harper & Row, Publishers, 1965.

VAN DUIJN, P. "The Interaction of Theories and Experiments in Science," in *Information and Prediction in Science*, eds. S. Dockx and P. Bernays. New York: Academic Press, 1965.

CHAPTER 2

SCIENCE AND
SCIENTIFIC EXPLANATION

I shall put forward a speculative hypothesis, which has in its favor only one argument—that it does connect and give tentative answers to all these questions.
—Julian Schwinger

THE NATURE OF SCIENCE: FURTHER EXPLORATION

Science as a Human Institution

Science is a product of human thought practiced by human beings as a means to human ends. Like the other thought systems, science is a tool men have developed to perform certain tasks. It is a powerful tool; science has radically restructured and reoriented the lives of most people in the world, yet it remains—for all that—a tool. Were we to invent a more powerful ordering system, we would abandon science in its favor in the same way that we abandoned portions of the theological systems which preceded science. From time to time, we will lapse into the habit of speaking of science as if it were a superhuman entity with a life and purposes of its own. This is inevitable because science is a supra-individual and supra-national institution. But to the extent that this habit shifts our focus from the fundamentally cultural

nature of science, it is a tendency we must guard against.

Science is a *megaconstruct*. We saw earlier that constructs are categories of experiences we fashion out of immediate, perceived events. Concepts, on the other hand, are non-empirical ideas lacking substantive content which we develop out of immense arrays of experience over very long periods of time. What, then, do we mean when we say science is a megaconstruct? We mean that science is a hierarchical structure built out of great numbers of concepts, constructs, and the relationships between them. Since science is a comprehensive ordering system, applicable to most regions of the P-plane, it is located in the zone of transition between constructs and concepts (Fig. 2–1). Like the other ordering systems, science has its origins in what we call the empirical world—that is, in constructs lying close to the P-plane. But it is also clear that science generates and incorporates concepts. Science is what a geographer calls a boundary-dwelling activity. It

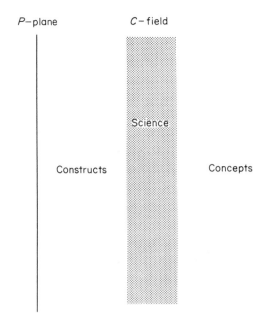

FIGURE 2–1

Science as a megaconstruct.

and so forth, but it is not necessary to do so to make our point. The substantive parts of the preceding scientific narrative can be divided in the following way:

Constructs	Concepts
car	mass
—member 1	velocity
—member 2	momentum
driver	pressure
brake pedal	friction
tires	deceleration
road	distance
etc.	etc.

While there might be some quibble over the assignment of a few terms to one or the other categories, this is not critical. What is important is the way science chooses concepts and makes them characteristics of constructs to produce explanatory narratives.

Because scientists are selective in the concepts they use (God, for example, is not permitted), it is obvious that the scientific method acts as a filter. Scientists, when practicing good science, usually ignore asserted C-field relationships for which they have no evidence. They are not seen as being relevant. Ideally, relationships postulated by theology or emotion are not perceived by the scientist because the scientific lenses through which he views the C-field do not permit him to see relationships between constructs which reflect light at their particular wavelengths. Science seems to be like a culture sometimes, and the argument that science is at least a distinct subculture has some validity. Scientists let their hair down emotionally and theologically as well as in other ways during their off hours. But so far as their professional activities are concerned, the way of life is total. One either believes that science is a valid and fruitful ordering system or one does not—there is rarely a middle ground.

That science is a subculture appears more and more convincing when we examine the development and diffusion of modern science, which is a peculiarly Western phenomenon. It is true that the scientific viewpoint has now spread to a large number of non-Western areas, especially in the last several decades. But it is also true that the spread of the scientific attitude is always an

persists and thrives because it enables us to connect constructs and concepts in an efficient and productive manner.

When we practice science we use concepts to build coherent networks of constructs. We attribute concepts to constructs in ways which produce satisfactory answers to our questions. For example, suppose we are explaining how a car struck a stalled vehicle at an intersection. First we attribute to the moving car a certain mass and velocity, and thus a certain momentum. The realization of the driver that something was amiss ahead led him to apply pressure to the brake pedal. This produced friction against the brake drums and thus friction between the tires and the road and thus deceleration. Because the momentum of the moving automobile was greater than the decelerative force which was applied in the distance remaining between the two cars, it was impossible for the driver to stop his vehicle before striking the stalled car. We could complicate this explanation by talking about human reaction times, attention spans, fluid mechanics,

element and index of the diffusion of Western culture. Of course, non-Western peoples often possessed accurate knowledge of the world before their contacts with Western people. But science as a distinct subculture having profound and extensive effects on all of society and on other realms of order is largely a Western innovation.

There is additional evidence that science is a subculture. Scientists exhibit the same concern for the preservation of science as they know it, as do apologists for any particular culture. Thomas S. Kuhn, in his fascinating book *The Structure of Scientific Revolutions*, has investigated the process of culture change which occurs within sciences, and has shown that science is very much like culture in that once a scientist has committed himself to practicing science in a certain way, using accepted theories and their consequences, it is very difficult and often impossible for him to adopt new ways of viewing old problems. Like political revolutions, scientific revolutions make exiles of most defenders of the old order.

Science is not a cumulative body of knowledge which is approaching an ultimate truth at an increasing or decreasing rate. Science is an order men impose on reality because it is convenient for them to do so. When an old form of science no longer enables scientists to reach the goals they set for themselves, it will be discarded in favor of new ideas which are more potent. Most geologists, for example, now accept the validity of the theory of continental drift and ocean floor spreading, because it explains the facts they experience more satisfactorily than older ideas. Yet one of us can remember a renowned professor of geology at a major university heaping scorn and ridicule on the idea in the early 1960s. Nothing is more illustrative of the fact that science is a culturally determined megaconstruct than the history of revolutions like this.

Kinds of Science

We have so far been talking about science as if the order which it imposes upon the world were a wholly unified body of knowledge. Yet we all have heard phrases like "pure and ap-

plied science" and "physical versus social science" used in discussions. Within Science—with a capital S—there are different kinds of sciences. But to the extent that they imply fundamental differences, most of the usual distinctions are artificial and nonproductive. We take the view that it is his *method* of imposing order which distinguishes the scientist from practitioners of other ordering systems, and that the method of science is common to all sciences, pure and applied, social and physical. We prefer to emphasize the features which scientists hold in common rather than the small differences that may exist. But before we examine the elements common to all sciences, there is one difference we must clarify because it bolsters another distinction we have made.

Carnap distinguished between the *formal* sciences and the *factual* sciences. Factual sciences are those which have empirical content. These would be all disciplines which deal directly with factual data, such as geography, anthropology, physics, history, and biology. The factual sciences comprise all disciplines—pure and applied, physical and social—except the formal sciences. The latter are those which have no empirical content but which are systems of thought. Thus the formal sciences are logic and mathematics, including geometry. Note that the two formal sciences are based on the two megaconcepts, *relationship* and *number*. Although the formal sciences may be applied to empirical phenomena and demonstrate their greatest values in such applications, neither of them concerns itself with any class of events or experiences. Logic and mathematics are thought systems which practitioners of the factual sciences use to order the empirical world of experience. Mathematics and logic are the only pure sciences in the sense that they are the only disciplines concerned with wholly non-empirical ordering structures. Because the formal sciences have no empirical content, they enter the construct portion of the C-field only when they are applied to constructs.

The distinction between formal and factual science mirrors our earlier distinction between concepts and constructs. Constructs are categories of experience which have empirical content. We can think of them as mental bins into which we sort experiences as we encounter them. Concepts are abstract, non-empirical ideas based on *number* and *relationship* or combinations thereof. The factual sciences produce knowledge about the world in the course of their operations. They are also concerned with explaining our experience by answering our *how* questions. The work of a factual scientist is subject to testing; his findings can be supported or weakened by empirical events. Practitioners of formal science are not concerned with explaining experience nor are their conclusions subject to empirical confirmation. The criterion against which we evaluate mathematics and logic is that of the absence of self-contradiction. Any logical or mathematical system is derived from certain fundamental ideas called axioms. Using these basic assumptions, some conclusions are valid according to the adopted rules of manipulation and others are not. The criteria by which formal sciences are evaluated are internal to the sciences and are based largely on consistency, although factors such as economy and elegance are often important also.

The way in which we use the formal and factual sciences may be illustrated by an example. The syllogism:

1. All men are mortal
   Socrates is a man

   Therefore: Socrates is mortal

is a valid argument. The conclusion that Socrates is mortal follows from the stated premises. Because the first premise is a universal statement, once we have made the second statement identifying Socrates as a man, he could not be nonmortal. If Socrates is a member of the class of men he must, by that fact, be mortal. Note that the conclusion that Socrates is mortal does not contradict the findings of the factual sciences. It is indeed our experience that all men die sooner or later. Here the formal science of logic provides a structure for presenting the results of factual science.

Let us, however, consider a second syllogism:

2. All carrots are elephants
   Socrates is a carrot

   Therefore: Socrates is an elephant

In the formal science of logic, this argument is valid. If, indeed, Socrates is a carrot and all carrots are elephants, then Socrates must be an elephant. Since the *form* of syllogism 2 is identical to that of syllogism 1, and since that particular form of the syllogism is one which our rules of logic hold to be valid, the argument stands. But unlike the argument in syllogism 1, that contained in syllogism 2 contradicts the results of experience and factual science. It is our experience that *carrot, elephant,* and *Socrates* are distinct, mutually exclusive constructs. The relationships hypothesized in the premises and conclusions of syllogism 2 can only be accepted by radically restructuring and changing the content of these constructs. In this case, we would argue that while the form of the argument may be valid, its factual content is not, and so we reject the conclusion (and the premises) as being contradictory to our knowledge of the world, however valid the form of the argument may be.

Geographers, like all factual scientists, use the formal sciences to help explain experience. By operating in the boundary zone between constructs and concepts, geography incorporates in its explanations ideas developed by the formal sciences. Neither geographers nor any other factual scientists can operate without the concepts produced by mathematics and logic. Geography has usually been logical and has never been wholly unmathematical, but the recent emphasis on using mathematics signals a recognition that more intensive use of that particular formal science is prerequisite to further advances in geographical knowledge and theory.

Because mathematical and logical structures can be evaluated without reference to the constructs to which they are applied, they greatly increase our power to manipulate those constructs. We can check the validity of our manipulations without being distracted by content. No factual science can make significant progress in today's world without intensive use of the fruits of the formal sciences. Throughout the rest of the book we will not be concerned with the formal sciences as such. When talking of scientists or science, we will refer to the sciences which have empirical content and which are devoted to explaining experience.

But we should never forget that the formal sciences are basic foundations of all factual science, and we cannot operate effectively without them.

SCIENCE IN THE SERVICE OF MAN

Science provides us with a means of determining what is happening in the world in which we live, as well as with a means of changing the world to make it more suitable for our own purposes. Science provides a more successful means of manipulating experience than the other ordering modes because scientists are the most adept at formulating and testing the answers they produce in response to the questions we ask.

The Capacity of Science for Manipulating Experience

Looking at the questions we ask scientists to answer for us and the things we ask scientists to produce for us, it is clear that we view science as a tool which enables us to take an active role in producing the events which impinge upon us. Since World War II, science has been viewed as an activity which will, at our direction, remake the world in which we live. Scientists have done more to transform our lives over the last century than any other group of people. Moreover, the transformation of life in the Western world in the last century has been absolutely and relatively greater than that which occurred in any previous period in history. When we discuss science and scientists, we are discussing the shapers of today's and tomorrow's world.

Perhaps our concern with manipulation is best demonstrated by considering the use we make of the health and medical sciences. Most of us, when we are ill, are very intensely concerned with being restored to a state of health. We expect the health professions to be able to explain to us why we are not feeling well and to take whatever actions are necessary to produce a more satisfactory state of affairs. Increasingly, we expect medical scientists to manipulate events in a preventive fashion.

Disease prevention—for example, smallpox vaccination and milk pasteurization—has been practiced for over a century, and the array of diseases against which we can take prophylactic action continues to increase. Thus our medical futures become progressively more secure and more satisfactory.

Ideally, a scientist would like to know what combinations of events produce a given condition before taking corrective or preventive action. This is not always possible; our friends in the medical profession tell us that they still do not completely understand how aspirin works. It is undoubtedly effective against some kinds of pain, but exactly how aspirin relieves pain remains obscure. In such cases, where there is an observed constant correlation between events, scientists usually use the remedy for the undesirable condition even if there are uncertainties involved. Often, the potential consequences of using the remedy can be demonstrated to be less damaging than not using it. Sometimes doctors have a very good idea of what causes a disease and what is required to cure it. In the case of polio, for example, we have progressed from a situation in which the cause of the disease was unknown in the early 1950s to the present situation wherein most types of polio are preventable through immunization.

In cases such as polio, scientists have succeeded in describing accurately the event chains which cause the disease. Such knowledge enables them to reverse the sequence in some instances and restore an ill person to health. Knowledge of the factors which produce a disease often enables us to interrupt the causal chain at some critical point so that the sequence is not completed and the disease does not develop. The most effective means of controlling malaria, for example, is to prevent mosquito larvae from hatching and thus eliminate the vector responsible for transmitting the disease to man.

In the same way that we want the health professions to manipulate our experience to prevent us from becoming ill or to cure us if we do, we ask other scientists to perform similar manipulative functions. We require economists to help us understand and adjust our corporate and national economic systems. We wish to manipulate economic activity to keep conditions compatible with chosen criteria of economic health. Aeronautical engineers are expected to provide us with more rapid means of travel. Physicists provide us with new sources of energy to heat and cool our houses and power our machinery. Geographers are expected to rectify existing spatial incongruities and to take preventive action against possible spatial incompatibility in the future. No matter what branch of science we consider, the demand for diagnostic, prescriptive, and preventive activity is similar. We want scientists to prevent us from experiencing unpleasant events and to structure the future in such a way that we will experience pleasant events in as great an abundance as possible.

The extent to which science has in the past been able to manipulate experience is only a fraction of the extent to which scientists will determine events in the future. Even taking into account the current declines in birth rates in the developed nations, our world, because of the enormous productive and consumptive power of the Western nations, becomes daily more intractable. Man and habitat are both degenerating, as is indicated by the rising incidence of diseases like cancer and the increasing pollution of the physical environment. If our species is to survive at Western levels of affluence, remedial work must be done to help non-Western peoples attain the same levels of living we enjoy, and we must ameliorate the undesirable side effects of technology on the environment. If by some miracle a state of worldwide affluence can be achieved, a tremendous amount of manipulation of the future will still have to be undertaken to keep such a finely tuned mechanism in order. We are accustomed to thinking of the human body as an extremely intricate system, and it is indeed the most complex form of life that 3.5 billion years of evolution have produced. We are now coming to realize that the entire complex of humanity is as complicated and delicate as is the human body, if not more so. Given the sizes of the organizations we are now capable of sustaining, keeping the whole of humanity in good social, political, economic, and spatial health requires much greater expertise than does keeping an individual healthy. Practicing "preventive medicine" for all of humanity will increasingly become the major focus of scientific activity.

Though not all scientists are concerned with finding solutions to immediate problems, society uses science as its major problem-solving institution. When we perceive the existence of unsatisfactory conditions, we expect scientists to be able to remedy them for us. We call upon them to perform diagnoses, to prescribe remedies, and to provide us with the means of avoiding problems which we think might arise in the future. Scientists have been rather successful in performing these tasks for the societies which support them, especially in the last several decades.

The Success of Science

A distinguishing characteristic of scientific ordering is the primacy of questions. Together with the institutionalized, systematic nature of scientific investigations, the emphasis on questions gives science a capacity for accurate explanation and prediction far in excess of that alternative systems can offer.

When we think of scientists at work, we almost always think of them attacking certain problems. The notion of a problem is a rather broad one, but one which deserves our attention because the absence of a focus on specifiable problems has often hampered the progress of geography and the other social sciences. What, then, constitutes a viable problem which can legitimately serve as the focus of the scientist's professional activity, and how do scientists go about formulating the problems to which they devote their energies? The answers to these questions are important because a strong "problem sense" is the key to success for the individual scientist, just as the ability to solve problems is the hallmark of a successful science.

In essence, problems are questions asked either by laymen or scientists. An unanswered question is a problem for the person who asks it. A person may have certain goals in mind but be unsure as to the most efficient means of attaining them. In such cases, laymen, if they are aware of the expertise possessed by different scientists, may consult the appropriate individual. For example, an entrepreneur may have put together a parcel of land in a suburb which he thinks is suitable for a major shopping center. In order to convince merchants to rent space in his center, he must demonstrate that it will be profitable for them to do so. Often entrepreneurs will hire a marketing geographer to delimit and analyze the trade area from which the proposed center will draw its customers and to make predictions concerning the volume of sales merchants can expect. Alternatively, a grocery chain may wish to open a store in a sector of the city but may be unsure as to where to locate the facility so that it will attract the maximum amount of business. Here, again, a geographer can find the optimum location within the selected region.

It would be easy to multiply examples of problems brought to science by laymen, both in geography and other disciplines, but the principle involved seems clear enough. Laymen usually address questions like those described above to practitioners of the various sciences. Whether we are ill or whether we wish to obtain advice on where to locate a supermarket, we normally turn to a practitioner first. The fact that laymen and society in general rely on practitioners to solve their problems emphasizes the critical role that well trained practitioners play in producing the success which a science enjoys. Without practitioners capable of solving the relevant problems laymen pose, a science is little more than a mutual admiration society. No science can possibly expect to achieve any measure of objective success until it has produced a corps of competent practitioners able to apply their specific expertise to significant problems raised by laymen.

The second source of problems to which scientists try to find solutions lies in the science itself. In the course of their practice, practitioners occasionally come across problems which cannot be solved using the current arsenal of tools which the science has at its disposal. If these are problems which the practitioner feels he should be able to solve, he, in conjunction with the science's methodologists and theoreticians, will set out to find a way of solving the problem in question. In this case, a problem which originates externally becomes an internal

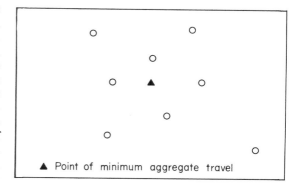

FIGURE 2–2
Point of minimum aggregate travel.

problem or question because it identifies a missing element in the current structure of the science.

Methodologists and theoreticians may, of course, discover such gaps themselves without questions being raised by the science's clients. For instance, methodologists and theoreticians in geography are now interested in finding a technique which will efficiently locate the point of minimum aggregate travel for a distribution of locations. That is, given a pattern of points (Fig. 2–2), there exists one point, not necessarily a member of the original set, which is closest to the entire distribution of points in the sense that if a person traveled from each of the points to the central location, total travel would be minimized. If you wish to think of this problem in a more concrete framework you can think of the distribution as that of small villages and think of the point we wish to find as the future location of a deep well to be drilled to provide all the villages with water. Because all the villages are equal in size, we wish to drill the well as close to all the villages *together* as possible. Geographers can solve this kind of problem with a technique known as the Weberian weight triangle, which we will discuss in greater detail later. The method can be programmed as a trial-and-error procedure for a computer. Because such trial-and-error methods tend to be costly in terms of time and money, methodologists and theoreticians would prefer to derive an elegant solution to this class of location problems—that is, a mathematical formula into which one could enter the locations of a distribution of points and then solve to yield the single optimal point immediately. Such a solution would save a great deal of computer and programming time and would reduce the aesthetic discomfort provoked by the use of iterative techniques.

Here, then, is an instance in which a solution to a problem exists which is used by practitioners. But the solution is not wholly satisfactory to methodologists and theoreticians on grounds of efficiency and aesthetics. The absence of an elegant solution is seen as an imperfection within the science. All sciences have internal theoretical and methodological gaps. If we think of a science as an elaborate, multilevel network in the C-field, there are always links between C's that we think should exist but whose existence has not yet been verified, and there are always elements of the structure whose simultaneous existence seems to be incompatible. Methodologists and theoreticians spend much of their time perfecting their structures so that they will be of greater utility to practitioners and more pleasing as well.

The importance of questions or problems in a science cannot be overemphasized. In many ways, asking the right questions is more difficult than providing answers to them once they have been formulated. Successful scientists make progress because they keep specific problems— problems which they have the capability of solving—foremost in their minds. A major factor which continues to hamper progress in geography and the other social sciences is either a difficulty in formulating problems or a concern with very general problems, or both. Some individuals have the ability to see problems which most scientists either do not perceive or do not think are important. Developing that elusive trait called "problem sense" is one of the most important aspects of the training of a scientist, whether he is destined to become a practitioner, a methodologist, or a theoretician. Practitioners need this skill as much as theoreticians, because laymen often come to the practitioner when they realize that they have a problem but can't specify what it is. In the same way that a medical practitioner is ex-

pected to diagnose an illness, geographers must be capable of identifying the exact nature of a problem before they can start attacking it. A feeling and sensitivity for problems and potential solutions seems to depend most on intuition and originality combined with rigorous training in theory and methodology. In this most critical area, the dispassionate and objective patterns of thought we normally associate with scientists can be drawbacks. In identifying and proposing solutions to problems, the intangible feelings in one's guts are often very valuable.

Science has traditionally been a luxury few civilizations could afford, but it is increasingly becoming absolutely necessary to the preservation of the social systems it has created. In any case, scientists who are supported with public funds have no right to carry on their activities in a social vacuum. A society which supports science financially has a perfect right to expect that scientists will give as good as they get. The massive financial support we are willing to accord science is evidence that in the last several decades science has often given much more than it has received. Investment in science and scientists has been very profitable indeed. Nonetheless, we cannot forget that scientists are useful because they devote themselves to solving the problems they can handle and to developing techniques for solving those which are currently intractable.

We can think of science as a massive *megastructure* knit together by innumerable communications channels. Conventions, monographs, journals, symposia, and disciplines are all important communications systems which unite scientists the world over. Most good scientists spend as much time reading about the research of other scientists as they do solving the problems which interest them. This being the case, scientists are apprised early of the results others working on the same problem have produced. Since scientists try to keep tabs on each other's work through these media, science as a whole is efficient at solving problems. Because scientists do work on problems of common concern and keep close track of each other's work, all aspects of scientific investigation and explanation are more objective and accurate than those encountered in other ordering systems. This does not mean to imply that scientists or science are objective according to any absolute criterion. Like other systems, science is order imposed upon the world, and even though we may feel that the order science produces is closer to whatever order exists in the world itself than that produced by other systems, this evaluation is itself subjective and teleological (based on the goals we have in mind). Even so, science has procedural safeguards which assure that the results of scientific analysis will be as accurate as is reasonably possible and more accurate than alternative systems. Here, again, the fact that the results of one scientist are replicable by another is extremely important. Without replicability, it would indeed be difficult to attain the degree of objectivity which science has achieved.

The nature of science and the focus on soluble problems have combined to produce a *method* of analysis and prescription which has been found to be very successful at fullfilling the tasks scientists set out for themselves. That the method of science is a valuable technique of problem analysis and solving is indicated by its continued use. Science earns its keep by using its method to answer our *how* questions more successfully than any other system can.

THE METHOD OF SCIENCE

The Formulation of Hypotheses

In order to solve a problem scientifically, one first needs a problem. Accordingly, the first and most important element of the scientific method to be considered is that of hypothesis formulation. Problem recognition and hypothesis formulation are two sides of the same coin, and intuitive ability is prerequisite to successful hypothesizing just as it is an important element of problem sense. Investigations often start with a kind of primeval rankling in the individual's mind which eventually produces a thought which starts with "I'll bet . . . ." Scientists rarely start an investigation cold, without any idea of what they are looking for. Sitting down with a pile of data and looking in the data for ideas about relationships is

very unproductive. It is equivalent to looking for something but having no idea of what it is you are seeking. Although a person may find something by engaging in this process, the chances of producing worthwhile answers to questions are much better if one has some questions to answer in the first place.

A hypothesis is basically a potential answer to a question or solution to a problem. As we saw earlier, questions may be raised by laymen or by scientists themselves. Hypotheses always deal with relationships between events. To take a current problem in social science as an example, suppose that we are worried about the frequency of urban riots by members of minority groups. As geographers, we are especially interested in the reason why riots explode at some places but not at others within metropolitan areas. Both as scientists and as citizens, we would like to know what causes such civil disturbances so that we can prevent them from occurring in the future. We might begin such an investigation into the locations and causes of urban minority riots in two ways. One way would be to consider a large number of situations which might result in riots. We could admit that, basically, we know little or nothing about the causes of such events, and start examining a great number of different characteristics of cities to see if we can discover a significant relationship between some factor which varies within the cities and the incidence of riots. This is not normally the way science operates, however, and it is not a productive way to begin solving this particular problem. Whether they are right or wrong, we do have some ideas about the things which cause riots. Rather than taking a scatter-gun approach, we would focus our attention on a restricted set of characteristics *which seem to us* to be logically related to the incidence of riots. We might suggest that riots are caused by poor police protection or by unemployment or by inadequate housing or by any combination of these variables alone or in conjunction with some other phenomena.

The assertion that some sort of relationship exists or may exist between two events or sets of events constitutes a hypothesis about the way our experience of these events is structured. More often than not, such hypotheses are

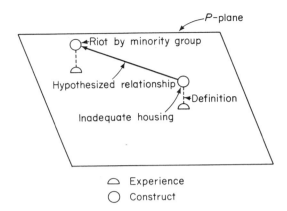

FIGURE 2–3

A hypothesis in the C-field: inadequate housing and urban riots.

causally stated. We are usually interested in whether an event acts in some way so as to produce another. Hypotheses are answers to *why* questions about our experience. Logically, a hypothesis that *X* causes *Y* can be stated in the form: *If X, then Y* (we think).

A hypothesis asserts that an interconnection between constructs exists in the C-field. Suppose, for the sake of argument, that urban riots are caused by inadequate housing; that is, when the ratio of people per residential room reaches some critical threshold, civil disturbances will almost always occur. But also suppose that we do not yet know that this is true. What we are asserting in our hypothesis that inadequate housing causes urban riots by minority groups is that there is a relationship in the C-field between the two constructs *inadequate housing* and *urban minority group riot* (Fig. 2–3). We can think of this hypothesis as being represented by a path from the P-plane through at least two C's, and back to the P-plane of event experience. Note that in our representation we have suggested that the relationship is one-way—that is, inadequate housing causes riots, but not all riots are caused by inadequate housing. Events rarely have single causes, and especially in the hypothetical case under discussion we would usually consider more than one set of relationships. Thus we could hypothesize that three or four things in conjunction cause riots (Fig. 2–4).

Or we might guess that any three of the four, taken together, will be sufficient to set a people to rioting.

Hypothesizing, then, is a fancy sounding term which is meant to obscure the fact that scientists do a great deal of educated guessing. Some guesses are accurate and confirmed by the scientists who put them forth; many hypotheses are found not to be accurate and are never heard from again. In reviewing what scientists produce, we must realize that we are looking at a very biased sample of all the hypotheses which have been formulated and tested. Because we have no *Journal of Negative Results*, scientists usually report to each other only those hypotheses they have been able to confirm. We will talk a little later about how we go about confirming or rejecting a hypothesis. For the moment, let us examine some of the more immediate steps the scientist must take before he can get to the stage of testing his hunches.

Observation and Description

Just as the origins of all order are found in experience of the world, all science has a basic concern with facts. Facts are selected experiences which a scientist judges to be particularly relevant to the testing of the hypothesis he has formulated. Getting the facts he needs to

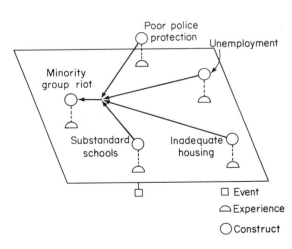

FIGURE 2–4

Hypotheses about multiple causation.

do this involves the scientist in a number of processes, all of them more or less influenced by the ideas he has already formulated.

The first problem in any scientific investigation is to decide which facts are relevant. When a scientist starts out to explain an event like the occurrence of racial disturbances in certain isolated locations in a city, he does not examine every possible fact that can be garnered about the city. Were he to try, at the rate information is being published these days he would never finish that task alone. Scientists observe very selectively and the selection process is based upon their preconceived notions about what might be relevant. Thus we encounter the first of the many "feedback loops" within the scientific method. Existing theory determines what facts will be gathered in any investigation. To help solve the problem of urban riots, then, we would probably gather facts on racial composition, median income, housing conditions, unemployment, literacy, ethnic composition, age/sex structure, and location of previous residence, to name a number of the relevant possibilities.

After a scientist has decided which facts might be relevant, he must describe them in a manner which is as unambiguous as possible. This involves two steps, the first of which is definition. Definitions themselves are of two kinds: nominal and operational. *Nominal definitions* are those which are found in dictionaries and those which we use when we make constructs. They express invariant connections between the P-plane and the C-field (Fig. 2–5). For normal discourse and activity, dictionary definitions are usually adequate, although we are all aware of the magnificently futile arguments which can occur when two people do not agree on the same definition of a term but think they do.

In science we usually wish to be as precise as possible. Some terms, such as "age" and "sex" or "address" are reasonably precise and need no further definition. Often, however, we are talking about hypothetical events, constructs, or concepts whose existence has not yet been firmly established. Thus scientists often use *operational definitions*. These are definitions which are produced by performing certain operations,

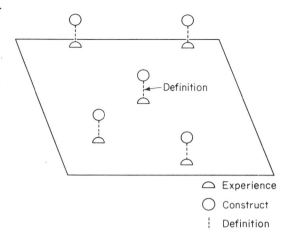

△ Experience
○ Construct
┊ Definition

FIGURE 2–5

Definitions are interconnections between experiences and constructs.

and the operations themselves are implicit or explicit parts of the scientific definition of the construct or concept in question. To determine a person's vital capacity, for example, operations must be performed to measure the volume of air he can exhale after taking a very deep breath. The entire set of operations by which "vital capacity" is measured is an implicit part of the definition of the term.

Suppose that we are still trying to explain why riots occurred in the ghettos of large metropolitan areas. The term "ghetto" is not one which is precisely defined for scientific purposes despite the careless abandon with which it is used in normal discourse. If our hypotheses about riots are going to talk about ghetto and nonghetto conditions, it is obvious that this term must be rigorously defined. In this particular case we might decide that we would use the term "ghetto" to designate any city block wherein blacks comprised 75 percent or more of the residents. The scientific definition of "ghetto" would then incorporate the operations required to measure the proportion of black residents. Now it is perfectly reasonable to raise questions about the figure chosen as the definition of "ghetto." You might contend that $66\frac{2}{3}$ percent black residents is a high enough proportion for an area to be termed a ghetto. We, on the other hand, might prefer to operationally define any block with between $33\frac{1}{3}$ and 75 percent black residents as a "transition neighborhood." What is important about operational definitions is that they are very useful despite the fact that the procedures upon which they are based are sometimes debatable. Once we have specified operational constructs, we can delimit ghettos accurately in a replicable fashion. Operational definitions are precise and replicable, and they allow us to get on with the business of science.

The second aspect of description with which a scientist concerns himself is measurement. Having decided which experiences are relevant to the question at hand, and having defined these experiences for his particular purposes, the scientist then begins to concern himself with the form in which the facts are to be collected and recorded. Facts may be collected and recorded in verbal, written, or numerical form. Scientists prefer to collect facts in numerical form or to render facts into numerical form wherever possible. There are several reasons why scientists insist on applying the concept *number* to experience. One reason is convenience; facts converted to numerical data are much easier to handle, store, manipulate, and reproduce than are facts recorded in written language. Second, facts recorded numerically are usually more accurate and objective than those recorded in other ways. For example, the ghetto resident's response that housing is unhealthy in block A is a very important fact because it is his evaluation of the environment in which he lives. But for many purposes, the knowledge that on the average, five families share a bathroom in block A would be more useful. Often, as we have seen, measurement is implicit in the definition of facts themselves.

Because of the scientist's inclination to use numerical data, observation usually involves measurement. At the present time, measurement is a major problem in the social sciences because techniques are poorly developed for measuring many of the facts which social scientists would like to measure. Geographers and other social scientists are making progress in developing these needed techniques, but we still require improved

means of observing social facts accurately. Measurement is a refinement of the normal descriptive process. By applying the concept *number* to experience we set up an alternative means of description. This alternative has proved to be very valuable and that is why scientists insist upon the measurement of experience whenever possible.

Numerical facts are more accurate, more objective, and more replicable than the alternatives. Measurements are performed with a standardized scaling device and according to standardized procedures. Thus the results of a measurement process are usually replicable by any other scientist. Equipped with a simple thermometer, each of us can determine whether in fact water boils at 100°C. at sea level. If we want to compare temperature at two places at the same time, we can state precisely how much warmer (on a particular scale) one place is than another, rather than relying simply on the ambiguous evaluation that one place feels hotter than the other. So far as scientists are concerned, measurement is the most important form of observation. Measurement is not only more accurate than other means of description in itself, it also helps make science more accurate and objective because measurements are usually repeatable at will.

Observation is a complex process. We do not simply go about collecting scientific data by some process of osmosis. Accurate and efficient acquisition of scientific data is a very difficult task, and methodologists devote a good deal of time to devising efficient data collection procedures. Scientists have always faced problems of selecting relevant data, but such problems will become increasingly serious in the future, especially in the social sciences. Geography and the other social sciences are now in a transition period between the decades when socioeconomic data were very scarce and the future decades when data are going to be superabundant. The observational problems of science of the 1980s and 1990s are not going to be those of collecting scarce data, but rather those of properly selecting the facts which will spew from data banks on command. More and more, attention must be given to techniques of data selection and manipulation, and to attempts to formulate more clearly the nature of facts and their relation to theory. In the same way that science creates an order to impose on the world of experience, theory creates relevant facts. Observation—data collection and recording—appears to the uninitiated to be a simple, fundamental operation. Although there is no denying that data collection is fundamental in every sense of the word, observation is by no means a simple element of the scientific method. And observation will become much more complex by the time those to whom this book is directed are full into their scientific careers in the decades after the year 2000.

Classification

After an investigator has obtained some data he thinks are relevant to a particular problem and after he has properly defined his variables, the next step is to manipulate the data to make them more useful. Often the step after observation is to generalize the facts by using a classification procedure. Classification is the first generalizing step that the scientist takes on his route to explanation. We have already encountered one form of classification in Chapter 1, when we noted that people construct classes of experience to which they relegate the events which impinge upon them. A construct is a class composed of certain kinds of events. The class-construct *rose* consists of all experiences which meet our criteria of "roseness." Most of us occasionally have the experience of encountering an object, say an unfamiliar flower or tree, which looks like it ought to belong to one of our existing mental classes. But we are not really sure. In such instances we are unsure whether the object or experience belongs to an existing class or to one which we have not yet established. In the case of more familiar experiences, recognition that the perceived event is a member of an often enountered class is almost automatic and completely unconscious. Constructs, then, are the results of taxonomic or classificatory procedures which were learned primarily in childhood but which continue to operate throughout life. Individuals must classify their raw material—their experiences—

and scientists also find it necessary to classify those particular experiences they have decided are relevant facts.

Taxonomy occupies a key position in the development of every science. The creation of a comprehensive taxonomy of the subject a science studies has often been a prelude to revolutionary change. For example, Linneaus' classification of life forms according to genetic relationships was a major stimulus to research which resulted eventually in the theory of evolution. By segregating organisms into genetically similar categories, Linneaus pointed biologists toward the idea of common ancestry. Similarly, Mendeleev's table of the elements stimulated rapid advances in chemistry. By arranging elements in order of their atomic weights, Mendeleev was able to go beyond the classification to predict the existence of elements which were not then identified. In both cases, a classification of existing knowledge produced the insight which led to hypotheses and predictions about relationships between the phenomena classified. Classifications are not fruitful ends in themselves. They are valuable because they enable us to discover relationships more easily than we can from unorganized data or facts.

At a more specific level, classification is necessarily an important element of individual research projects. For example, suppose a geographer interested in the location of urban civil disturbances had gathered some data on racial and ethnic composition, income, housing quality, and unemployment rates for census tracts in a large metropolitan area. If it is a large area, there may be several hundred tracts and for each tract we may have eight to ten pieces of information. Obviously, we have a lot of facts to feed into our analysis. The question is, what do we do with them? Sitting and examining the data in tabular form is likely to be confusing. With several thousand figures to look at, it is difficult to discern any relationships among the facts we possess.

In a case like this we would begin to "shrink" the number of things we have to think about by combining the observations into groups. For example, if our index of housing quality was average number of people per room, we might divide the array of several hundred census tracts into four or five groups, with the divisions based on values of this particular index. Tracts could be classified similarly on the basis of any of the other characteristics we have measured. The procedures by which one arrives at an optimal classification are not our concern at the moment; Chapter 6 is devoted to this topic. At this point we need only note that our classifications should contribute to the investigation which evokes them and that classes should be as homogeneous as possible. What *is* of interest here is the generalizing function which classification fulfills. We have several hundred ratios of people to residential rooms. By dividing this distribution of observations into five categories we "collapse" the several hundred items into five classes which we can think about much more efficiently. As an example, consider a hypothetical map (Fig. 2–6a). Imagine this as part of a map of census tracts in a metropolitan area with the number in each tract giving the ratio of people per room within that tract. Contrast this with a map produced by dividing these observations into several classes and shading the tracts on that basis (Fig. 2–6b). The map based on classes conveys information about housing conditions much more efficiently and clearly than does the plotting of raw data.

This illustrates a basic paradox. The map based on classes *contains less* information than the map with the raw data. When we make the class map we lose information since we cannot reproduce from the class map the data map which underlies it. The class map tells us which class any tract falls into, but it cannot tell us what the exact ratio of people to rooms is in that tract. Yet the class map *conveys* information more readily than does the data map. By abstracting and generalizing the relationships among facts, classification enables us to see patterns that are much more difficult to discern from the data themselves. The trouble with raw data, then, is that they usually contain too much information for us to comprehend. Trying to discover relationships among raw data usually produces a condition of *information overload*, a situation in which we cannot see the forest because there are so many trees in the way. By classifying, we summarize data and thus render it into a form which we

a

0.3	0.4	0.7	0.8	0.8	0.6	0.7	0.7	1.0	0.9	0.9
0.5	0.7	0.8	1.0	0.9	1.0	1.1	1.0	0.9	1.2	1.4
0.5	0.6	0.8	1.1	0.9	1.1	1.3	1.4	1.3	1.2	1.3
0.4	0.6	0.8	0.7	1.0	1.1	1.4	1.4	1.2	1.4	1.3
0.4	0.8	0.8	0.7	1.0	0.9	1.4	1.2	1.6	1.7	1.5
0.5	0.6	0.6	0.8	1.0	1.0	1.3	1.3	1.2	1.5	1.5
0.4	0.6	0.6	1.1	1.1	1.0	1.4	1.4	1.2	1.6	1.7

b

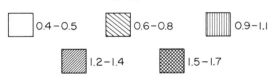

Average number of residents per room, by census tract

☐ 0.4–0.5   ◪ 0.6–0.8   ▥ 0.9–1.1

▨ 1.2–1.4   ▦ 1.5–1.7

FIGURE 2–6

A map as an areal classification.

can comprehend. As the first generalizing procedure performed on experiences, classification is an important step toward explanation.

The crossover here between classification and operational definition should be obvious. One might decide, for example, that a ratio of people to rooms higher than 1.0 (that is, on the average there is more than one person per room) defines a category or construct which we will label *unsatisfactory housing conditions*. This part of the process by which we make constructs. Thus if we were to look at our usual P-plane and C-field representation through a very powerful microscope, we would see that each construct is connected to a large number of experiences which we assign to that category (Fig. 2–7). At the micro level, classification is a process which occurs almost automatically for most experiences. We do not normally notice the process because it goes on at a scale we do not observe. The second way of characterizing classification is to think of selected data as very closely spaced constructs in the C-field which represent experiences a scientist deliberately encounters by collecting data. Classification can then be viewed as the regionalization of these constructs in the C-field (Fig. 2–8). We divide constructs into classes on the basis of their proximity to one another numerically. We can conceptualize numerical similarity as being represented by locational proximity in the C-field.

The fact that we can view classification in two ways illustrates two important principles. One is that scale of analysis is critical in science or any other ordering system. By looking very carefully with very powerful instruments, we see things that are not visible at higher levels of abstraction. We cannot overemphasize the importance of scale. As many futile arguments among geographers arise out of different opinions as to the scale at which phenomena should be analyzed as arise in normal discourse

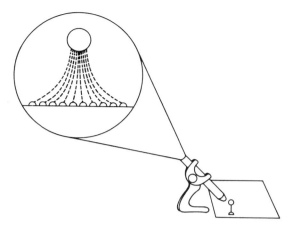

FIGURE 2–7

Classification at the experience-construct level.

might be the lowest of our categories, or it might contain the two lowest groups so that we could talk about *unsatisfactory* and *very unsatisfactory* facilities. Once we have made such a definition based on classes of data, we can proceed to try to discover meaningful relationships among the facts thus summarized and defined.

To return once again to our P-plane and C-field model of thought processes, we can characterize classification either as a process which occurs at a scale we have not previously discussed, or alternatively as a process of regionalization of the C-field. Classification is an integral

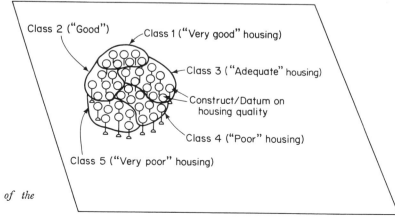

FIGURE 2–8

Classification as regionalization of the C-field.

because the discussants hold different definitions of critical terms.

The second principle evident here is that the categories into which we place observations are not precise and neat. Decisions about the location of the boundaries of classes are ultimately arbitrary, although they are usually defensible within the context of a particular investigation. In addition to being an important step toward explanation in itself, classification is also involved in what we normally think of as definition and observation. Whatever the level at which we choose to consider classification, it is obviously an important operation in science.

Hypothesis Testing and Law

Having gathered and summarized the data relevant to a hypothesis, the scientist is faced with the problem of deciding whether or not his assertion about a relationship between experiences is valid. Certifying the validity of hypotheses is a critical step. Scientists move with considerable caution in accepting a hypothesis, and such caution is necessary because at rock bottom the decision to accept or reject a hypothesis is a highly subjective one, although its subjectivity may be masked by the most impeccably objective language and statistical techniques. Confirmed hypotheses can acquire lawlike character and sometimes the status of scientific law. As laws, confirmed hypotheses become foundations of further investigations.

Basically, we test hypotheses by determining whether our observations are consistent with them, and by counting the number of times observations are consistent with facts if consistencies are not invariant. A hypothesis is part of a path which begins and ends in events and our experience of them (Fig. 2–9). Thus we can test a hypothesis against experience to see if the asserted relationship exists. If we are testing the hypothesis that poor housing conditions cause riots among minority groups in metropolitan areas, instances where racial minority groups lived in poor housing but did not riot would constitute evidence that our hypothesis was not valid. Instances in which poor housing was associated with riots by

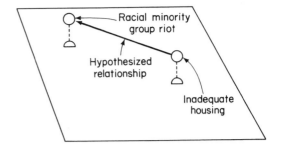

FIGURE 2–9
A hypothesis in the C-field.

minority groups would, of course, be evidence supporting our hypothesis. Thus we can test our hypothesis by examining all riots by minority groups to see if they were inevitably produced by poor housing conditions and by determining whether all cases in which minority groups have lived in inadequate housing were followed by riots.

It is necessary at this point to enter some warnings. For the sake of exposition, we are using a very simplified and somewhat hypothetical case. We mentioned earlier the fact that most events are caused by combinations of numerous events. Here, we are considering only a single possible cause. "Cause" is a very critical and complex notion and causes are usually multiple rather than singular. Our use of the term "cause" with such abandon would probably cause excruciating discomfort to many people inclined toward a more rigorous analysis of the scientific method. A second point to remember is that simple association *in itself* is never proof of causality. Even if we were to find that inadequate housing is invariably associated with urban minority group riots, the invariant association would not in itself be evidence of cause and effect. On the basis of invariant association alone, one could as easily argue that riots cause inadequate housing as the converse. Alternatively, it is possible—and indeed in this example probable—that both riots and poor housing may be effects of some still unrecognized third factor.

If we are inclined to argue that poor housing is the cause of riots on the basis of a strong correlation between the two events, we bring to such an argument a great deal more than the knowledge of invariant relationship. An assertion that inadequate housing causes riots is buttressed by a whole series of premises dealing with the inability of the occupants to find alternative housing, the psychological stress produced by residential crowding, the buildup of such stress, the stress-releasing function of riots, and so forth. We might indeed decide to accept the hypothesis that poor housing causes riots by urban minority groups, but the bases upon which we would accept such an assertion as valid are much broader than mere frequent or constant conjunction. In addition to frequent or invariant occurrence, we must be able to construct a logical relationship which makes it reasonable to accept the proposition. In the case we are considering, it seems reasonable that poor housing should engender frustrations which build up to the point where they are released explosively. Although these arguments are implicit rather than formally stated, there is a great deal more than mere frequent or constant conjunction in the statement that "inadequate housing is a cause of riots by racial minority groups in metropolitan areas."

For a number of years, there existed an almost perfect correlation between the number of divinity students graduated at Oxford and the number of arrests for prostitution in Sydney, Australia. Despite the fact that one could predict the value of one variable from the other very accurately, no one suggested the two were causally related. In addition to frequent or constant conjunction, then, we require the marshaling of a plausible argument that some functional relationship exists between two constructs before we will accept one as the cause or a causal factor of the occurrence of the other.

Although we would like to generate hypotheses which are *always* valid, we do not always require invariant relationships between events before we will admit cause. In most cases which a social scientist investigates, hypotheses are accepted that are considerably less than invariant. Suppose, again, that we are testing our hypothesis that inadequate housing is causally related to minority group urban riots. Suppose also that we had gathered data on a hundred

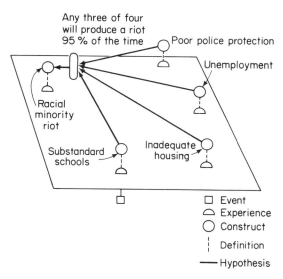

FIGURE 2–10
Hypotheses about multiple causation.

instances in which minority groups experienced poor housing conditions. Suppose that in 90 percent of these instances when poor housing reached a critical value, riots occurred, whereas in the other 10 percent of these instances riots did not develop. Should we accept our hypothesis as valid in this instance? Or suppose the respective percentages were 95 and 5; or 98 and 2; or 65 and 35. How much variation should we allow? Or suppose that we have a case where an event appears to have multiple causes. For example, suppose that when any three of four relationships were present (Fig. 2–10), a riot occurred in 95 percent of the cases. What are we to make of the hypothesis that the four factors in combination are the cause of a riot? Is it confirmed or not?

Unfortunately, there are no objective answers to these questions. Whether we wish to accept a hypothesis as valid when it accurately describes 95 percent of the cases to which we might apply it is a matter of individual judgment. Some scientists may feel that this is a sufficiently accurate description of the relationship, while others might reject the hypothesis because it does not meet their more stringent standards. It is simply a matter of personal choice. Often we would accept a hypothesis like this, but in our use of it we would qualify its applicability by phrases like "almost always" or "95 percent of the time," or "with a probability of 0.95," to indicate that the relationship is not completely invariant. But whether an individual scientist decides to build upon a hypothesis which is confirmed as being true 65 or 95 or 98 percent of the time depends on his own personal standards.

Basically, then, we test a hypothesis and decide whether to accept or reject it by looking into experience to see whether or not the hypothesized relationship exists, and by determining the proportion of times that it does hold if it is not invariant. When we find a hypothesized relationship to be invariant in a large number of cases, and when the relationship is one which has significant explanatory power, we accept the hypothesis as confirmed. When the relationship exists in a large proportion of cases examined, but is not invariant, we exercise greater caution in accepting its validity. Very rarely do we accept or reject a hypothesis on the basis of a single test case. Suppose, though, that a scientist hypothesizes the existence of some relation X whose existence is clearly incompatible with current scientific law, so that the hypothesis is at the same time an assertion that the law in question is invalid. If the scientist can demonstrate the existence of the relationship in a single instance, his hypothesis that the law is invalid will have been proved. Actually, however, cases like this are extremely rare. In the great majority of cases—especially those encountered in the social sciences—hypotheses which are accepted are not invariant and we must observe a large number of cases to determine the reliability of a hypothesized relationship. Proving a relationship to exist in a single case establishes only that it is potentially fertile; it must remain as merely an interesting research result until such time as it can be tested for reliability in a large number of cases.

Laws in science are hypotheses which have been tested and confirmed as being valid. Scientific law is critical because science depends on the laws it produces for its great explanatory power. The formulation of laws is most eagerly sought by scientists, although many would not phrase the goals of their research in those specific terms. To some extent, the term "law" is unfortunate because this analogy with jurisprudence sometimes produces confusion about the nature of scientific laws and the purposes they fulfill. Basically, the law of science is descriptive whereas civil law is proscriptive. Scientific laws describe what happens in individual cases or in large numbers of cases, and there are no enforceable penalties for noncompliance. Should water tomorrow suddenly become very perverse and decide to boil at 110°C., rather than 100°C., we would really have no choice other than to rewrite all our scientific textbooks or change our scale for measuring temperature, or both. There is no way we could punish or rehabilitate water that obstinately refused to boil at the proper temperature, whereas we do take such remedial actions when civil law is transgressed. So long as these distinctions between the nature and purposes of scientific law and legal proscriptions are kept in mind, no serious damage is done by the use of the same term in the two different contexts.

Scientific laws—as relationships between or among events—may be invariant or probabilistic. Usually, we feel that invariant laws are the most valuable because they offer us greater explanatory power. Invariant law enables us to make statements about individual members of a class of events. That is, if the one-way relationship between inadequate housing and riots were invariant—if every instance of very poor housing conditions resulted in a riot—we could state categorically that any observed instance of poor housing of this degree would soon be followed by a riot. This would be a useful thing to know, since we could then eliminate some riots by providing good housing. By monitoring housing conditions we could predict the occurrence of riots after each instance in which housing conditions crossed the causal threshold. Should we be lucky enough to discover a two-way relationship, we could, if we so wished, eliminate all riots. Suppose riots occurred if, *and only if,* racial minorities were subjected to poor housing. In that case, constructing adequate housing would completely eliminate the possibility of any riot occurring. We hardly ever find two-way invariant relationships, and even one-way invariant relationships are infrequent. Most events in this world—and especially those occurring in the realm of human behavior—are very iffy things, and we must usually make probabilistic statements about relationships between events.

Probabilistic laws describe nonconstant relationships. There is only a certain *chance* that the relationships will hold. Suppose that we know that 95 percent of the time poor housing will produce a riot. In 5 percent of the cases where we predict the occurrence of a riot based upon observation of inadequate housing, our prediction will not be fulfilled. When laws are probabilistic, we cannot make completely deterministic statements about individual events. Should we observe poor housing among a minority group, we could not say for certain that a riot would occur. We could only say that it was very probable that one would ensue. Probabilistic laws can be stated in the form:

$$\text{If } A, \text{ then } P(B) = 0.95,$$

where $P(B)$ means that the probability of B occuring as a result of A is 0.95. Invariant laws are deterministic; they hold without exception. Since the probability of a determined event—one which will certainly occur—is 1.0, we express a deterministic law as:

$$\text{If } A, \text{ then } P(B) = 1.0.$$

We cannot make sure predictions about individual cases on the basis of probabilistic laws, but this is not as serious a drawback as it might seem. Probabilistic laws are, in a sense, invariant with regard to *classes* made up of large numbers of the events to which they apply. Suppose we have established a probabilistic law which describes the fact that 1 percent of the people who walk by a hardware store will enter on impulse and make a purchase. And suppose that in order to make a profit, you, as the store owner, require a minimum volume of 10,000 impulse buyers per year. If you know that in the average year, 1,500,000 people will walk by your store, you need not concern yourself with whether or not the *next* person who walks by your store will enter and make an impulse purchase. You can plan a winter vacation in Florida assured that 15,000 impulse

buyers will patronise your establishment in a normal year, giving you a very tidy profit indeed. If a probabilistic law accurately states a relationship which is true for large numbers of events—which, by definition, it must in order to be accorded law status—we can afford to be unconcerned over the fact that such a law is relatively powerless to predict the outcome of any individual case.

Laws are tremendously useful in science because they enable us to predict future events. By the same token, law enables us to explain present and past experiences which raise questions in our minds. Our most satisfactory explanations of experience are those which can cite the experience as an instance to which a law applies. If we could demonstrate conclusively that all racial minority group riots resulted from poor housing, the next time we were queried as to why a riot had occurred, we would simply explain that very bad housing conditions were prevalent in the area in question, and that poor housing always produces riots. A more realistic example is the way we explain some things to children. If asked by a child why water boils when it gets hot we may try to explain the phenomenon in several ways, but somewhere along the line we will usually make some statement about pure water at sea level atmospheric pressure beginning to boil when it reaches a temperature of 100°C. After hearing about the temperature relationship several times the child soon understands that the relationship between temperature and boiling is invariant. When he sees water boiling he concerns himself with it no further because he recognizes it as an instance of a law that describes the relationship between temperature and the state of water.

In science we explain past and present experience and predict future experience by demonstrating that these actual or potential experiences are instances in which law is applicable. The two-way deterministic law reducible to the form

$$\text{If, and only if, } A, \text{ then } B$$

is the most comprehensive and satisfying form of law we can devise since it gives us power to predict and explain without qualification. But such laws are very rare. In both the physical and the social sciences, probabilistic relationships are by far predominant. Probabilistic law, though not quite so potent and somewhat less elegant, is almost equally as useful as deterministic law as a means of explaining and predicting our experience of the world.

Theory in Science

"Theory" is an overworked term. One of the uses of "theory" we shall *not* adopt is perhaps the most common usage, the treatment of "theory" as a synonym for "hypothesis." We often talk about "theorizing" to designate the kind of informal hypothesizing called "brainstorming." We put forth hypothetical relationships for evaluation and criticism. This use of the term is so common that it would be impertinent to call it incorrect. Like science, language is what most of the people who produce it say it is. We only note that this is not the kind of theory which interests us at this point, and that we would prefer that "hypothesizing" be used instead of "theorizing," leaving the latter term to designate the process of building formal conceptual structures in science.

As we intend to use the term, theories are structures composed of laws and the rules by which those laws are put together. Perhaps we can best illustrate this abstractly by referring again to our C-field model of thought, and adding another level to the structure previously built (Fig. 2–11). Although we represent laws as solid lines, the relationships which constitute laws are themselves "constructs" and they can be represented by the symbols we used for constructs. Thus we show *law* constructs at a higher position in the C-field, with connections between them to represent the rules which bind them together into theories. If events are the foundation of the sequence of steps toward explanation, we can see that experience is the first encounter with the world, and grouping constructs into classes is the first generalizing, abstracting process. Classes of constructs may be interconnected in various ways, with the interconnections forming a second level of structure. Finally, a number of laws taken together constitute the third level of structure, or theory.

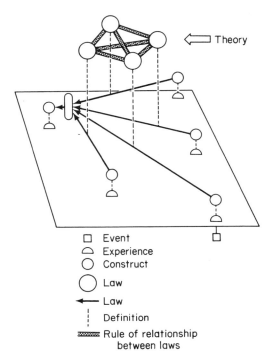

FIGURE 2–11
Theoretical structure in the C-field.

Symbol key:
□ Event
△ Experience
○ Construct
◯ Law
← Law
┊ Definition
▩ Rule of relationship between laws

Let us return to our attempt to explain urban riots. We saw earlier that if we could cite a riot as an instance to which a lawlike statement was applicable, the riot was usually considered to be adequately explained. But suppose we should ask, "Why does poor housing cause riots?" The statement about housing, in itself, says nothing about why the law—deterministic or probabilistic—holds. It simply states that it does. But you will recall our earlier observation that when we argue that one event causes another in such a fashion we always have a great deal more than constant or frequent conjunction in mind. The additional context—even though it may be very poorly formulated and almost subconscious—is a theory which relates a number of lawlike statements to each other.

Let us assume that we have five probabilistic laws of minority group urban riots (Fig. 2–12). Let us begin to speculate about the things which these five law constructs have in common and

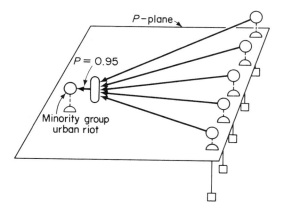

FIGURE 2–12

Probabilistic laws of minority group urban riots.

suppose we discover that all five of the causal constructs produce something we could call stress at a level sufficient to cause riots in 95 percent of all observed cases. At this point, we will have to elaborate our network (Fig. 2–13). We place a new construct, *stress*, in an intermediate position between our causes and the construct *riot*. Together, the five laws relating ghetto conditions to stress constitute a theory of ghetto stress. At the same time we expand our view of the C-field to consider other, previously unobserved elements. Assuming riots may be caused by a number of factors, we can add laws concerning the relationships between riots and mass hysteria, labor disputes, and religious friction. These laws, considered as a group, might constitute a theory of urban riots. It is obvious that by building and elaborating structures in this way, the networks in the C-field can easily become very complex—as intricate and detailed as are the networks of law and theory in science itself.

To focus more sharply on the theory of ghetto stress (Fig. 2–14), we can now ask questions about the nature of the relationships among the laws which comprise the theory, and about the utility of theory. The theory is a system of laws. It consists of the law statements which describe the factors which produce psychological stress in ghetto areas. The interconnections between these laws might be thought of as relationships of conjunction. Where

police protection is very poor, for example, crime rates are usually high. (This particular lawlike statement might itself be part of another theory.) Other interconnections might describe the additive or multiplicative effects of two or more of the stress-producing factors. Perhaps high crime rates and poor housing conditions produce far more stress in combination than each does separately. The rules which link laws together into theoretical systems are usually themselves laws of a higher degree of generality than those they interconnect.

By linking laws together in more general structures we can explain laws themselves. To return to an earlier question, that of why all five of the causal factors we cited in our laws of minority group riots produced riots, a theory of stress enables us to explain the causative action of poor housing or any of the other conditions by citing a law dealing with the stress-producing nature of poor housing and the stress-releasing nature of a riot. General theory of this kind is also fertile in that it often enables us to deduce specific laws and constructs from its very general statements. A statement such as "stress causes conflict" stimulates us to look for more specific relationships between special kinds of stress and alternative forms of conflict which we might otherwise overlook. Under so general a maxim, for example, we can subsume both the subtle conflicts which arise within the happiest of marriages, and the most horrendous battles fought in wars arising out of political and military stress. Theory has utility in science because it enables us to answer the questions we raise about the workings of laws, and because it stimulates our power to identify previously unsuspected general relationships.

The manner in which C-field structures in science grow more complex as general theories are constructed to explain both specialized theories and the laws which interconnect them is obvious. We have seen that as we become more sophisticated we can interpolate new theoretical constructs like *stress* in C-field structures which were once simpler. Because of such simultaneous lateral and vertical extension and intensification of the structure of science, the enterprise is indeed complex. Our examples have been oversimplified, we have ignored many

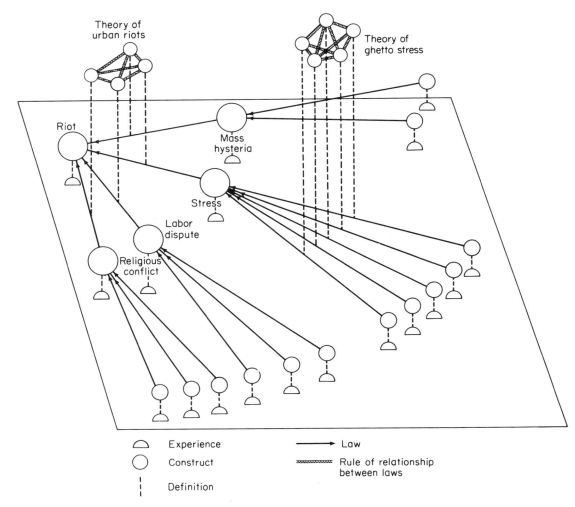

Experience ⟶ Law

Construct ▒▒▒ Rule of relationship between laws

Definition

FIGURE 2–13

Network structures of theory in the C-field.

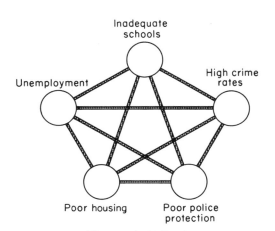

Theory of ghetto stress

▒▒▒ Relationships of conjunction, additive effect, multiplicative effect, etc.

FIGURE 2–14

Relationships among laws in a theoretical network.

of the implications of some of our rather naive assertions, and despite all this the structures we have discussed have grown like Topsy. The simple structures we have presented here in no way do justice to the actual complexity of the conceptual structure of science.

In contrast to the physical sciences, the social sciences have been relatively unsuccessful to date in developing theories composed of reliable laws. There are several reasons for this, not the least of which is the much greater intractability of the events with which the social scientist deals, although that is by no means the whole story. The relatively underdeveloped nature of theory in the social sciences need not concern us too much here, nor need social scientists be particularly apologetic over the limited progress they have made to date. Theory building is accelerating rapidly in all the social sciences. What should be emphasized at this point is the vital nature of continued theoretical work. It is simply impossible to produce good science without theory. Theory is operational at every level of the process of explanation, from the very basic determination of what constitutes a relevant fact to the explanation of theory itself. Theory is the matrix of all science, and the science with an underdeveloped theoretical arm is very much like a ship without a rudder; it drifts rather aimlessly and gets nowhere except by sheer good fortune.

Models

Because it is one of the terms which lends an aura of great *savoir faire* to those who use it, "model" has been virtually robbed of any specific definition. It is used to designate anything from a very simple relationship to a complex theory. Today, no social scientist would even consider submitting a research proposal which did not list as one of its goals the development of at least three "models" (preferably "mathematical models") of the process he was going to investigate. Now there is little doubt that each individual scientist who uses the term may have a very clear conception of what "model" means to him—and perhaps even to his colleagues in the discipline. But even within the geographical literature we encounter so many variant mean-

ings of the term that we are greatly tempted to suggest that it be eliminated altogether. But because the term is in common—if imprecise— use, and because the notion of a model is basically valid and is very useful, we shall attempt to set out a definition which will serve our purposes.

Philosophers of science would be happy if we restricted the use of "model" to those occasions when we wanted to talk about a structural isomorphism between two theories. Isomorphisms are one-to-one correspondences between two systems—for example, the correspondences between Arabic numerals and Roman numerals. Suppose we have two theories whose laws have the same form and whose laws are interrelated in identical ways. If there is a one-to-one correspondence between all elements of one of the theories and all elements of the second, either theory can be taken as a model of the other. Thus the model is not a theory itself. It is an abstraction of a theory, one which is stripped of all empirical content but which maintains the same structure. Philosophers prefer to restrict the use of "model" to those cases where both structural isomorphism and a one-to-one correspondence are actually or potentially in evidence.

The utility of models, thus defined, lies in their *heuristic* (revealing) potential. In instances where we have reason to believe that one theory's structure may be a model for another's, we may discover a lot of shortcuts in our attempts to build theory in an unfamiliar area of investigation. Geographers, for example, are currently very interested in the processes by which ideas and innovations diffuse through space, and they are expending a great deal of effort on the development of law and theory applicable to these processes. Now it so happens that there are some other spatial processes about which we know somewhat more than we know at the moment about the diffusion of ideas and innovations, processes for which we have relatively well developed theories. One such field is epidemiology (the study of the spread of diseases). Since we can find many one-to-one correspondences between the elements of the theory of epidemiology and the spatial diffusion of information, we can use the theory of epidemiology as a model of the theory of spatial diffusion. The formal structure of the theory of epidemiology will never exactly fit the structure of the theory of the diffusion of innovations and ideas, of course. But assuming that we know more about the theory of epidemiology than we do about the diffusion of ideas, examining the former—stripped of its empirical content—will provide us with some very useful insights into the structure of spatial diffusion processes.

The idea of a model has value when used as described above. But wishes of philosophers to the contrary, it is obvious after only a cursory perusal of the literature of any discipline in the social sciences that "model" has a far wider application than that described above. Geographers and other social scientists use the term to describe a great number of things. Generally speaking, "model" may refer to any simplification and abstraction of some relationship we believe to exist between experiences. Thus it is conceivable that "model" could be used to designate a theory, a law, a classification, a hypothesis, or even science itself—which certainly fits the characterization of "a description and abstraction of reality." When a term can be used to designate this many things, it is not very precise. Whatever meaning it has is restricted to whatever these designata have in common. Since all the things to which "model" is applied are abstractions and simplifications of reality, the social scientist's use of the term is not illegitimate, although it is not very precise as it stands. Nonetheless, even models defined in this manner have a great deal of utility.

Like philosophers, social scientists use models heuristically. We often take the structure of a theory in one realm of investigation and see if it is applicable in another. But perhaps the most important use of models at the present time is as *norms* against which we measure our experience of the world. When geographers and other social scientists talk about models they often talk about normative models. A normative model is one which is used as a standard or idealization of a particular relationship. Perhaps the most famous normative model in the social sciences is *homo economicus*—the entrepreneur in economic theory who al-

ways acts rationally in his own best interest, which he always tries to optimize on the basis of his complete knowledge of all factors relevant to any decision he must make. Now, no economist seriously believes that economic man really exists. But talking about what economic man *would* do *if* he existed is a very useful way to illustrate the intricacies of economic theory. And more importantly, economic man provides the standards or norms by which we can evaluate the economic behavior of other individuals.

One can always make rank-order comparisons of performance so long as we have some basis for measurement. If profits are a criterion, for example, it is easy enough to rank individuals according to the size of the profit they garner in a particular period of time. But after we have rank-ordered them, we have no absolute standard against which to compare their performances unless we adopt the performance of some admittedly unrealistic actor like economic man. If we know how much profit economic man could have made in similar circumstances, we have a criterion for evaluating the profits made by the actual entrepreneurs. Normative models are scale points on rulers which enable us to compare experiences to one another in precise ways. In many ways, a normative model can fulfill the same function as a 100 cm. mark on a ruler.

Normative models are not confined to economics. As we shall see later, geographers have developed a normative model of settlement process which produces a hexagonal network of evenly spaced villages, towns, and cities. In the same way that economists realize that economic man is an idealization, geographers do not really expect to find perfectly hexagonal patterns of evenly spaced settlements on the landscape. But analyzing the way settlements would be structured in space if the world were perfectly rational and undifferentiated makes it easier to understand geographical theory. And hexagonal lattices provide a basis for comparing the irregular settlement patterns we actually observe with each other.

Social scientists also use models for measurement in a non-normative fashion. In situations where we think a hypothesis or law might be applicable within a certain range of variations, we may "model" the situation within the context of the law or hypothesis from which the model is derived. For example, geographers and others have long been interested in spatial interaction, defined as flows of various kinds of commodities and information from one place to another. Investigating such problems, some scholars perceived a similarity between spatial interaction and Newton's law of gravity. The gravity law can be expressed in normal language as "The attractive force between two bodies is equal to the product of their masses and inversely proportional to the distance between them squared." Mathematically we can express this as:

$$F = K\frac{M_1 M_2}{d_{1,2}^2}$$

where: F = force of gravity,
   K = a constant of proportionality,
   M_1 and M_2 = the masses of the two bodies, and
   d = the distance between the two.

As it turns out, several kinds of human spatial interaction do conform fairly well to the gravity model. If, for example, we are interested in the flow of migrants between two towns, we can let the number of migrants correspond to the force of gravity and the populations of the two towns correspond to mass. Thus we have:

$$I = K\frac{P_1 P_2}{d_{1,2}^2}$$

where: I = interaction (migrants in this case) and
   P_1 and P_2 = the respective populations.

In the application of this model to actual interaction, we find in many cases that squaring the distance does not produce the best results. For a large number of pairs of towns, $d^{1.5}$ might produce more accurate predictions of actual interaction than d^2 if mail flow is to be predicted. On the other hand, $d^{2.3}$ may produce the best results for telephone calls. Thus we can change some of the terms slightly in order to make the model "fit" the empirical case we are investi-

gating more closely, but we would still call these investigations "gravity model approaches." All use the same structure as their basic organizing framework.

The similarity of this use of the idea of a model to the way philosophers use models is obvious. Generally speaking, the social scientist uses the term "model" to represent the structure of a wider variety of elements than the philosopher feels comfortable with. In either case, the heuristic role of models—whether those of theory, hypothesis, law, or some other element—is obvious. The way in which models fit into our own descriptive model of human thought should be obvious. When defined according to the philosopher's criteria, a model represents the perception on the part of a scientist that two identical hierarchical structures exist in the C-field. The constructs these structures interconnect are different but the frameworks are identical. When defined according to current social science usage, "model" represents a hypothesis about some simple or complex interconnection or structure of interconnections in the C-field.

We began our detailed discussion of the scientific method with more than a little trepidation for several reasons, the most important of which is the artificiality of the enterprise. Certainly there is a scientific method. Scientists do discuss problems of description and taxonomy, and they make conscious use of hypotheses, laws, theories, and models. Yet they do so in ways wholly at odds with the usual textbook discussions. The processes by which a scientist's mind operates are nowhere near as linear as the usual presentation makes them out to be. Scientists skip around between the various subprocesses which comprise the scientific method in very complicated ways (Fig. 2–15). Often they do not start with a hypothesis and basic observation of facts, but with a theory or a model. Often theorization precedes the formulation of the hypotheses and laws which we logically think support theory. Overall, the process of doing science is much messier than abstract expositions such as ours indicate. Rather than being linear and progressive, the method of science is actually highly interconnected and

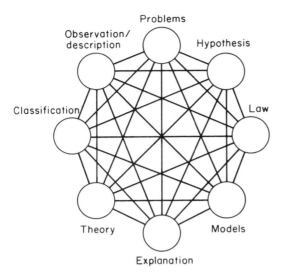

Problems

Observation/
description

Hypothesis

Classification

Law

Theory

Models

Explanation

FIGURE 2–15

Paths to explanation.

reflexive. All its elements affect each other.

When considering the method of science, we often fall prey to a process-product ambiguity. In discussing a particular science or science in general, it is obvious enough that descriptions, classifications, hypotheses, laws, theories, and models are all parts of science and are all used by scientists at one time or another. But when we consider the work of an individual scientist these elements are not readily identifiable or separable. When talking about elements of the scientific method we are in the rather unique position of being able to see the forest but not being able to see the trees. Components and relationships which are easily visible at the abstract level tend to disappear when individual scientists' thought processes are examined.

Nonetheless, considering the elements of the scientific method in the abstract is useful despite the artificial aura which suffuses such discussions. Logical exposition for pedagogical purposes and reality are always somewhat at loggerheads; that this is particularly true in the case of the material we have just covered need not concern us too much. Artificiality and abstraction are always the prices we must pay for whatever clarity we can provide.

EXPLANATION IN SCIENCE

Scientists are concerned with explanation more than with anything else. We have already described the way scientists explain events by citing them as instances to which law and theory are applicable. Other techniques of explanation are also used, however, and we shall therefore review the alternative forms of explanation.

The Deductive-Deterministic Approach

One form of explanation provides answers to questions we have raised about our experience by citing conclusions which must follow deductively from an argument based on law, theory, or both. For example, suppose our question is "Why did this bottle break?" (Remember again, that all *why* questions of this sort can be reduced to *how* questions where the *how* requests that we detail the sequence of events which have produced the event which interests us.) Our answer might go something as follows. "The bottle broke because the water inside it froze." This answer, which would satisfy most of us, masks a whole series of implied and understood statements which underlie the explanation. We could detail these somewhat as follows:

The water in this bottle was cooled below 0°C. Water cooled below 0°C. freezes. As the temperature of freezing water decreases, the water expands. Expansion produces pressure if freezing water is constrained in a bottle. Bottles made of the kind of glass this one was made of can withstand very little internal pressure. The expansion created greater pressure than the glass could withstand. Therefore, the bottle shattered.

All of the statements of this syllogism are either scientific laws which have been well established or declaratory sentences which narrate to us what happened in the particular sequence of events which interests us. If we accept the truth and accuracy of the premises, the conclusion deduced from them cannot be denied. It follows from the premises we have stated. The breaking of the bottle is explained because we have described a sequence of events which could have no other outcome.

We are particularly satisfied with such an explanation because it states both the *necessary* and *sufficient* conditions of an event. When discussing an event, we can talk about other events which are conditions of its occurrence. A necessary condition of the occurrence of the event X is one which, by its non-occurrence, prevents X from occurring. For example, having access to an adequate supply of water is a necessary condition of living for a pilot whose airplane is down in the middle of the Sahara. Without several quarts of water a day, he will die of thirst in short order. Having water does not assure that he will survive, however, for he may die of exhaustion, starvation, or snakebite even if he does have an adequate supply of water. A sufficient condition of an event is one which determines unequivocally that the event will occur. Suppose we consider the case of setting off a stick of dynamite to illustrate the concept of a sufficient condition. A stick of dynamite, a fuse, the proper burning of the fuse, and some means of lighting it, are the sufficient conditions of an explosion. If all these are present, the dynamite will explode—there is no other logical outcome. Thus a sufficient condition of an event is usually a set of events which together determine that the event will occur. In logical form we can state a necessary condition as:

$$\text{If not } X, \text{ then not } Y,$$

where: $X =$ the necessary condition of Y.

A sufficient condition is stated:

$$\text{If } X, \text{ then } Y,$$

which, you will recall, is the form of the deterministic scientific law. It is useful to be able to state the necessary and sufficient conditions of an event. We can look on the necessary conditions as "causal factors" and upon sufficient conditions, which may be plural, (that is, an event may have more than one sufficient condition) as "the cause." The deductive-deterministic form of explanation—the statement of both the necessary and sufficient conditions of an event—is generally held to be the most satisfactory form of explanation in science. This form ordinarily precludes further questions about the causes of the event in question.

Deductive-deterministic explanation—the mode in which we can predict individual events with certainty—is encountered very rarely, even in the physical sciences. The question is often raised of whether the social sciences, because of the nature of the events with which they deal, can utilize deductive-deterministic explanations. We think that in time they may acquire the ability to make deterministic statements. But even if geographers do not become capable of constructing deterministic explanations of human spatial behavior, there is no cause for concern since probabilistic explanation provides satisfactory answers to questions raised about human affairs.

Deductive-Probabilistic Explanation

We do not view probabilistic explanation as a different kind of explanation, distinct from a deterministic ideal. Our belief that probabilistic explanation is simply a slightly less precise form of the deterministic approach underlies our conviction that empirical science is a single, coherent body of thought. It is true that we have adopted the common convention of talking about natural and social sciences, and it is also true that the social scientists use probabilistic explanation somewhat more frequently than do those in the natural sciences. But the elegance of the deductive-deterministic model which is used in very limited portions of natural science has no doubt blinded us to the highly contingent nature of many physical events. Geneticists, for example, are no more able to make deterministic predictions about the outcome of many individual events than are nuclear physicists or sociologists. Most of our experience is inherently probabilistic and our explanations and predictions of such experience must be contingent also.

We are constructing a probabilistic set of sciences of man because man apparently has internal energies which permit him to act in accordance with individualistic needs and desires. It may or may not be true that those desires and needs themselves are determined in some ultimate sense, and that man only deludes himself when he thinks that he is capable of acting freely. Whatever the case, given the present state of the art in the behavioral and social sciences, we can provisionally accept the argument that man's will is free, and that his individual behavior is often very unpredictable. But despite the relatively low probability values we can attach to individual behavior, human behavior in the aggregate remains quasi-deterministic in that group behavior is often highly predictable. For classes containing large numbers of events, we are still able to state necessary and sufficient conditions even though these statements of relationship have probabilities of less than 1.0.

Whenever we are dealing with large numbers of recurrent events, the inability to predict accurately the exact nature of the next individual event is usually but a minor hindrance to our ability to order the world. We can predict relatively accurately, for example, what the demand for telephone circuits and operators will be on an average day and on a very busy day. Because we can predict aggregate demand we can design the network and schedule personnel so that sufficient service capacity is available to meet almost all of even an abnormally heavy load. Naturally, because the laws and theory upon which telephone operations are based are probabilistic, unexpected demand occasionally occurs and service cannot be provided. But these overloads occur very rarely —so rarely that it is not economical to try to design the system to accommodate them. Thus we have designed, and we operate successfully the largest and most complex machine in the world—our national telephone system—entirely upon probabilistic principles of supply and demand. That the telephone company cannot predict very accurately your individual telecommunications behavior is of little concern, because it can predict very accurately the intercommunications behavior of people like yourself taken in groups of 25,000 or so. The economical and efficient telephone service we enjoy in the United States is ample evidence of the power probabilistic law and theory gives us to explain and manipulate our experience.

Social scientists need feel no inferiority over their inability to attain the levels of explanatory and predictive precision achieved in a few portions of the natural sciences. As we have noted, invariant law and theory is confined to relatively restricted sectors of the natural sciences, and the phenomena which the social scientist studies are indeed less tractable than those which permit of deterministic explanation and prediction.

Genetic Explanation

There is a good deal of argument in some circles over whether or not what we usually call genetic explanation is "really" scientific explanation. By the criteria we set forth—the use of the scientific method and an ability to answer *why (how)* questions, genetic explanation qualifies as a member of the scientific community. Yet for at least one reason, genetic explanation is considerably less powerful than either of the deductive forms, and for that reason it is invoked with decreasing frequency by social scientists.

Genetic explanation, as the name implies, explains events by detailing their genesis—that is, by describing the sequence of events that produced the phenomenon being explained. Now there is nothing different here from the previous forms of explanation discussed. In order to explain we always describe an event sequence which has as its outcome the event about which we have questions. Moreover, we cannot fault genetic explanation for not using law and theory, for genetic explanations normally incorporate both, although it is true that in genetic explanation law and theory are almost always implicit and not explicitly stated. What, then, distinguishes genetic explanation from other kinds, and what is it about it that makes many suspect its scientific validity? Perhaps the answer to both questions will be easier to understand after we have examined an example of genetic explanation.

Suppose we ask: "Why are there large deposits of coal in southwestern Pennsylvania?" One way of answering this question is to give a more or less detailed description of the genesis of existing coal deposits in the region, as follows: "About 300 million years ago what is now southwestern Pennsylvania was the shoreline of a shallow sea. As the sea level rose and fell—which it did frequently—the area was intermittently a shallow sea bottom and a swampy lowland region which bore a very lush tropical vegetation which deposited considerable depths of organic matter as it flourished. Layers of this organic matter were interleaved with layers of sedimentary rock during the periods in which the region was submerged. After several tens of millions of years, the region was subjected to tectonic and gravitational forces which both squeezed the water and gaseous material out of the vegetative layers and at the same time uplifted the region to a higher altitude. After being subjected to such earth forces for several hundred million years, the vegetative strata were transformed into the coal seams found in the region today."

Note that this explanation cites no laws—deterministic or probabilistic. The entire explanation consists of declarative sentences about the past events in the region, which, taken together, tell us how the existing situation came to be. Yet underlying each of these declarative sentences is a law or theory which could be cited to explain why the events mentioned in that particular sentence occurred. The laws and theory are entirely implicit and understood in the explanation. The reason for this is that the question which evoked the explanation was asked about a particular unique case—that of coal in southwestern Pennsylvania. Now note that this question *could* have been answered in another way, as follows: "There is coal in southwestern Pennsylvania because what is now that area was once an intermittent, epicontinental sea during the Pennsylvanian geological epoch, and all such regions are areas in which coal was deposited. Where such regions have not been subjected to sufficient erosion to remove the coal deposits, the deposits remain today. Southwestern Pennsylvania is such an area. Therefore, there is coal in the region."

There is no question about the scientific validity of the second explanation. It is a good example (if an oversimplified one) of the deductive-deterministic form. The reasons for our doubts about the scientific value of genetic explanation should now be clear. The genetic explanation as given applies only to a unique case—southwestern Pennsylvania in this instance. There is no hint of generality in any of the statements in the genetic explanation. Nor, as we have pointed out, is there any explicit reference to any law or theory which could make these statements more general. It is our conviction that science is a generalizing activity which seeks explanatory statements of law and theory which are as universal as possible. The absence of universality in genetic explanation makes us uncomfortable in its presence. Genetic explanations, applied a million times over, will never produce law or theory. So long as the focus of genetic explanation remains restricted to unique cases, it will remain unproductive of the generalization we feel is the hallmark of science.

That such an approach is not without alternatives is indicated by the second explanation we gave, which does consist of several geological laws followed by a statement that southwestern Pennsylvania is an instance of the operation of these laws. The ` deductive-deterministic explanation could be applied to any one of a number of coal deposits in the world, with only the region cited in the last specific premise being changed.

Whether we choose to use genetic or deductive explanation in a particular instance like this is essentially a matter of choice and temperament. At some stages of science, choice may not exist. Constructing a number of genetic explanations and then noting the similarities between them and using such similarities as bases for formulating laws is a fairly common way of developing law and theory in science. But one will never develop law and theory unless he is consciously trying to do so and is searching for what is common in the individual explanations being constructed.

In explaining our experience, then, some people seem to prefer to emphasize what is common and general while others prefer to look for differences and emphasize the unique. Those who tend to argue that people and places are unique tend to rely heavily upon genetic explanations to explain both their human and physical environments. They also tend to argue that the function of the disciplines they practice is to "understand" our experience rather than to explain it. This approach to the world has been called *idiographic* and we sometimes speak of idiographic and *nomothetic* sciences. The former are those which investigate unique cases and are not law seeking, whereas nomothetic sciences try to develop universal statements of law applicable to all cases. In actuality, sciences themselves are not idiographic or nomothetic; it is their members who hold one attitude or the other. And we must emphasize strongly that this is a matter of attitude. Disciplines may be primarily idiographic or primarily nomothetic, but there are always individuals at odds with majority opinion in their field.

Were we discussing in detail the history of geography, this would be of more than passing interest since a considerable debate raged in geography about fifteen years ago over whether geography was an idiographic or nomothetic science. Debates of this kind are never won in the usual sense of the word; one opinion gradually comes to predominate by attracting most of the new entrants into the discipline. There are utilitarian criteria for evaluating the two approaches, and here the nomothetic approach seems to be clearly superior. If we wish to manipulate events and experience, the explanatory and predictive power produced by law and theory are prerequisite, for the genetic approach is relatively impotent in this particular area. It is for this reason that scientists feel so uncomfortable with genetic explanations. The idiographic approach which produces genetic explanation is unproductive of law, theory, and thus the ability to explain and predict *classes* of events.

Functional Explanation

The final form of explanation we wish to consider is one which has come to the forefront in the last decade after being unimportant for some time. The attribution of functions to phenomena is a very ancient habit of thought, having its origins in Greek cosmology. Functionalism has recently undergone a revival with the formulation of *systems theory*, a new general theory of social and physical organization.

An example of the kind of question answered by functional analysis is "Why do fish have fins?" A typical answer might be that "Fish need fins in order to swim." Or an anthropology student may ask: "Why do all cultures have an incest taboo?" His professor might answer: "Incest taboos are necessary to prevent the nuclear family from becoming indistinct." In both cases, and in many others which could be cited, events are explained by stating the functions they fulfill in some system. In the terminology we used in the first chapter, functional explanations answer *how* (utility) questions. In answering such questions functional explanations state, or more often imply, that the phenomenon in question is a component of a system, and that its existence is best explained by the function it fulfills within that system.

Functional analysis and functional explanation raise a number of conceptual problems with which we will not concern ourselves beyond noting their existence. One difficulty is the assumption that a system exists and that the phenomenon in question is a component of that system. Such an assumption is no better or no worse than many of the other assumptions scientists make, but it is often made unconsciously and not identified as an assumption by those who make it. Unrecognized assumptions are always sources of problems, but this is especially true in science. A further difficulty is raised by the fact that functionalists often not only assume that a system exists, but they often attribute goals and characteristics to the system on the basis of little or no empirical evidence that the system exhibits those characteristics or goals. None of these problems has really been thrashed out in any detail yet, as the renaissance of functionalism in the form of systems analysis is relatively recent. Of the social sciences, anthropology and sociology as well as fields like business administration have made use of systems concepts, usually with success. Functional or teleological explanation is especially relevant in the realm of human behavior, because people do have consciously perceived goals and they do make efforts to attain these goals.

Functional explanations can often be restructured into other forms with a little manipulation. The question as to why fish have fins could be interpreted as a question pertaining to the evolution of marine life and answered accordingly by citing certain laws of evolution. Similarly, the question about incest taboos could be answered developmentally. In some ways, functional explanation is similar to genetic explanation, especially when it answers questions about an existing state of a system developmentally. But we are less concerned with the scientific validity of functional explanations than we were with genetic explanations because those who produce functional explanations are very consciouly searching for universal statements of systematic relationships which will be applicable to many different kinds of systems. In this sense, systems theory is a very general theory of organization, and its specific functional explanations are usually products of individual laws (admittedly poorly formulated and often not explicitly stated) which comprise elements of the theory.

Since any detailed remarks we could make at this point on the validity of functional explanations and systems analysis would be very provisional, we will refrain from additional comment. We should only note that functional explanation becomes an increasingly valid way of ordering experience as the world becomes an increasingly manmade artifact, wherein many components of the physical and human environment have been designed with certain purposes in mind.

Explanation and prediction—with their applications—are the end products of science. We have confined our comments in this section to explanation alone. The reason for this should be obvious. Explanation implies prediction. If we develop the ability to explain our experience, we simultaneously acquire some ability to predict it. Explanation (retrodiction) and prediction are symmetrically arranged around the present, assuming conditions remain constant. Their forms are similar whether they are applied to past, ongoing, or future experiences. Explanations are produced by applying the method of scientific investigation to the past, and predictions are produced by applying the same method to the future.

Suggestions for Further Reading

BRODBECK, MAY, ed. *Readings in the Philosophy of the Social Sciences.* New York: The Macmillan Company, 1960.

BROWN, ROBERT. *Explanation in Social Science.* Chicago: Aldine Pub. Co., 1963.

CHORLEY, RICHARD J., and PETER HAGGETT, eds. *Models in Geography.* London: Methuen & Co., Ltd., 1967.

GREER, SCOTT. *The Logic of Social Inquiry.* Chicago: Aldine Publishing Co., 1969.

HARVEY, DAVID. *Explanation in Geography.* London: Edward Arnold, 1969.

KEMENY, JOHN G. *A Philosopher Looks at Science.* New York: Van Nostrand Rheinhold Company, 1959.

KUHN, THOMAS S. *The Structure of Scientific Revolutions.* Chicago: University of Chicago Press, 1962.

NAGEL, ERNEST. *The Structure of Science.* New York: Harcourt Brace Jovanovich, Inc., 1961.

PLATT, JOHN R. *The Excitement of Science.* Boston: Houghton Mifflin Company, 1962.

RUDNER, RICHARD S. *Philosophy of Social Science.* Englewood Cliffs, N. J.: Prentice-Hall, Inc., 1966.

Work Mentioned

CARNAP, RUDOLF. "Formal and Factual Science," in *Readings in the Philosophy of Science*, eds. Herbert Feigl and May Brodbeck. New York: Appleton-Century-Crofts, 1953, pp. 123–28.

CHAPTER 3

THE SCIENCE OF GEOGRAPHY

The geographer must conceive of the places of the earth as parts of a system, related to each other at different levels of interaction.

—Fred Lukermann

GEOGRAPHY AMONG THE SCIENCES

The Data of Science and Questions about Them

The essence of any science is a set of problems and a method for solving them. Problems in science are unanswered questions about our experience or about the way we explain experience. The scientific method is singular. Because one method is common to all sciences, method cannot be used as a distinguishing characteristic. Instead, distinctions must be based either on the constructs studied or the questions we ask about experience or both. Constructs and questions are both relevant, but questions are more useful than constructs since different sciences frequently study identical constructs.

We can, to a limited extent, divide science on the basis of the kinds of experiences to which scientists apply their expertise. The major division based on content is the traditional distinction between social and physical sciences.

This distinction is a valid recognition of interests; sociologists have little to do with subatomic particles. Physicists have no professional interest in urban riots. Despite the fact that it is possible to make gross distinctions like this on the basis of constructs, content does not provide distinctions finer than the very general categories of social and physical science. Different sciences often study exactly the same events with entirely different purposes in mind. Despite the fact that physicists, chemists, and engineers study identical phenomena, they often come to different, though complementary, conclusions about the nature of these phenomena. The same overlap is found more frequently in the realm of human affairs, where several sciences often study the same events and arrive at different conclusions about their importance and causes. Such disagreements are possible because an identical phenomenon studied by different sciences is defined differently by each separate discipline. Each scientist's theories about the world act as selective filters, forcing him to view experience in a way which scientists from other disciplines cannot. Different sciences appear to study the same phenomena, but in fact their respective viewpoints force them to define objectively identical phenomena differently. The same construct is, in effect, two or more different phenomena because it has more than one definition. A rose is *not* a rose is *not* a rose so far as several different sciences are concerned. A botanist, a landscape architect, and a geneticist will see the same rose as three distinctly different entities. Similarly, a riot will have different implications for a political scientist, a sociologist, a geographer, and an economist.

We can think of any event we experience as having a certain number of relevant contexts which correspond to the viewpoints held by different disciplines which are concerned with that event. Consider, for example, the construct *urban riot* (Fig. 3–1). Each axis which intersects the construct in the center of the drawing represents a context within which we may validly view that construct. Time and space are obvious dimensions of events such as riots, and their relevance gives rise to the interest of historians and geographers. But the social, political, cultural, economic, and psychological contexts of such an event are also important, and each gives rise to a distinct science which makes it its business to study and explain such events from its own particular viewpoint.

The contexts which intersect a given construct are analogous to different relationships between constructs in the C-field. Our argument that several contexts may intersect a construct repeats, in slightly altered form, an earlier argument. In the same way that a construct may be ordered within all four of the major ordering systems, a construct may be interconnected to other constructs by relationships produced by several different sciences. We have not previously differentiated the laws and theories which interconnect constructs, but different interconnections which compose the general structure of science are produced by different sciences. Several distinct scientific networks may knit together the same constructs and groups of constructs. *Poor police protection, high crime rates,* and *minority group riots* are interconnected by hypotheses, laws, and theories in sociology and political science just as in geography. In many cases, interrelationships are identical; that is, the same law may form a part of the theoretical structure of more than one science.

We can think of the C-field structure built by each of the sciences as a network which reflects light at a specific wavelength. Being trained in a science is much like being fitted with a pair of glasses which are opaque to the light reflected by networks built by other sciences. Those who are trained in a science wear theoretical "glasses" which make it difficult for them to see interconnections which reflect light at wavelengths other than the one for which their glasses are designed.

Now the parochialism of the sciences is not nearly as bad as we have made it out to be. Scientists in one discipline are not completely blind to the work done in related disciplines. But each discipline does have a distinct viewpoint on the structure of the C-field of experience, and each science usually concentrates on extending and intensifying its own C-field network. This explains why interdisciplinary research is often very fruitful. Relationships which are almost intuitively obvious to a scientist in one discipline may be very obscure or entirely overlooked by the practitioner of another. Focusing on a common problem from two or more viewpoints often helps people from several disciplines solve the immediate problem which brings them together more efficiently than they could if they worked separately. At the same time, they often make helpful contributions to each other's network structures.

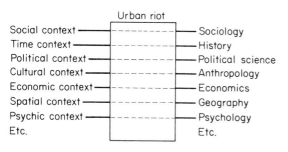

	Urban riot	
Social context ——		—— Sociology
Time context ——		—— History
Political context ——		—— Political science
Cultural context ——		—— Anthropology
Economic context ——		—— Economics
Spatial context ——		—— Geography
Psychic context ——		—— Psychology
Etc.		Etc.

FIGURE 3–1

Some construct contexts and related sciences.

Scientists are distinct from one another because they ask and answer different questions about the world. Unless we wish to get embroiled in the messy matter of operational definitions, we cannot differentiate sciences on the basis of the phenomena they study. Given the fact that all sciences have a single method, we are left with only one basis for making distinctions among the sciences—the questions scientists ask as they seek to solve problems. The fact that scientists ask different questions about a common stock of events and experience is what makes it possible to have several disciplines which study the same phenomena. By asking different questions, we create distinct but complementary structures of law and theory. It is the questions a particular science consistently asks about the world which distinguish it most clearly from the other sciences.

Locational and Spatial Questions

The distinctively geographical question is *"why are spatial distributions structured the way they are?"* This question—with its wealth of implications and elaborations—is the foundation of our science. Because spatial distributions are basic in geography, we must define the concept and the related idea of spatial process more precisely. A distribution is the frequency with which something occurs in a space. We can and do talk about distributions in one-, two-, three-, and N-dimensional spaces, and the underlying notion is similar in all cases. We often use histograms to describe distributions in one-dimensional space. By placing a cell along a line each time a value is observed, we produce a two-dimensional visual expression of the frequency of occurrence of a variable in one-dimensional numerical space (Fig. 3–2a). The nature of all distributions depends upon the scale at which we observe. If we use small divisions, we produce a different histogram than if we use larger divisions (Figs. 3–2a and 3–2b).

When we talk about distributions in two-dimensional numerical space, the same principles apply, except that we have two axes instead of one, and for each observation we will have two variables to plot. When we plot observations in this way, we produce *scatter diagrams* (Fig. 3–2c). These are useful for giving us some idea of the nature of any systematic relationship between the two variables. The frequency with which points occur in space over the diagram gives us information about such relationships. When geographers talk about distributions, however, they are usually interested in the frequency with which things occur in terrestrial space. Often, the two relevant variables are latitude and longitude. Just as the X and Y axes of the scatter diagram define a numerical space, latitude and longitude define terrestrial space. In the same way that a distribution of points in space on the scatter diagram is a starting point for further analysis, a distribution of phenomena in terrestrial space is the jumping off point for geographical analysis.

Distributions can be observed in three-dimensional statistical or terrestrial space also (Fig. 3–2d). Plotting an observation according to three variables (for example, latitude, longitude, and elevation) produces a distribution in three-dimensional space. It is difficult if not impossible for us to visualize more than three dimensions, but this does not prevent us from using spaces with four, ten, or N dimensions (*hyperspaces*) for some very useful purposes. For the moment, however, the two important things to note are that a distribution is always the frequency with which we encounter some phenomenon in a space, and that the scale at which a space is examined is a fundamental determinant of the nature of the distributions which will be observed there.

Spatial distributions may be composed of like or unlike things; we can fruitfully discuss the distribution of hogs and corn together or that of blast furnaces alone. Spatially distributed phenomena may be ubiquitous (available everywhere over the surface of the earth) or localized, but ubiquitous phenomena are rare; areal variation—spatial differences in occurrence and density—is characteristic of almost all distributions in terrestrial space.

It is not distributions themselves which excite geographers, but rather the fact that distributions vary in pattern and intensity from place to place. When we observe some-

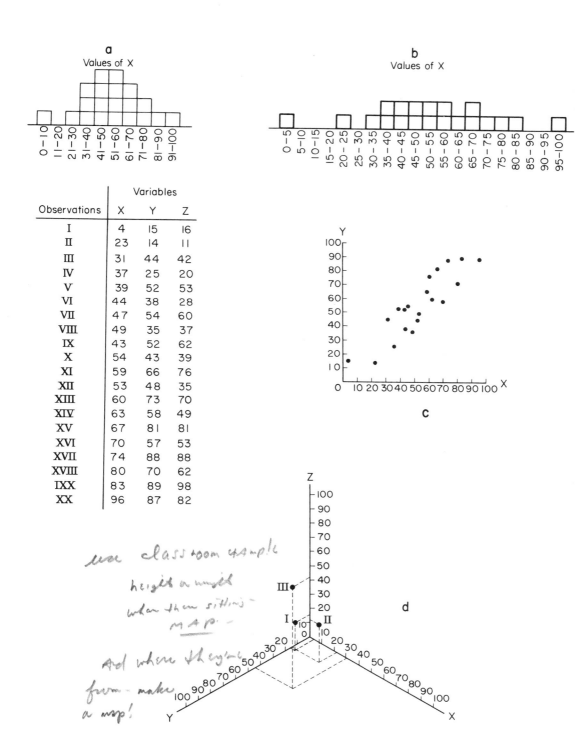

FIGURE 3–2

Distributions in spaces.

thing at one location but not at another, or when we note that densities of occurrence vary, we immediately begin to ask why this variation occurs. By asking such questions we create contemporary geography. Thus there would be no geography if all things were truly ubiquitous. Fortunately for geographers, there are few truly ubiquitous phenomena which are of any importance. Even air—the traditional "free good" of economic analysis—is no longer undifferentiated or costless; the prices of the hillside homes on the slopes of the Santa Monica Mountains are an index of the price people will pay for (among other things) air less polluted by the smog of the Los Angeles basin.

Spatial distributions and the processes which generate them are observable at several scales, some of which are within the geographer's purview and some of which are not. In size, the geographical scale of analysis and observation is bounded on the lower end by the architectural region—the area an architect usually considers when he designs a building—and on the upper end by the size of the earth. Thus geographers are not immediately concerned with the small spaces which architects investigate and the even smaller spaces of subatomic physics. Nor do they study distributions in the very extensive spaces of the astronomer. This does not mean geographers ignore spatial distributions and processes which occur outside these usual limits. Any investigation of spatial distributions or human spatial behavior is of interest to geographers because laws and theories from nongeographical space may be applicable to distributions in geographic space. Geographers derive many insights from investigations such as Edward T. Hall's and Robert Sommer's work on human spatial behavior on a very personal scale. Similarly, work in thermodynamics on the diffusion of heat through materials may provide useful models of geographical diffusion processes. There are analogs, for example, between the way heat diffuses from a source to a copper sheet and the way migrants to a new land might diffuse from their port of entry (Fig. 3–3). Formal sciences such as geometry, topology, and probability theory also investigate spatial distributions and produce

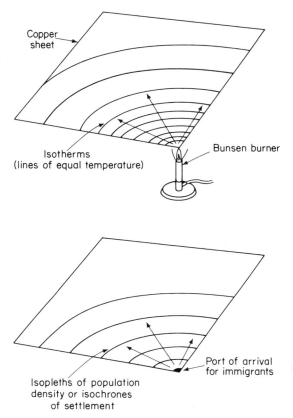

FIGURE 3–3
Spatial analogs at different scales.

techniques and theory which geographers find useful. The geographer's major concern is with distributions observable at the local to terrestrial scales, but he often finds it informative to pay attention to processes and structures which can be observed at other scales because of the homomorphisms which exist among spatial structures at several scales.

Because man has been unsuccessful in moving through the earth from one location on the surface to another, he still views the earth as flat. Thus the distributions geographers study are primarily two-dimensional. We recognize latitudinal and longitudinal extent, but rarely depth and height. This is not to say that we completely ignore the third dimension. Physical geographers and cartographers have always been sensitive to the third dimension, and because of the vertical growth of cities in the last several decades, human geographers have become increasingly aware of its existence also. With advances in movement technology, such as space travel and satellite communications, we shall soon have to think even more about the third dimension than we do now. At the present time, though, geography is still largely a science which asks and answers questions about distributions and processes in two-dimensional terrestrial space.

What kinds of questions are we asking then when we ask *"Why are spatial distributions structured the way they are?"* An important aspect of this question is that any discussion of the spatial distribution implies a *where* question concerning the location of that distribution. The matter of "whereness" is basic in all geography. Geographers ask *where* questions about all events with which they concern themselves, but they do not always provide the same answer, for there are two different kinds of whereness: *where* questions may be asked and answered in an absolute or in a relative way.

Kinds of Location

Absolute location is position in relation to a conventional grid system designed solely for locative purposes. Latitude and longitude are the most common means of describing absolute location. Washington, D. C., for example, is located at 38°50′ north latitude and 77°00′ west longitude. Another description of absolute location is a street address, such as 1146 Sixteenth Street, N.W. In both instances, location is fixed with respect to an abstract network of imaginary locations. Such locations are absolute in the sense that once a locational description of this kind has been adopted, it does not change over time.

Relative location is position with respect to other locations. The relative location of Washington, D.C., can be described in several different ways, one of which is to say that Washington is located on the Potomac River about 55 kilometers southwest of Baltimore and 175 kilo-

meters north-northeast of Richmond, Virginia. Or we might say that Washington is two hours driving time north-northeast of Richmond and one-half hour from Baltimore.

Relative location can be expressed in values other than the usual distance units. We could describe the distances between Washington and Richmond in terms of bus fare or the cost of a plane ticket. There is a large number of ways of describing distance and location in a relative context, but in the absolute context we are restricted to customary and unchanging units such as miles, kilometers, or degrees of latitude and longitude to measure distance. The relative locations of two places may change radically, even though absolute locations remain constant. An example would be San Francisco and New York City, which were about six months apart little more than a century ago, and which are now but six hours apart by jet aircraft. Today geographers are primarily concerned with the relative locations of phenomena. Traditionally, the major interest of geographers was accurately describing the absolute locations of distributions.

When we talk about the locations of sets of events rather than single events, description becomes considerably more complicated. What the relative and absolute locations of a spatial distribution mean is not clear. Obviously, we can locate a set of places by listing the absolute coordinate of each separate member. If we ask "where are the Caroline Islands?" or "where is the New York metropolitan area?" a list of the grid coordinates of each of the individual islands or of each county in the metropolitan region is not what we desire. We would prefer a description of some central location or perhaps the boundary which delimits the group of islands or the metropolitan region. We are interested in the location of the set as a whole, or alternatively, in the internal organization of the distribution—that is, the location of the elements of the distribution with respect to each other.

When we ask *where* questions about distributions we almost always desire an average location which represents the entire set. *Goode's World Atlas* gives the location of the Caroline

Islands as 9°30′ north latitude, 143°00′ east longitude, an arbitrarily selected location near the center of the island group. When choosing a point to represent the absolute location of a distribution, we might select the point of minimum aggregate travel (see Fig. 2–2). Alternatively, we could delimit a distribution on a map by drawing a line around it. These two ways of describing location can be thought of as external description. The distribution is *internally* undifferentiated. We take the Caroline Islands as a homogeneous portion of the earth's surface and are interested in the location of the whole. We can treat distributions similarly when we are concerned with relative location. We can describe the location of the Carolines as north of New Guinea, for example. So long as a distribution is internally undifferentiated, description of absolute or relative location is fairly straightforward.

Spatial Structure and Spatial Process

Today, geographers are more often interested in the internal organization of a distribution, the location of the elements of the distribution with respect to each other. This kind of location is always relative. Geographers frequently talk about the "pattern" of a distribution, using terms like "dense," "sparse," "agglomerated," "dispersed," and "linear." The way these terms simultaneously relate the locations of the elements of a distribution to each other *and* to the entire distribution are subtle but important. In recent years internal relative location has often been called "*spatial structure*," a terminology which we shall adopt. The spatial structure of a distribution is both the location of each element relative to each of the others, and the location of each element relative to *all* the others taken together. Describing spatial structure rigorously is not easy, and we shall postpone until later chapters a discussion of describing and analyzing the spatial structure of distributions of things. It is adequate for our immediate purposes to observe that spatial structure refers to the internal locational organization of a distribution in space.

Contemporary geographers devote a great deal of attention to spatial structures of all kinds. Almost any substantive problem a geographer tackles can fruitfully be considered to be a problem of describing accurately and explaining satisfactorily the spatial structure of a distribution. The emphasis in contemporary geography on spatial structure is somewhat misleading, because it overemphasizes distributions to the neglect of the *spatial processes* which interact causally with them. What we call spatial processes are mechanisms which produce the spatial structures of distributions. Reference to spatial process is inescapable in any explanation of spatial structure.

Why geographers are so much more aware of distributions than of the processes which produce them is not clear; it is probably because distributions of static things are easier to observe and record on maps than the processes which produce them. Their view that spatial structure is primary in their science often leads geographers to overlook important relationships, not the least of which is the causal effect of structure on process. Spatial structure and spatial process are *circularly* causal. Structure is a determinant of process as much as process is a determinant of structure. The existing distribution of supermarkets in a city, for example, is a fundamental determinant of the success of any new supermarkets established in the area.

In proper perspective, the distinctions we make between spatial process and spatial structure disappear because they are based upon a limited time perspective and are thus somewhat artificial. Processes are spatially variable and thus have distributions just as do "concrete" phenomena. Thus spatial structure is a concept applicable to both static distributions and to processes which appear to us to be dynamic. Process and structure are, in essence, the same thing. Whether we see process or structure when we look at a spatial distribution depends on the time perspective we adopt and the rapidity with which the process moves. Human movements in vehicles and on foot result in spatial structures of objects like roads, railroads, airports, and sidewalks on the surface of the earth. Similarly, human choices of agricultural, industrial, and commercial activities produce economic spatial structures.

Because we find it easier to make maps of

distributions of apparently static physical phenomena than of moving people, or their motives for moving or for making the choices they do, we tend to think that these distributions of physical phenomena are as static as they appear to be on maps. Yet upon reflection it is obvious that distributions and their structures constantly change, and when we consider apparently static distributions over periods of twenty-five or fifty years, their structures are usually very dynamic phenomena. Imagine, for example, a series of weekly maps of railroads in service in the United States and Canada over a period of fifty years. If a film were made of this sequence of 2,609 maps which showed twenty maps per second, the network would contract in some areas and expand in others in a very dynamic manner.

It is our limited time perspective which makes us designate some spatially differentiated phenomena as distributions of objects while we designate other, more rapidly changing distributions as processes. When we distinguish spatial process from spatial structure we are merely recognizing a difference in relative rapidity of change; both the obviously changing and the apparently static distributions on the earth's surface are components of spatial processes. "Spatial distribution" is a term we apply to spatial processes which appear to us to be static, and "spatial structure" is the term we apply to the internal spatial organization of these distributions of process elements. Properly considered, the spatial structure of a distribution is viewed as an index of the present state of an ongoing process.

The fact that spatial process and spatial structure are identical from one point of view does not necessarily negate the validity of the distinction we make between the two. There are processes which move rapidly and processes which move slowly, and it is useful to distinguish between them. Any attempt at explanation must segregate the events and experiences to be explained from those which caused them, artificial as this separation may ultimately be. We do this in geography by using spatial processes, causal event chains which are locationally differentiated, to explain spatial structures—relative locations with respect to each other of members of locationally differentiated sets of events. In turn, we use spatial structure to explain spatial process. Spatial process and spatial structure are two temporally defined aspects of process in general which are circularly causal. Structure always has explanatory power for subsequent process, and process always explains subsequent spatial structure.

The questions—implied and direct—which geographers ask about spatial process and structure are what makes geography distinct from other sciences. Our formulation of the geographical question conceals, as we have seen, two important ideas which are basic in geography. One, the questions about where things are in the absolute context, will be taken up in detail in the next section of this chapter, where we shall see how long it took geographers to describe the locations of the contents of our world. The second, the questions about why things are where they are, is a more current question which will occupy our attention for the remainder of this work. We could elaborate the geographical question to embrace these concealed aspects. We could ask "What is the spatial structure of events and how do their spatial structures and spatial processes interact?" But if we were to try to incorporate all possible facets of geographical inquiry into the question, it would soon attain unwieldy length. The question we have asked—"*Why are spatial distributions structured the way they are?*"—will serve as well as any so long as we remain fully aware of the riches it conceals.

Ultimately the form of the question or questions which distinguish geography from the other sciences is not too important. No other science consistently concerns itself with distributions of phenomena in terrestrial space; no other science consistently concerns itself with spatial structure. The questions about location, spatial structure, and spatial process which we ask and answer distinguish geography from the other sciences.

TRADITIONAL GEOGRAPHICAL QUESTIONS (250 B.C. TO ca. A.D. 1800)

For most of its 2,200 years of existence as a distinct body of thought geography has been concerned primarily with accurately describing the locations of places. Although more recent spatial questions have now relegated inquiries about abolute location to an implicit status, historically *where* questions have been preeminent. Until the basic task of accurately mapping places on the surface of the earth was completed, geographers had relatively little time for more detailed questions about what existed at places and why. Because so much of the world was unknown for so long, geographical manpower was largely devoted to the production of an accurate world map. In geography the first order of business is to fix accurately the locations of the elements of a distribution. Because the spatial horizons of Western man kept expanding up to 1800 and well beyond in a few areas, filling in the world map with places and names occupied most of the talent geography was able to muster until a relatively short time ago.

Origins of Western Geography

The question "where?" exists independently of geographers; people in the Western and Eastern world were certainly asking that question well before the first geographer bumbled onto the scene. The illustration on the cover of this book is a late Neolithic (ca. 5000 B.C.) village map which was chiseled in stone on a rock overlooking the settlement. This and other similar artifacts certainly indicate that spatial organization has been one of man's important concerns for a long time. Evidence of systematic geographical thought (as distinct from geography) is evident in Greek literature from its earliest beginnings. The works of Homer, Hesiod, the Greek dramatists, and Herodotus, among others, betray careful attention to the locations of places and peoples, and a curiosity about the characteristics of places. Although curiosity about the spatial context is a starting point, it is not in itself geography. In order to be a science, a body of thought must be the province of a group of scholars who are self-conscious about their inquiry, and who have methods for answering the questions they ask.

Of the early Greeks, Eratosthenes (276–196 B.C.) was the first to call himself a geographer and the first to devise a method which enabled

geographers to locate places with some accuracy. Eratosthenes was annoyed by the fact that the locations of places in the literature of his era migrated. What was Scythia in 650 B.C. was not the same place known as Scythia in 400 B.C., which in turn was different from the place designated by that name in 250 B.C. In a situation where definitions—spatial or otherwise—change unpredictably, the usual solution is to choose one definition of a term or place and formalize the choice by recording it in a dictionary or atlas.

Locating places unequivocally was not possible before about 200 B.C., because no accurate method of describing location on the earth's surface existed. In the absence of such a method of location, the drawing of an accurate world map was impossible. Eratosthenes devised a crude locative grid system. He divided the known world into rectangular regions with imaginary lines drawn through principal cities and important physical features (Fig. 3–4). The grid provided a framework for map construction and place location which helped make his map more accurate than previous efforts. Faced with his immediate locational problems, Eratosthenes devised a solution which looks crude to us, but one which is the foundation of the more sophisticated locational systems we use today. Geographers look to Eratosthenes as the first geographer because he provided the science with a method which enabled it to answer *where* questions satisfactorily.

Although there are some drawbacks in doing so, we will use the maps produced in each period as an index of the nature and quality of geographical knowledge during that era. As an indicator of the nature of geography at about 200 B.C., Eratosthenes' map signals a scientific interest in place location. By concerning himself with *methods* of location, Eratosthenes demonstrates his concern with all possible places and the utility of general place knowledge. Although Eratosthenes' map may look crude to us, it is, in terms of quality, superior to a number which were to appear in the succeeding centuries. The area around the Mediterranean is fairly accurately described and even such distant places as Britain and Ceylon are in

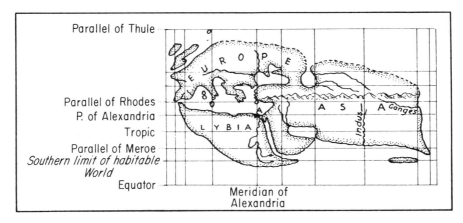

FIGURE 3–4

The world map of Eratosthenes (courtesy, Erwin Raisz).

approximately their correct relative locations.

Although Eratosthenes is claimed by geographers, he is most famous as a geometer. He was the first to measure accurately the size of the earth. His method was simple but ingenious. Eratosthenes knew that Syene (modern Aswan) was on the tropic because vertical objects cast no shadows there on the summer solstice. He also knew the approximate distance from Syene to Alexandria. Since the angle between a vertical stake and its shadow at Alexandria on the summer solstice was one-fiftieth of a circle, the linear distance between two cities had to be equivalent to one-fiftieth of the circumference of the earth (Fig. 3–5). We are not sure exactly what Eratosthenes' final measurement was because we are unsure about the length of the unit of distance (the stadium) he used; furthermore, there were a number of minor errors in the assumptions upon which the measurement was based. Nonetheless, the measurement was very accurate (say, within eight hundred kilometers), and the invention of the technique was magnificent. Both Eratosthenes' measurement of the earth and his map illustrate the interest of the earliest geographers in accurate measurement and description of location.

Several improvements in systems for locating places were made in the two centuries after Eratosthenes. Advances in geometry resulted in the convention of dividing a circle into 360 degrees. The application of the 360 degree division to the terrestrial globe and to two-dimensional representations thereof by Ptolemy

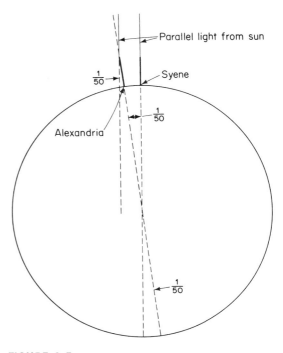

FIGURE 3–5

Eratosthenes' technique for measuring the earth's size.

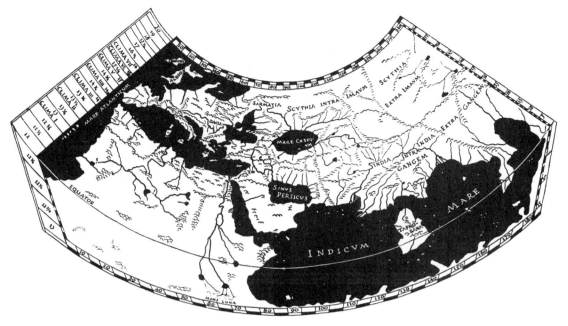

FIGURE 3–6

Reconstruction of Ptolemy's world map (courtesy, Erwin Raisz).

(died about A.D. 155) and some of his predecessors provided geographers of the later Hellenistic period with the same method of describing absolute location which we still use today. Ptolemy's use of latitude and longitude both for describing the locations of places on the earth's surface and for accurately plotting these locations on maps was a major achievement (Fig. 3–6). Although minor modifications of the Ptolemaic system have been made, we still use Ptolemy's basic locative system and we still use some of the map projections he invented.

Because mobility was limited, the geographers of the Hellenistic era could not provide a complete catalogue of terrestrial locations. This in no way detracts from their contribution in providing a method for geography, a method which put geography on a scientific footing at a very early date. Western philosophy is sometimes said to be a series of footnotes to Aristotle; we could also argue that much of our geography is a series of footnotes to Ptolemy. Such an assertion does not tell the whole story of con-

temporary geography, however, just as the former is misleading about the nature of modern philosophy. Geography is still very much concerned with whereness, but in a different context from the absolute locational framework emphasized by Eratosthenes and Ptolemy.

At the same time that techniques were being developed to provide satisfactory answers to *where* questions, geographers began to ask *what is where* questions. Such questions are certainly evident in the proto-scientific works of Homer and Herodotus, but Strabo (64 B.C. to A.D. 21) handled them in a more systematic fashion than earlier writers. More than any other geographer, Strabo demonstrates the Greek concern with the character of places:

. . . my first and most important concern, both for the purposes of science and for the needs of the state, is this—to try to give, in the simplest possible way, the shape and size of that part of the earth which falls within our map, indicating at the same time what the nature of that part is and what portion it is of the whole earth; for this is the task proper of the geographer [II, 5, 13].

Strabo's concern with the differentiated content of places, particularly his interest in cultural variations in space, is the beginning of a persistent trend in geography, but one which does not reach fruition until after 1800. Note that *what is where* questions are founded on prior knowledge of absolute location. Strabo's attention to the nature of the ecumene presumes the existence of maps which provide satisfactory answers to such questions.

Science, and especially geography, were as successful as they were during the 450 years between 250 B.C. and A.D. 200 because of the intellectual and political environment within which they existed. Geographers were able to answer practical and theoretical questions about events in the world. Moreover, both laymen and geographers had opportunities to travel during that era. Especially after the firm establishment of the Roman hegemony (but to some extent even before), movement throughout most of the Mediterranean region was easy and secure. Empire and geography have always been bedfellows. By producing settled conditions and curiosity about distant but important places, spatial hegemony stimulates *where* and *what is where* questions for geographers to answer.

Ptolemy was the last of the important ancient geographers. Even at the time he flourished, people in the Western world were beginning to turn their attention away from science, and the acquisition of new spatial knowledge was on the decline. Ptolemy's works were preserved on the eastern periphery of the Mediterranean area and further to the east, but in the Mediterranean region itself and in Europe scientific geography was neglected and soon disappeared.

The Christian West (A.D. 200–1400)

Because we feel that science is an outstandingly successful enterprise, we are imbued with the idea that science is progress and that science has always evolved in a positive direction. The history of geography and other sciences in Europe after A.D. 200 is adequate evidence that such is not necessarily the case. Geography degenerated in Europe after Ptolemy; it was not until 1200 years after his death that European geography and cartography began to

reach the levels of proficiency and accuracy attained during the Ptolemaic period.

The evidence available in the form of geographical expositions and maps from the era indicates beyond doubt that Christianity had much to do with the decline of geography. But what is probably more important is the fact that the habits of thought and the practice of traveling which promote geography both become increasingly rare as the Roman Empire deteriorated. As Medieval Europe closed in on itself to become a mosaic of political and social units, geography's questions were asked less often. Curiosity about what lies over the horizon is not compelling unless it is possible to travel.

At the same time that decreasing mobility was eliminating the need for scientific geography, increasing religious fervor was providing alternative answers to *where* questions in a form incompatible with those provided by science. Theological order contributed to the deterioration of scientific geography because it provided answers to *where* questions which, by modern standards, are not only erroneous but positively disastrous. Theology can at most be charged with contributory sciencide; the more basic reason for geography's demise is that it became increasingly superfluous as Europeans became increasingly immobile. Because geographical theory and practice were not tested against experience, and because they could not be tested given the immobility of people in the Middle Ages, a geography was developed which seems to us rather bizzare. The Bible contains a number of cosmological and geographical statements. Increasing respect for Biblically based theological ordering, developing at the same time that the need for experience-tested geography was decreasing, resulted in the former's supplanting the latter.

The replacement of scientific by religious cosmography is very evident in cartography. After Ptolemy there is a very obvious decline in the accuracy of world maps until the fourteenth century (Figs. 3–7 to 3–9). Whether we can fault the cartographers and geographers of the Middle Ages for the maps they produced is debatable. By our standards, their geographical lore and cartographical practice are obviously unsatisfactory. But given the

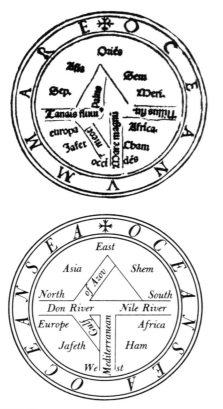

FIGURE 3–7

The world map of Isidore of Seville, A.D. *600 (from Lloyd A. Brown,* The Story of Maps, *facing p. 108, by permission of Little, Brown and Co.).*

exception of astronomy was as well formulated and methodologically sound as was Greek geography. The rejection of an established means of determining absolute location strikes us as foolishness, but the fact that it has happened more than once is evidence that scientific ordering is not omnipotent. As long as science produces the best answers to the questions people ask, it flourishes. If people stop asking the questions that science can answer and begin asking questions which can be better answered by other systems, science falls into disrepair. That geographical science in the Christian West died sometime after Ptolemy is clear. The dominance of geographical thought and cartography by theological cosmography is evidence of this. Even more compelling evidence is the loss of self-conscious geography. Ptolemy was the last self-conscious geographer to grace the Western world until the fifteenth century. The geographical literature which survives from the Middle Ages is geographical thought to be sure; man can never escape thinking about such a fundamental context of his existence. But geographical thought is not necessarily geography, and the absence of self-conscious investigators armed with an established methodology with which to practice geography during the European Middle Ages is the best evidence that geography was replaced by other methods of explaining man's spatial context during that era.

questions asked and the intellectual tenor of the era, what we judge to be erroneous beliefs about place location and the nature of the world were probably as satisfying or more satisfying than scientifically correct answers would have been. Theology replaced science as the primary method of ordering the world. Degenerative though it may appear to us, the change was simply a shift from a scientific ordering system to theological ordering, which many apparently found more satisfactory than Greek science.

There have been few instances in history in which scientific ordering has been scrapped in favor of an alternative system. Most of the other Greek sciences suffered similar fates, but no other empirical science with the possible

Medieval Muslim Geography (A.D. 800–1200)

When Muslim geography and cartography are compared with simultaneous developments in the Christian realm, they compare very well indeed. At any date between A.D. 800 and 1400, Muslim geographical knowledge more nearly matches our notions of what the world looks like. This relative superiority of Muslim geography and cartography is deceptive, however. By objective standards Muslim geography was no more successful than European geography and less successful than Chinese geography.

The Muslim intellectual world certainly had need of accurate knowledge of the locations of

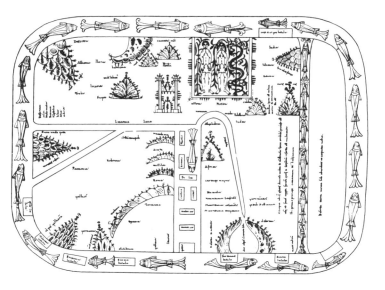

FIGURE 3–8

World map of Beatus, A.D. *787 (from Lloyd A. Brown,* The Story of Maps, *facing p. 126, by permission of Little, Brown and Co.).*

places. Only the later explosion of Europeans over the world surpassed the spatial conquests of the Muslims in the seventh and eighth centuries. By A.D. 800 the Abbasid Caliphate was sovereign over the entire realm from what is now Afghanistan to the Atlantic Ocean with the exception of France, Italy, the Balkans, and what is now Turkey. The extent of the Muslim Empire created administrative and military problems whose effective solutions often depended heavily on accurate knowledge of absolute and relative location. At the same time, travel and interest in other places were promoted. The religious requirement of *Hajj* (pilgrimage to Mecca) meant that the affluent and many not so affluent from all over the Muslim world traveled or anticipated traveling at least once during their lifetime. Moreover, a common language and religion throughout the realm and the charitable provisions made in many places for the support of indigent travelers led to seemingly casual journeying by literate adventurers such as Al-Biruni and Ibn-Battuta, to mention two famous Muslim travelers.

Thus the early Muslim period was one in which the spatial sciences could be expected to flourish. In A.D. 800 the Abbasid caliphs ruling from Baghdad reigned over a larger and more unified empire than the Muslims would ever

again achieve. Travel and commerce flourished, and enlightened monarchs subsidized the arts and sciences with a lavish purse. All of Ptolemy's spatial works (*Geography, Astronomy, Tetrabiblos*) were translated into Arabic. Thus at the very beginning of Muslim science, geo-

FIGURE 3–9

Bianco world map, A.D. *1436 (from Lloyd A. Brown,* The Story of Maps, *overleaf of page facing p. 126, by permission of Little, Brown and Co.).*

graphers were acquainted with the full corpus of Greek cartographic theory and technique. Unfortunately, rather than building on this foundation, with few exceptions Muslim geographers progressively ignored it as time passed.

For the most part Muslim geography deteriorated into a series of geographical dictionaries which even by contemporary standards were of limited value; the compilers of these works complain themselves of their inadequacies. Despite conscious attempts to describe accurately absolute locations, despite the fact that a method of measuring and recording location accurately was provided in Ptolemy's works, and despite the fact that almost all Muslim geographers cite Ptolemy as the founder of their science, very few Muslim geographers used latitude and longitude to describe locations. This is especially puzzling since Arab astronomers—certainly some of the best the world produced—continued to use latitude and longitude throughout the Muslim world. Because geographers rather than astronomers made maps, however, Muslim cartography declined steadily over the period. No extant Muslim world map attains the same degree of accuracy as can be had by simply following the directions in Ptolemy's *Geography* (Figs. 3–10 to 3–12). In general, we attribute the deterioration of Muslim cartography to the unavailability of Ptolemy's works to the later Muslim geographers, and to a lack of communication among geographers as the Muslim Empire fractionalized into small principalities after about A.D. 900.

As good as it appears when contrasted to the spatial knowledge in the Christian West, Muslim geography cannot be rated as successful by other standards. Given the body of knowledge and the method with which they started and the questions they were asking about absolute location, Muslim geographers must be given very low marks. In a relative sense, Muslim geography is poorer than Christian geography of the same era. European geographers provided what we consider to be poor answers, but they were asking questions in a context which led logically to those answers. Muslim geographers were able to provide objectively better answers to spatial questions than were the European geographers of the Middle Ages,

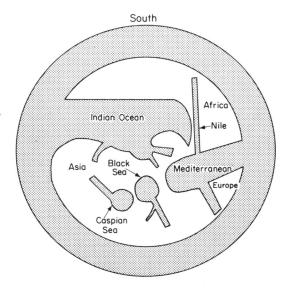

FIGURE 3–10
Outline of a map by Al-Istakhri, tenth century.

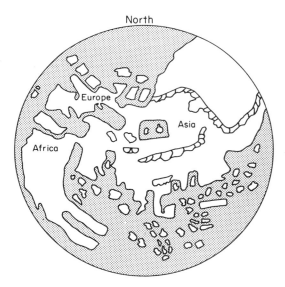

FIGURE 3–11
Outline of a map by Ibn-Said, thirteenth century.

but evaluated against the questions they were asking, these answers are relatively poor. Muslim geography and cartography come off well in comparison to European geography and cartography in the Middle Ages, but in

South

FIGURE 3–12

Outline of a map by Al-Idrisi, twelfth century (adapted from Ibn-Khaldun, The Muqaddimah: An Introduction to History, *trans. F. Rosenthal [Princeton, N.J.: Princeton University Press, 1967], Bollingen Series XLIII. Copyright © 1958, 1967 by the Bollingen Foundation, facing p. 110; courtesy also of Routledge & Kegan Paul Ltd.).*

terms of the potential it had and its subsequent ability to do its job properly, it was no more successful and possibly less so.

The Far East (A.D. 200–1600)

At the same time that geography in the West was becoming extinct, Chinese geography was developing into a sound spatial science. A self-conscious geography, described by that name, had developed in the first and second centuries A.D., and in the second and third centuries Chinese geographers began to use a rectangular grid and graduated scale to construct their maps. The grid system was not based on astronomical phenomena, as is ours, but it enabled Chinese geographers to construct extremely accurate maps nonetheless. Throughout much of the period between the third and the sixteenth centuries, Chinese cartography was superior to that produced by either Christian Europe or the Muslim Middle East. There were

evidently some transfers of knowledge from the West to the East and vice versa, but they are somewhat peripheral to our theme, and Joseph Needham has treated them far better than we could in any case. If you are interested in a detailed treatment of Chinese geography and cartography, you should consult Volume 3 of his excellent *Science and Civilization in China.*

The Age of European Exploration and Discovery

For reasons which are complex and not entirely clear, Europe underwent a revitalization after about 1200. Interest in travel and spatial knowledge in nontraditional terms increased rapidly. With the greater travel and important advances in maritime technology, accurate knowledge of absolute location again became fundamental to orderly human activity. At the same time, the rediscovery of Greek and Roman thought resulted in the translation of the Greek geographical literature. Ptolemy's *Geography* was first translated into Latin in 1410. The first printed edition was made in 1472, and between that date and 1500 six more editions were published. Ptolemy's *Geography* was one of the most widely printed books of the era, and for a century or more Ptolemy's maps were reproduced in the same atlases which contained more accurate maps of the world, which is some indication of the esteem in which Ptolemy was held. The Ptolemaic literature, the voyages of exploration, and the subsequent broadening of geographical horizons combined to produce a golden age. Geography and cartography have never been as useful and successful as they were between about 1450 and 1800.

It is difficult or impossible for us to understand the excitement which must have prevailed in informed circles during the era of European exploration. By all accounts, the interest in the new worlds being discovered was very intense. Our current interest in space exploration is only somewhat analogous. We knew what to expect when we got to the moon, and had some idea of the dangers involved; during the fifteenth, sixteenth, and seventeenth centuries, however, a whole new world was opened and there is probably no way we can exactly recapture the feelings of educated people of the time. But if we reflect upon the expansion of the European realm, and attend to the conceptual and practical problems this expansion involved, we can develop some appreciation for the ethos of the era and some indications of why geography was so successful during the era of exploration.

From a conceptual point of view, massive new inputs of locational knowledge had to be ordered and stored in an efficient manner. Maps are information ordering and storage devices particularly suited to handling spatial information, and map and atlas making flourished during this era. During the sixteenth century especially, new places were encountered at a furious pace. Reports of these discoveries were new data which had to be integrated into existing bodies of knowledge; simply assimilating new knowledge was a major project, and maps were almost continually revised as new information was added. Some very practical problems not too much divorced from conceptual difficulties were also encountered. Because the Age of Exploration was also the Age of Commerce, accurate knowledge of locations and thus of routes to different places (relative location) had actual monetary value. Shorter routes and accurate navigation were money in the bank to those who possessed them. By accurately describing absolute location, deriving relative location from such descriptions, and providing merchants and interested citizens with knowledge of what the new worlds were like, geographers made valuable contributions to the societies in which they lived and worked.

The question "Where?" was extremely important for several hundred years. It was the ability of geographers and cartographers to answer that and related questions which is responsible for the success of the geography of that era. Then, as ever, a science which answers people's most pressing questions is the one accorded the greatest prestige and financial support. Geography's utility peaked about 1500 to 1800. When contemporary geographers can offer their societies answers to practical problems as useful as the knowledge of accurate location was in the fifteenth to the eighteenth centuries, geography will again enjoy the position it held then. You might argue that this will be very difficult, because we have inconveniently run out of new worlds to discover. It is true that certain sciences are "adopted" by events during certain eras. An example is the stimulus given to the aerospace sciences, first by World War II, and then by the launching of the first earth satellite in 1957. It is also true that in one sense the world is a closed system, since our maps have been filled in and problems of absolute location are now passé. But we have only begun to explore problems of relative location and to map relative space, and there are worlds to discover in relative terrestrial space more exciting and potentially more valuable to humanity than the new worlds discovered up till now.

Modern Geography (ca. 1800–1950)

In about 1800 the nature of geography began to change somewhat from what it had been for the previous two thousand years. Increasingly, questions concerning absolute locations of places became outmoded. Having answered the question "Where?" to their satisfaction for most parts of the world, geographers became more concerned with describing *what* was where and with generalizing their knowledge of places.

"What is where?" is a question which is properly taken up only after the locational map of the earth is completely filled in. By 1800 an accurate world map had been produced, at least for the major outlines of the continents; the general configuration of terrestrial space was known. Even long before 1800, however, many people were more interested in describing than locating places. In the ancient world, writers such as Strabo and Pausanius wrote popular descriptions of familiar and far away places. We mentioned earlier the travel accounts of Ibn-Battuta and Al-Biruni, two Muslim travelers. The popularity of the tales of Marco Polo and other European voyagers are clear indications that late Medieval Europeans were curious about the quality as well as the locations of distant places. Nonetheless, throughout the period up to 1800, qualitative description was secondary in professional geography. Most descriptive geography was geographical thought, rather than geography per se; it was the product of talented amateurs and explorers. Self-conscious geographers continued to busy themselves primarily with locative cartography.

Discussions of the methods of describing places appear in Europe as early as the beginning of the seventeenth century, but it was not until the terrestrial sphere became a closed system about 1800 that geographers began to devote most of their attention to qualitative description and generalization of places. Immanuel Kant (1724–1804) added to the methodological foundation of descriptive geography begun earlier, but it was the works of Alexander von Humboldt (1769–1859) and Karl Ritter (1779–1859) which firmly established the methodology of descriptive geography and the place of geography among the modern sciences. Von Humboldt's investigations of specific topics such as types of vegetation and climates on a regional, continental, and worldwide scale resulted in a conception of geography as a systematic science—one which investigated the regional and worldwide distributions of phenomena like vegetation, population, and relief. Ritter's massive volumes of regional description and analysis firmly established the intensive investigation of regions as a geographical activity. The pupils and disciples of Humboldt and Ritter continued both traditions throughout the nineteenth century and beyond; the basic concepts of geography fostered by the works of Humboldt and Ritter remained dominant to the middle of the twentieth century.

Again, a good indication of the changing nature of geography is the maps geographers produced. After 1800, new innovations in cartography often involved maps of regions much smaller than the world, and more importantly, special purpose or topical maps of both the world and smaller regions. This shift in cartography mirrors the shift to topical studies and regional analysis, and away from the primarily locative work which dominated geography before 1800. Now we are not arguing that special purpose and small region maps were not used before 1800. Even maps of the Greek era showed locations of major cities and important features like rivers. Christian maps

of the Middle Ages were cluttered with symbols locating important religious places and mythical creatures. The maps and atlases produced by Flemish cartographic houses in the sixteenth and seventeenth centuries commonly used symbols to locate features such as mines, roads, and canals. But again, the earlier emphasis was *primarily* on describing absolute location. As the nineteenth century wore on, however, and as national censuses were established and scientific observations about the nature of physical distributions multiplied, the production of topical maps of features like population, transportation routes, vegetation, climatic elements, and topography became increasingly important, and increasingly the major concern of geographers.

Geography up to 1800 was primarily a locative enterpise. Fixing the locations of individual places is the first order of geographical business, and this had been largely finished by the time of von Humboldt and Ritter. Between 1800 and 1950, classifying places into sets on the basis of their contents or characteristics and dealing with groups of places became the major concern of the discipline. During this 150 years we can recognize several variations on this theme which held temporary sway, but all operated within the context of regionalism and with an emphasis on regionalization. If we think of terrestrial space as a gigantic C-field, we can designate the infinite number of places in the world as constructs. The emphasis in geography up to 1800 was largely on finding out which kinds of constructs were where. After 1800, with the major physical and cultural outlines of the world mapped, geographers could turn their attention to forming groups of C-field constructs by delimiting sets of places with similar characteristics, and they could attend to the interrelationships between the regions that sets of places form.

Emphasis on regions of like or dissimilar phenomena which are internally homogeneous goes beyond the simple "What is where?" asked by earlier descriptive geographers. Imposing order on the chaos of all possible places by segregating them into sets is a taxonomic process, which is a necessary and very important part of any science. Geographers between 1800 and 1950 sought to produce areal sets of places which were as homogeneous as possible, and to describe accurately the locations of the boundaries of the regions which resulted. This emphasis on producing internally undifferentiated regions is evident in both topical and regional geography. Topical geographers sought to delimit regions of particular kinds of places based on one or a few criteria. A physical geographer constructing a map of vegetation types, for example, would assign places with similar kinds of wild vegetation to the same region, probably on a worldwide or continental basis. Geographers who specialized in the intensive study of more restricted areas of the earth's surface tried to delimit regions based on larger numbers of spatially covarying characteristics of places. The American South, for example, is a region of uniformity in a large number of social, economic, political, and cultural characteristics which make regions like it internally homogeneous and distinct.

Although regional geographers rarely thought of it self-consciously, they were producing operationally defined regions in a C-field of geographical space. In fact, one of the major impediments to greater achievements by regionalists in geography was the failure to realize that their regions were operational definitions. For a long time, geographers tended to believe that internally homogeneous regions, based on one or a number of place characteristics, existed independently of the principles which defined them, and that regions themselves were intrinsically worthy of delimitation and study. Classifications must always have a specific purpose if they are to be fruitful.

By the time of World War II, geographers had compiled a large number of regional studies at all conceivable geographical scales, and while there is no denying that these investigations produced many valuable insights, it was at about that time that geographers realized that further progress in geography depended upon generalization and analysis at levels more advanced than classification. We have chosen 1950 as the date at which geography shifted from a concentration on regionalism to a greater emphasis on the contemporary geographical questions which evoke hypothesis, law, and theory. This date is arbitrary, of course, but the exact date is not as important as the recognition that it was at about this time that many geographers began to ask different questions.

The shift from classification to more advanced levels of analysis was accompanied by an immensely important change in the context in which geographers ask their questions and frame their answers. Throughout the period from 250 B.C. to 1950, geographical space was absolute space. *Where* questions and *what is where* questions were always answered in an absolute context, and the distances discussed by geographers were usually absolute distances expressed in stadia, miles, and kilometers. Since World War II geographers have more often measured space and location in the relative terms of time and cost. Maps of locations and distributions in relative spaces twist and truncate traditional spatial relationships in weird but exciting and stimulating ways. The more theoretical approach adopted in geography since the war is certainly important, but the concomitant change of context to relative spaces and distances is even more profound; it is the fundamental shift in geography since 250 B.C.

THE NEW SPATIAL CONTEXT

The Nature of Relative Space

Until 1950, geographers usually thought and hypothesized about distance and space in absolute terms. The measures of distance and location they used were the unchanging (absolute) units of miles, kilometers, and so forth. Since 1950, relative location and relative distance have been used to define new kinds of stretchable, shrinkable spaces. Please understand that we are not asserting that relative location and space were discovered only in 1950. They had both been examined in some detail and used to some degree for almost a century before that date. What we are concerned with here is their use by a large number of geographers, which we do feel is a relatively recent phenomenon. The shift to a relative spatial context is still in progress, and is probably the

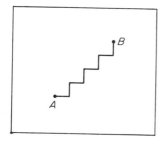

One of the many shortest routes between *A* and *B*. Diagonal movements are not permitted; only movements along cardinal directions are possible.

FIGURE 3–13

Manhattan space.

most fundamental change in the history of geography as it opens an almost infinite number of new worlds to explore and map.

Relative distance is the basis of relative space, since spaces are defined by distances along dimensions. The number of dimensions we use and the way we measure distance along them determine the nature of any space we construct. Numerical intervals along a single dimension produce the mathematical space we examined earlier when we discussed frequency diagrams. Numerical distance intervals along two orthogonal axes may define the two-dimensional space of a scatter diagram or two-dimensional terrestrial space. By choosing different distance measures, we can change space. In Euclidean space, the shortest distance between two points is a straight line. In the space of Riemann, the shortest distance between two points is a curved line. Riemann's curved space is appropriate for discussing the curved universe we believe we live in. A variant of Riemann's space is evident when we route a highway around a mountain because it is too difficult to go directly through or over it. Manhattan space is a variant of Euclidean space, in which the shortest distance between two points is a path consisting of line segments which meet at right angles (Fig. 3–13).

Obviously, in the latter two cases we have changed the measures of distance from those of absolute length, such as miles, to those of difficulty or cost. In terms of absolute distance,

it is longer to go around the mountain than to go over it, but in terms of relative distance, it is shorter. A graphic example will help make the relationships between absolute and relative spaces more clear (Fig. 3–14). Consider a set of seven towns mapped in absolute space (Fig. 3–14a). Distances are measured in absolute units and each outer location is one unit from the center town and its two nearest neighbors, and two units from the most distant town in the set. Assume now that travel from the center to each of the peripheral locations is measured in terms of time, and that distance on the map is scaled in travel time (Fig. 3–14b). Now let us assume that we introduce a new transportation system which makes travel in the region twice as fast. As measured by time, the structure of the distribution changes. Space is shrunk (Fig. 3–14c). On the other hand, congestion which doubled travel time would expand space (Fig. 3–14d).

Figures 3–14e to 3–14g are alternative ways of presenting the same information presented in 3–14b to 3–14d, respectively. In 3–14e to 3–14g we have used *isochrones* to illustrate the spatial relationships involved. Isochrones are lines of equal travel time from a single point. Isochronic mapping is a traditional method of presenting information about variable spatial relationships like these. Absolute terrestrial location is retained, and relative location is shown by lines of equal time distance. Figures 3–14b, 3–14c, and 3–14d are examples of another approach, that of drawing maps themselves on a relative scale, thereby ignoring absolute or "real" location.

There are several reasons why this might be done. Let us assume that we have three towns, *A*, *B*, and *C*, located in absolute space (Fig. 3–15a). Let us also assume that we have information concerning the costs of transporting a ton of freight between the places. It is possible to map the three locations in absolute terrestrial space using *isotims* of equal cost distance (Fig. 3–15b). But even with only three locations, the map is very complex and extremely hard to read. The isotim map produces information overload. A much simpler way of showing the relationship between the three points would be to map them in a space measured by cost

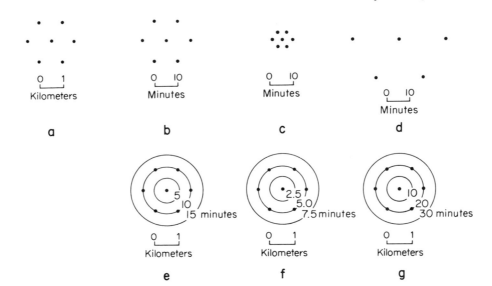

FIGURE 3–14

Absolute and relative space.

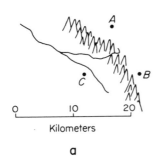

Cost of shipping one ton of goods (cents)

	To	A	B	C
From A		–	75	165
B		75	–	125
C		165	125	–

a

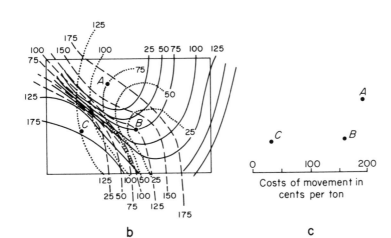

b

c

Costs of movement in cents per ton

FIGURE 3–15

Three towns in transportation space.

(Fig. 3–15c). Although a cost-space map does not contain as much information as an isotim map, it conveys more information than the latter because it is simpler. Moreover, it represents the true spatial structure of the distribution of towns more accurately. In transportation space, *A* is much closer to *B* than to *C*.

Mapping places in different relative spaces appears to "distort" spatial relationships because we think that absolute space is normal and other spaces are somehow deviant. Because such contexts are new and unusual, however, does not mean that they are deviant in any pejorative sense of the word. Maps in relative space do deviate from the traditional techniques of expressing spatial relationships, but they are no less valid for that reason, and they are considerably more stimulating. Maps in relative space are, in fact, considerably *more* valid for many purposes. Attempts to explain human spatial behavior are an increasingly important concern of geographers, and such attempts will be more successful if they are cast in relative spatial contexts. People shipping goods or taking trips between towns *A*, *B*, and *C* are not as much concerned with absolute distance as they are with cost and time; they make their decisions in cost- and time-space, not in absolute space.

The spaces in which people live are much more psychological than absolute. If we are concerned, as we often are, with explaining spatial interaction, what is important is not how far two interacting places are from each other in absolute space, but rather how far the people at the two places *think* they are apart. Affluent members of our society give little thought to journeys between the east and west coasts of the United States because the costs to them of making such trips are small. Less wealthy individuals make such trips far less often. If a trip from New York to San Francisco costs 6 percent of your annual income and requires six days driving time because you think you cannot afford to fly, you are far less likely to make such a trip than is a person who expends only 2.5 percent of his annual income and twelve hours of his time to make the same journey. Time and cost are far more powerful determinants of perceived space than are absolute

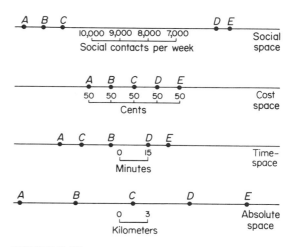

FIGURE 3–16
Relative locations in relative spaces.

distances, and they are thus better explanatory variables of spatial behavior.

Places have a number of relevant contexts, each of which is a different space. A given set of places may have a different spatial structure within each of these various spaces. Consider five hypothetical places at equal intervals along a line in absolute space (Fig. 3–16). In time-space, *C* might be closer to *A* than *B* if these two places were connected by an expressway which bypasses *B*. In cost-space, as measured by bus fare, the places might be equidistant also. In social space, *A*, *B*, and *C* might be close to each other while *D* and *E* are distant from the other three but very close to each other. This could be the case if *A–E* were sectors of a city and *D* and *E* were inhabited by blacks, whereas *A*, *B*, and *C* were occupied by whites. In this case increasing distances represent decreasing social contact. These are three examples of the kinds of space in which distributions can be located and in which an identical set of locations may have very different spatial structures. It is often useful to speculate on the number of different spaces which are relevant contexts in any area you are studying.

Identifying and mapping different spaces is a very important part of contemporary geography. Whenever a geographer encounters a distribution determined wholly or in part by human choice—and the structures of almost all geographically relevant distributions are functions of human spatial decisions—accurate mappings of distributions in the appropriate relative space will sometimes do more than could anything else to answer his questions. Human spatial behavior is determined by the interplay of decisions made in political, economic, cultural, sociological, psychological, and other spaces. People seeking to fulfill different goals in different relative contexts set in motion the spatial processes which produce spatial distributions.

Recognition of the existence and importance of relative spaces has come about only recently. Although a few pioneering works antedate World War II, investigations of distributions in relative spaces and of the nature of these spaces have been frequent only since about 1950. Watson, more than anyone else, is responsible for directing the attention of geographers to the importance of distance and relative distance. He is also one of the first geographers to draw a map of relative space. His map of cost distance in Hamilton, Ontario (Fig. 3–17), looks rather unimpressive, but it is of tremendous importance because it is indicative of the abandonment of the absolute space which was the context of geographical theory and practice for the previous two thousand years. Since the early

1950s, more and more geographers have presented the cartographic results of their analyses in relative spaces measured by cost and time. We are thus only beginning to explore the kinds of relative spaces in which human activity can be analyzed. We have concentrated on cost and time spaces up to now because they are two of the most obvious non-absolute measures of distance. But there remains a large number of spaces whose metrics are social, psychological, and other kinds of distance, which we have barely begun to explore.

Cartograms and Maps of Non-absolute Space

Intuitively, geographers have recognized the utility of non-absolute space for several decades. Areal cartograms were popularized by Erwin Raisz in the early 1930s. Cartograms retain a few of the properties of absolute space, such as partially accurate relative location, and they are to some extent maps in relative space (Fig. 3–18). They are based on areal scales (for example, one square unit equals one million people), and on *static* distributions. Areal cartograms are useful and stimulating ways of presenting information concerning the spatial structure of static distributions. But geographers are now as much concerned with spatial process as with its static manifestations,

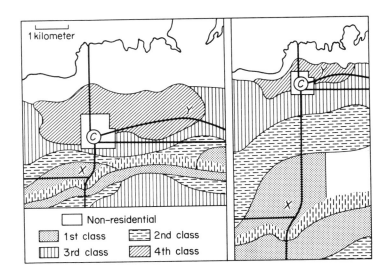

FIGURE 3–17

Cost distance in Hamilton, Ontario. Cost distance (right) is greater than absolute distance (left), resulting in sharp residential gradients (courtesy J. W. Watson, Fig. 1, p. 7).

Legend within figure:

1 kilometer

Non-residential
1st class   2nd class
3rd class   4th class

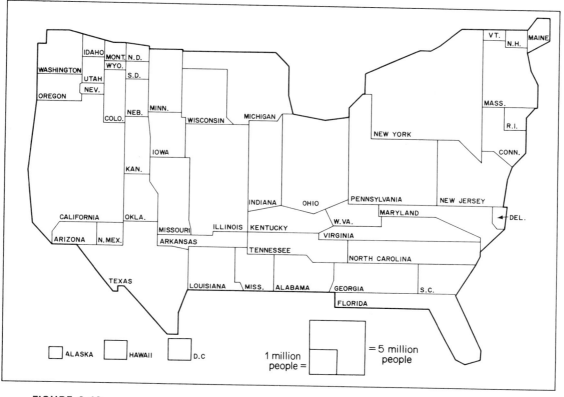

FIGURE 3–18

Population space: United States in proportion to population, July 1, 1967 (courtesy, Division of Research and Statistics, Ohio Bureau of Employment Services, Chart E–500).

and for that reason they have developed maps of relative spaces defined by measures based on actual or potential movement.

Maps in true relative space appeared almost simultaneously with Watson's recognition of the importance of relative distance. Many geographers began to try to represent relative spaces in useful and informative ways. Edgar Kant constructed the logarithmic projection used by Torsten Hägerstrand in his studies of migration to and from Asby, a district in Sweden (Fig. 3–19). On a logarithmic map, distance away from the center of the district decreases in proportion to the logarithm of "real" (absolute) distance.* Such maps are useful for presenting data about migration

*The logarithm of a number is simply the *power* to which ten must be raised to make it equal the number. For example $10 = 10^1$ so the log of 10 is 1; $100 = 10^2$ so the log of 100 is 2; $57.4 = 10^{1.7589}$ so the log of 57.4 is 1.7589.

because most moves are short, and since the near areas which contain many origins and destinations are relatively enlarged, many symbols can be shown in these areas without crowding. Peripheral areas of limited importance as origins or destinations of migrants are small and do not detract attention from the intensity of movements at the center. Cartographic design aside, such a map is useful in another way, because it gives us an idea of the "mental map" of the typical resident of Asby, who, like the resident of any other place, usually has a good deal of detailed information about his local region and less abundant information about more distant places.

Isochrone maps, while drawn to the traditional absolute scale, do illustrate some of the peculiarities of relative space, and geographers continue to use this technique. One of Kok's maps of 's-Gravenhage (The Hague), for example, shows the involuted space produced

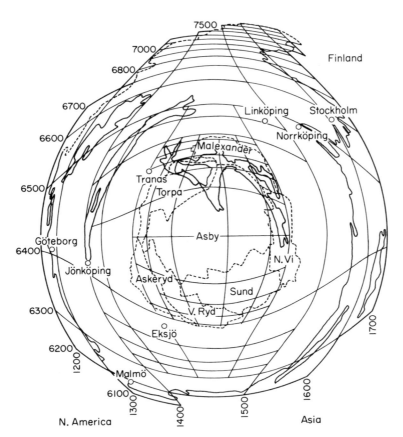

FIGURE 3–19

Hägerstrand's migration map on logarithmic projection (courtesy, Torsten Hägerstrand, Fig. 38, p. 73).

by public transportation systems (Fig. 3–20). Gouda is much closer to the center of 's-Gravenhage in temporal distance than many places which are much closer in absolute distance, where trains do not stop. In several areas, small circular areas of accessibility appear ahead of the general waves which spread out from the center. Central Leiden is as close as Central Delft in time despite the fact that it is about eight kilometers further away from 's-Gravenhage than Delft. While isochrone and isotim maps do give us some idea about the structure of relative space, mapping distributions in relative space itself is, as we have seen, a superior means of describing the nature of the spaces which most strongly influence human decisions. Thus it is preferable to translate isochrone and isotim maps into their relative counterparts whenever it is

possible to do so. Figure 3–21, for example, consists of two maps of Seattle, one in absolute and the other in relative space. Both convey similar ideas about the time it takes to reach various places in the city during the rush hour. One presents isochrones in absolute space (Fig. 3–21a). The other transforms the city into transportation space and presents the isochrones as concentric circles (Fig. 3–21b).

There are instances in which mappings in relative spaces are not possible. Imagine three places located at the following cost distances from one another:

$$A \text{ to } B = \$7.00$$
$$A \text{ to } C = \$3.00$$
$$B \text{ to } C = \$1.00$$

Because it is impossible to construct a triangle with sides that are seven, three, and one unit

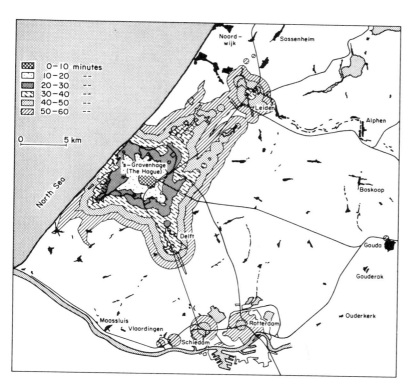

FIGURE 3–20

Isochrone map of 's-Gravenhage (The Hague) (Courtesy, T.E.-S.G., *Afb. II, p. 263).*

FIGURE 3–21

Seattle in absolute and relative space (courtesy, Traffic Engineering Division, City of Seattle and William Bunge, Figs. 2–13 and 2–14, p. 55).

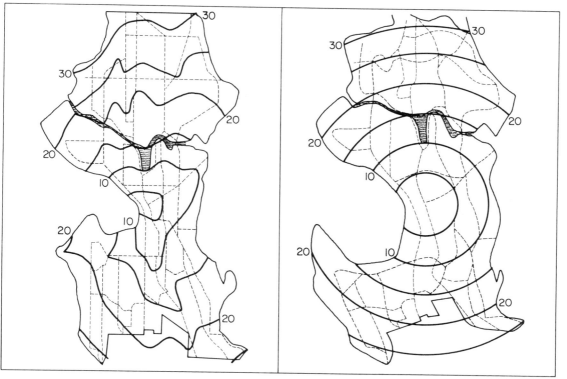

a   b

long, this spatial relationship could not be presented in the kind of relative space we have been discussing. There exist other techniques for handling such cases, however, one of which is a matrix. The set of cost relationships could be summarized thus:

		To	
Cost from	A	B	C
A	—	7	3
B		—	1
C			—

Matrix representation is especially useful when distances are not symmetrical. Often the relative distance between the same two points is different in opposite directions. A flight from New York to Paris by modern jetliner, for example, requires about six hours, whereas the return trip from Paris to New York takes eight hours because on the return trip the jetliner must fly against the high altitude westerly winds which boost its speed on the eastbound journey. Thus:

	To	
Hours from	New York	Paris
New York	—	6
Paris	8	—

Matrices are also efficient ways of presenting complex networks which would be confusing if mapped, and they offer the additional advantage of computational simplicity, as we shall see in Chapter 7.

Even though they are not suitable for presenting all spatial relationships, maps in relative spaces can often provide useful insights into human spatial behavior. Professor Waldo Tobler has been our most active explorer of the new worlds of relative space. One of his numerous transformations is reproduced in Fig. 3–22. Map *a* shows the boundaries of postal rate zones in absolute space. Map *b* shows the transformation of the spatial relationships between Seattle and other cities into cost space. Note the scale, given in cents, and the graph of cost distance against absolute distance. If one wishes to send a parcel via United States mail, all places in the nation between 3000 and 5000 kilometers away are equidistant from Seattle.

Tobler has not sketched in the coastline in the transformation although this could be done with some difficulty. Tobler's efforts to transform absolute space into the relative spaces of greater use to contemporary geographers are of monumental importance.

If we are to make worthwhile contributions to the explanation of human spatial behavior we must explore to the limits the nature of relative space. People make decisions in the context of relative rather than absolute space. If relative space had no influence on human behavior, we would still be living the spatially circumscribed lives our ancestors did thousands of years ago instead of the highly mobile existences we enjoy today. Absolute space has not changed to any significant extent in the last several thousand years, but relative space has changed enormously, and our spatial behavior has evolved along with it. Because relative space does change over time, and because the nature of relative space affects our behavior to the extent that it does, such changes are important explanatory variables. Man's continued manipulation of relative spaces creates new spatial contexts which in turn produce new behavior patterns. *Space-adjusting* techniques such as transportation and communication enable us to restructure space by changing the relative distances which separate places and sets of places.

There are other, less obvious, means of adjusting space which also must be taken into account. A geographer might look upon education, for example, as a space-adjusting technique of sorts in that being or not being highly educated often has a great bearing on mobility and migration. Increasing a person's knowledge and worth to society also increases his income and mobility. We can promote greater spatial interaction directly by reducing the friction of distance relative to the resources potential travelers command, or more indirectly by increasing the financial resources they can

FIGURE 3–22

The United States in parcel post space (courtesy, Waldo Tobler and William Bunge, Figs. 8–17 and 8–18. [a] is based on data contained in a map issued by Parcel Post Unit 5553, Seattle).

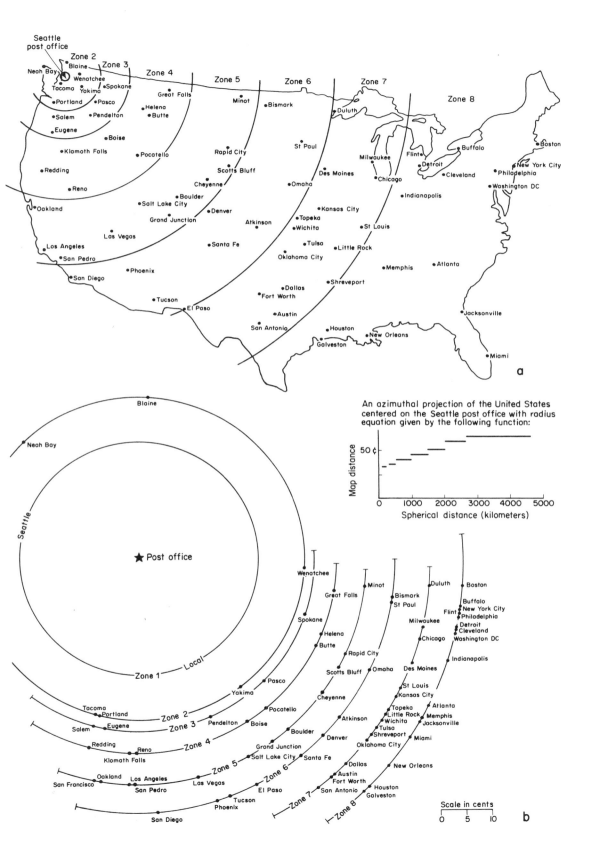

Seattle post office

Zone 2 Zone 3 Zone 4 Zone 5 Zone 6 Zone 7 Zone 8

An azimuthal projection of the United States centered on the Seattle post office with radius equation given by the following function:

50¢

Map distance

0   1000   2000   3000   4000   5000
Spherical distance (kilometers)

Post office

Zone 1 Local

Zone 2
Zone 3
Zone 4
Zone 5
Zone 6
Zone 7 Zone 8

Scale in cents
0   5   10

a

b

expend on travel. Either will effectively change patterns of spatial interaction and adjust space. Any activity we undertake which makes it easier or more difficult for people, ideas, or objects to move through space has significant effects on spatial processes and the structures they produce. In very subtle ways, man adjusts space when he takes such apparently irrelevant actions as saving money for his child's college education. The realization that many seemingly innocuous and seemingly aspatial decisions have significant effects on spatial process and structure provides even more grist for contemporary geographical mills.

Individuals and groups of people live at intersections of numerous relative spaces. The structures of many distributions in those spaces determine the goals people have, the information they receive, and the choices open to them in fulfilling their goals. At the same time, human decisions constantly alter and restructure relative spaces. It has taken geographers a long time to challenge the pervasive tyranny of absolute space, and we still have difficulty escaping its conceptual clutches. Shown a map drawn in relative space, most of us describe it as a distortion of "real space." Relative spaces are just as real as traditional absolute space. Indeed, they are more powerful explanations of human behavior, because they are the spaces within which people really operate. It is only our limited mental agility which leads us to characterize absolute space as real and relative spaces as a distortions of reality. In fact, the opposite is the case, and recognition that relative spaces constitute the truly relevant context of human behavior is the hallmark of contemporary geography.

TRADITIONAL GEOGRAPHICAL QUESTIONS IN A NEW SPATIAL CONTEXT

Although the context within which geographers work and view the world has changed recently, we have not abandoned traditional questions; "Where?" and "What is where?" are basic in any geography and they still occupy much of our time. Of necessity, the answers to these questions are not the same as those which were developed in the context of absolute space.

Locational Questions

Asking "Where?" in the context of relative spaces presents no major difficulties, although the question itself requires more attention than it did in the past. Answers to *where* questions in a relative context are always provisional. Locations in absolute space have always been considered immutable, although the recent confirmation of ocean floor spreading and continental drift indicate that even things like continents are located in a relative space if we consider very long periods of time. Over the short run, however, once we have accurately determined the terrestrial coordinates of a place, absolute location becomes a given. We do not need to check every decade to see if Washington, D.C. has moved since we last measured its location. But because man is incessantly fiddling around with space-adjusting techniques, we must constantly repeat locational observations in relative spaces, because places do change location frequently in those contexts. We must be constantly alert to identify these locational changes and measure their extent and effects if we wish to produce acceptable explanations of spatial structure and process.

One way we can monitor locational changes in relative space is to measure time-space convergence, a concept developed by Donald G. Janelle. In our informal discussions of relative space, we frequently talk about the "shrinking world" in which we live. We can monitor this shrinkage by measuring the rates at which places on the surface of the earth approach one another in time distance. Janelle has calculated the rate at which London and Edinburgh are converging in time-space (Fig. 3–23). The rate at which two places converge upon each other in time-space is given by the formula:

$$\frac{TT_1 - TT_2}{Y_2 - Y_1}$$

where: TT_1 and TT_2 = the travel times between the two places in year 1 and some

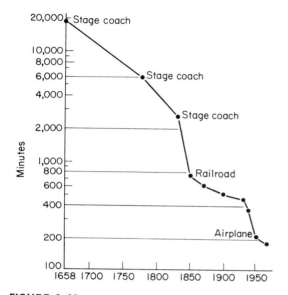

FIGURE 3–23

Time-space convergence between London and Edinburgh (courtesy, Donald Janelle, Fig. 1, p. 6).

later year 2, respectively, and

Y_1 and Y_2 = the two dates in question.

In the case of Edinburgh and London, the rate of time-space convergence between 1776, when the journey was made by coach, and 1966, when the trip can be made by airplane, is 29.4 minutes from the calculation:

$$\frac{TT_1 - TT_2}{Y_2 - Y_1} = \frac{5,760 - 180}{190} = \frac{5,580}{190}$$
$$= 29.4 \text{ minutes per year}$$

You should be able to think of cases from your own experience where time-space convergence has occurred. The building of limited access highways, for example, has produced time-space convergence of places in many parts of the world. Any place which you can reach more quickly now than you could five or ten years ago has converged toward you at an average rate you can calculate with the formula above. A peculiarity of time-space convergence is that distant places converge on each other at a greater rate than close places. Janelle illustrates

this with a hypothetical example (Fig. 3–24). When travel speed between all places along the route from *A–F* is doubled, *F* approaches *A* at a rate five times that at which *B* approaches *A*. The way this convergence might affect your spatial behavior should be obvious. Note that in 1950 it took 24 minutes to go from *A* to *B*, whereas in 1970 it took 12 minutes. Note also that by traveling for 24 minutes in 1970 you could reach *C*. Now suppose that you live in a very small town *A*, and that *B* is a larger town than *A*, and that *C* is even larger than *B*. Suppose also that you have customarily done your shopping in town *B*, where an adequate but not abundant array of merchandise and services is available. Suppose, in addition, that there is a significantly larger array of goods and services available at *C* than at *B*. With the improvement in travel you may choose to continue to shop at *B* and to use the 12 minutes you now save on each one-way trip for some other purpose. What is more likely, however, is that you will continue to spend 24 minutes traveling each way when you shop, but that you will go instead to Town *C* to do your shopping, where a greater selection is available. You should be able to think of examples from your own experience where you have used the time saved by

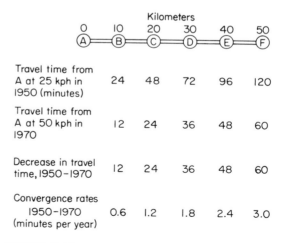

	Kilometers				
0	10	20	30	40	50
(A)	(B)	(C)	(D)	(E)	(F)
Travel time from A at 25 kph in 1950 (minutes)	24	48	72	96	120
Travel time from A at 50 kph in 1970	12	24	36	48	60
Decrease in travel time, 1950–1970	12	24	36	48	60
Convergence rates 1950–1970 (minutes per year)	0.6	1.2	1.8	2.4	3.0

FIGURE 3–24

Increase in time-space convergence rates with increasing distance (courtesy, Donald Janelle, Fig. 3, p. 9).

more efficient transportation for traveling farther rather than for nontravel purposes, because this allocation of "saved" time is very common.

Time-space convergence has significant effects on human behavior whether we consider the consumers at *A* or the merchants at both *B* and *C*, who will also be significantly affected by their new locations in time-space. A merchant at *B*, unaware of the nature of relative space, might have decided to invest a large sum of money to upgrade his store in hopes of increasing his business, only to discover his sales volume declining rather than growing after the transportation innovation is instituted. Similarly, an unperceptive merchant at *C* might not realize that the innovation would increase his business and thus might not be able to capitalize on the new customers from *A* before they decide to shop at the establishments of more alert entrepreneurs who expanded their facilities in anticipation of the new volume.

These direct and indirect effects of space adjustments demonstrate the necessity of being aware of changing relative location by monitoring rates of change. We have dwelt at some length on the example of time-space convergence, but it is obvious that we must also concern ourselves with cost-space convergence, social-space convergence, and all the ways places are approaching one another as we continue to adjust spaces.

"What Is Where?" in Relative Space

"What is where?" is a question which is also producing different answers from those given in the past. To a significant degree this is because of a new context itself, but it also follows more directly from the fact that geographers are now interested in the "whats" which define these relative spaces. In the past geographers were primarily interested in gross patterns of spatial organization and usually in static distributions of the phenomena they studied. More recently, geographers have concentrated more on dynamic spatial phenomena in small regions. Thus the subject matter of the discipline has changed considerably over the past several decades. Increasingly, geog-

raphers view themselves as social rather than physical scientists, or as a combination of the two. Accordingly, geographers are now more likely to ask "What is where?" about sociocultural constructs than about aspects of the physical environment. When he asks "What is where?" today's geographer is usually inquiring about the location of some humanly determined distribution of phenomena in relative space. This being the case, geographers —like other social scientists—now encounter problems of data collection and aggregation which are somewhat different from those confronted by earlier practitioners of the science.

In order to obtain the answers to "What is where?" geographers must increasingly rely on data collected by census bureaus and other governmental agencies. But like other social scientists, geographers can also generate their own data by conducting surveys and interviews. Both sources produce relatively limited quantities of data about spatial process and spatial behavior. Self-generated data is frequently more useful in this regard, because the geographer who designs the survey will take pains to insure that the facts he collects are suitable answers to questions he is asking. Data collected by censuses and by government agencies, on the other hand, are usually designed to provide answers to other questions. In general, we feel we do not have many of the data we need to answer our questions about human spatial behavior, nor do we command the financial resources we need to acquire it.

In situations where we wish to explain events but lack adequate data, we are forced to apply more than the ideal amount of deductive theory to the limited stock of facts available. Thus our theories and models of spatial process are not all that we could hope for; they could be improved considerably, if we had the data needed to refine them. Hopefully, data to answer our *what is where* questions will soon become more abundant.

Since most of the "whats" geographers are now interested in are things which move over the surface of the earth or are static manifestations of the movement of ideas over the earth's surface, satellite technology offers one possibility for monitoring movements and distribu- tions of many kinds with unprecedented speed and accuracy. With the techniques presently available, geographers could have daily reports on crop distribution, crop production, com- muting patterns in urban areas, seasonal population migrations, construction progress in metropolitan areas, and a host of other relevant distributions. Unfortunately, most of the technology available in this particular realm of data gathering is not readily available to civilians. Nonetheless, such information systems will eventually become more accessible to all and as they do, some of our data problems will be solved. We can also expect significant improvements in conventional data gathering systems in the next decade.

Uniform and Nodal Regions

As locational information on the "whats" of concern to geographers becomes more abun- dant, the difficulties of organizing spatial data into areal classes will become more severe. Taxonomic problems are serious enough now, even with the limited amount of data geog- raphers presently enjoy. Constructing taxon- omies of moving objects in relative space produces regions which are different in nature from the ones geographers formerly relied upon most heavily. Areal classifications of static phenomena usually produced regions which were composed of homogeneous phenomena and which were relatively *uniform* throughout. The regions geographers require today as inputs to their analyses are more often *nodal*. Nodal or functional regions are not uniform throughout; they are delimited on the basis of spatial interaction and are thus regions of spatial organization.

To clarify the distinction between uniform and nodal regions, consider the following hypothetical case. Suppose we have two items of information about some counties in the American Midwest (Fig. 3–25). One is the percentage of agricultural land devoted to grains for consumption by livestock, and the other is the percentage of the labor force in each of the counties which works in a city which is located in a central county. Note that a

superhighway runs east-west through the re- gion (Fig. 3–25a). Figures 3–25b and 3–25c are histograms of the two distributions along the west to east traverse, R to S. Slightly more land is devoted to feed grains in the west than in the east, and there is an expected dip in the central county where the city is located. Employment is very high in the urban county itself, and drops off relatively rapidly with distance from the town. Figures 3–25d and 3–25e graph these percentages along a north-south traverse. There is a slight decline in the importance of feed grains toward the south and a more rapid drop in commuting to the central town for employ- ment than was observed along the west-east axis, reflecting the fact that the latter is favored by the superhighway.

As shown by isolines (Fig. 3–25f), the pro- portion of agricultural land devoted to feed grains in the region varies little. We would de- scribe such an area as a uniform region, prob- ably as part of the uniform region known as the Corn Belt. Employment in the central town is quite different, however. The region within the isoline with value 0 (Fig. 3–25g) is homo- geneous in the sense that it is the region which contributes commuting labor to the city. But it is not uniform in the contributions it makes, inasmuch as the percentages of the local labor force which commute from counties in this region vary between 0 and 78 percent. The commuting region is nodal because the flows of people through space converge on and dis- perse from a central location, and variations in the intensity of commuting decline with distance from this node. We have used the ex- ample of commuting employees in this case, but we could just as well have chosen the ex- ample of the trade area from which the city draws its customers, or the circulation intensity of the weekly newspaper published there. Uniform regions are composed of static phenom- ena and there is relatively little variation in the density or intensity of occurrence of such phenomena throughout the uniform region. Nodal regions, on the other hand, depend on phenomena in motion, and significant variations in intensity of flow within the region are usually the rule.

The heights of the histogram bars which

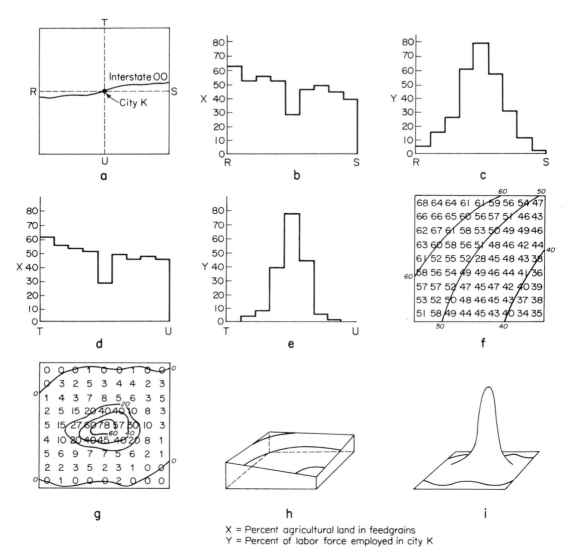

X = Percent agricultural land in feedgrains
Y = Percent of labor force employed in city K

FIGURE 3–25
Uniform and nodal regions.

represent the percentages for each of the counties define two *statistical surfaces* over the region. If we think of each of the counties in this example as having an *x* and a *y* coordinate which define its location in two-dimensional space, we can also think of each county as having a large number of *z* coordinates, each of which defines the height of a statistical surface in the third dimension. If we connect the *z* values for a number of subregions (for example, the counties), we produce the statistical surface defined by that particular variable. Histograms are cross-sections through three-dimensional statistical surfaces (Figs. 3–25b-e). Figures

3–25*f* and 3–25*g* are contour maps of the statistical surfaces sketched in Figs. 3–25*h* and 3–25*i*. Uniform regions are statistical surfaces which are flat, sloping, or gently rolling. Nodal regions are statistical surfaces which are usually somewhat similar in form to volcanic peaks. Nodal regions are produced by spatial interactions which link places with each other.

Classifications of things have usually been based on similarities among the things classified, the most similar objects being placed in the same class. Nodal regions are areal classes of similar places, but "similar" is defined in a slightly different way than before. Rather than being similar in *form* to other places in the same region, as is usually the case with uniform regions, the places assigned to the same nodal region are related in *function*. It is true that counties which comprise the trade area of a central place are formally similar in the sense that more people from these counties shop in the given central place than in any other town. But the *functional* interrelationships between the consumers in these counties and the providers of goods and services in the central place, and the movements through geographical space involved in securing and providing these goods and services are the most important characteristics of nodal regions. Regions defined by the flows of people, money, messages, goods, political power, and a host of other phenomena which vary in intensity with distance from nodal places, are important explanatory variables of human behavior, and these are the kind of "whats" which geographers are interested in locating and grouping into areal classes at the present time.

Classification of sets of locations into regions is the point at which description and analysis overlap. The answers to simple *what is where* questions are essentially descriptive in nature, but the extension of the descriptive process to classification is the beginning of the abstraction and generalization process which ultimately culminates in explanation. We emphasize again that geographers today are still asking the same questions that they have asked for the past two thousand years, in addition to the questions which have come to the fore more recently. But because they are now more interested in dynamic distributions and movement systems, and in the relative spaces and nodality that space-adjusting techniques create, nodal regions are the usual results of attempts to answer the question "What is where?"

CONTEMPORARY GEOGRAPHICAL QUESTIONS

It would be difficult and foolish to try to list all the possible questions we can ask in the context of relative space. The combinations of *where*, *when*, *what*, *how*, and *why* questions which can be devised are numerous, and the spatially relevant experiences to which they can be applied are almost infinite. At one time the answers to *where* and *what is where* questions were almost ends in themselves. We now view answers to these questions as preliminary steps toward the explanations we produce by answering *why* (actually *how*) questions. We no longer see the foundation of our science as description of the spatial organization of the world. Our view now is that the explanation of classes of events by demonstrating that they are instances of widely applicable laws and theories is the function of geography. The classes of experience to which we apply our explanatory expertise are now primarily those which are related to the locations of people and their activities.

The immense variety of distributions available to serve as grist for our analytical mills is a function of the fact that almost all things man encounters are variant in density and occurrence. This is especially true of man himself, and his activities. Of all the elements of the ecosystem, man and his activities are the most variable in occurrence and density over the surface of the earth. Some features of man's world are more worthy of analysis than others, at least insofar as we view geography as a diagnostic and prescriptive science. No matter what distribution we choose to try to analyze, however, we need not be overly concerned with the variety of potential and actual locational questions which exists. The analysis of the location of any single event or of any spatially distributed set of events is a similar process. We may have to devise slightly different techniques to handle different cases, but the underlying strategies are the same no matter what distribution concerns us.

Our interest in distributions is always focused on their spatial structures and the processes which produce these structures. Movements of phenomena through space result in the establishment of movement systems, which in turn facilitate further movements through space. But movement systems favor some places at the expense of others, so that the relationship between movements (process) and transportation systems (structure) is not one-way, but is reflexive and circularly causal. Similarly, movement systems or space-adjusting techniques are very powerful determinants of the location of human activities, while at the same time places which are nodal and have already attracted intense human activity exert powerful influence over the structure of transportation and communication systems. Through these circularly causal mechanisms and relationships, man—consciously or unconsciously — invents spatial organization. The interaction between process and structure is a fundamental aspect of any locational problem, so that whether we are trying to explain the distribution of neolithic agricultural sites in Europe or whether we are trying to decide where to locate a number of hospitals to serve a spatial distribution of people, we can break the problem down into terms of process, structure, and their interaction.

Thus the contemporary question *"Why are spatial distributions structured the way they are?"* is a convenient shorthand which represents all the questions which have to be answered in order to provide satisfactory explanations of spatial process and structure, and of the relationships between the two. The theory and methodology of a science are the strategy and tactics it brings to bear on the problems which confront it and the questions it raises. By becoming thoroughly familiar with these aspects of our discipline, you will be much better equipped to solve successfully the problems which you encounter in the next fifty years, whether you choose to be a theoretician, methodologist, practitioner, or lay student of geography.

In any case, explanation and manipulation are the focus of our discipline today and will continue to be in the future. Because of our wish to manipulate events, we must explain both process and structure to ourselves so we can intervene in process to produce the spatial structures of activities we desire. It is our conviction that as we continue to expand our knowledge of the operations of process, space itself will come to be recognized as a major explanatory variable of human behavior. The kinds of space and distance which we will come to recognize as the causes of much of man's spatial behavior will not be absolute space and distance, which, as we have seen, are increasingly losing what validity they have as determinants of human activity. Rather, we shall come to explain human spatial behavior as the product of the relative spaces which man himself creates by his space-adjusting activities.

Suggestions for Further Reading

Contemporary Trends

ACKERMANN, EDWARD A. "Where Is the Research Frontier?" *Annals of the Association of American Geographers*, LIII (1963), 429–40.

BLAUT, JAMES. "Space and Process," *Professional Geographer*, XIII: 4 (July 1961), 1–7.

BUNGE, WILLIAM. *Theoretical Geography*, rev. ed. Lund: Gleerup, 1966. Lund Studies in Geography, Series C, no. 1.

BURTON, IAN. "The Quantitative Revolution and Theoretical Geography," *The Canadian Geographer*, VII (1963), 151–68. Reprinted in Brian J. L. Berry and Duane F. Marble, eds. *Spatial Analysis: A Reader in Statistical Geography*. Englewood Cliffs, N. J.: Prentice-Hall, Inc., 1968.

HAGGETT, PETER. *Locational Analysis in Human Geography*. London: Edward Arnold, Ltd., 1965.

HARTSHORNE, RICHARD. *Perspective on the Nature of Geography*. Chicago: Rand McNally & Co., 1959.

JANELLE, DONALD G. "Spatial Reorganization: A Model and Concept," *Annals of the Association of American Geographers*, LIX: 2 (June 1969), 348–64.

KIRK, WILLIAM. "Problems in Geography," *Geography*, XLVIII (1963), 357–71.

KOHN, CLYDE F. "The 1960's: A Decade of Progress in Geographical Research and Instruction," *Annals of the Association of American Geographers*, LX: 2 (June 1970), 211–19.

SAUER, CARL ORTWIN. "The Morphology of Landscape," *University of California Publications in Geography*, II: 2 (1925), 19–54. Reprinted in *Land and Life: A Selection from the Writings of Carl Ortwin Sauer*, John Leighly, ed. Berkeley and Los Angeles: University of California Press, 1963.

The Science of Geography. Washington, D. C.: National Academy of Sciences, National Research Council, 1965. Publication No. 1277.

TAAFFE, EDWARD J., ed. *Geography: The Behavioral and Social Sciences Survey.* Englewood Cliffs, N. J.: Prentice-Hall, Inc., 1970.

TOBLER, WALDO. "Geographical Area and Map Projections," *Geographical Review*, LIII: 1 (January 1963), 59–78.

WATSON, J. W. "Geography: A Discipline in Distance," *Scottish Geographical Magazine*, LXXI (1955), 1–13.

Foundations of Contemporary Geography

BAGROW, LEO. *History of Cartography*, rev. and enlarged by R. A. Skelton. Cambridge, Mass.: Harvard University Press, 1966.

BUNBURY, E. H. *A History of Ancient Geography.* London: John Murray, 1879. Reprinted by Dover Publications in 1959.

DICKINSON, ROBERT E. *The Makers of Modern Geography.* New York: Frederick A. Praeger, Inc., 1969.

———, and O. HOWARTH. *The Making of Geography.* Oxford: The Clarendon Press, 1933.

FREEMAN, T. W. *A Hundred Years of Geography.* Chicago: Aldine Pub. Co., 1962.

GLACKEN, CLARENCE J. *Traces on the Rhodian Shore.* Berkeley and Los Angeles: University of California Press, 1967.

HARTSHORNE, RICHARD. *The Nature of Geography.* Lancaster, Penn.: Association of American Geographers, 1939 and later editions.

NEEDHAM, JOSEPH, and WANG LING. "Geography and Cartography," in *Science and Civilisation in China*, Vol. III: *Mathematics and the Sciences of the Heavens and the Earth.* Cambridge: The University Press, 1959.

TAYLOR, GRIFFITH. *Geography in the Twentieth Century.* New York: Philosophical Library, 1957.

WRIGHT, JOHN K. *Geographical Lore at the Time of the Crusades.* New York: American Geographical Society, 1925. Reprinted by Dover Publications in 1965.

Works Cited or Mentioned

BROWN, LLOYD A. *The Story of Maps.* Boston: Little, Brown & Co., 1949.

BUNGE, WILLIAM. *Theoretical Geography.* Lund: Gleerup, 1966. Lund Studies in Geography, Series C, No. 1.

HÄGERSTRAND, TORSTEN. "Migration and Area," in *Migration in Sweden.* Lund: Gleerup, 1957. Lund Studies in Geography, Series B, No. 13.

HALL, EDWARD T. *The Hidden Dimension.* Garden City, N.Y.: Doubleday & Co., Inc., 1969 (Anchor Books).

IBN-KHALDUN. *The Muqaddimah: An Introduction to History*, trans. F. Rosenthal. Bollingen Series XLIII. Princeton: Princeton University Press, 1967.

JANELLE, DONALD G. "Central Place Development in a Time-Space Framework," *Professional Geographer*, XX: 1 (January 1968), 5–10.

KOK, R. "Isochronenkaarten Voor het Locale en Regionale Openbaar Personen Vervoer Van 's-Gravenhage," *Tijdschrift voor economische en sociale geografie*, XLII (1951), 261–78.

NEEDHAM, JOSEPH, and WANG LING. *Science and Civilisation in China.* Vol. III: *Mathematics and the Sciences of the Heavens and the Earth.* Cambridge: The University Press, 1959.

RAISZ, ERWIN. *Principles of Cartography.* New York: McGraw-Hill Book Company, 1962.

SOMMER, ROBERT. *Personal Space.* Englewood Cliffs, N. J.: Prentice-Hall, Inc., 1969.

TOBLER, WALDO. "Map Transformation of Geographic Space." Unpublished Ph.D. dissertation, University of Washington, 1961.

WATSON, J. W. "Geography: A Discipline in Distance," *Scottish Geographical Magazine*, LXXI (1955), 1–13.

PART II

Measurement, Relationship, and Classification

CHAPTER 4

THE PROBLEM OF
MEASUREMENT AND SCALING

The phrase "more or less" is a fault much in evidence in kings and geographers.

—Strabo

The problem of measurement and scaling is the most fundamental one faced by geography and other factual sciences. In very general terms, measurement is the application of the concept of number to the constructs in the C-field that we invent to order our experiences in the P-plane. If the problem of measurement cannot be tackled and resolved in a satisfactory way, anything vaguely approaching scientific inquiry is impossible. This, of course, is quite a strong statement, and not one you will accept easily the first time you meet it. Does it mean, for example, that if we cannot measure the things we are interested in as geographers, the path of inquiry is closed? Frankly, the answer is yes. But it all depends upon what you mean by measurement. It is precisely to the meaning of measurement that we now turn.

QUANTITY AND QUALITY

It is paradoxically eerie, yet comforting, to see the many broad similarities in the development of all scientific disciplines; the same early gropings, the same years of patient factual recording, the same heated arguments, and the same basic methodological problems appearing, no matter which physical, biological, or social science we consider. Since these similarities help us see our own position in the whole history of scientific investigation, it is worth looking for a moment at some of these basic underlying patterns of development. Most sciences, and certainly geography, started when men tried to satisfy naive and innocent curiosities about the things around them. Experiences of peoples, places, objects, thoughts, and sensations were described in story and later in writing. Often similar events and experiences were described, and collections of like things made by grouping the most similar ones together.

It really does not matter what we take as an example. We can imagine a group of medieval vintners sitting around a banquet table and commenting upon the vintages, evoking the bouquets of 1245, 1259, and 1261 to label these wines as *excellent,* while wrinkling their noses at the thin and flabby wines of 1250 and 1263, labeling them *poor.* As the evening wears on, and as their judgments are reinforced by judicious tasting, our vintners might make more refined judgments, ranking the '61 ahead of the '59 but behind '45 (Ah, my boy, what a year that was!). Now, while agreeing upon the superiority of the '45, not all might agree about the order of the last two, some preferring the '61, some the '59. In the course of resolving such a crucial question voices are raised and tempers flare until one vintner, the youngest and crassest, says "Why not measure them in a different way and see?" Of course you could have heard a pin drop. "Measure which is better? Preposterous! Absurd! The fellow must be out of his mind! You can't measure the quality of wine. I mean there are some things you just can't measure—and thank goodness for that, eh! Or it won't be long before we have these alchemists running the winery!" But with the boldness of youth, the youngest vintner persists. "Well why not? You've been arguing half the night and getting nowhere; why not agree on some sort of measuring scale and see which comes out ahead? Why not take the number of crowns paid for a hogshead at the spring auction, and then, allowing for the infla-tion caused by the Duke of Tuscany's latest rampage, use these numbers to measure which is best?"

Although somewhat homely and apocryphal, this is a familiar story. In order to refine the description of things it is necessary to measure, and nearly the whole history of measurement is contained in this little fable. Consider the problem of measuring temperature, probably the most famous example in the whole history of science. Long ago describing temperatures as hot or cold was quite sufficient, perhaps on the simple grounds of skin sensation. Later the in-between notion of warm was developed: not too hot, not too cold. Then came the necessity for ranking different levels of hotness and coldness: frosts do different orders of damage; some metals melt much more quickly than others in the charcoal furnace.

Next came the invention of the thermometer, which means literally the "heat measurer."

But this was quite a jump, because the thin glass tube and its moving column of mercury was a very sophisticated measuring device, and its invention raised a difficult question. If we are no longer going to be satisfied with ranking things by putting them in order, but feel now that the interval between things is important, where are we going to start measuring? Where will our base line, our zero point, be? And what intervals shall we use? If we are only inter-ested in the level of a particular temperature, say a child's body in fever, or in comparing the intervals between two readings, it makes no difference where we put the zero point. 0°C. is the freezing point of water; 0°F. is the coldest temperature artificially achieved under precisely defined conditions at the time of Herr Fahren-heit's experiments; both are quite arbitrary, but both are equally useful for comparing the temperatures of different people or successive days. However, the arbitrary nature of the zero point means that we cannot say that 150°F. is twice as much heat as 75°F. Think about it for a minute, and ask yourself what meaning such a statement has. Similarly, we set an arbitrary starting point when we measure calendar time (sometimes the birth of Christ), because we are interested in the intervals between events. But it obviously makes no sense to say that the year A.D. 2000 is twice as much time as the year A.D. 1000.

Finally, as men pushed temperatures back and back, liquifying and finally solidifying gases at extremely low temperatures, they gradually learnt that there was a limit below which no temperature could go; a zero point not arbi-trarily set by man, but a natural limit where every atom was at rest. Thus the Kelvin scale, which measures temperatures from a true zero point, is used increasingly for scientific work, and upon such a scale it does make sense to say that 200 degrees of heat Kelvin are twice as hot as 100°K.

Paradoxically, one of the most serious obstacles to the development and use of measurement in geography has been the sophisticated nature of the earliest forms. We seldom think of measuring distance as a terribly difficult task employing highly sophisti-cated ways of measurement. From earliest childhood we have seen people measure distances with rulers and tape measures, with the result that many of us assume that this is the only real form of measurement possible. But as children we started at the top, at the most advanced level of measurement and scaling, where not only intervals between marks on the ruler can be compared, but ratios between measurements have an obvious meaning. This is because distance scales have natural, as opposed to quite arbitrary, zero points—namely, no distance. So familiar and obvious is such measurement that simpler forms of scaling are often regarded with great suspicion, and even denied to be measures at all, particularly where such approaches are used in areas of research where qualitative, as opposed to quantitative notions formerly reigned. Great misunderstanding has been generated by putting quality and quantity in opposition to one another, instead of regarding them as opposite ends of a long continuum on which many forms of useful measurement are placed. The difference between quantity and quality is one of degree not of kind. The theory of measurement has come a long way in the twentieth century, and there are few areas of scientific research where the answers to important and imaginative questions can be obtained without appropriate measurement techniques.

WHAT IS MEASUREMENT?

Strictly speaking, before we look at ways of measuring things, we should define what we mean by the act of measurement. But this is difficult to do in a consistent way, and for the moment we shall deliberately dodge the problem and take it up again after we have some actual examples to help us. Let us start by reversing the order of childhood, and take the simpler measurement scales first.

Binary and Nominal Measures

It may be difficult to think of assigning events to classes as an act of measurement, but the *binary* scale simply consists of putting objects into one of two pots labeled Yes or No, Present or Absent, One or Zero. With such a simple

scale, it is hard to think of anything which could not be measured, for it is usually possible to *dichotomize*—that is, literally "divide into two" —anything that varies over the surface of the earth. For example, once we have decided upon a point of decision, climates might be measured simply as Wet or Dry, assigning a 1 to wet and a 0 to dry. Of course, with such a simple measure we would lose large amounts of information compared to what we could get from more sophisticated approaches, but classification and numerical assignment constitutes a genuine and quite common scaling procedure. Questionnaires frequently contain items to which a yes-no, one-zero, dichotomized answer can be given, and there are many areas of geographic research where the sheer presence or absence of something is worth measuring in this very simple way. Often such a scale is the only form of measurement possible. As with pregnancy, the only variation between things may appear in this rather stark yes or no form.

Supposing, as geographers attached to a World Health Organization medical team, we were studying the distribution of blind people in a rural area of Africa (Fig. 4–1). We take a random sample in the area and find out the proportion of people who are sightless. In the course of sampling the small family compounds in the field, the local people tell us that the high level of blindness is caused by evil water spirits who live in the flood plain of the nearby river. Rejecting such theological explanations, we are surprised in mapping our results to find that there does indeed appear to be a higher incidence of blindness in the compounds near the river. Are flood plain living and blindness associated? Or is our cartographic pattern just an accident of the sample? How can we relate blindness to location on or off the flood plain when our measurements are simply 1's or 0's? Note that it is here, after we have measured, that we raise for the first time the question of relating something to something else—blindness to location. Thus we have our first clue for the reason we try to measure in the first place. To explain a set of experiences in the P-plane, as opposed to recording and describing them, we must as the minimal step relate them to another set of events. Relationships and ex-

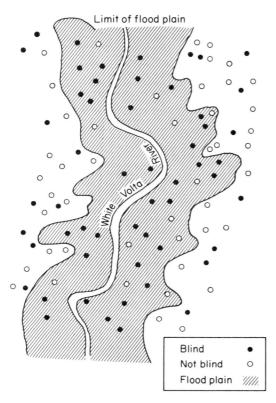

FIGURE 4–1

Distribution of family compounds with blind members near a river in Africa.

planation are big enough topics to require considerable discussion on their own (Chapter 5), but measurement itself, considered as an attempt to refine description, frequently raises questions about those relationships between things that constitute explanations.

But the question remains: how can we see if our two dichotomized variables, blindness and flood plain location, are associated or related one to another? Furthermore, is the degree of the relationship strong or weak? Suppose we transfer our field results from the map to a two-by-two table (Fig. 4–2). Each of the one hundred sample observations fits into one of the four cells, and simply from a careful inspection of the proportions of observations in each cell we can see that the local people may well be on the right track. To examine the relation-

ship in a more precise way, we can label each of the cells and the marginal totals, and calculate a number (usually called a phi coefficient) varying between -1 and $+1$ indicating how strong the overall relationship is. If:

$$\phi = \frac{a \cdot d - b \cdot c}{\sqrt{A \cdot B \cdot C \cdot D}}$$
$$= \frac{36 \cdot 34 - 18 \cdot 12}{\sqrt{48 \cdot 52 \cdot 54 \cdot 46}}$$

then we have:

$$= \frac{1224 - 216}{2490}$$
$$= 0.40$$

If the relationship between blindness and location were exact, all the observations would be in cells a and d, or in b and c, so that the numerator would be the same as the denominator, and the phi coefficient (ϕ) would be $+1$ or -1. On the other hand, if the observations were equally divided among a, b, c and d, then the term $(a \cdot d - b \cdot c)$ would be zero and phi would also be zero. In our example, phi is 0.40, so some degree of relationship may exist, and it might be worth following up this clue, investigating the hypothesized and partially supported relationship further.

It is possible that such a value may have arisen simply by chance, by the observations falling in the cells because of an unfortunate quirk of sampling, leading us to think that there

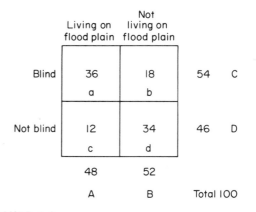

FIGURE 4–2

A cross-classification of blindness and location.

is a relationship when there is really none. In this case, the probability of getting a value as high as 0.40 when no relationship exists is very small indeed, but we shall not explore such questions further. We know today that "river blindness," as it is called by the local people, is carried by the simoleum fly which lives and breeds in the brush near water. But it is just this sort of careful observation, measurement, field work, recognition of spatial pattern, and structuring of relationships that uncovers those vital clues leading to satisfying explanations.

Working at the simplest levels of measurement does not confine us to binary or dichotomized variables. It may be appropriate to work with a Yes-Maybe-No, or $+1$, 0, -1 scale, or even more refined scales based upon five or seven values. Such scales are termed *nominal*, and still involve classificatory principles in which a number is assigned to a particular type of thing. For example, semantic differential scales are used extensively in the social and behavioral sciences, and require people to classify objects by placing them in seven categories along a continuum.

Suppose in a geographical study on residential desirability we asked African university students to evaluate their personal feelings about districts in their country. The question could be posed in terms of assignments as government servants or teachers after graduation. The participants could be asked to consider each district on a number of scales measuring their perception of the facilities and people in the districts. The scales might look like this:

ZONGO MACHERI DISTRICT

Facilities: Many things to do, spare time no problem   Not much to do in spare time, far from the center of things

$$+3 \quad +2 \quad +1 \quad 0 \quad -1 \quad -2 \quad -3$$

Local People: Friendly and very easy for me to get on with   Very difficult for me to fit in and get on with people there

$$+3 \quad +2 \quad +1 \quad 0 \quad -1 \quad -2 \quad -3$$

and a person could record his own, quite personal feelings about Zongo Macheri on each scale. Good feelings about the district would be recorded high on the plus side; bad feelings would mean a shift to the negative side.

Similarly, a geographer working at the micro-spatial level in a city might ask a sample group to evaluate a series of carefully chosen routes on semantic differential scales whose opposite ends are labeled open-closed, noisy-quiet, bright-dark, and so on. Special methods, some of them quite complex, are needed to analyze and combine scales of this sort, but the major point here is that such a difficult thing as the human perception of the urban environment can be measured effectively today.

The Ordinal Scale

As its name implies, the *ordinal* scale rests upon our ability to put things we are trying to measure in a certain order. To use ordinal scales we must distinguish between things on the basis of some criterion—size (bigger than, less than), distance (closer to, farther from), and so on—and then assign the ranks 1, 2, 3 . . . up to N, the number of observations. This is often a requirement that can be met in geographic research, and it offers us very powerful measurement tools. For example, we can often get people to rank or order their preferences for towns or states by asking them to compare each area with the others in turn, saying to themselves "Now, if I really had to choose between these, which would I prefer?"

Upon their first acquaintance with ordinal scales people are often very uncomfortable. This is because the requirement of simple order does not support our intuitive ideas of what real measurement should be. Our commonsense views of measurement rest upon notions that a mathematician would term *metric*—that is, ideas rooted in the notion of distances between things. As a child you stood by the door and successive pencil marks recorded the distance from the floor to the top of your head; downtown and school were a certain distance from home, and so on. In contrast, the simple business of putting things in order seems a much less satisfactory way of going about the business of measurement. There is some conflict here between the commonsense and scientific modes of inquiry. Yet a powerful and growing body

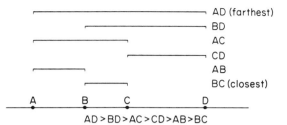

AD > BD > AC > CD > AB > BC

FIGURE 4–3

The order of six observations represented by the distances between four beads on a string.

of theoretical work is based upon just such a simple idea, for it turns out that ordinal scales actually contain a considerable amount of metric or distance information. The formal proofs of such an assertion are difficult, but let us see if we can get an intuitive feeling for what is involved.

Let us say that we have six observations and "all" we can do is order them from smallest to largest. But now suppose that we can represent the six observations by four beads on a string (Fig. 4–3). We slide the beads, labeled *A*, *B*, *C*, and *D*, back and forth until the intervals between all the pairs match the ordering of our objects. But if we can do this think what it means. We have represented an ordering of observations by intervals between beads—that is, distances between beads—so we have uncovered some properties here that are very close indeed to the metric properties required by our intuition and traditional, commonsense thinking.

But, you say, I can still move the beads around without changing the order of the distances between pairs. And, in a sense, you are right: but how much actual freedom do you have to move the beads back and forth before you break one of the ordered relations? Let us take a closer look, and considering bead *A*

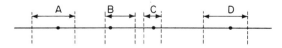

FIGURE 4–4

The ranges over which the beads can be moved before the order of a pair of observations is broken.

slide it slowly to the left (Fig. 4–4). The distance *AD* will always remain bigger than *BD*, but look what happens to the order *BD* > *AC*: bead *A* cannot move very far to the left before *AC* > *BD* and our order is broken. Similarly, if we slide *A* to the right, it is not long before the order *AC* > *CD* is reversed. All the other beads have rather narrow ranges over which they can move before some ordered relation is changed, so it turns out, even in this small example, that we really have very little freedom to alter positions and keep the ordering of our six observations. With any realistic number of observations the freedom to slide the beads is drastically reduced, for only very small shifts will be required to break one of the orders. Thus ordinal scales may well contain a high degree of metric information, giving our intuition some comfort as to their validity, and some assurance that they have the power to measure in a very refined way.

In certain situations, ordinal scaling can raise very uncomfortable problems, particularly when ordered preferences of people have to be combined under democratic, or equal weighting conditions. Suppose in a study on a footloose industry, where "non-economic" factors such as pleasant climate, good schooling for employees, and recreational facilities have a strong influence on locational choices, six decision-makers rank three areas (Fig. 4–5). Unfortunately, there is

	Areas		
	A	B	C
Merrill	1	2	3
Lynch	2	3	1
Pierce	1	2	3
Fenner	3	1	2
Smith	2	3	1
Beane	3	1	2

FIGURE 4–5

The order of locational preferences for six decision-makers.

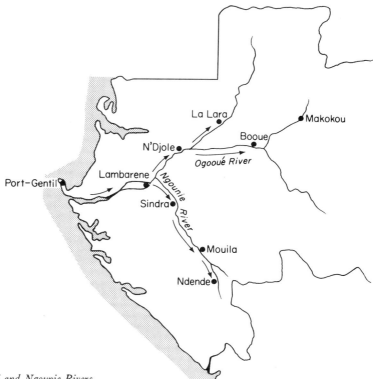

FIGURE 4–6
Flows of early trade goods on the Ogooué and Ngounie Rivers.

no way to come to a decision in this case, because a majority (four out of six) prefers area A to B, a similar majority prefers B to C, but a majority also prefers C to A! This is a famous paradox and much has been made of the fact that under certain circumstances choices may be intransitive, or that A may be preferred over B, B over C, but C may be preferred over A. Fortunately, such paradoxical situations appear more frequently in the minds of ingenious theorists, rather than in actual choice situations.

When we try to structure relationships between things measured upon ordinal scales, we can bring a well-developed body of methods to bear. Suppose, for example, you are a geographer reconstructing the trading patterns on the Ogooué and Ngounie Rivers around the turn of the century (Fig. 4–6). Records of commercial trading firms based on Port-Gentil have long been lost, but turning to oral tradition you establish the order of shipment values at various landing stages through interviews with old and now retired river pilots. Patient, cross-checked field work in the interior establishes the order of population around each river port so that you have the following ordinal information (Fig. 4–7). What sort of relationship was there between size of population and shipment in those days? Thinking of population as a measure of demand for trade goods, do shipments reflect such demand, or were other

River port	Rank of river port	Rank of shipment
Lambarene	4	2
N'Djole	1	1
Mouila	5	4
Sindra	8	6
La Lara	6	7
Booue	7	8
Ndende	2	3
Makokou	3	5

FIGURE 4–7

Ordinal data for shipments and population on the Ogooué and Ngounie Rivers.

forces at work shaping the flows of such commodities? A number indicating the degree of relationship between two ordered lists is Spearman's rank correlation coefficient. It varies from $+1.0$ to -1.0 and is calculated by:

$$r_s = 1 - \frac{6\Sigma D^2}{N(N^2 - 1)}$$

where: $\Sigma D^2 =$ the sum of the squared differences between each pair of ranks in the list, and

$N =$ the number of observations (in this case 8).

Working the example through:

$$r_s = 1 - \frac{6 \cdot 16}{8(8^2 - 1)}$$
$$= 1 - \frac{96}{504}$$
$$= 0.81$$

Since the value of r_s is close to 1.0, we now have evidence that the spatial pattern of trade shipments was evidently influenced by the demand generated by the people around the small river ports.

Although they only require a geographer to rank events from the largest to the smallest, ordinal scales are very powerful measuring devices. Very frequently we can get information about geographic events that lets us order them in the form of a list. If we can do this, it means we are very close to powerful and precise metric scales. Many events, previously considered qualitative and immeasurable, can now be handled in a scientific way. Thus we can clarify and test a number of relationships between geographic events that were formerly suspected and handled in a descriptive fashion.

The Interval and Ratio Scales

We come finally to the most familiar forms of measurement—the interval and ratio scales. The *interval* scale is a true metric scale. Intervals between successive measurements may be represented by distances between points. Only

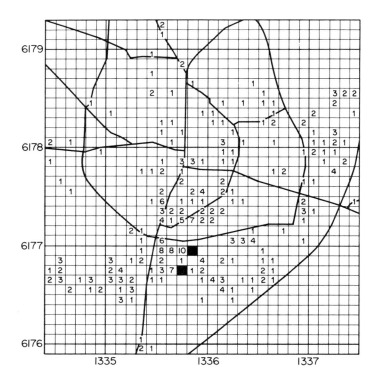

FIGURE 4–8

Computer plotting on the Swedish national grid of children attending the Järnåkra School, Lund (adapted from Nordbeck, Fig. 7, p. 85).

he starting place is arbitrarily set. The most
amiliar use of the interval scale in geography is
he precise characterization, or measurement, of
ocation upon the earth's surface using latitude
nd longitude. Both the equator and the merid-
an of Greenwich are arbitrary starting places
·om which distances are measured to locations.
imilarly, some countries, like the United King-
om and Sweden, superimpose national grids
·ver their territories, and upon such coordinate
ystems, with arbitrary origins or zero points,
he locations of all the features are specified.
.ven census information can be located using
he national grid systems, and today computers
·lot many distributions needed in spatial
·lanning and analysis. In planning new schools,
·or example, it is important to know where the
hildren are (Fig. 4–8). If every child is located
n spatial coordinates on the national grid,
·s well as in the more traditional temporal
·imension, it is easy to get a computer to print
naps at any scale required.

The *ratio* scale is also a true metric measure.
Because the zero point is not set in an arbitrary
·vay, we can use it to compare ratios as well as
ntervals between successive measurements.
Many spatial variables are measured in this way:
·listances or travel times between places; areas
·f counties, states, and countries; densities of
·eople, and so on.

If the ordinal scales contain so much metric
nformation, why bother to measure things on
nterval or ratio scales at all? What additional
nformation is gained by trying to refine our
neasurements, often at greatly increased cost?
The answer is that ordinal measures tell us
nothing about the rate at which things vary
ogether, only that two variables move up or
lown together. As an illustration, let us suppose
hat the records of old trading firms are suddenly
liscovered in the files at Port-Gentil, so that we
now have the actual shipment values and
·opulation counts for early trade (Fig. 4–9).
Ne know from our analysis of the ordinal data
hat a close relationship existed around the
urn of the century between the two. But what
·dditional information do the ratio measures
give us?

If we plot the new metric data on a graph
·Fig. 4–10), and then fit a straight line to the

River port	Population of river port	Value of shipments landed
Lambarene	3900	3400
N'Djole	5500	3900
Mouila	3400	2100
Sindra	1900	1500
La Lara	2900	800
Booue	2400	100
Ndende	5000	2300
Makokou	4300	1800

FIGURE 4–9

Interval data for shipments and population on the Ogooué and Ngounie Rivers.

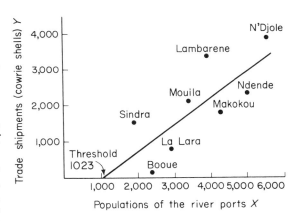

FIGURE 4–10

A graph of shipments and populations of river ports.

points representing the river ports (we shall
discuss the question of line-fitting extensively
in Chapter 5), we can see the rate at which
shipment values increase with population.
Roughly, the population has to go up 1,000 for
there to be an increase in shipment value of
753 cowrie shells. Notice also where the line
crosses the horizontal axis of the graph. We can
think of this as a rough estimation of the threshold population (1023)—that is, the minimum
population needed before any shipments of
trade goods are generated. Finally, we can

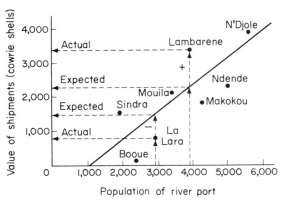

FIGURE 4–11

The deviations of the river ports from the general relationship between populations and shipments.

TABLE 4–1

Categories of Measurement

Scale	Basic Empirical Question	Geographic Examples
Binary	Are things in one of two categories?	Land use types
		Locations in or out of defined areas
Nominal	Are things in one of several categories?	Semantic differential scaling
		Intensity of answers on questionnaires
Ordinal	Is something greater or less than something else?	Ranking of place preferences
		Richter scale for earthquakes
Interval	Are two intervals the same?	Calendar time
		Contour lines on maps
Ratio	Are two ratios the same?	Distances between places
		Bushels per acre
		Per capita incomes

---

measure the vertical distance each town deviates from the line (Fig. 4–11). The line represents our general statement about the overall relationship, to which there are exceptions of varying degrees. Some towns, for example Lambarene, actually received somewhat greater shipments than we would expect from the general relationship represented by the straight line. Others, like La Lara, got less than the values expected on the basis of their populations. We can plot the distance each town deviates from the line on a map (Fig. 4–12). Towns close to Port-Gentil got much more trade than we might expect, while those far away got less. From the way these values change regularly in space, it would seem that the use of ratio measures has uncovered another variable, distance from Port-Gentil, whose effect before was obscured by the weaker ordinal information.

If we summarize what we know now about measurements (Table 4–1) and look for the single thing that describes all the binary, nominal, ordinal, interval, and ratio scaling procedures we have used, we are finally in a position to answer our original question: What is measurement? Underlying all forms of measurement is the ability to assign numbers to things according to a clear and well-defined rule. The rule may be as simple as assigning 1 or 0 to the answer of a question, or as sophisticated as giving a numerical value to an object, so that

its interval from another can be determined. No matter where the problem lies on the continuum from quality to quantity, all we require is an unambiguous rule and we can fulfill the task of measurement.

RULERS AND SURROGATES

The minimal requirement for interval measurement is the use of some sort of ruler marked off in clearly defined units. The common meter ruler, the surveyor's chain, and the electronic depth sounder are all instruments which determine how numbers of units (meters, feet, and fathoms) are assigned to distances. But the problem is not always so easy, and we may have to construct our own measuring stick against which a particular object is to be scaled. Sometimes crude rulers can be made after we have numerical information by noting how far individual values are away from the average of the group. Using the average, or mean, as a base point on a ruler is a very common procedure, the most familiar being the scaling of intelligence tests. Large numbers of tests are scored, and only then is the average score given the arbitrary value of 100. This becomes the

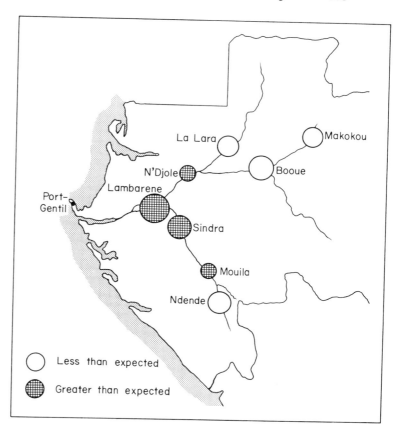

FIGURE 4–12

Deviations of trade flows from those expected according to the flow-population relationship.

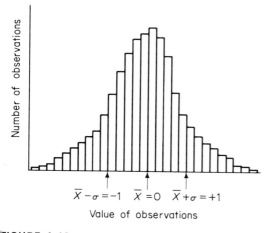

FIGURE 4–13

A scale with the mean ($\overline{X}$) as the arbitrary origin and the standard deviation (σ) as the unit on the scale.

point on the ruler against which all the other observations are measured according to their degree of dispersion, or distance, above or below this most typical value.

Often standardized units of dispersion around the mean are used when the original scores are reasonably symmetric (Fig. 4–13). The units used for measuring dispersion depend upon the overall degree of scatter around the mean. We start the calculation by subtracting the mean value from all the scores. Since some are negative, while others are positive, they are squared before they are added together. The sum of the squared values is then divided by the number of scores to get the average squared scatter. Finally, because we are working with squared units, we take the square root. This common measure of dispersion around the mean $\overline{X}$ has been given the symbol σ (sigma).

It is called the *standard deviation*, so that:

$$\sigma = \frac{\Sigma (X - \bar{X})^2}{N}$$

As an example, suppose we had the following values:

| 1 | 4 | 8 | 9 | 9 | 12 | 17 | 20 |

The mean is:

$$\bar{X} = \frac{80}{8} = 10$$

Subtracting this from each of the individual observations, we have:

(1 − 10)	(4 − 10)	(8 − 10)	(9 − 10)
−9	−6	−2	−1
(9 − 10)	(12 − 10)	(17 − 10)	(20 − 10)
−1	+2	+7	+10

These are now squared and added together to give:

$$81 + 36 + 4 + 1 + 1 + 4 + 49 + 100 = 276$$

which is then divided by the number of observations:

$$\frac{276}{8} = 34.5$$

Finally, since we are still in "squares" we take the square root, so that:

$$\sigma = \sqrt{34.5} = 5.87$$

Thus, with the mean $(\bar{X})$ as our starting point, and the standard deviation (σ) as our unit, we can construct a ruler to measure any individual observation. If we call an individual observation z, then:

$$z = \frac{X - \bar{X}}{\sigma}$$

If the original, or raw value of a particular observation coincides with the mean of all the observations, the *standardized* or z value will be 0, while scores a single standard deviation above or below the mean will be measured as +1.0 or −1.0. Providing the distribution is reasonably symmetrical, observations lying within one of the standard deviation units above or below the zero point on our ruler will be quite com-

mon. Values closer to the tails will appear much less frequently.

Sometimes making rulers is difficult because we do not have large samples to work with. In such cases, we must try to define some value as clearly and unambiguously as we can simply as an ideal against which we can compare the individual observations we make. For example, if we wanted to measure the degree to which family incomes in a *municipio* of Guatemala could provide minimal requirements for clothing, shelter, and a nutritionally adequate diet, we might be able to determine the minimum income necessary for each area under such well defined conditions. We would then be in a position to measure the departures of the actual incomes from the clearly defined, minimum values.

Setting up ideal or normative values is a common procedure when very difficult things like spatial patterns and shape have to be measured. If we compare patterns of settlement, for example, we can construct a ruler by finding the average distance between settlement points on the map to their nearest neighbors (Fig. 4–14). Let us call this actual average distance to nearest neighbor $\bar{r}_a$ and define it as:

$$\bar{r}_a = \frac{\Sigma r}{N}$$

That is, we measure the distance of every point to its nearest neighbor, add all the distances together, and divide by the number of points.

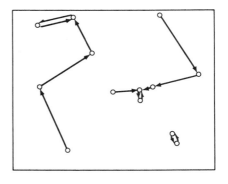

FIGURE 4–14

Measured distances (r) of each point in a pattern to its nearest neighbor.

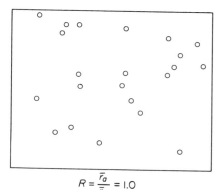

$$R = \frac{\bar{r}_a}{\bar{r}_e} = 1.0$$

FIGURE 4–15

A random pattern of points.

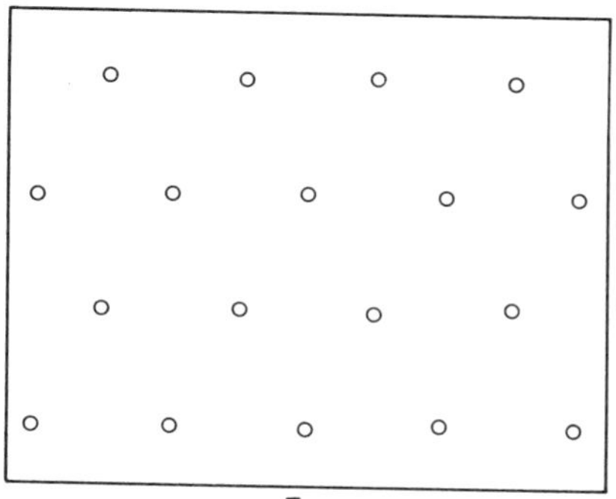

$$R = \frac{\bar{r}_a}{\bar{r}_e} = 2.149$$

FIGURE 4–16

A hexagonal pattern of points.

Now it can be shown that if the pattern is random, we would expect the average distance ($\bar{r}_e$) to be:

$$\bar{r}_e = \frac{1}{2\sqrt{\rho}}$$

or the reciprocal of twice the square root of the density ρ (rho). The density is simply the number of points per square unit—for example, villages per square kilometer.

When an actual settlement pattern is random (Fig. 4–15), the ratio (R) of the actual average distance to nearest neighbor ($\bar{r}_a$) to the expected ($\bar{r}_e$) will be 1.0. On the other hand, if all the villages push each other away, they will form a hexagonal lattice (Fig. 4–16), and R will take the maximum ratio of 2.149. Notice that if the settlements huddle together at a single point the distances between the nearest neighbors will all be zero, so $\bar{r}_a$ and R will also be zero. Thus, by setting up an ideal pattern, we have generated a ruler (Fig. 4–17) for measuring actual patterns of settlement starting at 0 (complete agglomeration), moving through 1.0 (random), to 2.149 (ideal or normative hexagonal lattice).

On occasion the task of measurement in geography seems almost insuperable. At the heart of many of the problems that geographers find most intriguing are those will-o'-the-wisp things like accessibility, modernization, eco-

nomic development, social distance, and intervening opportunity. These are all ideas that seem intuitively obvious, until we try to pin them down with a ruler. The result is that we often have to cast around for *surrogate* measures— measures of something else, not quite what we were after, but perhaps close enough to stand in place of the actual thing until we get greater insight or more money to improve our rulers.

Suppose we were investigating the way the number of amphibian and reptile species varied on islands, and hypothesized that the numbers increased with "habitat diversity." How can we measure such a difficult idea as this? From what definition can we even start? When does one habitat diverge sufficiently from another so that we can count it as a different one? Such questions are extremely difficult, and we might well give up the whole idea of trying to measure such an obvious and elusive variable. But suppose we cast around for reasonable surrogates, variables that might be quite easy to measure that would "stand in" for the one that is causing so much

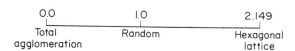

FIGURE 4–17

The ruler for measuring point patterns.

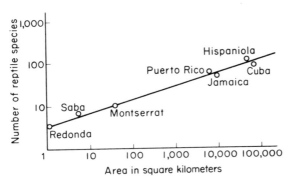

FIGURE 4–18

The area-species relationship for West Indian Herpeto-fauna (adapted from Robert MacArthur and Edward O. Wilson, The Theory of Island Biogeography, *Fig. 2, p. 8. Copyright © 1967 by Princeton University Press).*

trouble. Perhaps we can use the area of the island instead, on the reasonable grounds that the thing called habitat diversity might well be so closely related to area that it will not make much difference which we use. Substituting area for habitat diversity (Fig. 4–18), it would seem that our surrogate measure is not so bad after all. The relationship is so close that if we know the area of an island, we know almost exactly the number of amphibian and reptile species we are going to find.

PROPERTIES AND PRIMITIVES

In one very real and profound sense we cannot measure geographic things. This does not contradict what we have just said, but simply recognizes the obvious fact that we cannot measure directly the actual objects we deal with. Flows of goods and people and ideas, cities and settlements, cultural and political barriers are never measured directly. Rather, it is their properties that we try to scale with our measurement methods. Thus, although a city or region may be truly immeasurable, the properties of these things may be more susceptible to our rulers.

It is at this point that we enter a very difficult area. If we really think hard about the properties of the things we measure in geography we are

forced back to consider the basic primitives of the field. That is, what are the most essential, most primitive things we have to deal with, the basic blocks from which we build all our other constructs in the C-field? So little thought has been given to the question of geographical primitives, that it is easy to convince ourselves that we seldom have the faintest idea what we are measuring or even investigating. Sobering up after a measurement binge is no fun, and the headache can be so severe that we may decide never to touch the subject again. On the other hand, the experience of thinking about the fundamental constructs may so clear our sight that a judicious and experienced approach to measurement questions will mark our efforts in the future.

Dimensional Analysis

Let us consider some of the very primitive notions we deal with. Probably the most basic one is *length*. Indeed, for some geographers the notion of distance, the concept of number applied to length, characterizes the discipline which has as its core a concern for spatial pattern, arrangement, and juxtaposition. Absolute or relative spaces are usually considered to be geographic areas with two dimensions, but we can build the notion of area from the more basic notion of length since it is simply $(L \cdot L)$ or L^2. If we want to distinguish between length in two different directions, say longitude and latitude or the horizontal X and vertical Y directions on a graph, we can break the L^2 of area down into two components $L_x^1 L_y^1$. What other dimensions do we deal with? Clearly time T is crucial whenever we work with geographic processes, and from this additional primitive we can build the further notion of velocity. This is simply a length L per unit of time T or L/T. We usually write this as LT^{-1}, the exponent -1 implying that we are dealing with $1/T$. We may not be able to measure directly an early settlement wave or pulse down a river like the Ohio, but we can perhaps measure its property of velocity, which in turn is composed of even more primitive notions. If a settlement process accelerates across a new pioneer region, it simply means that an area L^2 is being covered by people at a

ster and faster rate so that settlement accelera-
on has the dimension L^2T^{-2}.

Let us consider a very simple example, in
hich we relate some familiar geographical
nings to each other, remembering that at the
enter of any relational expression is the equality
gn—a familiar, but by no means a trivial
bject. The equality sign implies that something
n one side of an equation is equal to the
nings on the other side. The question of balance
implicit here, so that we can think of an
quation as a child's seesaw with the equality
gn as the fulcrum. Thus if we write:

$$A = B + C$$

implies that B and C together balance A (see
ig. 4–19). The notion of equational balance
vital when we consider the basic dimensions,
r primitive terms, that make up the two sides
f an equation because they must also balance
ne another. This is simply another way of
aying that dimensional equality requires all
f the geographical primitives on either side
f an expression to be compatible.

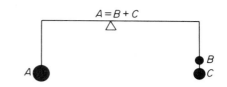

FIGURE 4–19

A dimensional equation with A *balancing* B + C *around
the fulcrum or equality sign.*

Suppose we take a classical geographic
problem on the size and spacing of central
places, and examine the primitive notions from
which we may be able to build higher-level
expressions. Assume, for the purposes of simplification, that we have a homogenous plain
(Fig. 4–20). Since it is an area it will have the
dimension L^2. Now we may not be sure, at the
first stages of our thinking, whether it will be
important to consider the separate length dimensions, so we can distinguish between them just
to be on the safe side making our area $L_x^1 L_y^1$. In
the plain we have some towns and villages, and

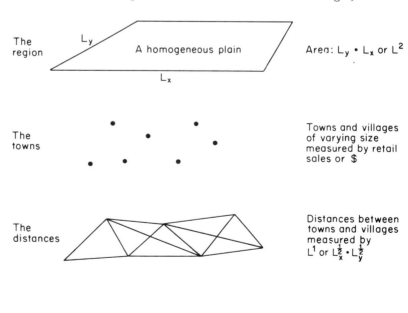

IGURE 4–20

*he dimensions of a central place
roblem. The space* L^2; *the towns* $;
e distances L^1; *and the purchasing
ensity* $\cdot L^{-2}$.

we face the problem of specifying their primitive properties that we can actually subject to measurement. The property we are particularly interested in is size, symbolized by s, but what are its basic dimensions? Since we are trying to measure the influence of towns upon others, the basic dimensions are hard to pinpoint. Population might be one way of measuring the size of central places, but if incomes vary greatly over our region such a measure may not be the best. Since towns are centers serving the surrounding area with goods, recreational facilities, and so on, let us take total retail sales as a measure giving it the dimension *dollar* or *$*.

The distances between central places are easy to specify in primitive terms for they simply have the dimension L. Since we are on a plain of two dimensions, we can split distance into two components and write it as $L_x^{1/2} L_y^{1/2}$. We can always add the two exponents $\frac{1}{2}$ and $\frac{1}{2}$ if we want to reconstitute the basic dimension of distance L^1.

What else might influence the size and spacing of our central places? If we look at settlement patterns across the Midwest and into the Great Plains—for example, eastern and western Kansas—we can get some clue. As population thins out in the drier, western areas, the towns serving the population also thin out. Clearly density of population or purchasing power may have something to do with the size and distance relationship of the central places. The density of purchasing power p is simply dollars spread over area, so the dimensions will be $\$L^{-2}$ or $\$L_x^{-1}L_y^{-1}$. Summarizing, we have:

Size of central places in retail sales $s = \$$.
Distance between central places $d = L$
$$= L_x^{1/2} L_y^{1/2}$$
Density of purchasing power $p = \$L^{-2}$
$$= \$L_x^{-1}L_y^{-1}$$

Now suppose we were interested in the relationship of distance d to the other things, size s and purchasing power p. We could write:

$$d = f(s, p)$$

and read "d is some function of s and p." But we can be a little more specific and say:

$$d = k \cdot p^\alpha \cdot s^\beta$$

The exponents α (alpha) and β (beta) we have given to p and s respectively can vary. The value of the constant k can also change, but in any particular problem, it only depends on the units we used for measuring our variables: miles or kilometers; dollars or francs, and so on. If we want to make the variable d on the lefthand side balance p and s on the right, we must find values for α and β. Only by choosing the right values can we make the dimensions on one side equal, and therefore balance the other side around the equation's pivot—the equality sign.

Instead of writing our equation with the symbols d, p, and s, let us substitute their dimensions, so that:

$$L_x^{1/2} \cdot L_y^{1/2} = (\$ \cdot L_x^{-1} \cdot L_y^{-1})^\alpha (\$)^\beta$$

We now face the task of finding the values of α and β to make the dimensions on one side balance those on the other. Any equation must be in dimensional balance, and we must find the appropriate values of the exponents α and β simultaneously. We do this by considering each of the fundamental dimensions one at a time. Let us look at the dimension L_x first. On the lefthand side, L_x has the exponent $+\frac{1}{2}$. In the dimensional expression of p on the righthand side L_x appears with the exponent -1. This means α must be chosen so that when -1 is multiplied by α it is changed to $+\frac{1}{2}$, which is the exponent of L_x on the left. Of course we already know the answer (α must be $-\frac{1}{2}$), but we shall continue our analysis formally because sometimes the answer is not nearly so clear.

In the second term s on the righthand side L_x does not appear at all. As far as we are concerned β could be zero. Thus, with respect to L_x we can write:

$$-1\alpha + 0\beta = \frac{1}{2}$$

Similarly, with respect to L_y:

$$-1\alpha + 0\beta = \frac{1}{2}$$

And, finally, with respect to $\$$:

$$+1\alpha + 1\beta = 0$$

From these it is quite easy to see that:

$$\alpha = -\frac{1}{2} \text{ and } \beta = +\frac{1}{2}$$

and when we substitute them back into our original expression:

$$d = k \cdot p^{-1/2} \cdot s^{1/2}$$

$$d = k \cdot \sqrt{\frac{s}{p}}$$

Does the expression make sense? Examine it carefully: when the size of central places (s) goes up, the distance (d) between them will increase if purchasing power (p) is held constant. This means that big towns, with very specialized functions, will compete for customers supplying the necessary retail sales to keep them going. They will have to be far apart, for if they are too close together they will cut each other's throats, and the weakest will decline to a smaller size in the urban hierarchy. On the other hand, as an area develops economically, and purchasing power increases, we would expect distances between central places of a given size to decrease—all other things being equal. As the density of purchasing power goes up, some small towns lying halfway between the bigger ones will grow to supply the increased needs of the people. A richer area needs more places to serve it, so more places will come into the system, and the distance between places of a similar size will decline. As for the value of k, it is simply a constant of proportionality varying with our measurement units; dollars per square mile, rubles per square verst, or kronor per square kilometer.

As a final example, to stretch ourselves a little, consider the primitive things that might have to be measured in a dynamic spatial study such as the pioneer settlement of Pennsylvania by the early Scotch-Irish. Suppose a pioneer wave moves out from a source point like Philadelphia, and we think about the things that might explain the velocity of the wave at some point on the landscape. What things might enter the problem? As a start, we are dealing with people P, distance L and time T. First, velocity or speed s is length per unit of time or LT^{-1}. Next, the energy of the pioneer population p may explain the speed of a settlement process, for we might expect very large numbers of highly motivated farming immigrants to push out into new lands with greater vigor than would be shown by just a few. What dimensions will such a concept

have, and how can we measure it? Perhaps by noting the ability of the pioneers to cover area at an increasing rate, so population energy may take the dimension PL^2T^{-2}. But what about the terrain over which the pioneer settlement moved? Some landscapes may be "stickier" than others. Perhaps we can think of the viscosity of a pioneer area v, measured in terms of the ability of people to move over a distance in a certain length of time, possibly settlers per mile per week or $PL^{-1}T^{-1}$. Finally, the speed of the settlement waves may depend on the distance d from the original source area which has the simple dimension L.

Following the same formal analysis of dimension, we have

velocity of the settlement wave $s = LT^{-1}$
population energy of the settlers $p = PL^2T^{-2}$
viscosity of the landscape $v = PL^{-1}T^{-1}$
distance from source $d = L$

Now suppose that the velocity of the settlement wave depends on the population energy, viscosity of the landscape, and distance from the source, or:

$$s = f(p, v, d)$$

In slightly less vague terms:

$$s = k \cdot p^\alpha \cdot v^\beta \cdot d^\gamma$$

or in dimensional terms:

$$LT^{-1} = k \cdot (PL^2T^{-2})^\alpha (PL^{-1}T^{-1})^\beta (L)^\gamma$$

Now what values will α, β and γ (gamma) take to ensure that the expression is in dimensional balance? Considering P first, we have:

$$1\alpha + 1\beta + 0\gamma = 0$$

while for L:

$$2\alpha - 1\beta + 1\gamma = 1$$

and finally for T:

$$-2\alpha - 1\beta + 0\gamma = -1$$

Working the equations through to find the values simultaneously:

$$\alpha = 1 \qquad \beta = -1 \qquad \gamma = -2$$

Substituting back:

$$s = k \cdot p^1 \cdot v^{-1} \cdot d^{-2}$$

Or:

$$s = \frac{k \cdot p}{v \cdot d^2}$$

As we might expect, as the population energy p increases, so does the speed s of the settlement wave. On the other hand, the more viscous the landscape v, and the further the pioneer fringe is from the source d, the slower the wave s. Notice that we may well be able to measure in some crude fashion population energy, distance from source, and velocity of the pioneer fringe much more easily than we can measure "landscape viscosity." Thus, we may want to rearrange our expression to:

$$v = k \cdot \frac{p}{s \cdot d^2}$$

to put all the easier things to measure on one side so that we can estimate values of the most difficult thing—landscape viscosity.

In these examples of dimensional analysis we have obviously stretched things too far, for crude physical analogies will not solve the problems we face in specifying and measuring the properties of geographical things. Nevertheless, such an analysis forces the geographer to think very hard about the subject of his inquiry, and it lays bare in an uncomfortable and even embarassing way the roots of his ignorance. Most of the time we are measuring, or even worse writing about, things we know so little about that we are hard-pressed to specify the primitive building blocks from which they are constructed. Nor, let it be clearly understood, will dimensional analysis solve all the problems, for nothing comes out that did not go in originally. But dimensional analysis does help us to think about difficult measurement problems, exposing to a distressingly bright light the things we know and those we do not. This is usually a pretty good place to start.

Suggestions for Further Reading

BRIDGEMAN, P. W. *Dimensional Analysis*. New Haven: Yale University Press, 1922.

GOULD, P. R. "Computers and Spatial Analysis: Extensions of Geographic Research," *Geoforum*, I (1970), 53–69.

GREENSHIELDS, BRUCE D. "Traffic and Highway Research and How It May Be Improved," *Science*, CLXVIII (1970), 674–78.

HÄGERSTRAND, TORSTEN. "The Computer and the Geographer," *Transactions of the Institute of British Geographers*, XLII (1967), 1–19.

KEMENY, JOHN G. "Measurement," in *A Philosopher Looks at Science*, Chap. 8. New York: Van Nostrand Rheinhold Company, 1959.

STEVENS, S. S. "Measurement, Psychophysics, and Utility," in *Measurement: Definitions and Theories*, eds. C. West Churchman and Philburn Ratoosh. New York: John Wiley & Sons, Inc., 1959.

Works Cited or Mentioned

CLARK, P. J., and F. C. EVANS. "Distance to Nearest Neighbor as a Measure of Spatial Relationships in Populations," *Ecology*, XXXV (1954), 445–53.

HUNTLEY, H. E. *Dimensional Analysis*. New York: Dover Publications, Inc., 1967.

MACARTHUR, ROBERT, and EDWARD O. WILSON. *The Theory of Island Biogeography*. Princeton: Princeton University Press, 1967.

MAXWELL, A. E. *Analyzing Quantitative Data*. New York: John Wiley & Sons, Inc., 1961.

MILL, JOHN STUART. *Philosophy of Scientific Method*, Book III. New York: Hafner Publishing Co., Inc., 1950.

NORDBECK, S. *Barnens Skolvägar och Trafikvanor*. Lund: Institutionen för Byggnadsfunktionslära, 1967.

SHEPARD, R. N. "Analysis of Proximities as a Technique for the Study of Information Processing in Man," *Human Factors*, V (1963), 19–34.

TORGERSON, W. S. *Theory and Methods of Scaling*. New York: John Wiley & Sons, Inc., 1958.

CHAPTER 5

STRUCTURING GEOGRAPHIC RELATIONSHIPS

*In our description of nature the purpose is not to disclose the real essence
of the phenomenon but only to track down, so far as it is possible, relations
between the manifold aspects of our experience.*

—*Niels Bohr*

MEASUREMENT, RELATIONSHIPS, AND EXPLANATION

Measurement is seldom undertaken for its own sake. Rather, it forms a crucial link in the chain leading from initial hunches and questions to intellectually satisfying explanations. On a map, for example, it might appear that tuberculosis decreases with distance from a city. But to move beyond such a geographic speculation we need accurate values of tuberculosis rates at carefully measured distances from the center. Only then can we investigate the relationship between the two variables, and perhaps uncover other factors, such as poverty, that explain the spatial variation in the disease. Of course, poverty must also be measured accurately. Thus additional links in the chain, binding measurement and explanation together, are constructed from relationships.

If we want to explain something that is geographically intriguing, as opposed to reporting upon it with an accurate description, we can do so only by measuring it and relating it to other measured things. It is hardly surprising, then, that relationships emerged in the chapter on measurement, for the two are almost impossible to separate. Relationships are the fibers that bind constructs together in the C-field. Measurement takes on purpose when explanatory relationships are suspected; but our ability to examine relationships rests upon our prior ability to measure. Thus to explain is to relate is to measure.

What Is a Relationship?

In a sense we could stop right here, except that we would be pushing the real question into the background. For what is a relationship? What basic thing do we imply when we say that one thing is related to another? A common thread in all the relationships we examined in the last chapter was the simple but profound idea that as one thing changed so did another in a fairly consistent manner: a change in location onto the flood plain seemed to change the possibility of blindness; larger river ports received larger shipments of trade goods; the bigger the island the more reptile species we would be likely to find. And so we have come to the most fundamental idea of all—change or variation. We investigate relationships by examining the way things vary together. It is the changes and variations in things that are the subject of our inquiries, and our explanations of why, how, and where things vary are always made in terms of the way they change together or co-vary with others. Thus we have come back to the elementary, but important notion of a function:

$$Y = f(X)$$

which says that Y is some function of—that is, it co-varies with—X.

Other things being equal, blindness Y varies with location X; size of shipments Y changes with population X; interaction Y is a function of distance X. We could continue these geographical examples for pages, but to do so would be merely to indulge in an irresponsible exercise so long as we absolve ourselves from the responsibility of specifying, examining, and testing the form of the function. Until we do that, the equation above remains a thoroughly pretentious piece of symbolic trivia. In contrast, a clearly specified function is anything but trivial. It may even stand as a concise, explanatory model, a simplifying and compressing statement through which we can look at reality with greater understanding.

The accurate specification of a function, describing the way things vary together in a regular way, may place us in the enviable position of being able to *predict*. Quite apart from the usefulness of prediction in spatial planning, the ability to predict implies a reasonably profound knowledge of a problem, allowing us to anticipate the effect of some event before the event has taken place. But notice that we anticipate—that is, we hazard a guess; prediction, in the way we shall use the term, does not mean gazing into crystal balls and foretelling the inevitable course of the future with gypsylike omniscience. It is most unlikely that human spatial behavior will ever be completely predictable, for we live in a probabilistic rather than a deterministic world. We must be content with a more modest requirement from our predictive statements, being grateful for those perspicacious and frequently lucky occasions when we come reasonably close to the mark.

In our groping for geographic understanding we shall often be more than content if we can *postdict*—that is, find relationships and models that clarify our understanding of developments in the past. If, for example, we can specify the very general relationships that characterize and explain the spatial growth and development of cities from 1900 to 1950, we can test the appropriateness of the relationships by examining their ability to postdict the patterns of 1960—patterns which we have already recorded, but whose explanation was previously obscure or even unknown.

How Are Relationships Suspected?

Suspecting interesting and important relationships between things is usually a creative act difficult to formalize into a set of investigative rules. All scientific inquiry is pattern- and relation-seeking activity, and the perception of pattern and order in a seeming chaos of events is a creative act founded upon a touchstone of faith that orderly relationships are there in the first place. Indeed, so deep is the apparent need for order and pattern that we frequently impose a structure of C-field relationships upon reality, forcing the world to conform to a conceptual framework of our previous making. This, of course, is the great danger of acquiring prior experience and knowledge of simplifying structures and intriguing analogies, for we may end up dealing only with those relational questions that fit this year's model. On the other hand, and as Michael Polanyi has pointed out, prior experience and knowledge may enhance the possibility of perceiving patterns and relationships, and those with avaricious minds and large funds of analogy often have accounts bearing rich creative interest. There is a balance between forcing the world into preconceived models and total methodological virginity.

The act of creative suspicion starts with a

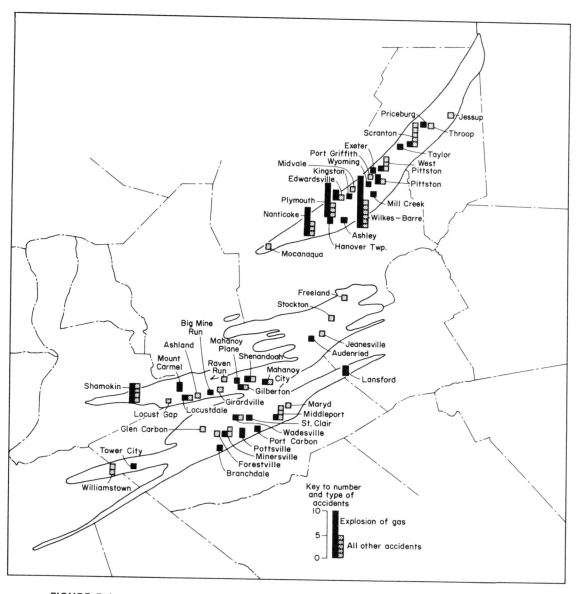

FIGURE 5–1

Distribution of gas explosions in the anthracite mines of Pennsylvania (adapted from Deasy and Griess).

question to be answered. It is nurtured by the apparent perception of pattern in a series of events, and sustained by evidence that the pattern of relationships is more than a figment of the imagination. Many creative leaps have been made in science by men who have seen patterns of relationships that were there for all to see—once the "obvious" was pointed out. Thus the perception of pattern may be likened

to extracting a signal that is blurred by noise. Seemingly random perturbations in predicted planetary orbits had been recorded for years until the French astronomer Leverrier saw relational patterns in the "error" terms, and predicted the position and very existence of Neptune. Sitting before the fire, the chemist Kekulé dreamt of whirling serpents and imagined one grasping its own tail. Immediately

113

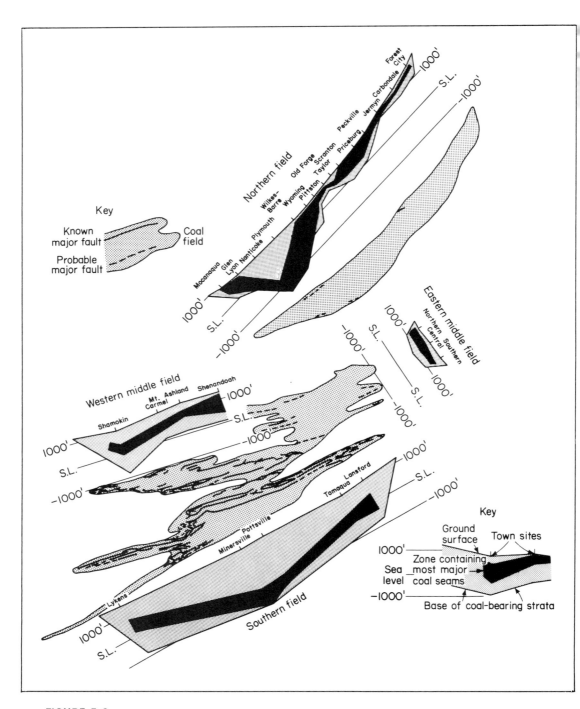

FIGURE 5–2

Depth of coal seams and degree of faulting in anthracite mines of Pennsylvania (adapted from Deasy and Griess).

he awoke with a burst of creative certainty that the benzene atoms were related to one another in the form of a ring.

Although such acts of creation cannot be taught and formalized, we can examine ways of determining the form and strength of relationships once an act of suspicion has led to the measurement of the variables involved. What are some of the relationships examined by geographers? Remember that many of the relations deal with the way in which things vary together over geographic space. Perhaps the simplest spatial relationship is one of regular locational juxtaposition and proximity. That is, if two things are regularly found together at the same location we may strongly suspect that there is a relationship between them. An example is the distribution of gas explosions in the anthracite mines of Pennsylvania (Fig. 5–1). Despite extensive mining over all the fields, mine accidents are highly concentrated in just a few areas. When the distribution of gas explosions is matched against maps and diagrams showing the depth of the coal seams and the degree of faulting (Fig. 5–2), the high degree of spatial coincidence is obvious. Shallow seams have hardly any gas explosions; deep seams, with extensive faulting allowing gases to escape rather than accumulate, have a moderate number; while deep, unfaulted seams are clearly the most accident prone.

Often the degree of spatial regularity of events in time is of considerable geographic interest: that is, are the times at which events occur related to, and therefore predictable from their locations? An example of a single measure changing or covarying in a regular way with location is the time of first settlement in western New York State. The simplest description of the relationship between space and time is a plane tilted in a east-west direction (Fig. 5–3), so that we can think of our pioneer settlers moving westward up a flat time gradient. The further west a location is, the later it was probably settled. But the relationship between space and time may be more complicated, for western New York was anything but a uniform transport surface. Thus, the relationships may be described by more complicated time surfaces that are curved over geographic space, emphasizing the

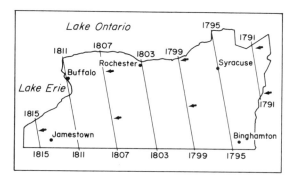

FIGURE 5–3

A settlement time plane over western New York; the simplest description of variation in time over geographic space (adapted from Gould, 1966).

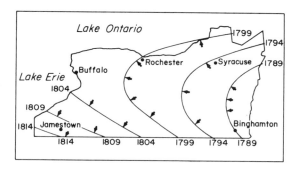

FIGURE 5–4

The settlement surface over western New York warped in two directions: a more accurate, but more complicated relationship (adapted from Gould, 1966).

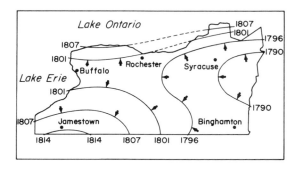

FIGURE 5–5

The settlement surface over western New York which has been warped in three directions: the most accurate, but most complicated relationship (adapted from Gould, 1966).

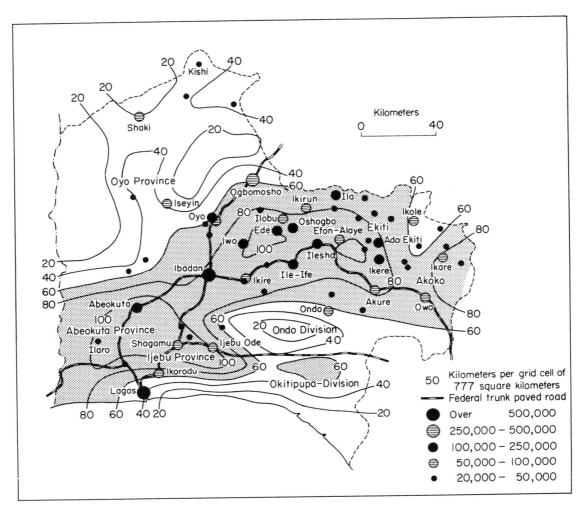

FIGURE 5–6

Road density surface for the western region of Nigeria (adapted from Ola).

corridor effect of the Mohawk-Buffalo alignment (Fig. 5–4), and the steep time gradient along the shore of Lake Ontario (Fig. 5–5) as the early settlers bypassed it on their way west because of military forays by the British from Canada.

Two or more variables may also move up and down together over an area indicating a strong relationship between them. In the western region of Nigeria high road densities mark the main, crescent-shaped core of the area where most of the major towns are found (Fig. 5–6). If we ask young teachers about their residential preferences in the area, we can combine and map their answers as a surface undulating over the region (Fig. 5–7). High peaks indicate areas that are viewed very favorably, while low valleys are places that are disliked. The surface appears to covary with the road densities to a very marked degree, possibly because the densities are surrogate measures for variations in information flows that mold and shape these mental images of geographic space. A graph of values at sample points on the maps (Fig. 5–8) confirms the visual impression of a regular relationship between the two.

As we saw even in the case of simple spatial juxtaposition, where explosions varied with depth of coal seam *and* with the degree of fracturing, a number of things may vary together

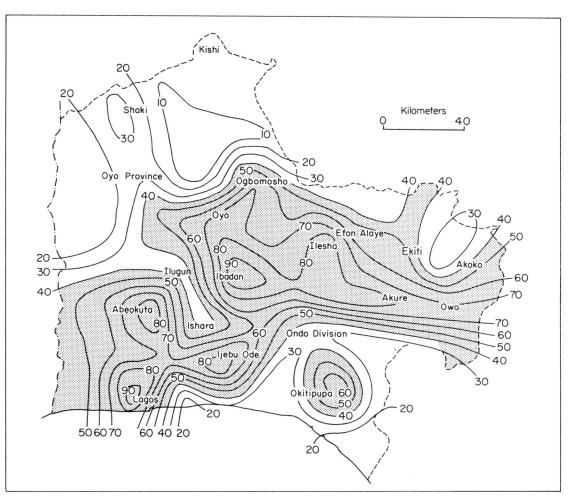

FIGURE 5–7

Residential preference surface for teachers (adapted from Ola).

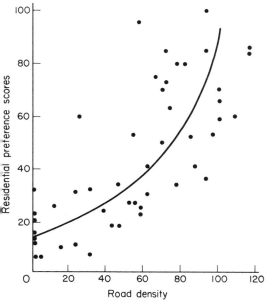

FIGURE 5–8

A plot of road density and residential preference measures at randomly selected locations on Figs. 5–6 and 5–7.

in extremely complex associations over a region. The associations may be so complicated that special methods may have to be used to simplify and disentangle them. For example, an air photo of the ocean may appear to be a hopelessly complex surface, but if we record the heights of the waves at regularly spaced intervals we can break up the highly convoluted surface into a sequence of very regular sine waves (Fig. 5–9). Each of the sine wave surfaces accounts for a different proportion of the irregularity in the original ocean wave surface, and when they are all added together their sum reproduces the original surface almost exactly. Sometimes it is even possible to interpret the individual pieces that go to make up the whole. Thus a

wave surface in the middle of the Atlantic W may be made up of waves from a Caribbean hurricane C, waves from a storm north of Iceland A, and a long swell from the Indian Ocean I. Thus we could write the relationship:

$$W = C + A + I$$

The notion of decomposing complex spatial patterns into constituent parts is a common one in all the sciences, and where spatial periodicities are suspected in human patterns such methods, imposing simple additive relationships, may be helpful. In the same way we could think of decomposing a population surface into separate urban cones accounting for the effect of each city and town on the overall population

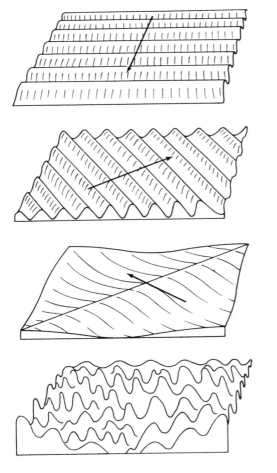

Waves from a storm
in the North Atlantic
(A)

Waves from Caribbean
hurricane
(C)

Long swell from Indian
Ocean
(I)

Total wave surface
observed in mid–Atlantic
(W)

A + C + I = W

FIGURE 5–9

The decomposition of a wave surface into independent parts.

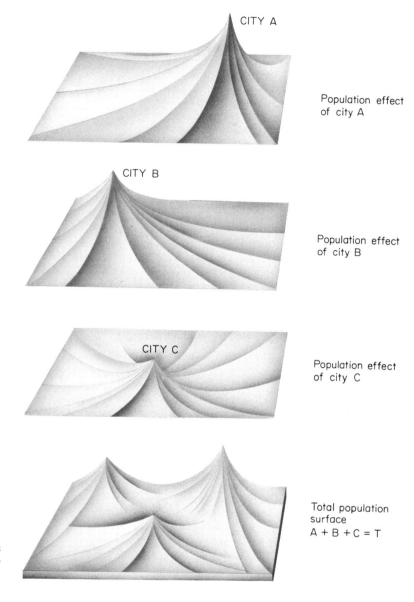

CITY A

Population effect of city A

CITY B

Population effect of city B

CITY C

Population effect of city C

Total population surface
A + B + C = T

FIGURE 5–10

The decomposition of the population surface T *into independent parts due to the effects of cities* A, B *and* C.

distribution in the region (Fig. 5–10). The urban cones are located so that they always account for the greatest amount of the remaining irregularity in the surface. When the separate pieces are added together, in much the same way that we added the wave surfaces, they reconstitute the original population distribution in the region.

THE DESCRIPTION OF RELATIONSHIPS

Whether our concern is for simple relationships arising out of locational juxtaposition, regular trends of a single variable in a region, the covariation of two or more variables in geographic space, or the decomposition of complicated surfaces for the purpose of simpli-

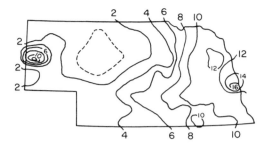

FIGURE 5–11a

Rural farm population density surface, Nebraska (adapted from Robinson, Fig. 1, p. 415).

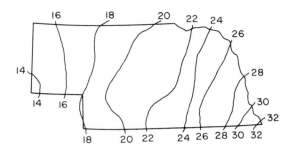

FIGURE 5–11b

Surface of annual precipitation for Nebraska (adapted from Robinson, Fig. 1, p. 415).

fication, the question of the strength and form of relationships is a critical one. Traditionally, geographers have compared patterns on maps, either by placing them side by side or overlaying them upon one another, to examine the degree of correspondence and to make some subjective judgments as to the strengths of the relationships between them. In the initial and exploratory stages of many types of geographic inquiry such methods are still valuable, and map overlays are often an extremely effective way of presenting simple relationships to laymen interested in spatial problems. Geographers also have bulging mental files of spatial patterns, and with their personal information retrieval systems they compare and evaluate maps constantly. But the human eye is not always a very precise assessor of the strengths of spatial relationships, and it can be an extremely misleading instrument. Relationships suspected on theoretical and intuitive grounds, or from an examination of empirical cartographic evidence, must be examined much more rigorously.

If we have enough information to construct maps and tables, it is usually possible to plot the same data upon a graph. For example, we can sample points on the population density and rainfall surfaces in Nebraska (Figs. 5–11a and b). Here we assume that the rainfall and population density vary continuously over the state, tending to be high in the east and low in the west. From a simple comparison of the two maps, we can see that the correspondence may be close, but certainly not perfect. If we take the values of both variables at sample points selected at random we can plot them on a graph to form a *scatter diagram* (Fig. 5–12). The distinct trend confirms our "visual hunch" of a strong relationship between the two.

Some spatial patterns are not continuous, and we may have values of two variables only at specific and quite discontinuous points. The flow of trade along the Ogooué River was an example in which values came from selected points that were not on a continuum. You will remember that when we were considering measurement questions we avoided the problem of fitting a line describing the general trend of the relationship. It is time we thought about it more deeply.

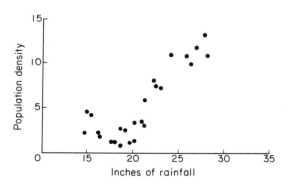

FIGURE 5–12

The scatter diagram of rural population density and precipitation at 25 randomly selected points on Figs. 5–11a and b.

Fitting Lines to Points

Consider, once again, the scatter diagram of trade shipments and populations (Fig. 5–13), and the problem of trying to describe the general trend of the relationship between the two. Our generalization is that an increase in population generally produces a regular increase in the trade shipments, so we want to find a straight line that best fits the points. The crux of the problem is what we mean by "best-fitting line," and this raises in turn the question of the criterion we choose to define such an optimal property. At the outset, let us be quite frank: there is no ultimate, "correct" answer to the general question, What is the line of best fit? There are an infinite number of lines we could choose, each of them describing the relationship in a slightly different way. The very notion of best fit is itself a construct of the human mind in the C-field. Which is the ideal line can never be known with certainty, indeed the question is quite metaphysical, although we may argue in the context of a particular problem that some lines are more plausible and intuitively more satisfying than others.

As you look at the scatter of points in Fig. 5–13, and try to fit a line by eye, think about what you are thinking about. Your brain is superimposing a straight line on the point pattern so that it is as close to all the points as

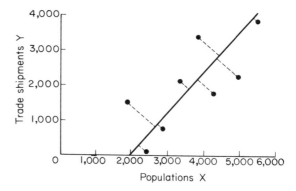

FIGURE 5–14

The line as an axis of rotation, with each point considered as a weight balancing around the line.

possible. This is a good spatial and geographical problem, for we could think of locating a straight road to some villages of equal size, or serving towns with equal populations with a straight gas pipeline. There are many practical problems in geographic space involving the idea of best-fitting lines. A line we could fit would be one around which the points balanced: that is, we could think of each point as an equal weight, so that our line would be an axis of rotation with the distance of each point measured perpendicular to it (Fig. 5–14). The only trouble is that an infinite number of lines can be found which

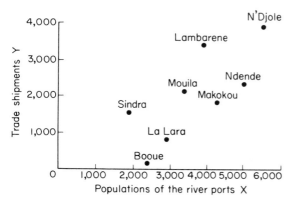

FIGURE 5–13

The scatter diagram relating shipments on the Ogooué River to the populations of the river ports.

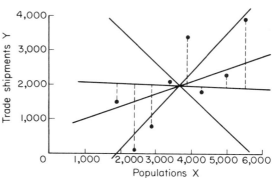

FIGURE 5–15

Other axes of rotation, each fulfilling the criterion of best fit. An infinite number is available.

all conform to such an axis of rotation criterion (Fig. 5–15), so that it hardly represents an unambiguous rule upon which to build further scientific work.

Suppose we simply require that the best-fitting line should be the one which minimizes the sum of all the perpendicular distances to it (Fig. 5–16). We may be able to find such a line, one which will pass directly through two of the points in the scatter diagram But finding a line to meet this criterion is extremely difficult. We can only do it by trial and error, and even with computers doing the trial and error (or iterative) work, the line we end up with may not satisfy our intuitive feeling for "best-fittingness." Usually it seems off-center from the cluster of points, and it fails to go through the mean values of the two variables—a condition which we may feel should be met, again on purely intuitive and aesthetic grounds.

One way out of our difficulty would be to square all the distances before we added them together, trying to find a line that minimized the sum of all the squared distances measured perpendicular to the line (Fig. 5–17). Such a line has very interesting properties, and in one, only slightly ambiguous sense it is the best of all "best-fitting" lines. In the first place, no other line reduces the squared deviations of the points from the line as much, and as it also passes through the mean values of the two variables, it satisfies our intuitive feelings on two counts. Finally, it is very easy to find the slope of such a

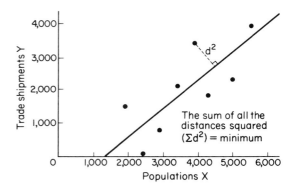

FIGURE 5–17

The line minimizing the sum of all the squared perpendicular distances—known as "the line of least axis."

line, and with the slope and the mean point it is easy to plot it on the scatter diagram.

Unfortunately, the line does have a serious drawback; it is not stable when we change the units of measurement on one or other of the axes. For example, if we plotted the scatter diagram on a thin piece of rubber graph paper and stretched it, the lines from the points to the axis would no longer be perpendicular. In fact, on our stretched graph paper, another line could be found which now minimized the sum of the squared perpendicular distances. A mathematician would say that the line is not invariant under a simple linear transformation. Such a line, called the line of least axis, has seldom been used in geographical research, although it may be more appropriate when variables are mutually dependent upon one another and we wish to describe an overall *structural*, rather than *predictive* relationship between them.

If a relationship is not a two-way affair, and our concern is to predict as best we can the value of one variable when we know the value of another, it can be shown mathematically that the line of least axis does not represent the best-fitting relationship. Our example of trade shipments and populations of river ports might be such a case. Here we feel that the demands for trade goods are measured by the sizes of the populations, and that from these independent values we can predict the size of the shipments

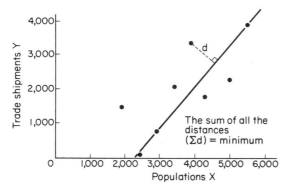

FIGURE 5–16

The line minimizing the sum of the perpendicular distances.

which are dependent upon them. We assume that the relationship is one-way, since it is unlikely that we would be interested in predicting the populations from the shipments. Now if our task is to predict trade goods given the populations, we may wish to concentrate our attention on minimizing the vertical rather than the perpendicular deviations from the line. That is, we want to minimize the deviations of the trade shipments, which are the very things we are trying to predict. Suppose we want to find a line that minimizes the sum of all the distances measured parallel to the shipment Y axis (Fig. 5–18). Once again, we are in a situation where we have no mathematical way of finding it, and so are reduced to trial and error methods. But even if we let a computer do the tiresome iterative work, we have not solved the problem because there are an infinite number of lines to choose from if there are an even number of points in the scatter diagram. If we happen to have an even number of observations (Fig. 5–19), we shall have a range over which our line can move parallel to its original position. Any upward movement towards half the points, decreasing the sum of the distances to them, is exactly balanced by the increasing distances from the lower half. With an odd number of points such a range would not occur, but we can hardly rest our case of best fit upon a criterion that works for odd but not even numbers of observations.

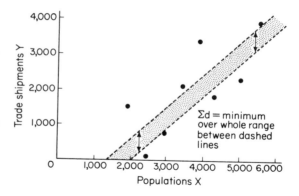

FIGURE 5–19

The range over which a line may move when fitted according to the criterion of minimizing the sum of the vertical distances—an infinite number of solutions to the problem.

Once again, we can get around our problem by squaring all the distances before we add them together (Fig. 5-20), and take as our criterion the minimization of this quantity—the sum of these squares. This is the way you will usually see a line fitted to a set of points, and for obvious reasons it is called the "line of least squares." It has become almost the only method of line fitting in many sciences because of some very useful properties that not only satisfy our intuition about what a best-fitting line should be, but also meet some well proven notions about the accuracy of any predictions we want to make.

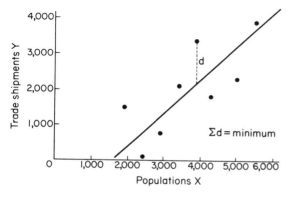

FIGURE 5–18

Fitting a line by minimizing the sum of the vertical distances.

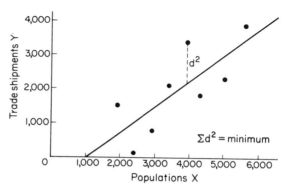

FIGURE 5–20

Fitting a line by minimizing the sum of the squared vertical distances—known as "the line of least squares."

It is also easy to compute lines of least squares, and computational ease is a very practical criterion not to be sneered at.

The Simple Linear Equation

The straight line can be described by the equation:

$$Y = a + bX$$

By convention we measure the variable that depends upon the other along the Y, or vertical axis, and the independent variable along the X, or horizontal axis. In our example, the shipments are the values of Y and the populations the corresponding values of X. Let us consider a general line described by such a linear equation (Fig. 5–21), remembering that an equation of this sort is essentially a balancing or equality statement—one side must exactly equal the other. Substituting $X = 0$ into the equation means: what is the value of Y when X is zero? Clearly when X is zero and multiplies b, Y will take the value a. This gives us a clue as to the general interpretation of a: it is simply the value of our dependent variable Y when the independent variable X is 0. Thus it is the place where our line cuts the Y axis. But what about b? Suppose we let X take the value 1. Now the value of Y will be $(a + b \cdot 1)$ or $(a + b)$. Starting at a on our graph, we go out one unit on the horizontal axis. But such a unit increment in X means the corresponding value of Y goes up by $b \cdot 1$, or simply b. If we went out ten units on X we would go up $b \cdot 10$ units on Y. Obviously b is the slope of the line, telling us the rate at which Y changes with a change in X.

The two values a and b are called *parameters* and define a particular straight line completely. That is, if we know that a line starts at a on the Y axis, and that it goes up at the rate of b, we need no other information to plot the line and write its descriptive equation. Of course, a and b can take the values of any real numbers positive or negative. If a is a negative number, the line cuts the Y axis below the X axis. If b is negative the line will slope down from left to right. Thus positive b values imply direct relationships between variables, while negative slopes imply inverse relationships.

Least Squares and the Minimization of Error Terms

The line we obtain by minimizing the sum of the squared vertical deviations is simply another way of describing the relationship between two variables. But the line has an additional property: any predictions or estimates we make of Y, from plugging particular values of X into the equation, will be the best that can be devised. When we use such a line for predictive purposes we assume that all the errors of measurement are in the Y observations. That is we assume the values of X are free of error, and only the Y observations contain error terms. If we want to be very strict about it, we should write the linear equation derived from the method of least squares as:

$$Y = a + bX + e$$

The equation states that if we put in a value of X, to estimate a corresponding value of Y, we recognize that we may have to add or subtract a small amount (e) to make the equation balance because of measurement and other errors in the Y values. We also assume that the e, or error terms, cancel one another out because they are symmetrically distributed around a mean of zero. What the method of least squares

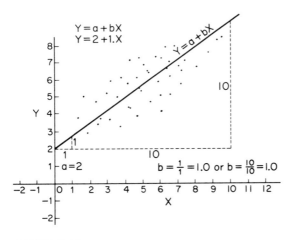

FIGURE 5–21

A linear relationship fitted to a set of observations. The intercept on the Y *axis is* a; *the slope of the line is* b.

does as it fits the line to the points is minimize the effect of error terms on the predicted Y observations, giving us the line that is best for estimating purposes. The assumption about error only being in Y is always incorrect, because we know we can never measure anything with complete precision. Indeed, in the assumption is the unspoken implication that we can manipulate X, setting it to different values, and then observe the corresponding values of Y. But this is an ideal experimental situation approached in the laboratory of the physicist but more difficult to meet in the social and behavioral sciences. In geographic studies there is always some error in both the variables, and ideally we would like to minimize their effects simultaneously upon our estimating line and equation. In other words, we would like to write:

$$Y + e_X = a + bX + e_Y$$

There is only one snag. No one knows how to find the parameters a and b which will minimize the effect of both error terms at the same time, for it remains an unsolved mathematical problem. Nevertheless, the line of least squares is still a most useful standard form of line fitting, and it is the best we can devise when equations and relationships are required for predictive purposes.

Finding the Parameters

So far we have discussed the linear equation in very general terms, giving its parameters the symbols a and b. A further question is: how do we estimate the values of these parameters from a set of actual data to get our best fitting line

TABLE 5–1

Sample Table of N Observations for X and Y Values for Plotting a Scatter Diagram

Observation	Independent Variable (X)	Dependent Variable (Y)
1	X_1	Y_1
2	X_2	Y_2
3	X_3	Y_3
.	.	.
.	.	.
.	.	.
N	X_N	Y_N

and equation? When we plot a scatter diagram we always start with a table of N observations, each observation consisting of a pair of X and Y values (Table 5–1). Now we could consider each observation separately, and link the X and Y values together in the form of a linear expression. Thus:

$$Y_1 = a + bX_1$$
$$Y_2 = a + bX_2$$
$$\vdots$$
$$Y_N = a + bX_N$$

If we add all these together: $\Sigma Y = Na + b\Sigma X$

This is called a normal equation, and we can easily find three of the values in it, namely ΣY, N, and ΣX. The other two, a and b, we do not know yet for they are precisely the parameter values we are trying to discover. Unfortunately, one equation is not terribly helpful when we are trying to calculate two unknown quantities. We obviously need another. Suppose we look at our equations again, and multiply each one right the way through by its own particular X value. Thus our first equation would become:

$$X_1Y_1 = aX_1 + bX_1X_1$$
Which is the same as: $X_1Y_1 = aX_1 + bX_1^2$
While the others are: $X_2Y_2 = aX_2 + bX_2^2$
$$\vdots$$
$$X_NY_N = aX_N + bX_N^2$$

Now, if we add all these together: $\Sigma XY = a\Sigma X + b\Sigma X^2$

In this second normal equation we can find ΣXY by simply multiplying each X in the table of data by its corresponding Y value, and then adding all the XY's together. We have already found ΣX and can easily get ΣX^2. Notice this is the sum of all the X values after they have been squared. ΣX^2 is not the same as $(\Sigma X)^2$. Now we have two equations, so we can solve them simultaneously to get the two unknowns a and b:

$$\Sigma Y = a \cdot N + b \cdot \Sigma X \quad (1)$$
$$\Sigma XY = a \cdot \Sigma X + b\Sigma X^2 \quad (2)$$

Let us take the example of trade shipments and population once again, labeling the shipments our dependent variable Y, and the populations our independent variable X (Table 5–2).

TABLE 5–2

Data for Plotting a Scatter Diagram: Trade Shipments and Population of River Ports in Gabon

River Port	Population (X)	Shipment (Y)
Lambarene	3900	3400
N'Djole	5500	3900
Mouila	3400	2100
Sindra	1900	1500
La Lara	2900	800
Booué	2400	100
Ndende	5000	2300
Makokou	4300	1800

While the numbers are a bit cumbersome, we can find all the values in our normal equations. They are:

$$N = 8$$
$$\Sigma X = 29,300 \qquad \Sigma Y = 15,900$$
$$\Sigma X^2 = 118,290,000 \qquad \Sigma Y^2 = 42,610,000$$
$$\Sigma XY = 66,500,000$$

Now we know:

$$Na + b\Sigma X = \Sigma Y$$
$$a\Sigma X + b\Sigma X^2 = \Sigma XY$$

So substituting in:

$$8a + 29,300b = 15,900 \qquad (1)$$
$$29,300a + 118,290,000b = 66,500,000 \qquad (2)$$

Multiplying (1) by 3662.5, and subtracting it from (2)

$$29,300a + 118,290,000b = 66,500,000 \qquad (2)$$
$$29,300a + 107,311,250b = 58,233,750 \qquad (1)$$

$$\overline{ 10,978,750b = 8,266,250}$$

$$b = \frac{8,266,250}{10,978,750} = 0.7529$$

Substituting b in (1)

$$8a + (0.7529 \times 29,300) = 15,900$$
$$8a + 22,060 = 15,900$$
$$8a = 15,900 - 22,060 = -6,160$$
$$a = \frac{-6160}{8} = -770$$

Solving these, the value of a is -770 while b is 0.753, so we can write:

$$Y = -770 + 0.753X$$

and plot the line on the scatter diagram. Our equation, and its graphical expression in the form of a line, is a mathematical model linking two variables together in an abstract and generalized fashion. When Y is zero in the equation, we get the point where the line crosses the X axis. This is of particular interest since it is an estimate of the population needed to trigger any trade shipments at all—a threshold value that a river port or village must reach before the small trading boats will bother to stop and offload goods. Furthermore, the equation states that generally an increase in population of one person will produce an increase in the value of trade shipments by 0.753 of a cowrie shell.

Measures of Variation and Covariation

How strong is the overall relationship between the two variables, one of which we have described as a simple linear function of the other? First, let us consider the more general question of measuring the strengths of relationships, starting with some graphical examples to enhance our intuition. If points or observations on a scatter diagram all happen to fall exactly on a straight line (Fig. 5–22a), it is obvious that there is a perfect relationship between the two variables. Our Y variable changes or covaries in a perfect linear fashion with the X variable so we could predict one from the other exactly. On the other hand, if there were no relationship between the two, our scatter diagram would indicate no overall trend to the points, but either a completely circular cluster (Fig. 5–22b), or a pattern of points whose general trend is parallel to one of the axes of the graph (Figs. 5–22c and d). Providing we have not foolishly hypothesized some degree of relationship where in fact there is none, most of our examples will lie between the extremes of a perfect relationship and no relationship at all, with scatter diagrams showing elliptical clusters trending to the corners of the graph (Figs. 5–22e and f). Fat "footballs of points" will indicate quite weak relationships, while thin "cigars of points" will tell us we are dealing with strong ones.

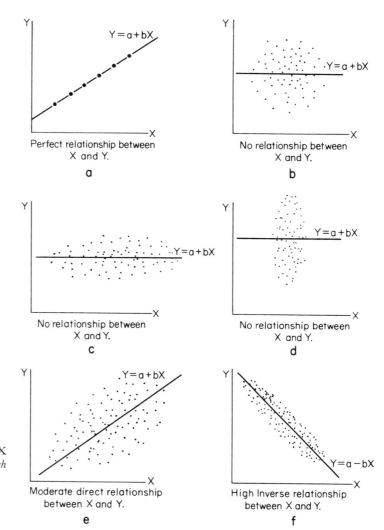

FIGURE 5–22

Scatter diagrams between two variables X *and* Y, *ranging from a perfect through a moderate to no relationship.*

Now our judgment about the strength of a relationship from a visual inspection of a scatter diagram rests upon an intuitive, almost unconscious appraisal of the degree to which one variable covaries with another. We have a certain amount of variation in X, for the values of our X observations do vary rather than being all the same, and some of this variation may account for the fact that our Y observations also vary in turn. That is, when our X values vary up or down the Y values also vary with them. But before we can tackle the task of measuring

the strength of a relationship, or the degree of *co*variation between two things, we must construct three building blocks. Our building blocks will be measures of variation within our two variables considered separately, and a third measure of the covariation, the "changing togetherness" between them.

Three Building Blocks

What sort of measure can we devise for the variation in a set of observations? Let us look again at the populations of the river ports (Table 5–2), which are obviously not all the same, so there must be some variation within the list. We could take the average or mean value, 3662.5, and subtract it from each of the individual values in turn (Table 5–3). The first column of numbers indicates the degree to which each of the individual ports varies from the average of all of them. Perhaps we could simply add up all the individual variations from the average, and use this as a measure of the total amount of variation. The only trouble is that some of our observations like Sindra and Booué lie below the mean, and have minus values, while others like Lambarene and Makokou lie above and have plus values. When we add them all together the sum is zero, so our measure tells us we have no variation at all—which is obviously not true! To get around the difficulty of values above and below the mean canceling out, let us square all of them so that they become positive, and after adding them together use this quantity, the sum of the squared deviations from the mean, as our measure of the *total variation* in X. We usually give this the symbol Σx^2, using the lower case x for our observations whenever they have had the mean value $(\overline{X})$ subtracted from them, rather than capital X when they are in their original or raw form. We can also do exactly

the same for the shipments, and so obtain the *total variation* in Y or Σy^2. For our third building block we require a measure of the covariation between X and Y, a measure of the variation across the two variables. For this we can simply take the two values associated with each observation after they have had their own mean values subtracted from them, or all the little x's and their corresponding y's, multiply them and add them together to get Σxy, or the *total covariation*.

Buckets of Variation and Variable Sponges

What can we do with these three quantities, the total variations Σx^2 and Σy^2, and the covariation Σxy? Consider first of all the total variation in the shipment variable Y that we assume is dependent upon the river port populations X. Surely the total variation in Y, or Σy^2, is precisely what we are trying to explain. We have made the educated guess, or hypothesis, that the reason the values of the shipments change and vary is because the populations also change and vary. Thus Σy^2 is the total variation in the shipments that we want to explain by relating it to variation in population size, or Σx^2. The real question is, *how much* of the total variation in Y is accounted for by its covariation with X? We can think of a bucket

TABLE 5–3

Calculating the Total Variations and Covariation in a Set of Observations

River Port	Population $-\overline{X}_{population}$ (x)	Shipment $-\overline{Y}_{shipment}$ (y)	x^2	y^2	xy
Lambarene	237.5	1412.5	56,406.25	1,995,156.25	335,468.75
N'Djole	1837.5	1912.5	3,376,406.25	3,657,656.25	3,514,218.75
Mouila	−262.5	112.5	68,906.25	12,656.25	−29,531.25
Sindra	−1762.5	−487.5	3,106,406.25	237,656.25	859,218.75
La Lara	−762.5	−1187.5	581,406.25	1,410,156.25	905,468.75
Booué	−1262.5	−1887.5	1,593,906.25	3,562,656.25	2,382,968.75
Ndende	1337.5	312.5	1,788,906.25	97,656.25	417,968.75
Makokou	637.5	−187.5	406,406.25	35,156.25	−119,531.25
	$\Sigma = 0$	$\Sigma = 0$	10,978,750.00	11,008,750.00	8,266,250.00
			Σx^2	Σy^2	Σxy
			Total Variation in X	Total Variation in Y	Total Covariation between X and Y

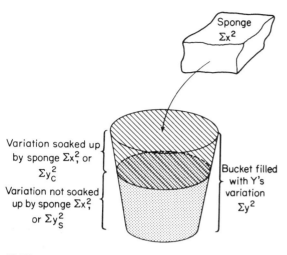

Variation soaked up by sponge Σx^2, or Σy_c^2

Variation not soaked up by sponge Σx^2, or Σy_s^2

Bucket filled with Y's variation Σy^2

FIGURE 5–23

The bucket and sponge analogy; the independent variable X *as a sponge soaking up a part of the total variation in the dependent variable* Y.

filled to the brim with Y's variation (Fig. 5–23), the very thing we want to explain. The variation in X, or Σx^2, is a sponge that we can dip into the bucket to see how much of Y's variation it will soak up. If Σx^2 is a very good sponge it will completely soak up Σy^2, which would be the graphical equivalent of all the points of our scatter diagram lying exactly on a straight line. Or the sponge might be completely impervious to the Σy^2 in the bucket, and come out quite dry. This would be the equivalent of no relationship between the two variables whatsoever, and a formless scatter diagram. Usually our independent variable sponge will soak up a certain proportion of the Σy^2 variation, and we call this amount the *explained variation* or Σy_c^2. But some will remain in the bucket after we pull the sponge out, and will be *unexplained variation* or Σy_s^2. The subscripts c and s are simply conventional symbols for these quantities. Thus we can split up the total variation that we want to explain in Y into two parts: (1) that explained by the covariation with X, and (2) that which remains unexplained. Symbolically:

$$\Sigma y^2 = \Sigma y_c^2 + \Sigma y_s^2$$

The next question is, how we can calculate the explained variation Σy_c^2, the amount of variation soaked up by our sponge? For this we need the slope of the line b which happens to be the ratio of the covariation Σxy to the variation in the independent variable Σx^2. Or,

$$b = \frac{\Sigma xy}{\Sigma x^2}$$

Rather than letting your eyes skim over this expression, examine it carefully because it makes good intuitive sense. As the scatter, or total variation Σx^2 along the X axis of our graph gets bigger and bigger we would expect any best fitting line to become flatter and flatter. Thus the slope b gets smaller as Σx^2 gets bigger relative to the covariation between X and Y, or Σxy. With the slope of the line it is now easy to find the explained variation Σy_c^2 from:

$$\Sigma y_c^2 = b\Sigma xy$$

This, of course, is the same as:

$$\Sigma y_c^2 = \frac{\Sigma xy \cdot \Sigma xy}{\Sigma x^2} = \frac{(\Sigma xy)^2}{\Sigma x^2}$$

So it turns out that the variation in Y explained by its covariation with X is simply the ratio of the squared covariation to the squared variation in the independent variable. If we take the explained variation Σy_c^2, and express it as a decimal percentage of the total variation we originally set out to explain Σy^2, we have an excellent measure of the strength of the overall relationship between the two variables. This would be the equivalent of taking the water soaked up by our sponge, and expressing it as a percentage of all the water originally in our bucket. Giving this percentage the symbol r^2, and calling it the *coefficient of determination*, we have:

$$r^2 = \frac{\Sigma y_c^2}{\Sigma y^2}$$

Obviously this expression will vary from zero, when we have explained nothing and no relationship exists, to one when the explained variation equals the total and we have a perfect, or 100 percent explanation and relationship. In our example (Table 5–3), the coefficient of determination is:

$$\Sigma y^2 = 11,008,750$$

$$\Sigma y_c^2 = \frac{(8,266,250)^2}{10,978,750} = 6,233,404$$

$$r^2 = \frac{6,233,404}{11,008,750}$$

$$= 0.57$$

So 57 percent of the variation in the shipments is accounted for by variations in the size of the river ports. Another measure sometimes used is the square root of r^2, or simply r, called the *coefficient of correlation*. This number always looks better because it is bigger, in this case 0.75. For most purposes it is less readily interpretable, although as we shall see later it has its own uses.

Assigning Residual Variation to the Map

While we have accounted for 57 percent of the variation in trade shipments, we can also say, from a less optimistic point of view, that 43 percent remains to be explained. What are some other things that might help us account for the residual variation? Or, using our analogy of the bucket and sponge, what other sponges

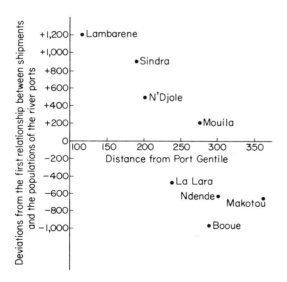

FIGURE 5–24

The residuals, or deviations above and below the line relating shipments to population, plotted against distance from the entrepôt Port-Gentil.

might soak up the variation left in the bucket after the first sponge has done the best it can? If you turn back to Fig. 4–12, you will remember that we plotted upon the map the degree to which each of our river ports deviated from the overall relationship. This is virtually the equivalent of taking the residual variation left over in our bucket and assigning it to various locations on the map. From such a map of residuals it is quite evident that ports above the line receiving more shipments than we might expect from their size all seem closer to Port-Gentil, while those below the line, receiving fewer shipments, are all farther away. If we plot the amount each port deviates above and below our first line against its distance from Port-Gentil (Fig. 5–24), it seems that the new distance sponge is going to account for much of the variation that was still unexplained after considering population size alone.

A Further Look at the Coefficient of Correlation

We gave the correlation coefficient r rather cursory treatment when we introduced it along with the coefficient of determination, although it is a measure of the strength of a relationship often reported in scientific work. Compared to the coefficient of determination it is more difficult to interpret directly, but against this disadvantage it does have a very clear geometrical expression. Geometrical expressions can be extremely helpful when we try to visualize more complicated problems in which many variables are involved in complex relationships with one another. In the next chapter on classification it will be particularly helpful to have such geometrical images, for geographers appear to have unusually visual and strongly spatial minds and find such graphical and geometrical representations particularly helpful.

Suppose we represent any variable, say trade shipments, population, distance, road density, and so on, as a line whose length is unity or one. Such a line is called a unit *vector*. Now suppose we have two variables that are related to some degree. We can represent them as two unit vectors joined at one end with a certain angle

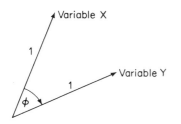

FIGURE 5–25

Vector representation of two variables with a moderate relationship between them.

FIGURE 5–26

Vector representation of two variables with a strong relationship between them.

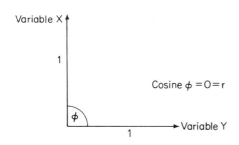

FIGURE 5–27

Vector representation of two variables with no relationship between them.

between them (Fig. 5–25). If the relationship is very high, so that the coefficient of correlation is nearly one, the two vectors will lie almost on top of one another making a very small, almost zero angle between them (Fig. 5–26). On the other hand, if there is hardly any relationship between them, so the coefficient of correlation is almost zero, the vectors will be roughly at right angles to one another (Fig. 5–27). Now the interesting and useful thing about these vectors is that the cosine of an angle between two vectors varies from zero to one just like the coefficient of correlation, so that we can think of two variables as unit vectors with the strength of the relationship between them shown by the cosine of the angle they make with one another. (Remember that in a right-angled triangle the cosine of an angle is the ratio of the length of the adjacent side over the length of the hypoteneuse —see Fig. 5–28.) If we think in terms of our two vectors or variables X and Y (Fig. 5–29), we can see that the cosine of the angle between them is simply the right angled projection of one vector onto the other. But since the vectors are both of unit length the cosine is simply the length of the projection. In this case the length of the projection is 0.8192, so the cosine of the angle is 0.8192/1.0000 or 0.8192. The extreme cases are: (1) when the two vectors lie exactly together, so that the angle is zero, the projection and the coefficient of correlation are both one and the relationship perfect; and (2) when the vectors are at right angles, so that the projection and the coefficient of correlation are both zero and the relationship nonexistent.

This geometrical view of a correlation co-efficient measuring the strength of a relationship

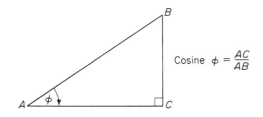

FIGURE 5–28

Cosine of ϕ (phi) as the ratio of the adjacent side over the hypotenuse.

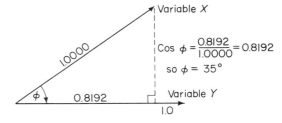

FIGURE 5–29

The right-angled projection of one unit vector upon another as the cosine of the angle between them.

TABLE 5-4

**Data for Three Observations
of Two Variables**

Places Interacting	Distance X	Interaction Y
$A \to B$	3	4
$P \to Q$	4	2
$L \to M$	5	3

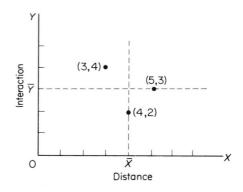

FIGURE 5-30

Scatter diagram of three observations in Table 5-4.

between two variables also falls right out of our previous discussion of variation and covariation. Suppose we take only three observations on two variables X and Y, which might stand for the interaction between pairs of places and the distance between them (Table 5-4). We can plot them as a scatter diagram (Fig. 5-30), and see that there is a general trend down and to the right indicating an inverse relationship. This is what we would expect: interaction gets smaller as distance between the two places gets bigger. We can subtract from the observations their respective means (Table 5-5), and make a scatter diagram of the little x's and y's rather than the raw values or big X's and Y's. This is simply the equivalent of moving the origin of our graph to $\bar{X}$ and $\bar{Y}$ (Fig. 5-31).

With our observations in the form of deviations from their own means, we can calculate b, the slope of the line, and r, the coefficient of correlation, in the conventional way. Thus:

$$\Sigma x^2 = 2 \quad \Sigma y^2 = 2 \quad \Sigma xy = -1$$

$$b = \frac{\Sigma xy}{\Sigma x^2} = -\tfrac{1}{2} \quad \text{and} \quad \Sigma y_c^2 = b\Sigma xy = +\tfrac{1}{2}$$

so that:

$$r^2 = \frac{\Sigma y_c^2}{\Sigma y^2} = \frac{\tfrac{1}{2}}{2} = 0.25$$

By convention, because the relationship is an inverse one, we give the value of r in this case a minus sign, or -0.50. The important

TABLE 5-5

**Subtracting the Means from
the Observations in Table 5-4**

Places Interacting	Distance x	Interaction y
$A \to B$	-1	$+1$
$P \to Q$	0	-1
$L \to M$	$+1$	0

thing to recognize is that everything we have done and shown here in the conventional way can also be represented by vectors. Instead of showing our observations, after they have had their means subtracted from them, as three points in a graphical space of two dimensions, we can show our variables as two points representing the ends of two vectors in an "observation space" of three dimensions (Fig. 5-32). As our vectors are drawn they are not of unit length (Fig. 5-33), but rather $\sqrt{1^2 + 1^2}$ or $\sqrt{2}$. We must shrink the vectors back to unit length without giving them any rotation that might change the angle between them. We can shrink them back to unit length by dividing each coordinate value by the present length of the vector, or $\sqrt{2}$ (Fig. 5-34). Since they

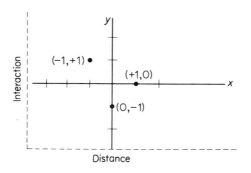

FIGURE 5-31

Scatter diagram of three observations after subtraction of the respective means.

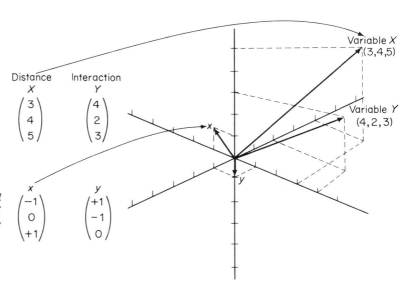

Distance X
$$\begin{pmatrix} 3 \\ 4 \\ 5 \end{pmatrix}$$

Interaction Y
$$\begin{pmatrix} 4 \\ 2 \\ 3 \end{pmatrix}$$

Variable X (3,4,5)

Variable Y (4,2,3)

FIGURE 5–32

Two variables in a three-dimensional observation space. They are X and Y before subtracting the means; x and y afterward.

x
$$\begin{pmatrix} -1 \\ 0 \\ +1 \end{pmatrix}$$

y
$$\begin{pmatrix} +1 \\ -1 \\ 0 \end{pmatrix}$$

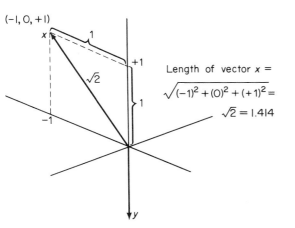

$(-1, 0, +1)$

Length of vector $x =$
$$\sqrt{(-1)^2 + (0)^2 + (+1)^2} = \sqrt{2} = 1.414$$

FIGURE 5–33

One of the vectors x as a hypotenuse of a triangle. Its length is the square root of the sum of the squares on the other two sides.

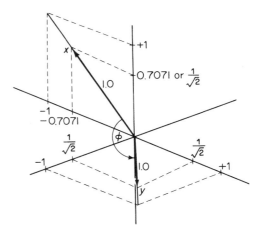

0.7071 or $\frac{1}{\sqrt{2}}$

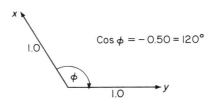

$\text{Cos } \phi = -0.50 = 120°$

FIGURE 5–34

Shrinking the vectors back to unit length.

133

now are both of length one, the cosine of the angle between them is simply the cross-product of the coordinate values, or:

$$\cos \phi = \frac{-1}{\sqrt{2} \cdot \sqrt{2}} = -\frac{1}{2} = -0.50 = r$$

Thus we have ended up with exactly the same measure as the coefficient of correlation. Of course, in trying to give you the geometrical image we are confined by the two dimensions of this book. The maximum number of dimensions we can portray with a perspective drawing is three, so we are limited to two vectors defined by three observations. Our Gabon example, with eight observations, would require a space of eight dimensions in which the two shipment and population vectors would be located. This is difficult to visualize, but the geometric perspective we gain from simple examples in spaces of two and three dimensions will be of considerable conceptual help later on.

Residual Analysis

In our most optimistic and sanguine moments we would hardly expect the world in which man orders his spatial and temporal activity to be described satisfactorily by simple linear relationships between pairs of variables. Even our example of trade flows indicated that shipments were not simply a function of population, so that:

$$S = f(P) \quad \text{or} \quad S = a + bP$$

but that distance from the point of shipment was also of explanatory value, so that:

$$S = f(P, D)$$

We might have suspected that distance had a frictional effect upon movement, simply on grounds of common sense and our own personal experience of effort required to overcome distance. Nevertheless, the way we uncovered this additional variable was by discovering spatial regularities in the residual, or "left over" variation from our first relationship to population. Examining the anomalies or "misfits" from overall trends and generalizations is a very old method in science, and one commented upon extensively by the philosopher John Stuart Mill in his "Method of Residues."

Like many powerful ideas, residual analysis is very simple and obvious once it has been pointed out. In essence, we first try to establish generalizations about patterns, processes, and relationships, knowing full well that anything as complex as human spatial behavior is seldom going to be accounted for completely by very simple models and statements. However, having established overall trends and relationships of some explanatory value, we are in a position to examine the exceptions to the generalizations or the "other things being equal" statements that we have postulated. Sometimes we can only account for the exceptions by a series of reasons unique to each case or observation. When this happens scientific inquiry with some generalizing and theoretical significance must stop, and we must be content for the time being with a certain, perhaps unsatisfactory degree of general explanation, ascribing the variation we cannot account for to uniqueness and error of measurement. But we must also remember that a scientific explanation is never a closed and final one that becomes encrusted with authority and elevated to dogma. By its very nature science is always open. Explanations are provisional, and the residuals with no pattern or discernable structure for one generation may become the source of further fruitful inquiries for the next.

SPATIAL TRENDS AND REGULARITIES IN A REGION

The simplest geographic relationships involving more than two variables are evident whenever we try to establish the areal regularity or general trend of some observations in a region. What degree of spatial regularity, for example, was there to the dates of pioneer settlement in Pennsylvania, a state notorious for its rough, barrier topography? Was there a sufficiently high degree of regularity so that if we know the midpoint location of one of the 1500 townships we can predict the time of first settlement? In other words, we are asking about the strength of the relationship described by the function:

$$T = f(X, Y)$$

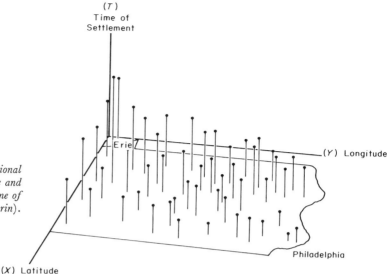

FIGURE 5–35

Pennsylvania lying in three-dimensional space, with two spatial axes (latitude and longitude), and one temporal axis (time of township settlement) (adapted from Florin).

or Time of Settlement $= f$(Latitude, Longitude)

Let us think of Pennsylvania lying in a space of three dimensions, with each township represented by a point located by coordinate values on the spatial X and Y axes and the temporal axis T (Fig. 5–35). The simplest description of the trend would be a plane that best fits the points, and this raises once again the "best fit" question. We have variation in the time of settlement (Pennsylvania was not settled instantaneously), so we can think of a bucket of time variation that we want to relate to two locational sponges, X and Y. If the sponges are dipped into the bucket together, they will account for about 63 percent of the variation in settlement time with a plane that trends upward from southeast to northwest (Fig. 5–36). This is equivalent to a coefficient of determination of 0.63, or a correlation coefficient of 0.79. Measuring the locational coordinates on a grid with an origin in northwest Pennsylvania, the plane of best fit has the equation:

$$T = 1855 - 0.0004(X) - .0001(Y)$$

which implies an intercept of 1855 near Erie, Pennsylvania, diagonally across from the Philadelphia starting point, and a drop in the north-south direction four times as severe as in

the east-west with every change in a unit distance. Thus, making one variable a linear function of two others is simply a matter of adding another term onto the equation. However, we can also specify a slightly more complicated function, allowing our plane of best fit to warp to see if such freedom to bend can give a much better fit. A warped plane can describe a *quadratic* relationship between the variables. Thus:

$$T = 1788 + .0003 \,(\text{Lat}) + .0002 \,(\text{Long})$$
$$- .0007 \,(\text{Lat})^2 - .0015 \,(\text{Lat} \cdot \text{Long})$$
$$- .0014 \,(\text{Long})^2$$

Note that both the latitude and longitude variables appear in squared, or quadratic, form together with a (Latitude·Longitude) or interaction term. To get the plane of settlement to warp, we have had to use three more sponges. However, with five spatial sponges at our command we can now soak up 77 percent of all the time variation in the bucket, ($r^2 = 0.77$), and we now have a slightly curved "settlement surface" over Pennsylvania (Fig. 5–37). It seems that despite such rugged topography, and the unique character of the land and people, the pioneer movements were extremely regular, although the time waves of settlement were virtually parallel to the strong

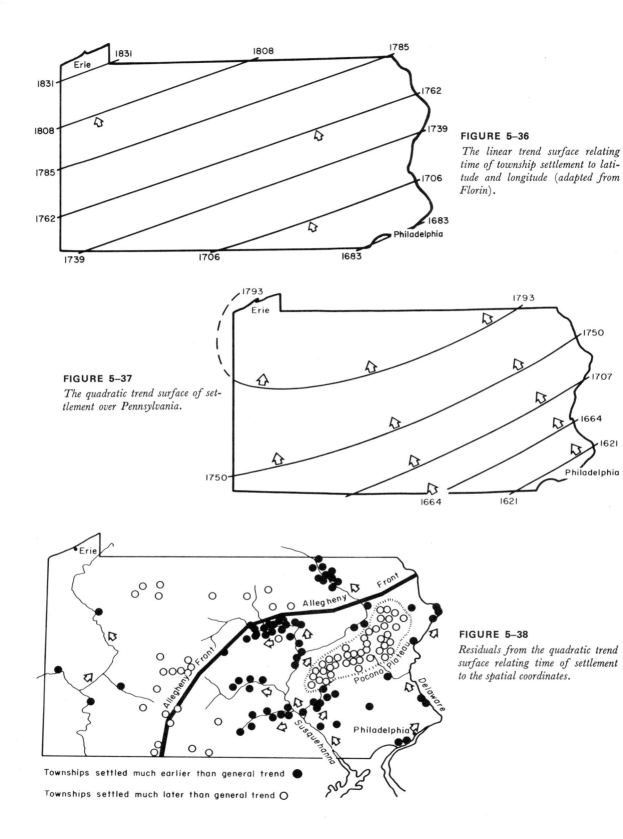

FIGURE 5–36

The linear trend surface relating time of township settlement to latitude and longitude (adapted from Florin).

FIGURE 5–37

The quadratic trend surface of settlement over Pennsylvania.

FIGURE 5–38

Residuals from the quadratic trend surface relating time of settlement to the spatial coordinates.

Townships settled much earlier than general trend ●

Townships settled much later than general trend ○

.lignment of the ridge and valley section that
lices diagonally across the area.

Despite our reasonable success, we must not
orget that we still have 23 percent of the time
·ariation yet to account for. The extreme anom-
.lies, the very large residual distances of the
ownship points above and below our curved
ime and space plane, produce a distinctive
·attern on the map (Fig. 5–38). Nearly all the
ow exceptions, or townships that were settled
·efore the overall trend, lie on the Susquehanna
.nd Delaware river systems. Clearly the his-
orical clichés about river transport and water
aps have some merit, but these are highlighted
·nly after an overall trend accounting for nearly
0 percent of the variation has been slipped out.
·imilarly, the late anomalies lie in a distinct
luster on the rugged and less fertile Pocono
·lateau and west of the most famous of barriers
—the Allegheny Front.

·TAYING WITH LINEAR RELATIONSHIPS
OR ESTIMATING PURPOSES

Sometimes we know that the relationships
·etween variables are not linear ones. For
xample, the effect of distance upon movement
; seldom described by a simple, straight-line
·elationship, but by a curvilinear one repre-
·ented by a line falling quickly at first, and
·hen more slowly as distance increases (Fig.
·–39a). Travels, communications, migrations,
·nd urban influences seldom seem to conform
· the simple linear hypothesis. Nevertheless,
near relationships are so easy to handle, and
·he procedures for estimating the parameters
·re so simple, that we often try to straighten out
· nonlinear relationship if we possibly can. In
·ig. 5–39a, showing the relationship between
·teraction and distance, we could fit a straight
·ne to the obviously curved scatter of points,
·ut in the process of establishing a simple,
·verse relationship we would lose much of the
·formation that the scatter diagram contains.

One way of straightening the scatter would
·e to use the logarithms of the interactions and
·istances instead of the numbers themselves
·Fig. 5–39b). By using the logarithms, small
·teractions and distances—say between ten and

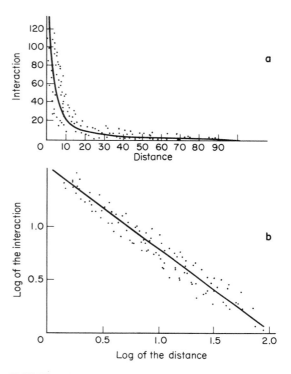

FIGURE 5–39

*An exponential relationship between interaction and distance,
with a transformation to logarithms for the purpose of linear
estimation of parameters.*

a hundred—would be pulled back to the ones,
while big values in the thousands would be
radically transformed back to the threes (see
footnote to Chapter 3, p. 77). The word
transformed is the key here, for we are under-
taking a logarithmic transformation of the data,
so that:

Interaction (or X_1) = Log interaction
$$\text{(or Log } X_1)$$
Distance (or X_2) = Log distance (or Log X_2)

Such a transformation straightens out the
curved scatter diagram and allows us to fit
an equation with a simple linear form.

Let us consider as an actual example the
question of migration to Guatemala City
(Fig. 5–40). Like most underdeveloped areas,
Guatemala is dominated by its capital city,
which forms a major attraction for the people

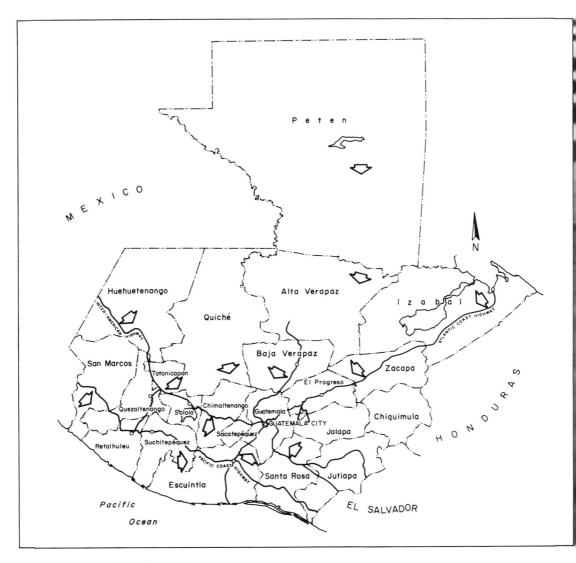

FIGURE 5–40

Guatemala City and the major routeways for migration (adapted from Thomas).

of the country. The hundreds of *municipios* that make up the country do not all send equal numbers of migrants, so we have, once again, a bucketful of variation to account for, this time variation in migration. Suppose we relate the logarithm of the number of migrants to the logarithm of the population in the *municipios* on the commonsense grounds that heavily populated *municipios* will probably send more migrants to the capital than those with sparse populations. If we dip the population sponge into the bucket of migrant variation, we can soak up about 49 percent with this single variable, and express the relationship in the form:

$$Y = 0.0336 + 0.2073X_1$$

Log of number migrants

$= a + b_1$ {log of population of the generating *municipio*} his would be a straight line with a positive lope expressing the expected direct relationhip. What happens, though, when we take the ariation we have left in the bucket after withrawing our population sponge (51 percent), nd use it to construct a map of residuals (Fig. -41)? Mapping only the highest and lowest fth of the residuals to simplify a very complex 1ap, it is clear that most of the large positive esiduals are close to Guatemala City, while enerally the very negative ones are far away.)bviously migration to the capital is a function ot only of the size of the generating *municipio* ut also of distance from Guatemala City s well. Thus we can take a second sponge, the og of the distance, and dip it in the bucket of esiduals to see what ability it has to account or the still unexplained variation. If we do 1is, we find that 60 percent of the variation in 1igration can now be explained by population nd distance together. With three variables e would have a three-dimensional scatter iagram, with a best fitting plane whose equaon is:

$$Y = 0.8811 + 0.1649X_1 - 0.4934X_2$$

og of number migrants

= $a + b_1$ {log of population generating *municipio*} − b_2 {log of distance from Guatemala City}

emember that our variables are really the ogarithms of the values. When we add logathms we are really multiplying the original alues, and when a logarithm is multiplied by constant it is the same as raising the number to 1at power. In other words, our equation could e restated as:

$$\text{Migrants} = 7.605 \cdot \frac{(\text{Population})^{.1649}}{(\text{Distance})^{.4934}}$$

hus our transformation allows us to estimate 1e exponents of the well known potential 1odel formulation in which interaction is irectly proportional to the size of population nd inversely proportional to distance. With the ffect of varying population distribution taken 1to account, the distance exponent (.4934) an be thought of as a measure of the friction f distance upon migration flows. If this value 1creases, the effect of distance in the expression

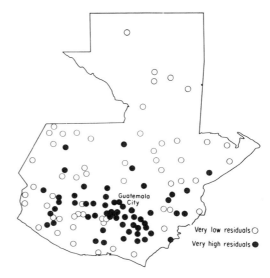

FIGURE 5–41

Residuals from the relationship: Migration = f (population of municipio), with only the highest and lowest fifths recorded (adapted from Thomas).

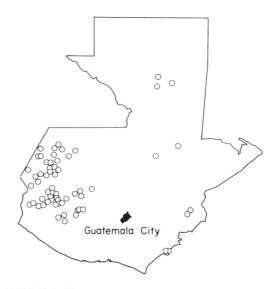

FIGURE 5–42

Lowest fifth of residuals from the relationship Migration = f (population of municipio, distance from Guatemala City) (adapted from Thomas).

increases and migration is lowered accordingly. In this particular case the migration from a *municipio* appears inversely proportional to approximately the square root (.5000) of the distance from Guatemala City.

Once again we can take the unexplained variation of 40 percent and this time construct two maps of residuals, plotting the lowest and highest fifths separately to clarify two points. First, the lowest residuals (Fig. 5–42) are nearly all in the westernmost part of the country. If we compare this map to another showing the major areas of Indian population (Fig. 5–43), it would seem that this area, quite distinct from the more *ladino* or latinized parts of the country, produces a clear cultural barrier to migration. Second, there is a remarkable correspondence of the highest fifth of the residuals (Fig. 5–44) with *municipios* containing, or adjacent to a *cabacera*, a provincial capital. It would seem that there is some evidence for a steppingstone effect to the pattern of migration in Guatemala, with people often moving first to the provincial towns from the countryside, and only then to the capital city. If migration is basically a response to information flows, these

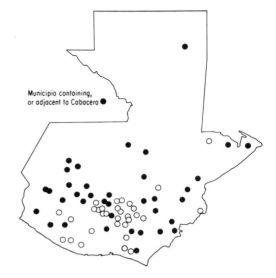

Municipio containing, or adjacent to Cabacera ●

FIGURE 5–44

Highest fifth of residuals from the relationship Migration = f (population of Municipio, distance from Guatemala City) High residuals containing, or adjacent to, a cabacera shown in black (adapted from Thomas).

would appear to be mainly between the urban nodes of the country. We can measure the sheer presence (1) or absence (0) of a *cabacera* in a *municipio* on a binary scale, and when this third variable is included our explained variation increases to 66 percent. In this final expression our model relates migration to three variables simultaneously:

$$M = f(P, D, C)$$

which becomes:

$$Y = 1.0009 + 0.1211X_1 - 0.5266X_2 + 0.5338X_3$$

Log of number migrants
$= a + b_1$ {log of population of the generating *municipio*} $- b_2$ {log of distance from Guatemala City} $+ b_3$ {presence of *cabacera* in the *municipio*}

Of course, we cannot be content with 34 percent remaining unexplained, but migration patterns are extremely difficult things to account for with simple relationships of this sort. They often appear to develop as the result of strong flows of information feeding back to the

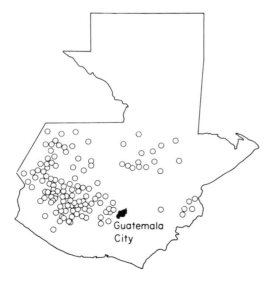

Guatemala City

FIGURE 5–43

Municipios *containing major portions of Indian population* (*adapted from Thomas*).

source areas, so that the earliest patterns formed from mere trickles of migrants become more and more accentuated through time. With our very simple linear model, such feedback, or *recursive* features cannot be taken into account. Moreover, very difficult relationships of this sort quickly become impossible to handle with the mathematical tools presently at our command. In such cases, geographers turn to simulation models, in which complex, nonlinear relationships with recursive features are specified and solutions estimated by long trial and error runs on a computer.

EXPLORING WITH LINEAR RELATIONSHIPS

Although linear models may sometimes appear inappropriate, and even naive, their very simplicity makes them very useful tools for exploratory probes into new areas of inquiry where patterns are suspected upon intuitive grounds. One of the most profound examples of spatial pattern-seeking comes from the work of the animal behaviorist Glenn McBride. Starting from a hunch that patterns of dominance in chickens might have crucial implications for their spatial behavior, he sat for hours above a chicken house taking hundreds of photographs at random. He knew from previous periods of observation the dominance relationships between the chickens. Every photograph was analyzed by drawing direction lines through the heads of the chickens (Fig. 5–45), and then joining the heads of a dominant and subordinate chicken with a third distance line (X_1).

Careful measurements were made of the angles Y and X_2, and the distance X_1. Using hundreds of observations, the relationships between the variables were explored using a simple model of the form:

$$Y = f(X_1, X_2)$$

While the relationships were probably not linear, a simple linear model was deliberately chosen for an initial exploration of the signs of the parameters controlling the relationship. The signs turned out to be:

$$Y = a + b_1 \cdot X_1 - b_2 \cdot X_2$$

and in other experiments, with different sorts of animals, no exception has ever been found. The signs mean that as the distance X_1 decreases, the head angle of the subordinate chicken Y also declines so that he ends up facing the dominant bird. Conversely, an increase in the head angle of the dominant chicken X_2 immediate produces the opposite movement in his subordinate.

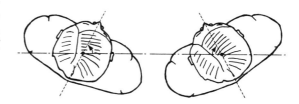

A permissible angle of conversation

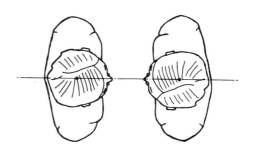

A forbidden angle of conversation

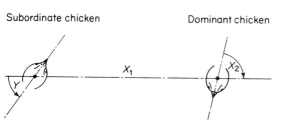

Subordinate chicken   Dominant chicken

FIGURE 5–45

A man's eye view of birds: The measurements of angles and distance made by McBride to investigate the spatial movements of chickens.

FIGURE 5–46

A bird's eye view of men: Angles and distances between human beings in conversation.

This apparently simple analysis with linear functions is in the great tradition of scientific discovery for its implications are profound. What formerly appeared to be a random, patternless movement of chickens in space turns out to be a tightly ordered and structured movement governed by spatial relationships that are highly predictable. Now that a man has discovered pattern where others only saw chaos, the dynamic spatial arrangements formed by animals are seen to have all the formality and predictability of a ballet. The implications for human behavior have yet to be explored, although we all have experiences with angles and distances daily. As you get closer to someone you are talking to you increase the angle you make with them (Fig. 5–46). When engaged in conversation with someone a meter away, it is perfectly permissible to face him; but it is forbidden when he is only thirty centimeters away. There are exceptions, of course, as every man and woman knows.

MORE COMPLICATED RELATIONSHIPS

So far, with the exception of the quadratic terms that bent our flat trend surface, and the logarithmic transformation that let us estimate parameter values in an exponential migration model, we have only considered straightforward linear relationships. These are simple relationships, and we have just seen that simplicity has considerable merit as we try to unscramble and structure the complex spatial world around us. Linear relationships are also conservative, so that in fitting straight lines to data we never overestimate the strength of a relationship. Rather we lose information when we fit a linear function to a nonlinear relation, and therefore underestimate its actual strength. In this way we chose a conservative, even timorous approach to many of our problems, although all too seldom do we think about the implications of this fact explicitly. This is a pity, because no matter what area of geographical inquiry we take, linear relationships can seldom be defended on conceptual grounds. We live in a nonlinear world, and our linear structures, even when they are carefully interpreted and

hedged about with warnings, only tend to be rough, first approximations.

To ask, What nonlinear functions are there? is absurd, since literally an infinite number can be devised. We shall consider only a few, simply to emphasize that careful argument should be marshaled for the form of a particular function whenever possible.

Learning about the Local Area

Suppose we consider how people learn about the local geography when they come into a new area. We might ask a group of university students to write down all the street names they can remember, and then plot the numbers of streets against the time they have been in residence (Fig. 5–47). If we simply fit a linear function (Line *A*), we not only get a poor fit, but it is very difficult to argue that students learn about street names, or other sorts of important spatial information, at a constant rate throughout their college careers. It would seem more likely that they need a certain amount of local information quickly to get around, but then their needs taper off. The simplest curvilinear function is probably the quadratic (Line *B*), in which we simply add a squared term to the linear function. In this case the negative parameter multiplying the squared term tells us that the line will slope up but at an ever-decreasing rate. The only trouble is that a time will come when the squared term gets bigger than the linear term, and the function begins to turn down. While we could interpret this as an example of professors becoming increasingly absentminded after many terms of university residence, it would be kinder to reject the function as the relevant one describing the actual relationship in a plausible fashion.

Instead of calculating in a frantic hit and miss fashion, let us think for a moment what we require of a function that relates local information to time in $I = f(T)$. It seems likely that learning takes place very quickly when students first arrive and require a few basic names to find their way around the areas between their residences, bookstores, lecture halls, and restaurants. But after a while there is no need to remember more names because they are seldom

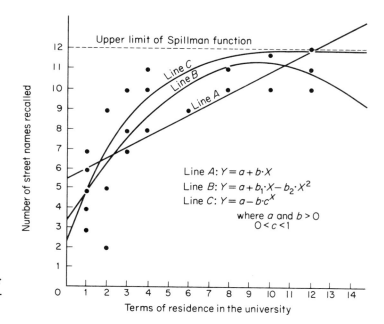

Upper limit of Spillman function

Line C
Line B
Line A

Line A: $Y = a + b \cdot X$
Line B: $Y = a + b_1 \cdot X - b_2 \cdot X^2$
Line C: $Y = a - b \cdot c^X$

where a and $b > 0$
$0 < c < 1$

Number of street names recalled

Terms of residence in the university

FIGURE 5–47

Local space learning as a function of time: the linear, quadratic, and Spillman functions fitted to sample data.

used as reference lines in the local space. From inspection the upper bound appears to be roughly twelve, so our function should level off around this limit.

A function with the properties we require is called the Spillman function, and it has the form:

$$Y = a - b \cdot c^X \qquad a \text{ and } b > 0$$
$$0 < c < 1$$

For this function we have three parameters a, b, and c linking the number of street names Y to the terms of residence X. In this case, our independent variable appears as the exponent X, and if we think about the values of Y when X changes we get some clue about the interpretation of the parameters. Clearly when X is zero c becomes 1 and Y takes on the value $(a - b)$. This then is the intercept on the Y axis, similar to the a value in our now-familiar linear equation. But as X gets bigger, raising the decimal value c to ever greater powers, then $b \cdot c$ will get smaller and smaller. Thus c is the rate at which Y changes with X, while a is the largest value that Y can ever take—and strictly speaking, even this only happens when X is infinite. The problem is, how can we

estimate the values of a, b, and c in this nonlinear exponential function? We know how to find the parameters of a simple linear relationship, so if we can squeeze the Spillman function into a linear form we can use our familiar least-squares method. A little algebra shows that:

$$Y = a - b \cdot c^X$$
$$a - Y = b \cdot c^X$$
$$\text{Log } (a - Y) = \text{Log } b + X \cdot \text{Log } c$$
$$\text{Log } (a - Y) = \text{Log } b + \text{Log } c \cdot X$$

Notice this is in the form:

$$Y = a + bX$$

We have found a linear form for the exponential function by having $\text{Log } (a - Y)$ as our dependent variable, and X as our independent one. Now you will remember that we can solve for the linear parameters, a and b, with two simultaneous equations. Unfortunately we have three unknowns here, so we must set one of them before we start our calculations as sensibly and as plausibly as we can. Setting a, the upper bound to twelve in our example, we can then estimate b and c (Line C), so our function is:

$$Y = 12 - 10 \, (.7)^X$$

The intercept when X is zero is two, implying that students generally have a pair of local coordinates when they arrive—perhaps the crossroads where they will live.

The Diffusion of Radio and Television

One of the most important and common nonlinear functions in geography is the logistic function, an S-shaped or sigmoid curve that characterizes many diffusion processes (Fig. 5–48). It rises slowly at first as innovations begin to diffuse through an area, then more quickly, and finally slows down once again as it flattens out to an upper bound of saturation. It has the less familiar equation:

$$P = \frac{U}{1 + e^{(a-bT)}}$$

an expression relating P, the proportion of people accepting a new innovation, to T, the time since the process of innovation diffusion started. The two variables are linked by three parameters; U the upper limit, b the rate at which P changes with T, and a, which determines the value of P when T is zero. The symbol e is simply a mathematical constant with a value 2.7183, and forms the base for natural or Napierian logarithms, in just the same way that ten forms the base for the common logarithms we used before.*

Before we worry about estimating the parameters, let us consider a few values of T and think about corresponding values of P. When T is zero, e is simply raised to the power a, and if this is quite large the denominator $(1 + e^{a})$ will be big relative to U, the upper limit, and this means in turn that P will be very small. This is what we require at time zero, the start of a diffusion process when only a very few people have accepted an innovation. But notice what happens as T gets bigger. Gradually the exponent term $(a - bT)$ gets smaller and smaller,

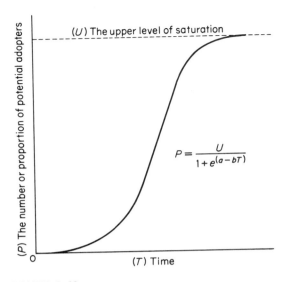

FIGURE 5–48

The logistic curve of acceptance.

decreasing in turn the whole denominator, which increases P more and more quickly. Eventually T will get so large that the exponent term $(a - bT)$ becomes negative, implying that we are now taking reciprocals of powers of e, rather than raising e to powers. Thus as T gets very large, and much time passes for the innovations to diffuse, the e term goes to zero, the denominator gets closer and closer to one, and P converges to the upper limit U.

To estimate the parameters we must set one arbitrarily, say the upper bound U, and manipulate the exponential expression into a linear form. Once again, a little algebra shows that:

$$P = \frac{U}{1 + e^{(a-bT)}}$$
$$P(1 + e^{(a-bT)}) = U$$
$$P + Pe^{(a-bT)} = U$$
$$Pe^{(a-bT)} = U - P$$
$$e^{(a-bT)} = \left(\frac{U-P}{P}\right)$$
$$\mathrm{Log}_e\left(\frac{U-P}{P}\right) = a + bT$$

which is the same form as:

$$Y = a + bX$$

By treating the whole left hand side of the equa-

*For example, the natural logarithm of 5 is the power to which e must be raised so that it equals five. In other words, in $e^{x} = 5$, x is the natural logarithm, or 1.6094. Logarithms to the base e can be found from the usual tables of common logarithms to the base 10 in the following way. If $e^{x} = 5$, then $X \cdot \log_{10} 2.7183 = \log_{10} 5.0000$, or $x = \log_{10} 5.0000 / \log_{10} 2.7183 = 0.6990 / 0.4343 = 1.6094$

tion as our dependent variable in a simple linear regression, we can estimate the remaining parameters a and b.

In the United States, radio and television broadcasting were two innovations with dramatic and far-reaching geographical consequences (Fig. 5–49). Making some assumptions about the upper bounds of saturation, we can plot the two logistic functions and compare both their graphical forms and parameter values. Radio broadcasting started in the early 1920s and diffused quite slowly at first. Television started two decades later and although it was interrupted by the Second World War it increased dramatically during the 1950s. However, it appeared to be slowing down in the late 1960s, and the upper bound is lower than for the earlier innovation. This is because the costs of installing and running a television station are much higher than for radio, and questions of channel availability and control by the Federal Communications Commission also arise. A comparison of the two functions is revealing:

Radio	*Television*
$P = \dfrac{5000}{1 + e^{(5.6091 - 0.1618T)}}$	$P = \dfrac{700}{1 + e^{(6.3598 - 0.3232T)}}$

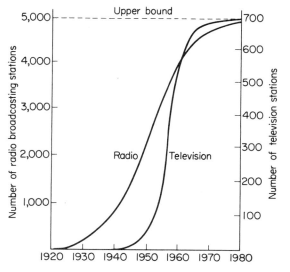

FIGURE 5–49

Innovation curves for the acceptance of radio and television broadcasting in the United States.

The b parameter controlling the rate of increase for television (0.3232) is twice that of radio (0.1618), indicative of the way the diffusion of large-scale innovations have changed in the past half century in the United States.

COMBINING SPACE AND TIME IN A NONLINEAR RELATIONSHIP

As our final example of geographic relationships, let us consider a quite complex integration of space and time variables in a diffusion problem. Suppose we have a number of regions, one of which contains a center of technological innovation (Fig. 5–50). Within the innovative Region A we can relate the percentage of potential adopters P to the time T, with the logistic function:

$$P = \frac{1}{1 + e^{(a - bT)}}$$

Here 1.0 is the upper limit because we are considering decimal percentage values rather than absolute numbers. Other regions, such as B and C, surround the area where the innovation started, and eventually they also begin to adopt the innovation. We can describe each of their individual adoption curves by separate logistic functions. Note, however, the way the parameter values for curves in regions farther away from the innovation center will differ from those of the logistic function characterizing the course of diffusion within Region A itself.

Remember that the parameter a locates the start of the adoption curve on the time scale, because it determines the value of P when T is zero. Then regions like C, far away from the center of innovation, will start the adoption process later, so their a values will be bigger than the a value of the region starting the innovation wave. These larger values simply slide the logistic curves to the right. What about the parameter b? Remembering that this value controls the steepness of the curve, and assuming that economic innovations get adopted first in places like Region A where they are the most profitable, we might expect the b values to decline away from the innovation pole. We have shown this schematically with the arrows becoming less and less steep (Fig. 5–50).

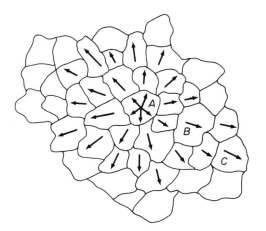

or:

$$P = \frac{1}{1 + e^{(a_1 + a_2D + a_3D^2 + b_1T + b_2DT + b_3D^2T)}}$$

In linear form this would be:

$$\mathrm{Log}_e\left(\frac{1-P}{P}\right) = a_1 + a_2D + a_3D^2 + b_1T$$
$$+ b_2DT + b_3D^2T$$

Notice that we have six parameters to estimate. Fortunately we also have six variables, so we can obtain six simultaneous equations to estimate them. Notice that some of these variables are "pure" like P, D, D^2, and T, while others are derived or "interaction" variables

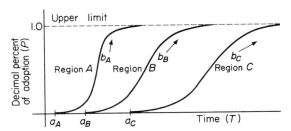

FIGURE 5–50

The innovation curves for the three regions A, B, *and* C, *indicating the way in which the parameters* a (*controlling the starting point*) *and* b (*determining the rate of increase of adoption*) *are functions of distance from the innovating region* A.

What we have implied in our discussion above is that the parameters a and b are themselves functions of distance from the growth pole. Suppose, instead of considering the diffusion curves in each separate region, we took a logistic function for the country as a whole. We could make the a and b parameters of this curve quadratic functions of distance so that:

$$a = a_1 + a_2D + a_3D^2$$
$$b = b_1 + b_2D + b_3D^2$$

Now we can substitute these back into the logistic function for the whole country, so that the percentage of potential adopters P is made a nonlinear, logistic function of both distance and time. This means that:

$$P = f(D, T)$$

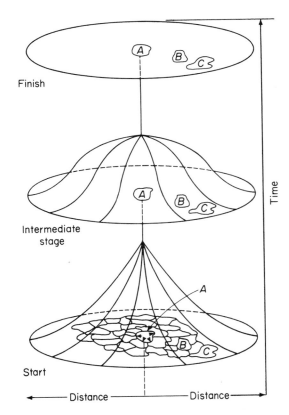

FIGURE 5–51

A schematic representation of the logistic-innovation function integrating both space and time effects. The innovation starts in the center Region A; *then slows up as peripheral regions* (B *and* C) *catch up; until finally saturation is reached at the top.*

between distance and time like DT and DT^2.

The graphical expression of such a complicated model of spatial diffusion is relatively simple (Fig. 5–51). Before the innovation appears, we can think of a flat disc overlaying the country because the percentage of potential adopters is zero everywhere. As the process starts, and we move up the diagram along the vertical time scale, the proportion of potential adopters rises first in the center immediately around Region A. Notice that the innovation has yet to seep out to the edges. Later the proportions of adopters in regions like B farther from A rise, and even the peripheral areas like C have some adoption going on. Finally, at the top of the diagram, the proportion of potential adopters is the same everywhere, or 100 percent, so we have a flat disc once again. As an analogy, we can think of the logistic function as a flapping sea gull's wing, or a rope tied to a tree which a child twitches to make a pulse run through it.

Consider an actual technological innovation like the tractor, which had dramatic consequences for land use patterns in the United States (Fig. 5–52). A major area of innovation was North Dakota, where flat land and a grain economy were ideal for mechanical cultivation. If we relate the proportion of farms in a state with tractors to time and distance from North Dakota, the logistic function becomes:

$$P = \frac{1}{1 + e^{(1.85 + 0.0000034D^2 - 0.114T)}}$$

Notice that four of the terms have disappeared, since their parameters did not differ significantly from zero. The fit is extraordinarily good, for we can account for 92 percent of the variation in the proportion of adopters in each state if we know only the square of the distance from North Dakota and the number of years that have passed since the state started the tractor adoption process. In essence the geographer Casetti said: tell me the time and the place, and I will predict (or rather postdict) almost exactly the proportion of farmers using this innovation.

Suggestions for Further Reading

COLE, JOHN P., and CUCHLAINE A. M. KING. *Quantitative Geography*. New York: John Wiley & Sons, Inc., 1968.

FIGURE 5–52

The diffusion of tractors in the United States from the innovation center of North Dakota. The logistic model with time and space variables accounts for 92 percent of the variation in the proportions of adoptions.

$$P = \frac{1}{1 + e^{(1.85 + 0.0000034D^2 - 0.114T)}}$$

Proportion of
adopters

GOULD, PETER. "Methodological Developments since the Fifties," in *Progress in Geography*, Chap. 1, eds. C. Board et al. London: Arnold Publishers, Ltd., 1969, pp. 1–49.

KING, LESLIE J. *Statistical Analysis in Geography.* Englewood Cliffs, N. J.: Prentice-Hall, Inc., 1969.

POLANYI, MICHAEL. "Experience and the Perception of Pattern," in *Modeling of Mind*, eds. K. M. Sayre and F. J. Crosson. Notre Dame, Ind.: University of Notre Dame Press, 1963.

YEATES, MAURICE. *An Introducton to Quantitative Analysis in Economic Geography.* New York: McGraw-Hill Book Company, 1968.

Works Cited or Mentioned

CASETTI, E., and R. K. SEMPLE. Concerning the Testing of Spatial Diffusion Hypotheses," *Geographical Analysis*, I: 3 (July 1969), 254–59.

————. *A Method for the Stepwise Separation of Spatial Trends.* Michigan Inter-University Community of Mathematical Geographers, Discussion Paper No. 11. April 1968. Mimeo.

CROXTON, FREDERICK E., and DUDLEY J. CROWDEN. *Applied General Statistics.* Englewood Cliffs, N. J.: Prentice-Hall, Inc., 1955.

DEASY, GEORGE F., and PHYLLIS R. GRIESS. "Major Gas Explosion Accidents in the Anthracite Fields," *Proceedings of the Pennsylvania Academy of Science*, XXXVII (1963), 235–46.

FLORIN, JOHN. "The Advance of Frontier Settlement in Pennsylvania, 1638–1850: A Geographical Interpretation." Unpublished master's thesis, The Pennsylvania State University, 1965.

GOULD, PETER. *On Mental Maps.* Department of Geography, University of Michigan, 1966. Michigan Inter-University Community of Mathematical Geographers, Discussion Paper No. 9. Mimeo.

OLA, DANIEL. "Perception of Geographic Space in Lagos and the Western States of Nigeria." Unpublished master's thesis, The Pennsylvania State University, 1969.

POLANYI, MICHAEL. "Experience and the Perception of Pattern," in *Modeling of Mind*, eds. K. M. Sayre and F. J. Crosson. Notre Dame, Ind.: University of Notre Dame Press, 1963.

ROBINSON, ARTHUR H. "Mapping the Correspondence of Isarithmic Maps," *Annals of the Association of American Geographers*, LII (1962), 414–29.

THOMAS, ROBERT NELSON. "Internal Migration to Guatemala City, Guatemala, C. A." Unpublished doctoral thesis, The Pennsylvania State University, 1968.

CHAPTER 6

CLASSIFICATION

*We would never have learned anything if we had never thought:
"This object resembles this other, and I expect it to manifest the same
properties."*

—*Bertrand de Jouvenel*

THE NEED FOR CLASSIFICATION

If every object and event in the world were taken as distinct and unique—a thing in itself unrelated to anything else—our perception of the world would disintegrate into complete meaninglessness. The purpose of classification is to give order to the things we experience. We classify things so that we may learn more about them. If we did not classify, for example, we could not give names to constructs like "cabbage," "Saturday," or "rabbit." If we did not classify objects and experiences in commonly accepted terms we could not transmit information because no one would understand what we were talking about. Nor could we make inductive generalizations because each event could be understood only as a unique experience. Thus, language is itself an important classificatory device invented by the human mind.

Because it helps us form hypotheses and guides further investigations, classification is the first big step taken by most sciences. Some sci-

entists go so far as to argue that the state of classification in a science is a measure of its level of development. Chemistry reached its first important classification stage with Mendeleev's periodic table of the elements; botany has just recently passed beyond classification to other critical problems. Sociology and psychology—indeed most of the social and behavioral sciences—still devote an important fraction of their talent to classification.

Consider the way Americans assign unofficial and official names to the parts of their country. Unofficially we speak of New England, the South, the Midwest, and so forth. Such names are widely used but seldom do they have a precise definition. Despite the fuzzy meanings, however, the names are extremely useful and are used every day. Effective government, on the other hand, requires precise areal classifications for administrative purposes. Thus we have clearly demarcated boundaries between the United States and Canada and between the United States and Mexico. When a question arises about the exact location of a boundary, a boundary commission is created to settle the matter. Inside the country, boundaries between states, counties, townships, towns, and cities are precisely defined and jealously guarded. Without sharply delimited areal classes, taxes could not be levied, mail could not be delivered, nor could the other business of everyday life be transacted.

The common names for things are seldom satisfactory for the scientist. Botanists seldom use local names for plants when discussing them in technical journals. If they did there would be mass confusion because the same name is often applied to different plants at widely separated localities. A geographer studying migration must define the phenomena he wants to observe. Does a person "migrate" if he moves from Illinois to California only to return to Illinois a year later? Can we apply the term "migrant" to a person who moves from the central city to the suburbs and never returns? Some moves are over long distances and for long durations; others are short. Some are one-way and permanent, whereas others involve temporary stays and a return trip. Can there be any laws of migration until usage is made more precise? Perhaps we should forget the term "migration" and start to think of "circulation" in terms of distance, duration, and periodicity. Unless we invent precise terminology, we are forced to lump together the travels of an immigrant arriving in New York from the West Indies, and the journey to work of the suburban commuter. For purposes of analysis it would make more sense to separate such events.

If a science is to make progress, students and scientists must be able to exchange knowledge. For example, no person can know all the facts about all the different neighborhoods and land use zones in each of the cities of the United States. Yet if similar urban parts are grouped together, then statements can be made about land use classes which will apply to the members of that class. Consider the urban areal class "black ghettos." No one in the United States or elsewhere has first-hand facts about the black ghettos of every American city, yet certain statements can be made about such places

which seem to apply with some degree of generality.

Specialized vocabulary enables us to make novel and useful distinctions. After our first psychology course we all begin labeling (often inaccurately) acquaintances and friends as neurotic, or manic-depressive, or suffering from some complex or other. Each generation of adolescents, aware that profound differences in ability and personality exist from one person to another, gropes for words to express the distinctions it feels must be made. This is why parents never seem able to understand the jargon of the teenager. The college experience standardizes speech to some extent and attaches more precise meanings to our words so that by the time we are in our twenties we can communicate with the world.

Generalization through the naming of things is a form of intellectual shorthand which makes the transmission of knowledge easier. But perhaps the most important purpose of classification systems is to permit inductive generalization about the objects we want to study and understand. Thus, every classification in science is an implicit theory. Suppose we want to know more about urban places and how they grow. First we have to decide what an urban place is. We might say an urban place is a cluster of buildings and people. This definition is too broad, including as it does a prison, a university, a carnival, a Girl Scout camp, or an agricultural village. Alternatively, we might say an urban place is a nucleated settlement of the permanent homes and work places of a diverse and predominantly nonagricultural population. This definition eliminates non-nucleated settlements, nomadic camps, seasonal settlements, and agricultural villages. It includes the tiniest hamlet at a crossroads and the largest metropolises in the world.

Once we have identified our object of inquiry, penetrating investigation can proceed. The first step is classification. For a long time urban researchers tried to understand urban places by classifying them according to principal function—for example, administration (capital cities), defense (fortress and garrison towns), institutionalized culture (university, cathedral, or pilgrimage towns), communications (collecting, processing, and distributing centers at transport nodes), production (manufacturing towns and craft centers), and recreation (health spas and tourist resorts). Such terminology has become common usage, but has never enjoyed any scientific acceptance because the areas of overlap are great and borderline cases are hard to assign, as well as because these classes are not very meaningful. They do not suggest any important hypotheses, nor do they aid analysis in any way. Perhaps no implicit theory exists. Instead of classifying urban places, attention has recently turned to classifying their employment structures in terms of the Standard Industrial Classification (S.I.C.) prepared by the U.S. Bureau of the Budget. Instead of describing each urban place in terms of its principal function, it is described according to the fraction of its employed residents who work in each S.I.C. class, such as construction, durable goods manufacture, retail trade, finance, government, and so forth. Using such information we can make inductive generalizations about why certain cities grow and others decline. After we have classified the members of the employed labor force we may discover that fast growing places have a high fraction of their employees working in growth industries, whereas the employee ranks in declining areas have an excessive fraction of their workers in stagnant or declining activities like mining and railroading. Such an inductive generalization is useful and valid. The Bureau of the Budget devised the classification scheme to help us see things which would otherwise escape us in the face of large amounts of unorganized information.

There are no natural classifications that everyone agrees upon. Instead, classifications are contrived either for general purpose use or for specialized uses. They are not devised for the sake of classification alone. There is no reason to place an object into a class unless something is implied in addition to membership in that class. This is why some young men are classified by the Selective Service System as 1–A (fit for military conscription) instead of 1–A–O (conscientious objector but physically fit) or 4–F (physically unfit). The classification implies something more than simple membership in the class. A prime aim of classification is to enable inductive generalizations to be made about the objects under study. When we make a frequency distribution to classify a nation's families according to income, we do so in order to relate income to other variables and to assess income change through time.

Every good classification has an aim beyond the classification alone. Zoning laws in cities are classification systems contrived to help the city function better. Intelligence tests are administered so that a student's curriculum can be matched with his abilities. The census taker asks you how much income you earned last year because your income class implies something about your behavior. Social scientists feel that income is one of the best predictors of level of living. Grouping people by income class enables us to identify the poor so that programs can be developed and administered on their behalf.

As a rule, *general purpose* classifications are developed to give names to things as members of groups and to transmit important information. *Special purpose* classifications, such as those developed in every branch of science, are the first formal step in science itself. Classifications allow inductive generalizations, and general statements allow the study of relationships among events. Regularities in relationships suggest hypotheses which when tested and refined become empirical laws. New knowledge is thereby created.

HOW WE CLASSIFY

Classification is the systematic grouping of objects or events into classes on the basis of properties or relationships they have in common. Classification must follow a definite plan and the grouping can be accomplished by two routes: by logically subdividing a population, or by agglomerating like individuals. In logical subdivision, the subdivision procedure is accomplished in a series of steps using carefully defined criteria, usually the absence or presence of one or more attributes. The agglomeration method takes a number of individuals and assembles them into classes according to some grouping procedure.

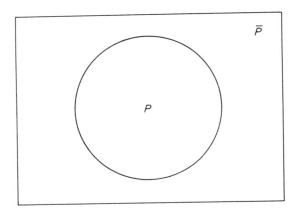

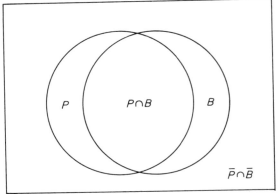

FIGURE 6–1

The universe of all college students includes the set P—students who like pizza, and the complement of P, *which is* P̄—*students who do not like pizza.*

FIGURE 6–2

Adding a second set to Fig. 6–1 increases the number of possible categories from two to four. B is the set of students that like cold beer. What are the other sets?

Classification by Subdividing a Population

Let us consider several examples of logical subdivision of the population of American college students. The universal set of all college students may be represented by a rectangle, and the set of students who like pizza can be designated by a circle inside the rectangle (Fig. 6–1). Such a representation is termed a Venn diagram. Points outside the circle but inside the rectangle represent college students who do not like pizza. Points inside the circle represent people who are college students and who also like pizza. Each American college student can be assigned to either set P or to its complement set $\overline{P}$. Through such a procedure we logically divide students into one set or the other on the basis of a stated criterion.

The classification example based on pizza preference reveals two important attributes of successful classification schemes. First, no matter how a classification is achieved—that is, whether it proceeds by logical subdivision or by grouping like individuals—it must be exhaustive. Every individual must fit into one of the classes. Secondly, the classes must be mutually exclusive. Classes must be designed so that an individual cannot fall into more than one. In the pizza example, the classification is

exhaustive because there is a category for every student; it is exclusive because a student ends up in one and only one category. We assume he either likes pizza or he does not.

The simplest classifications involve only one criterion such as "likes pizza." The classification can be refined by adding an additional criterion such as "likes cold beer." The introduction of a second criterion produces a revised Venn diagram as a basis for a more elaborate classification of American college students (Fig. 6–2). The Venn diagram discloses four groups of students: those who like pizza and do not like beer (P), those who like beer and do not like pizza (B), those who like both pizza and beer $(P \cap B)$, and those who like neither pizza nor beer $(\overline{P} \cap \overline{B})$. Again, the classes were defined *a priori* as logical constructs so that each student could fit one and only one of the four classes.

Two attributes (e.g., "likes pizza" and "likes beer") define two sets which divide a universe into four classes (e.g., Fig. 6–2). Three attributes define 2^3 or eight classes; four attributes define 2^4 or sixteen classes, and so forth. In addition to food preferences, college students hold strong views about where they would like to live and work after they have graduated. Consider the following criteria: "likes New England," "likes the Midwest," and "likes

California." The three preferences, which may be held or not held in any combination may also be expressed as a Venn diagram (Fig. 6–3).

Some students may prefer New England and only New England (N), whereas others like New England or California but would never live in the Midwest ($N \cap C$). A distinct preference pattern is represented by each region in the Venn diagram. Some students, including many in the South, prefer none of the three regions listed. They are, therefore, represented by points outside the three intersecting circles ($\bar{N} \cap \bar{M} \cap \bar{C}$).

Venn diagrams are a way of representing a division of a population in a concise and often highly useful way. Consider the census tracts of Los Angeles in terms of conditions thought to underlie the pathologies that contributed to the Watts riot of 1965. Five conditions or attributes define five sets: (1) median family income in the tract less than \$5000; (2) blacks were 75 percent or more of the population; (3) the tract contained the highest population densities in the city; (4) school drop-out rates were at a maximum; and (5) crime rates were the highest in the city. The five attributes define five sets which intersect one another. Where all five

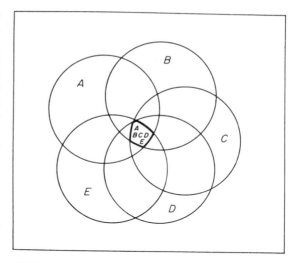

FIGURE 6–4

Five intersecting sets identify locations with maximum potential for violent civil disorder.

sets intersect we would expect living conditions to be the worst (Fig. 6–4). Turning to census tract data and other social area indices we discover that only in the Watts area of Los Angeles did all five conditions prevail at the same time (Fig. 6–5).

When we classify using Venn diagrams we must take care to define our sets and arrange their intersections so that we avoid contradictions. For example, geography students can be grouped into nine categories according to their likes, dislikes, and neutrality about the subject's two major divisions: spatial analysis, and man-environment studies (Fig. 6–6). Some students like all geography ($A \cap B$); some dislike it all ($C \cap D$). Students represented by points outside the circles are either indifferent to these issues or ignorant about them.

Another procedure used in logical subdivision of a population is a stepwise disaggregation into a hierarchy of classes. If the students in a university are the population, the first order of classes might be the colleges; the second the departments; the third the major areas of specialization within the departments; and the fourth the concentrations within the specializations (Fig. 6–7). If, instead of college students, we considered the parcels of land in a county as the

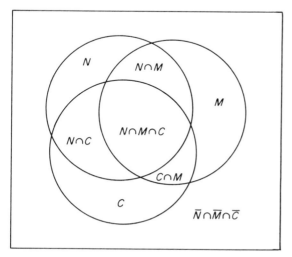

FIGURE 6–3

Residential preferences of college students: N = "likes New England"; M = "likes the Midwest"; and C = "likes California."

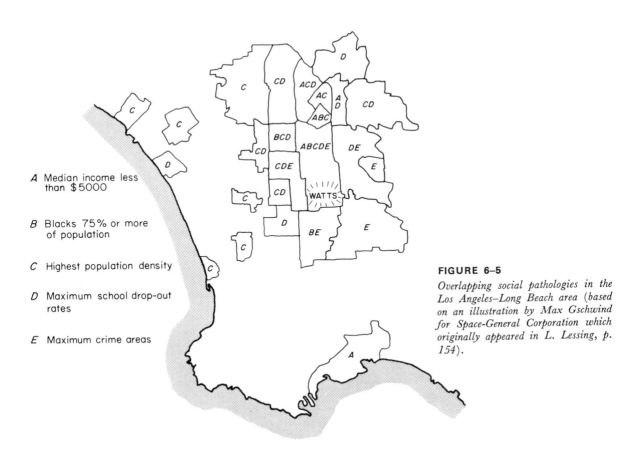

A Median income less than $5000

B Blacks 75% or more of population

C Highest population density

D Maximum school drop-out rates

E Maximum crime areas

FIGURE 6–5

Overlapping social pathologies in the Los Angeles–Long Beach area (based on an illustration by Max Gschwind for Space-General Corporation which originally appeared in L. Lessing, p. 154).

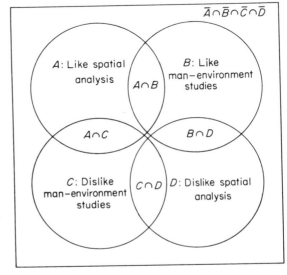

$\bar{A} \cap \bar{B} \cap \bar{C} \cap \bar{D}$

A: Like spatial analysis

B: Like man–environment studies

$A \cap B$

$A \cap C$

$B \cap D$

$C \cap D$

C: Dislike man–environment studies

D: Dislike spatial analysis

FIGURE 6–6

If geography students like, are neutral to, or dislike the branches of geographical studies, they can be represented by a Venn diagram of intersecting sets.

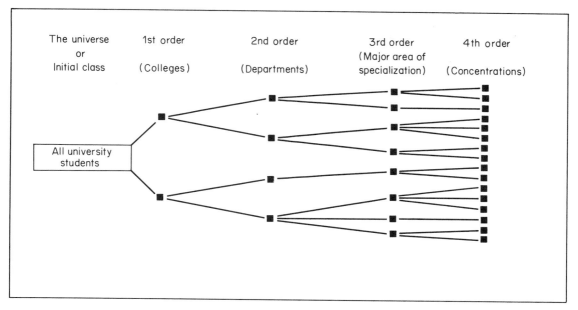

FIGURE 6–7

Logical subdivision often is a stepwise a priori *disaggregation of a population into a hierarchy of classes.*

universe and subdivided them according to the uses to which they are put we would follow exactly the same procedure (Fig. 6–8).

Classification by Grouping Similar Observations

Agglomeration of similar individuals into classes is the second approach to classification and it is this approach which will concern us in the remainder of this chapter. Grouping similar observations is simply the other side of the coin from subdividing a population. Only the perspective at the starting point is different. If we subdivide a university's student body—

first into colleges, then into departments, and so forth—all students should fall into classes with the same members as if we had examined the transcript and curriculum of each student, one student at a time, and slowly assembled the students into classes with students following similar academic programs.

At this stage in our discussion it might appear that the logical subdivision procedure is neater and more elegant. Yet as we shall see, the grouping of like individuals is operationally far superior. It is faster and easier, and all classes that we define have at least one member. In a logical subdivision, we can define any number

FIGURE 6–8

Partial illustration of how land parcels in a country can be classified according to a hierarchy of land uses. Only farm land is subdivided into second-order classes; and only orchard land is subdivided into third-order classes (adapted from Grigg, Fig. 2).

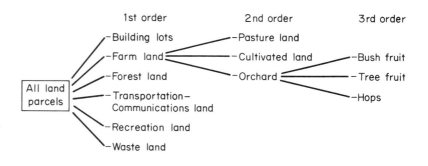

of classes, but many may have no members. In a university, for example, a department may offer a major concentration which no student currently has chosen, yet the class may still be defined.

Once we have decided on the agglomerative approach, the next choice to be made is the method by which individuals will be compared. We may use discrete attributes like male vs. female; born in Wisconsin vs. born elsewhere; or Protestant vs. Catholic vs. Jew. Alternatively, our observations may be measured in terms of continuous variables like income, intelligence, health, or height. When we use continuous variables we are forced to define the limits of each class before we start grouping our observations into classes. If we want to classify families according to a continuous variable like annual family income we first examine the range of observations and then set up classes like $0–$500; $501–$1,000; $1,001–$1,500; $1,501–$2,000; and so forth. To the question: how many classes should we specify and what should be the

class limits, we can say only that the choice is up to us, and proper choice depends on the aim of the classification.

Let us consider some different ways of grouping the fifty states and the District of Columbia in terms of the percent of the selective service draftees examined for military service in 1967 and subsequently disqualified for medical reasons, for failing mental tests, or for both reasons. A column of fifty-one percentages ranging in magnitude from 50.8 percent (Alabama) to 22.8 percent (Oklahoma) is much too detailed and cumbersome for convenient study and interpretation. If the fifty-one figures were plotted on an outline map of the United States, all the meaningful spatial patterns would be swallowed up in a sea of detail. One alternative to mapping the fifty-one percentages is to shade all the states where disqualification rates exceeded the national average of 40.3 percent (Fig. 6–9). By grouping states according to whether they were above or below the national average a surprising thing happens. We elimi-

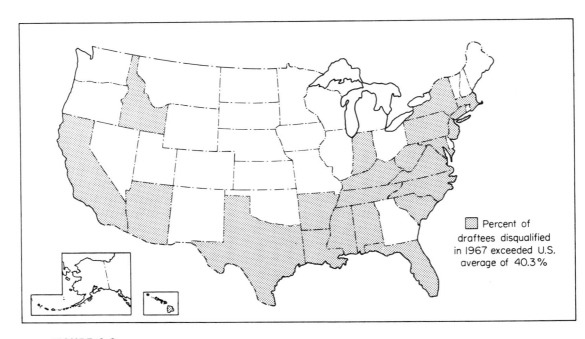

Percent of draftees disqualified in 1967 exceeded U.S. average of 40.3%

FIGURE 6–9

A two-way classification of the states based on draftee rejection rates (from Statistical Abstract of the United States: 1968, *p. 264).*

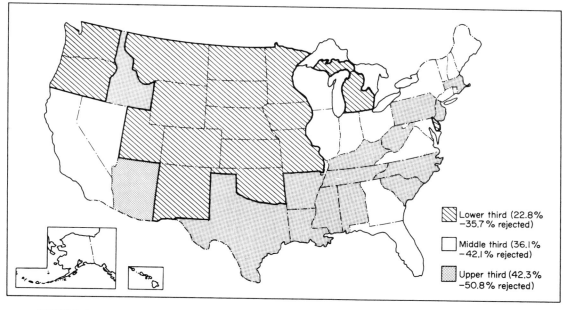

FIGURE 6–10

Grouping the states and the District of Columbia into three equal classes according to draftee rejection rates (*from* Statistical Abstract of the United States: 1968, *p. 264*).

nate details but begin to see patterns. Too much information is almost as bad as no information at all. A classification reveals that above average disqualification rates seem concentrated in the East and South, while below average rates are common in the interior and North Central states.

If a map of two classes can reveal a pattern, perhaps three classes can tell us even more. For example, we could rank the observations and divide them into three groups of seventeen observations each: high rejection rates; average rates; and low rates (Fig. 6–10). The high rates are now more clearly seen as a southern and Appalachian phenomenon, with the lowest rates concentrated in the Upper Midwest and the Plains states. Such patterns raise a number of hypotheses about the geographical distribution of physical fitness, general health, child care, climate, and medical conditions in America.

Each time we group observations into classes we lose detail in order to gain ease in interpretation. Going in the other direction, when we subdivide broad classes we gain detail but our

sharp spatial patterns on the map begin to disintegrate in the same way that a newspaper photo under a magnifying glass dissolves into dots of various sizes, or a mosaic under close inspection becomes so many colored tiles.

We classify to clarify. What is the best number of classes? It depends on why the grouping is being done in the first place. We try to pick the class intervals and the number of classes that will give the greatest amount of useful information about the observations we are examining. Every classification or "taxonomy" is a simple but implicit theory. If by trial and error we are able to produce classifications and maps of spatial phenomena which suggest or expose important relationships, then our choices are good choices. Our classifications have served their purpose.

As an example of a classification scheme using two continuous variables let us turn to East Africa and consider the economic development potential of several regions in terms of their *natural resource base* and again with respect to *overhead facilities* for development. Natural resources include arable land of different qualities, irrigation possibilities, and sporadically occurring minerals such as coal, copper, lead, petroleum, salt, soda, tin, zinc, and other items. Overhead facilities for development include, in addition to large urban centers, abundant water, power, railways, and main roads.

Regions may be ranked in terms of their resource endowments; and they may also be ranked in terms of the facilities available for development. As we can see from a contingency table, the two variables are strongly related to one another (Fig. 6–11). The contingency table helps us identify four different combinations of resource endowment and development facilities. Development zones of the first type show considerable promise for further development. Not only do they enjoy excellent resource bases but most of the facilities needed (railways, roads, power, and water) are immediately at hand. Development zones of the second variety are rich in natural wealth but at the moment they do not possess the full range of facilities needed for regions to realize their development potential. Development zones of types three and four include the drier and tsetse-infested areas of only moderate to very limited promise. Such areas are sparsely populated with natural conditions permitting mixed farming, combining cultivation and stock grazing at one end of the scale down to primitive animal herding at the other. They might have some mineral wealth but the facilities for its exploitation are practically nonexistent. Overall, the identification of distinctive types of development zones allows government planners to deal with regional development problems as classes of problems, thereby avoiding the need to confront each small region as a unique event.

Classifications (taxonomies) based on one or just a few characteristics are called "monothetic," and all the items allocated to one class must share the trait under consideration. Thus the members of the class of "native speakers of French" must in fact be people who learned French as a mother tongue. Classifications based on several characteristics are termed "polythetic." Membership in a class does not depend on just one property or characteristic shared by all members. Instead, a given "taxon" or class is established in terms of a set of attributes, and some of the members of the class, while they share most of the attributes appropriate for the class's members, do not share all of them. As a result most members of the taxon "birds" have wings, yet there are some birds, such as the kiwi of New Zealand, which lack wings. Similarly, some vertebrates lack red blood, and some mammals such as the platypus lay eggs instead of bearing live young.

Establishing classifications based on many attributes or characteristics is a complicated business. The human mind has a hard time tabulating and processing many items in terms of many attributes. Traditionally, therefore, each science's taxonomists (the people who set up the taxons or classes) have favored one aspect over another in their classification efforts. Geography is no exception. In the development of geographic thought, the whole idea of the natural region—physiographic or otherwise—has had a debilitating stranglehold on many geographers' thinking about areal classifications. The arrival of the computer has reversed this trend and has opened a whole new field of possibilities for objective, explicit, and replicable classifications of the world's areas.

Numerical Taxonomy

A well-known taxonomist, Robert Sokal, has written that the purpose of classification or taxonomy is to group events into "natural" classes or taxa. What constitutes naturalness is often disputed, but the underlying idea is that the members of a natural taxon are more similar to the other members of the taxon than they are to nonmembers. Another way of saying this is to say that once objects or events are assigned to classes, the variation among individuals within each class is minimized and the variation between each group and members of other groups is maximized.

Groups are commonly understood as clusters of events or objects. They are defined in terms of the similarity or proximity of their members. Before the grouping of the operational taxonomic units (OTU's—the events or objects themselves) we must specify what we mean by proximity or distance of each OTU from every

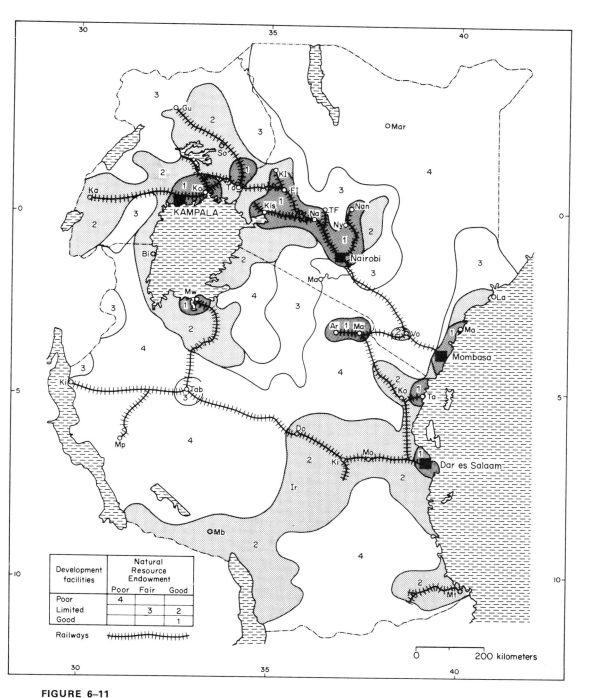

FIGURE 6–11

Four types of development zones in East Africa. Each type is defined in terms of two variables (adapted from T. J. D. Fair, Fig. 5).

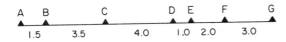

FIGURE 6–12

Seven service plazas on a road and the distances between each plaza and its nearest neighbors.

other OTU. Let us assume that our objects are seven service plazas which lie along a toll road (Fig. 6–12). Once the distance between each adjacent pair of plazas is specified, we can construct a distance matrix, often called a similarity matrix, which records the distance of each plaza to every other plaza (Table 6–1). If we treat the service plazas as OTU's and begin to group them together according to their distances from one another, we may proceed in any one of three ways.

The *single linkage* method groups nearest neighbors. It starts with the shortest distance

TABLE 6–1

A Similarity Matrix Records the Distance from Each Operational Taxonomic Unit (OTU) to Every Other

	A	B	C	D	E	F	G
A	0.	1.5	5.	9.	10.	12.	15.
B	1.5	0.	3.5	7.5	8.5	10.5	13.5
C	5.	3.5	0.	4.	5.	7.	10.
D	9.	7.5	4.	0.	1.	3.	6.
E	10.	8.5	5.	1.	0.	2.	5.
F	12.	10.5	7.	3.	2.	0.	3.
G	15.	13.5	10.	6.	5.	3.	0.

between any two OTU's. The two OTU's separated by the shortest distance become the first to join and form the first group. In Fig. 6–13 D and E join over a distance of 1.0. The next shortest distance is between A and B. These two OTU's join over a distance of 1.5. The

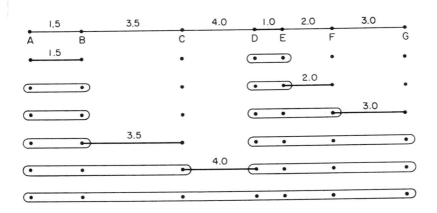

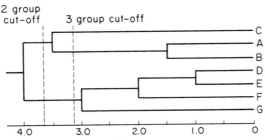

FIGURE 6–13

The single linkage method whereby OTU's and clusters of OTU's are linked together a step at a time over ever greater distances. The clustering can be stopped at any desired degree of uniqueness or generality. Optimal division of the seven plazas into two groups and three groups is shown.

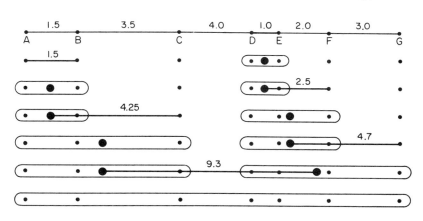

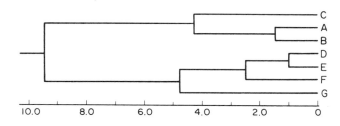

FIGURE 6–14

The average linkage method whereby distances are measured between an OTU *(small dot) and the center of gravity (large dot) of the cluster it joins, or between the centers of gravity of the clusters that join.*

next shortest distance is between group cluster *D-E* and *F*. In the single linkage method the distance between a group and an OTU outside group is measured in terms of the group member that is closest to the outside OTU. According to this rule the distance between group *D-E* and *F* is 2.0. Clustering continues until all OTU's are joined. As clustering proceeds step by step the criteria for joining a group become less stringent. OTU's *D* and *E* joined over a distance of 1.0; the last linkage joining groups *A-B-C* and *D-E-F-G* spans a distance of 4.0. One way to summarize how OTU's join and the distance over which they join at each step along the way is to construct a linkage tree (Fig. 6–13). Before grouping begins we have complete uniqueness; after all OTU's have been connected we have complete generality, which means complete loss of detail. Finally, the linkage tree enables us to divide the OTU's in a rational and objective way into two groups or three groups or a greater number according to the criterion that members of a group are more

similar (closer) to one another than they are similar (closer) to outside members. Where we draw our cutoff on the linkage tree determines the number of groups.

In the *average linkage* method the first group is the same as in single linkage. In later steps, an OTU will join a cluster if the distance between it and the center of gravity of the cluster is (1) less than any other cluster to OTU distance, or (2) any other OTU-OTU distance (Fig. 6–14). In the average linkage example shown, the sum of the distances between a cluster's center of gravity and points on one side equals the sum of the distance between the center of gravity and the points on the other side. Accordingly, centers of gravity and distances must be recomputed after each linkage step. As a result, with average linkage the order in which OTU's group may differ slightly from the order in single linkage. The distances over which OTU's and clusters link also differ from those in the single linkage method. The differences are easiest to see in the linkage trees. In the example presented here, cluster *D-E-F-G* forms before *C-A-B* under the single linkage method (Fig. 6–13); but under the average linkage method this order is reversed (Fig. 6–14). Moreover, clustering distances vary substantially between the two methods. Using the average linkage method, the last two clusters to join had to span the distance of 9.3 which separated the centers of gravity of the two main clusters: *A-B-C* and *D-E-F-G*. Under the single linkage method these same clusters joined over a distance of 4.0.

In the *complete linkage* method a distance threshold criterion is used to specify which OTU's in the distance matrix will join in each

of several successive passes over the matrix (Fig. 6–15). In the first pass, for example, let us join all OTU's separated by distances up to and including 1.0. According to this threshold criteria, OTU's *D* and *E* will cluster at this level. Now let us see what happens if we raise the threshold by 1.0 for each successive pass through the distance matrix. At the level of 2.0, *B* joins *A* and *F* joins the *D-E* cluster because *F* is 2.0 units away from *E*. At the level of 3.0, *G* joins the *D-E-F* cluster. On the next cycle, as the distance threshold is raised to 4.0, clusters *A-B* and *D-E-F-G* are joined by *C*, the only remaining OTU. The linkage tree based on the complete

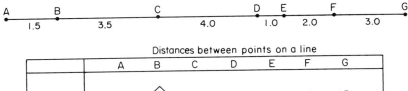

Distances between points on a line

	A	B	C	D	E	F	G
A	0	1.5	5	9	10	12	15
B	1.5	0	3.5	7.5	8.5	10.5	13.5
C	5	3.5	0	4	5	7	10
D	9	7.5	4	0	1	3	6
E	10	8.5	5	1	0	2	5
F	12	10.5	7	3	2	0	3
G	15	13.5	10	6	5	3	0

Clustering thresholds:

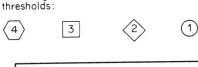

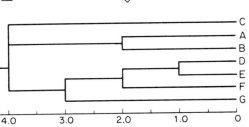

FIGURE 6–15

The complete linkage method. Clustering proceeds at each successive threshold, which starts here at 1.0 and increases by 1.0 after each pass through the distance matrix. After four passes all seven places are grouped into one cluster.

linkage procedure differs from the other two linkage trees both in the order in which OTU's combine and in the distances over which the grouping occurs. The complete linkage method can be used with the original distance matrix as we have done, or the distance matrix can be recomputed after each cycle, replacing each cluster with its center of gravity and the appropriate distances.

Regardless of the linkage procedure used, the initial clustering step is the same and the final cluster—namely the cluster including all the OTU's—is necessarily the same. The intermediate clustering steps, however, can be quite different at roughly the same stage. At the center of each procedure is a working definition of the term "similarity." A pair of OTU's combine because they are more similar to one another or closer to one another on some scale than they are to any other OTU's.

The nature of "similarity" is fundamental in classification problems in geography and other sciences. It is an ancient philosophical problem which has become acute recently as we apply computer technology to classification procedures. We have to answer the question, What do we mean when we say that A is similar to B? Only if we are able to respond that A is similar to B in such and such a respect, can we make any progress. In numerical taxonomy—classification on the basis of interval measurements of similarity—we assume that we can measure degrees of similarity on an interval scale. This means that we must locate the objects or constructs we want to classify in a taxonomic space and obtain the distances between them. In the examples we looked at so far we used a space of only one dimension—points (service plazas) were located along a line (a road). In that one-dimensional space we were able to make statements like "D is closer to E than it is to C," or, "F is just as close to D as it is to G," and so forth. Establishing comparative similarities of the type: A is more similar to B than it is to C, is the basic principle underlying any attempt at clustering objects into groups.

Throughout this discussion we have defined similarity in terms of distance in a taxonomic space. Our examples so far have used only a one-dimensional taxonomic space because we wanted to deal with (1) the fundamentals of grouping procedures and linkage trees, and (2) the idea that as grouping proceeds in a classification, detail is systematically sacrificed for generality and understanding. Before we begin grouping our OTU's we have complete uniqueness—maximum detail. After all OTU's have been grouped into one super cluster, all detail is lost and complete generality prevails. Between the two extremes lie useful classifications. Some are more generalized with few classes; some are refined with many classes and only a few OTU's in any one class. What a classification is to be used for determines the point at which clustering should halt. A choice must be made. The clustering procedures are objective; the decision about when to stop clustering is a subjective matter.

Multivariate Classification

There is no reason why our classification efforts should be based on a single characteristic. Two, three, or more variables can be used to define taxonomic spaces of two, three, or more dimensions. Usually the addition of a second characteristic changes the way in which OTU's cluster. Let us consider the seven OTU's used in the one-dimension example and measure each of them in terms of a second characteristic (Fig. 6–16). Since each OTU now occupies a position in terms of two distinct dimensions we can consider the distances between OTU's in a two-dimensional taxonomic space.

Points F and G are the closest. They are three distance units apart. They both have the same Y value of six, so they differ only with respect to their respective X values, 12 and 15. All other pairs in the two-dimensional taxonomic space have different X values and different Y values. We can compute the distance between any pair of points by using the Pythagorean theorem from plane geometry which states that in a right triangle, the square of the hypotenuse is equal to the sum of the squares of the other two sides. Consider points B and D. By drawing lines parallel to the two axes and by connecting the two points B and D, we can construct a right triangle whose hypotenuse can be computed easily. The base of the triangle equals

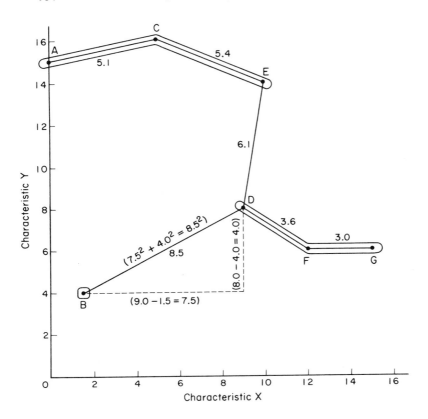

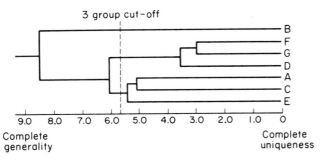

FIGURE 6–16

Classification according to the distances between OTU's in a two-dimensional taxonomic space. Distances are computed according to the Pythagorean theorem: the hypotenuse of a right triangle (e.g., B–D) squared equals the sum of the squares of the other two sides of the triangle.

D minus B on the X axis or $9 - 1.5 = 7.5$. Similarly, on the Y axis, D minus B is $8 - 4 = 4$. The sum of the squares of these two sides equals $7.5^2 + 4^2 = 56.25 + 16 = 72.25$. The square root of 72.25 is 8.5, which is the length of the hypotenuse and thus the distance between B and D in our two-dimensional taxonomic space. The distance between any pair of OTU's can be computed in the same way, and a similarity matrix describing the distance from each point to every other point is constructed accordingly.

In our example (Fig. 6–16) the first pair of OTU's to join are F and G. They cluster over a taxonomic distance of three units. Using the single linkage method, whereby a cluster joins additional OTU's on the basis of the nearest neighbor criterion, we find that in the next linkage, point D joins the F-G cluster because its distance of 3.6 units from that cluster is smaller

TABLE 6-2

Computing the Distance between Two OTU's
in a Two-dimensional Taxonomic Space

OTU	Position on Dimension X	Position on Dimension Y
B	1.5	4.0
D	9.0	8.0
Difference:	7.5	4.0
Difference squared:	56.25	16.0
Distance between B and D = $\sqrt{56.25 + 16.0} = \sqrt{72.25} = 8.5$		

than the distance between any ungrouped OTU pair, and smaller than the distance between any other OTU and members of the *F-G* cluster. The clustering continues and as each linkage is established the tree gradually grows. The first pair to join are *F* and *G*; the last linkage in the linkage tree occurs when *B* joins cluster *F-G-D-A-C-E* over a distance of 8.5 units. If we want to group our seven OTU's into three classes so that members of each taxon are more like one another than like OTU's outside the taxon we would stop grouping just after cluster *A-C* joins *E* over a distance of 5.4 units. At that point three taxa remain: *B*, which stands alone; *F-G-D*; and *A-C-E*.

If we add a third characteristic to our taxonomic scheme we create a three-dimensional taxonomic space. In such a three-space OTU's are described in terms of their position along three different dimensions. Although three dimensions are a little harder to visualize, the distance between pairs of OTU's remains the criterion of similarity (Fig. 6-17). For example, consider *B* and *D* and the distance between them in the three-space. How can we compute the distance? By using the Pythagorean theorem

once more. Remember that when *B* and *D* were plotted in two-space we were able to make the computation shown in Table 6-2.

When *B* and *D* are plotted in three-space the procedure is exactly the same except that we must take account of each OTU's position on the third dimension, as shown in Table 6-3. The distance between *B* and *D* in three-space is 9.39. All the other distances between pairs of OTU's can be computed in the same way since each OTU has three values attached to it—one for each dimension in the taxonomic space. Once all distances are computed and the similarity or distance matrix is prepared, grouping can proceed according to any one of the three linkage methods.

There is no limit to the number of dimensions we can incorporate into a classification scheme. We cannot draw pictures showing points plotted in four, five, ten, or *N* dimensions but the idea is the same. In the U.S. Census of Population, for example, places are described in terms of numerous dimensions—median age of men; median age of women; median family income; median number of school years completed; percent nonwhite; and so forth. If we want to

TABLE 6-3

Computing the Distance between Two OTU's
in a Three-dimensional Taxonomic Space

OTU	Position on Dimension X	Position on Dimension Y	Position on Dimension Z
B	1.5	4.0	4.0
D	9.0	8.0	8.0
Difference:	7.5	4.0	4.0
Difference squared:	56.25	16.0	16.0
Distance between B and D = $\sqrt{56.25 + 16.0 + 16.0} = \sqrt{88.25} = 9.39$			

TABLE 6–4
OTU's Described by Five Variables

OTU's (Tracts)	Position on Dimension V (Median Age of Men)	Position on Dimension W (Median Age of Women)	Position on Dimension X (Median Family Income)	Position on Dimension Y (Median School Years)	Position on Dimension Z (Percent Non-white)
A	34.5	32.1	54.6	10.7	15.7
B	28.3	26.2	57.9	11.5	11.6
C	31.5	30.1	61.0	12.0	8.0
.	.	.	.	.	.
.	.	.	.	.	.
.	.	.	.	.	.

group the census tracts in a city according to such attributes we proceed as shown in Table 6–4.

Usually the raw data matrix is converted to z-scores by subtracting the column's mean from each column entry and by dividing the difference by the standard deviation of that column's entries. If there are N dimensions to be considered, each OTU occupies a position in an N-dimensional taxonomic space. The distance between any pair of OTU's in N-dimensional space is computed by simply extending the

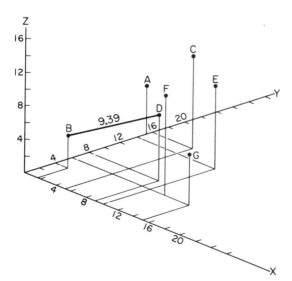

FIGURE 6–17

Plotting OTU's *in a three-dimensional taxonomic space and computing the distance between pairs of* OTU's; *for example, between* B *and* D *the distance is 9.39.*

Pythagorean expression to include the N squared differences.

Classification with Several Somewhat Redundant Variables

Sometimes when we want to classify events in geographical research we find that we have more information than we really need. In metropolitan area analysis, for example, we know that certain census tract attributes are highly correlated (Table 6–5). If this is so, the variables used for the classification are not completely independent, which is simply another way of saying that they are measuring the same thing to some degree. It is for this reason that we say correlated variables contain redundant information. When variables are not independent we can predict something about the magnitude of one variable by knowing the magnitude of another. In most American cities, for example, the correlations between median education, median income, and housing values in a tract are such that by knowing any one of them we can form a pretty good idea about the levels of the others.

TABLE 6–5
Correlation of Census Tract Variables

Variable	(1)	(2)	(3)
1. Median school years completed	—		
2. Median family income	.699	—	
3. Median value of owner-occupied housing	.698	.829	—

TABLE 6–6

Original Data Matrix

Observations (Counties)	Mules/Tractors (1)	Signs/Mile (2)	Mint Juleps (3)	% of Restaurants (4)
1				
2				
3				
.				
.				
.				
N				

If our variables are closely correlated and measure highly interrelated phenomena we need a convenient procedure for squeezing the maximum amount of information out of them while leaving the redundancies behind. How do we combine variables? How do we weight them so as to overcome the problem of redundant information? In Chapter 5 we discussed procedures for measuring the correlation between variables in a set of observations. Let us take an example and review what correlation really means.

Suppose we want to classify the counties in a region on the basis of four criteria. For each county observation we have information on four variables:

1. The ratio of mules to tractors
2. The number of "Keep America Beautiful" signs per mile of county highway
3. The annual consumption of mint juleps per capita
4. The proportion of restaurants that serve automatically a side dish of grits

Since we cannot draw four-dimensional graphs we cannot portray our counties as dots in a four-dimensional classification space. Yet if some of these variables measure only slightly differing aspects of the same characteristics they will be interrelated, and our observations will carry redundant information. We may well ask if we need four dimensions after all to plot our observations so that grouping of counties may proceed according to some linkage method.

We start our analysis with N counties and four variables (Table 6–6). Recall that we can show the correlation between any two variables as two vectors of unit length drawn from a com-

mon origin so that the cosine of the angle between them equals the correlation coefficient between the two variables. In this example the four variables are interrelated and we can describe the correlations in matrix form (Table 6–7). The same correlations can be represented by the cosines of the angles formed by a set of vectors, each of unit length. Let us see how this is done.

TABLE 6–7

Correlation Matrix

	(1)	(2)	(3)	(4)
1. Mules/tractors	1.0	–.79	.45	.95
2. Signs/mile	–.79	1.0	–.93	–.94
3. Mint juleps/capita	.45	–.93	1.0	.70
4. % of restaurants	.95	–.94	.70	1.0

A right triangle is formed when one unit vector is projected vertically onto another (Fig. 6–18). The vectors meet at some angle θ (theta). If θ equals 0°, the projection of one vector coincides with the other and the cosine of θ (equal to the adjacent side A divided by the hypotenuse H) equals unity. When θ equals 60°, cos θ equals 0.5. At 90°, the projection has shrunk to zero length so A/H, the cosine of θ, drops to zero. For larger angles, A is negative but H remains positive. Thus, when θ is 120° cos θ is -0.5, and when θ is 180°, cos θ equals -1.0. The cosines for all angles between 0° 180°, when plotted, form a continuous curve with values ranging from $+1.0$ when the two vectors coincide, to -1.0, when they head in exactly the opposite direction (Fig. 6–19). When two vectors are perpendicular (that is

$\theta = 0°$
$\cos \theta = +1.0$

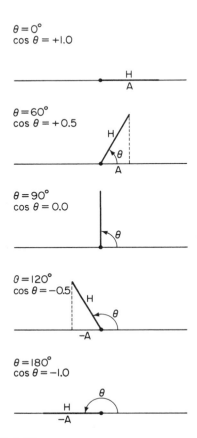

$\theta = 60°$
$\cos \theta = +0.5$

$\theta = 90°$
$\cos \theta = 0.0$

$\theta = 120°$
$\cos \theta = -0.5$

$\theta = 180°$
$\cos \theta = -1.0$

FIGURE 6–18

The vertical projection of a line of unit length onto the horizontal base defines a right triangle. As θ increases from $0°$ to $180°$ the cosine of θ goes from $+1.0$ to -1.0.

orthogonal) to one another, θ equals $90°$ and $\cos 90°$ equals 0.0.

Correlation coefficients also range in magnitude from $+1.0$ to -1.0. A correlation coefficient of $+1.0$ means that two variables are perfectly correlated. They also head in the same direction. A correlation coefficient of 0.0 means that the two vectors of variables head in orthogonal directions. If variables represent two vectors heading in exactly opposite directions, so that by knowing one variable in an observation we can predict perfectly the value of the other in the same observation, then the correlation coefficient between the two vectors is -1.0.

Let us return now to the data matrix des-

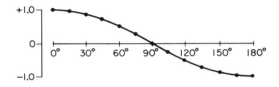

FIGURE 6–19

Cosines of angles between $0°$ and $180°$.

cribing N counties in terms of four vectors of variables (Table 6–6). How well can we predict the mule/tractor ratio for the first county by knowing the mule/tractor ratio for that county? We can predict it perfectly; the correlation coefficient between vector 1 and vector 1 is $+1.0$. We represent vector 1 perfectly correlated with itself by a vector of unit length (Fig. 6–20).

Now notice in the correlation matrix that vector 1 correlates with vector 4 to a very high degree. The correlation coefficient is $+.95$ which means that vector 1 and vector 4 are very close together and headed in the same direction. If we draw vector 4 so that it intersects vector 1 at an angle of $18°$, the cosine of the angle formed equals $+.95$, the correlation coefficient between variable 1 and variable 4. Continuing, variable (that is, vector) 4 has a correlation of $+.70$ with variable 3. If we draw vector 3 so that it intersects vector 4 at an angle of $45°$, the cosine of the angle formed equals $+.70$, the correlation coefficient between variable 4 and variable 3. After vector 3 has been drawn, we see that it forms an angle of $63°$ with vector 1 which we drew earlier. The cosine of an angle

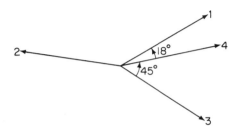

FIGURE 6–20

Vector representation of the correlations between four variables. Cosines of angles formed equal correlation coefficients between the respective variables (vectors).

of 63° is +.45 which we see in our Table 6–7 is the correlation coefficient between variable 1 and variable 3. As an exercise it would be instructive to verify that the cosines of the angles formed by the four vectors in Fig. 6–20 equal the respective correlation coefficients.

Especially interesting is vector 2 representing the second variable—the number of Keep America Beautiful signs per mile of county highway. Vector 2 forms angles of more than 90 degrees with each of the other vectors. Accordingly, the cosines are all between 0 and −1.0. In the table of correlation coefficients we see that the only negative coefficients involve variable 2 which is highly correlated with each of the other variables, but always in the opposite direction.

The four variables which we thought we could represent only in four dimensions can be shown in a two-dimensional space—on a page in this book. The reason we can represent four different variables this way is that they are interrelated. In fact, they contain so much redundant information that we need only two scales or dimensions to construct a classification space based on the four original variables. How do we do this? How do we represent four vectors with only two dimensions?

The Construction of a Multifactor Taxonomy

We want to classify the counties in a region on the basis of four variables and we have already discovered that the four variables are sufficiently interrelated that they can be represented as a cluster of four vectors. The cosine of the angle between each vector pair equals the correlation coefficient between those variables.

We can represent our vector cluster in a still more compact form. Each of our unit vectors passes through a common vertex. Let us now find the axis through that vertex which is located so that the sum of the squares of the right angled projections of the unit vectors on it are maximized. We will define such an axis as the axis of best fit (Fig. 6–21). The projections of each of the four vectors onto this main axis are the lengths l_1, l_2, l_3, and l_4. Since the vectors

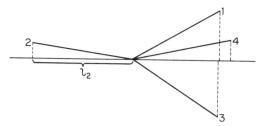

FIGURE 6–21

The principal axis passes through the common vertex and is located so that the sum of the squares of the projection lengths is a maximum. Each length is measured from the vertex out along the axis as shown for vector 2 and l^2, the length of its projection.

are each of unit length (1.0), the projection lengths are the same as the cosines of the angles formed between the vectors and the main axis. Therefore the lengths l_1, l_2, l_3, and l_4 are also the correlations of each of the four variables with our major axis or *scale*.

Once the lengths of the projections onto the first axis have been measured, we can put a second axis through the vertex at right angles to the first axis (Fig. 6–22). Once more we can measure the projection lengths. This time projections represent correlations of the four variables with the second axis which was drawn perpendicular to the first so that the second set of projections would be independent of those on the first axis. The second axis forms a

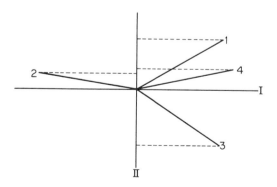

FIGURE 6–22

When a second axis, perpendicular to the first, is passed through the vertex, the projections of each vertex onto it produces a second set of projection lengths.

TABLE 6-8

**The Original Data Matrix Weighted by the Matrix of Loadings
to Score Each County according to the Two Principal Axes**

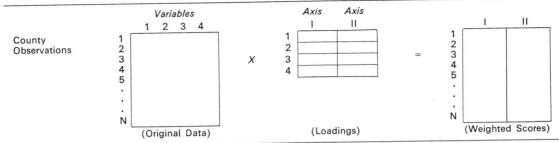

County Observations (Original Data) X (Loadings) = (Weighted Scores)

second scale. Many of the things we try to measure cannot be completely scaled on a single ruler. Rather, we may need several dimensions or scales for a particular measurement problem. But notice that the two new scales contain all the information that was originally held in the four variables.

The projection lengths are called *loadings*. They may be used as weights to combine the original four variables into two (Table 6-8). Once four variables are distilled to two we have the ingredients for the two dimensions or axes of a simple taxonomic space within which each county observation can be represented by a single dot. Note in Table 6-8 that the data in the original data matrix must be in comparable units. Rank order data are acceptable, or z-scores may be used for each variable where

$$z = \frac{X_i - \overline{X}}{\sigma}$$ as discussed in Chapter 4.

Once the four variables have been collapsed to two independent axes or reference scales, we can construct a taxonomic space with two independent dimensions containing no redundant information (Fig. 6-23). Since each county occupies a location in the taxonomic space, the last classification step is to begin to group the counties according to some acceptable linkage procedure.

CLASSIFICATIONS IN ACTION

Classifying Households, Housing Units, and Urban Census Tracts

We shall turn to the city for a practical example of how classification can shed light on important questions about households and housing usage patterns. Each urban household can be described according to socioeconomic attributes like (1) income, (2) educational level, and (3) the occupation of the principal breadwinner. The same households also can be described in terms of a second dimension called stage in the family life cycle. The second dimension is independent of the first and is based on the predictable sequence people pass through as they are born, grow up and

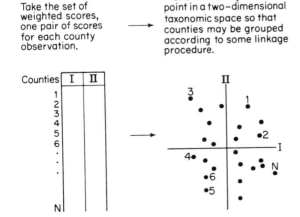

Take the set of weighted scores, one pair of scores for each county observation. → Plot each county as a point in a two-dimensional taxonomic space so that counties may be grouped according to some linkage procedure.

FIGURE 6-23

Weighted scores plotted in a taxonomic space of two dimensions.

pass through adolescence, reach maturity, marry and raise a family, watch the children mature while they themselves retire, grow old, and finally die. Stage in the family life cycle is reflected in measures like (4) size of the household, (5) age of the head of the household, and (6) number of dependent children in the household.

A data matrix describing household according to these six variables can be subjected to correlation analysis to reveal how the six variables are intercorrelated and therefore the degree to which they report redundant information (Fig. 6–24a). In this example the first three are highly intercorrelated and the last three are also highly intercorrelated. If each of the six variables is treated as a unit vector plotted in two-space, according to the principal axis (factor analysis) procedure discussed in the last section, then loadings (lengths of the vector projections on the axis) can be computed for each variable indicating its degree of correspondence with the principal axis (Fig. 6–24b). A second axis orthogonal to the first can be passed through the vertex and a second set of loadings are thereby provided. The loadings on the two axes or factors contain the information that was contained in the original matrix of correlation coefficients. The first factor represents socioeconomic status; the second factor is stage in the family life cycle. When the original data matrix is multiplied by the matrix of factor loadings, each household is scored in terms of its position on each of the principal axes (Fig. 6–24c). The households can then be plotted in a *social* space of two dimensions (Fig. 6–25). Socioeconomic status ranges from low to high along the vertical axis; stage in the life cycle is measured along the horizontal axis. A grouping procedure can then start from this point.

After we have classified households, we can classify the housing units that the households occupy. Each urban housing unit has value and quality attributes that correspond to the socioeconomic status of households. Included are things like (1) persons per room; (2) condition of the housing unit; (3) its price. Housing units can also be measured in terms of style features such as (4) building type—from high rise

apartment to single family house; (5) the size of a housing unit's private lot and yard; and (6) the number of dwelling units per acre. Each style of housing unit corresponds to a type of housing generally needed at a certain stage in the family life cycle.

A data matrix describing housing units can be evaluated using the same procedures as those employed with the household data. This time, however, the weighted scores describing the housing units can be plotted in a *housing* space of two dimensions. Housing value and quality is represented on the vertical axis, and housing type is scaled along the horizontal axis (Fig. 6–26). In the same way that each household occupies a position in social space determined by socioeconomic status and life cycle status, so the dwelling units occupied by the households occupy analogous positions in housing space.

We can carry the example one step further and consider the households and their housing units simultaneously by focusing on the housing process itself. The housing process can be defined as a set of households in a place living in the housing units located at that place. Each household has a set of attributes which relate to its housing needs and wants. A household's housing requirements vary systematically throughout the household's life cycle. The housing stock in a place depends on the construction history of that place. Consequently the attributes of the housing units are created to a large extent independently of the present households. Yet when several household classes purchase and occupy the different kinds of housing available in a city, a series of distinctive housing usage patterns is the outcome. Using the household and housing unit data which is available from the census of population and housing, we can classify census tracts into housing usage patterns according to either of two procedures.

Following the first procedure we can take household and housing data by census tract, convert it to z-scores, compute correlation coefficients between all pairs of variables, compute the loadings of each variable on two orthogonal axes, and use these loadings to compute weighted scores for each census tract

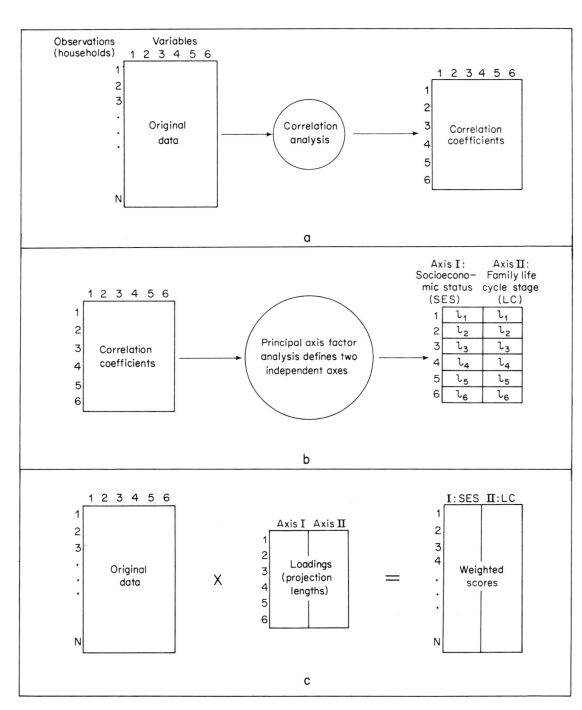

FIGURE 6–24

Reduction of an original data matrix to a matrix of weighted scores.

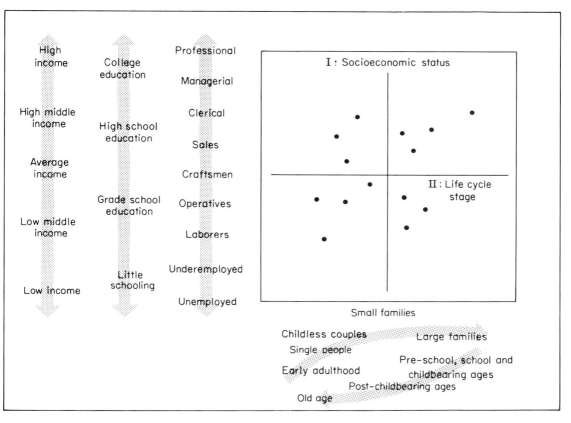

FIGURE 6–25

A social space of two dimensions. Each household occupies a position inside the social space (adapted from B. J. L. Berry and P. H. Rees, "The Factorial Ecology of Calcutta," American Journal of Sociology, LXXIV:5 [March 1969], Fig. 14, p. 464. Copyright © 1969 The University of Chicago Press).

according to the two principal dimensions of variation. Since each tract would occupy a position on each dimension we could plot each tract in a two-dimensional *community* space and then group the tracts according to any acceptable grouping or linkage procedures (Fig. 6–27). In Midwestern American cities it appears that census tracts which score high on dimension I, such as tracts 2, 1, and 6 in Fig. 6–27, usually lie in the wealthy socioeconomic sector or sectors of the city. Alternatively, tracts scoring low on dimension I are concentrated in other sectors. Tracts 8, 9, and 11 define such a low socioeconomic sector where household needs and tastes are provided for with housing units

of appropriate quality, style, and price. The variation of tracts along the second dimension, on the other hand, reflects housing use patterns for families at different stages in the family life cycle. Tracts in the innermost ring score lowest on dimension II. These tracts comprise households at the beginning and the end of their life cycles and housing requirements. The very old and the very young have the most modest housing requirements and the least ability to pay for them. Tracts scoring highest on dimension II such as 5, 6, 10, and 11 are those with households at the peak of their housing requirements. Between the two extremes are the families still growing or households in the process of

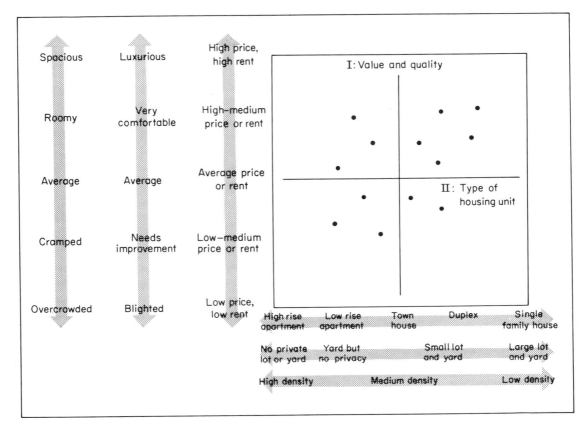

FIGURE 6–26

A housing space of two dimensions. Each housing unit occupies a position inside the housing space (adapted from B. J. L. Berry and P. H. Rees, "The Factorial Ecology of Calcutta," American Journal of Sociology, LXXIV:5 [March 1969], Fig. 14, p. 464. Copyright © 1969 The University of Chicago Press).

shrinking. Thus, the socioeconomic status of households and their related housing use patterns appear to define a sectoral patterning of census tracts. Variations due to changing housing requirements throughout the family life cycle coupled with the spatial arrangement of housing units by size, density, style, and price, promote what is called an annular or concentric zonation of housing use patterns.

A second approach to defining the separate housing usage patterns in the metropolis is a refinement of the first and also uses census tracts as observations and household data and housing data as variables. It examines five housing usage clusters, how households pass

from one cluster to another, where the clusters are located, and how they compete with one another. With few exceptions each household in a city belongs to one of five household

FIGURE 6–27

Defining a community of space of two dimensions. Each census tract occupies a position according to its fundamental household and housing attributes. Inside the physical space of the actual city, a tract's sectoral location depends on its dimension I score in the community space; ring location is related to its dimension II score (adapted from B. J. L. Berry and P. H. Rees, "The Factorial Ecology of Calcutta," American Journal of Sociology, LXXIV:5 [March 1969], Fig. 14, p. 465. Copyright © 1969 The University of Chicago Press).

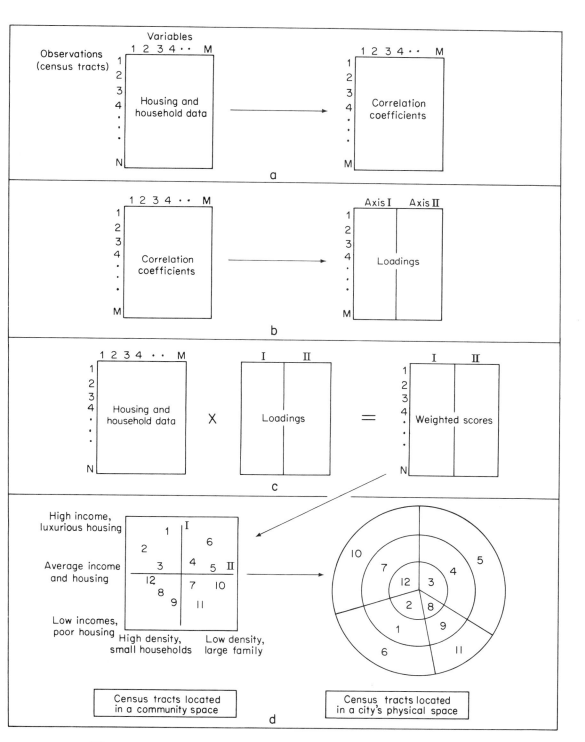

groups: (I) young, footloose cosmopolites; (II) blue-collar working-class families; (III) rising young families; (IV) mature established families; and (V) aged declining households. Each of the five groups forms a distinct usage cluster. Some households in group I pass in time to group II and remain there. In these families the principal breadwinner's earning capacity reaches its peak relatively early in his working years and housing consumption patterns stabilize accordingly. Not so with the remainder of the group I households. They proceed to group III and then on to group IV as upwardly mobile middle-class families who reach their peak earning power and housing consumption relatively late in their careers (Fig. 6–28).

To determine the location within the city of each of the five usage clusters we take household and housing data by tract, compute z-scores and correlation coefficients, and then compute the loadings of each variable on the required number of axes—five in this case. (Every well run computer center maintains a library of standard computer programs to

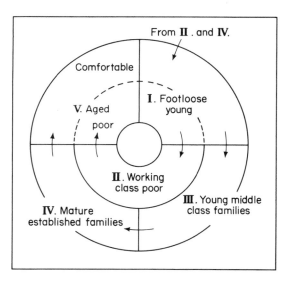

FIGURE 6–28

Five household groups within the family life cycle. Most single people and young marrieds in group I enter group III. Many families in group II remain there. Groups I, II, and V often compete for the same kinds of housing units in the same places in the city.

execute each of the steps of such an analysis. These programs allow the user to specify how many orthogonal axes should be defined so that factor loadings and factor scores may be computed.) When the original data matrix is multiplied by the matrix of factor loadings, each tract receives a score based on its position along each of the five separate dimensions.

Such an analysis was performed for the Twin Cities of Minneapolis and St. Paul and nearby suburbs using data from the 1960 Census of Population and Housing. In this real, and very complex example, five orthogonal axes were needed to define and measure the five major housing usage patterns. Each census tract received a weighted score for each of the five axes. The tracts scoring high on each dimension were mapped separately and the five maps revealed in sharp relief the five expected usage clusters (Figs. 6–29a to 6–29e).

I. Young, Footloose Cosmopolites. Young, unattached working and university populations constitute the first distinctive housing usage pattern which displays peak intensities around downtown Minneapolis and downtown St. Paul, and a third peak around the campus of the University of Minnesota midway between the two downtowns (Fig. 6–29a). Additional corridors of housing usage by these footloose cosmopolites extend along major transportation arteries. Persons in this usage cluster are often full- or part-time college students. They have traveled abroad, often walk to work, live with friends rather than relatives, and occupy rented quarters.

II. Blue-collar Working-class Families. The second housing usage pattern (Fig. 6–29b) is created by a blue-collar or working-class population clustered near the downtown with two important outlyers, one a black ghetto area immediately to the north and west of downtown Minneapolis, and the other area to the northwest in an old streetcar suburb which recently has been engulfed by the postwar suburbanization wave. Households in the usage cluster are mobile and usually rent their housing; if owned, housing is often substantial. In the same cluster live recent migrants from the South. The number of persons per housing unit is high and many households comprise single persons

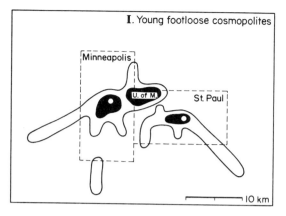

FIGURE 6–29a

Young single people and childless married couples—workers and university students—cluster near the downtowns, the university, and along major transport corridors.

living alone or groups of unrelated individuals. Very few people in this cluster attend high school, implying that they have been graduated or have dropped out.

The location of this unconsolidated working-class cluster around the downtowns occasionally brings its households into competition with the young middle- and upper-middle-class unattached cosmopolites. To the extent that the latter group has superior purchasing power, the working-class households are left with what amounts

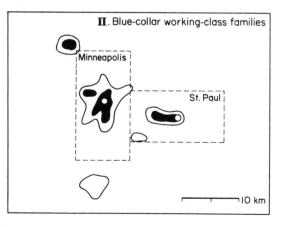

FIGURE 6–29b

Low-income working-class families concentrate around the downtowns with additional clusters in older inexpensive suburban housing.

to a housing residual. Accordingly, some working-class households prefer equivalent housing remote from the downtown. The young cosmopolites prefer locations easily accessible to the downtowns, or the university, or both.

III. Rising Young Families. The third kind of housing usage (Fig. 6–29c) comprises the young middle-class family, which is well educated, owns its own home, has plenty of young children, and makes heavy use of the automobile. Such clusters dominate four large suburban zones at the edge of Minneapolis and five others overlapping the edges of St. Paul. In the type III usage zone, tract housing on a large scale is common over large areas. At the central city edges, between the downtown and the rising young families' usage zones, are older developed areas, either run down, too expensive, too crowded, or otherwise unsuited to the needs and tastes of young middle-class families.

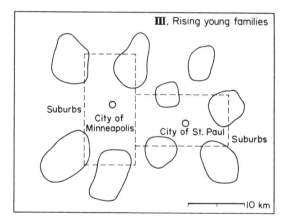

FIGURE 6–29c

Young middle-class families predominate in intermittent tract housing zones at the fringes of the central cities.

IV. Mature Established Families. The fourth usage class, comprising mature families (Fig. 6–29d) represents the next stage in the family life and income cycle when a household's housing needs are at a maximum. Housing is expensive whether it is owned or rented. Families are prosperous and well educated. Typical households live a significant distance from the downtown and deteriorating or dilapidated

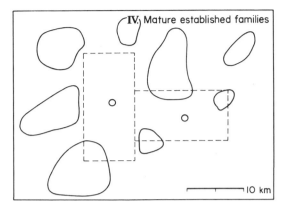

FIGURE 6–29d

In the more exclusive suburbs prosperous middle-aged families predominate. Instead of tract housing, residences usually are designed and constructed individually.

housing units are a rarity. Minneapolis and St. Paul each has four such zones for mature, established families. Occasionally the zones overlap the central city. Normally they are confined to a few select suburbs. Presumably there is little competition between the households in classes III and IV. Income and life cycle stage mark the main differences between them. When a class III family's income and age finally place it in class IV, the family usually moves from its old house to a new residence in a class IV neighborhood.

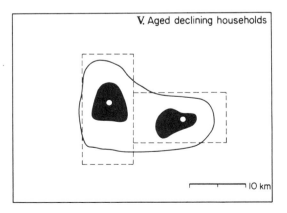

FIGURE 6–29e

In 1960, the suburban apartment boom was not yet under way. The elderly, living on reduced incomes, concentrated in rental quarters near the downtowns.

V. Aged Declining Households. The last of the five important usage clusters (Fig. 6–29e) comprises elderly men and women, living as retired couples or alone in older rental quarters near the downtown. Children are virtually absent. So is new high quality housing. Families use public transportation instead of private automobiles and they usually do not change their residence unless forced. Within this elderly group, the foreign-born constitute a significant fraction. The housing requirements and income status of this usage group somewhat resemble that of Class I: the young footloose group at the other extreme of the household life cycle. Thus, at central city locations near the downtown, Classes I and V compete with one another and with Class II, the unconsolidated working class. A housing problem created in one of these three usage classes quickly becomes a problem for all.

To summarize, we may classify (1) households, or (2) housing units, or (3) housing usage patterns. In the third case our observation unit is the census tract. Starting with a data matrix of housing and household data, census tracts may be scored with respect to two or more independent dimensions of variation. Once the census tract observations are located in a two-, three-, or *n*-dimensional taxonomic space, clustering may proceed according to any acceptable method. Grouping may cease at any point and the separate groups may be mapped to reveal the spatial expression of the taxonomy. Another procedure, employed in the Twin Cities example, stops short of the linkage analysis. Instead of classifying census tracts as OTU's in a five-dimensional taxonomic space, a separate map was prepared to describe the pattern of scores along each dimension of variation. The five maps representing five independent dimensions of variation really represent five separate classifications of the census tracts.

Classifying Weather Stations and Defining Climatic Types

The classification procedures available in human geography can be used with equal effectiveness in grouping places according to their physical attributes like temperature, humidity, precipitation, weather variability, wind velocity, and so forth. The traditional classifications of climate are based mainly on average annual temperature and on average annual precipitation with some attention paid to their seasonality. The general idea is that places vary from hot to cold and from humid to dry. If these two dimensions of variation were independent, this approach toward classifying the places in the world might have been quite satisfactory. But two major problems cloud the issue. In the first place, temperature and humidity are not independent. It takes substantially more rain to make a hot place humid than it takes to make a cool place humid. If we define a dry place as one where there is not enough precipitation for plants and man, threshold water requirements cannot be specified until we know what temperature regime we are referring to. The second problem concerns the seasonality of temperature and rainfall. Some places have a wet season and a dry season as well as a cool season and a warm season. If the rain comes in the summer when it is warm, it might be inadequate to water crops and vegetation or supply man's needs. On the other hand, if the same amount of rain were to fall in the cool season when evaporation and plant transporation rates are reduced and water has time to soak in and recharge underground storage, then we might argue that precipitation is adequate.

Climate is the average course or condition of the weather at a particular place over a period of many years. Climate is multidimensional in character. Every weather station in the world gathers a different set of weather data each day. The task of the climatologist is to classify the stations into a smaller number of manageable classes. In other words, since the climate at each place as measured by a set of climatic elements is always unique, climatic stations can be grouped only on the basis of their similarities; no two of them are identical. But how should similarity be measured? Dieter Steiner posed this question using climatic data from sixty-seven weather stations across the coterminous United States. For each of the sixty-seven observations he assembled the following sixteen pieces of information:

1. Mean annual temperature (in degrees Fahrenheit) for the period 1931–1960
2. Mean January temperature: 1931–1960
3. Mean July temperature: 1931–1960
4. Mean annual precipitation (in inches): 1931–1960
5. Mean January precipitation: 1931–1960
6. Mean July precipitation: 1931–1960
7. Mean annual percentage of possible sunshine with length of record varying from 2 to 67 years depending on the station
8. Mean January percentage of possible sunshine, variable record length
9. Mean July percentage of possible sunshine, variable record length
10. Mean annual relative humidity at 1:00 P.M. in percent, length of record variable from 2 to 72 years
11. Mean January relative humidity at 1:00 P.M. in percent, variable record length
12. Mean July relative humidity at 1:00 P.M. in percent, variable record length
13. Temperature range (July/January in degrees Fahrenheit): 1931–1960
14. Precipitation ratio (July/January): 1931–1960
15. Sunshine ratio (July/January), variable record length
16. Humidity ratio (July/January), variable record length

These sixteen variables were measured instead of others because traditional definitions of climate include them. (Some contemporary definitions also include things like wind speed and certain kinds of radiation.) The aim of the classification is to group the weather stations on the basis of similarities in their climatic records. Using the principal axis factor analysis method described earlier, four basic uncorrelated axes were identified. No less than 88.6 percent of the variation in the original data matrix of sixty-seven observations and sixteen variables was accounted for by the four principal axes or components. This means that a great deal of redundant information is present in the sixteen variables chosen for the classification. Thus, the climate of the United States, described by sixteen original variables and recorded at sixty-seven stations can be measured with a high degree of accuracy by four new and independent components, which may be termed: humidity; atmospheric turbidity; continentality;

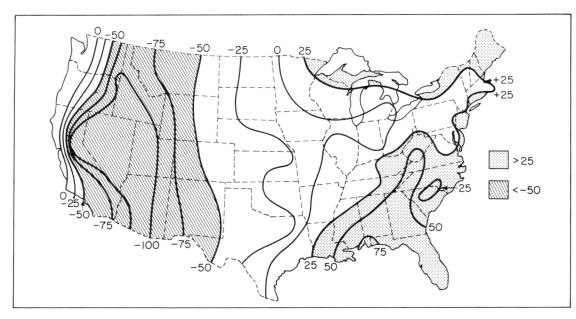

FIGURE 6–30

Isoline map of scores on component I: Humidity (adapted from D. Steiner, Fig. 2).

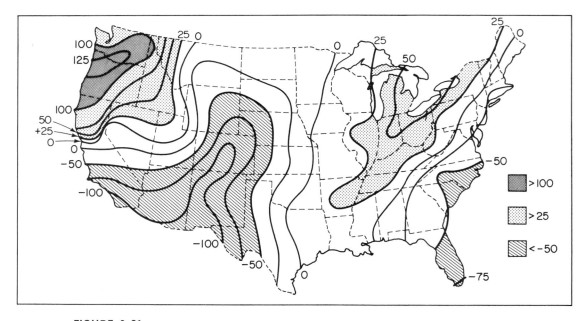

FIGURE 6–31

Isoline map of scores on component II: Atmospheric Turbidity (adapted from D. Steiner, Fig. 3).

and thermality. Since each of the sixteen original variables is associated to a greater or less degree with each of the four principal components, and since the closeness of the association in each case is reported by the loading of each variable on each component, the 16 by 4 matrix of factor loadings can be used to compute four weighted scores for each of the sixty-seven weather stations.

A map was constructed for each component using the set of sixty-seven scores from each. The first component, called humidity, is the most important because it accounts for 28.8 percent of the variation in the original data matrix (Fig. 6–30). This component expresses the dominant role of summer rainfall. Highest scoring places include the region southeast of the northern Great Lakes. The lowest scoring places include most of the Great Plains and the Rocky Mountain areas.

Stations with high scores on the second component have high relative humidity and very little sunshine, especially in the wintertime (Fig. 6–31). Maximum atmospheric turbidity, indicated by the highest scores, occurs in the Pacific Northwest and around the Great Lakes. The greatest absence of these conditions is recorded in the clear and arid Southwest and in the south eastern Atlantic region.

The third component reflects station to station variation according to thermal continentality (Fig. 6–32). Stations with high factor scores have large temperature ranges and high precipitation ratios. The same stations have extremely low January temperatures and receive their maximum rainfall in the summer.

The fourth component reflects thermality and weather station scores vary latitudinally (Fig. 6–33). The highest scoring stations are located across the southern part of the country and the lowest scores are found adjacent to the Canadian border and along the West Coast.

Once the scores are assigned to each weather station it is possible to group the sixty-seven stations according to their scores on the four dimensions. Imagine that each of the stations is a point in a four-dimensional taxonomic space. The distance between points is a measure

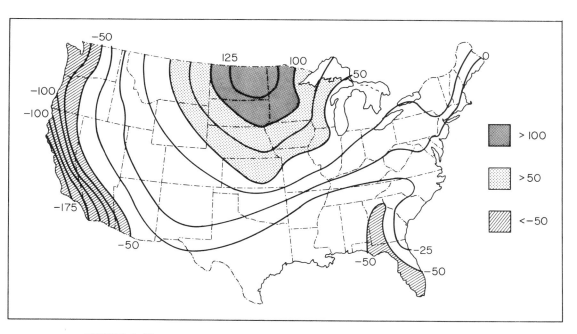

FIGURE 6–32

Isoline map of scores on component III: Continentality (adapted from D. Steiner, Fig. 4).

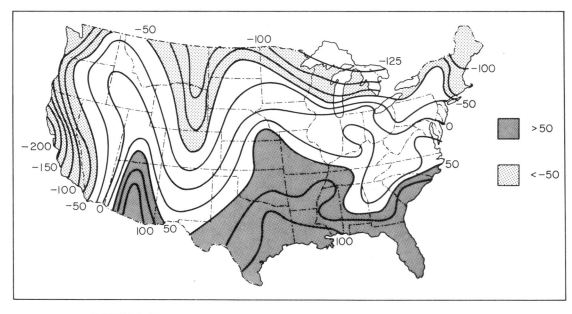

FIGURE 6–33
Isoline map of scores on component IV: Thermality (adapted from D. Steiner, Fig. 5).

of the similarity of the climate recorded at those stations. The points can be grouped according to a systematic linkage procedure on the basis of their proximity in the four-dimensional taxonomic space. Steiner began his grouping with a 67 by 67 distance matrix. Then he put the two stations separated by the smallest distance into one group, calculated the group mean for each of the four dimensions to determine the center of gravity of the group. Then, treating the cluster as one observation, a 66 by 66 distance matrix was computed. The process was repeated and the linkage continued until all stations had been joined into one cluster of sixty-seven stations. Moving backward up the main trunk of the linkage tree, we can obtain the first optimum branching of the weather stations into two groups. Another movement up the tree provides three groups; another step means four groups and so forth. There is one optimum solution for each level of detail required. If we want to divide the sixty-seven weather stations into five different climatic classes, then we stop clustering when we have branched five times in the tree leaving five

clusters yet unjoined. Optimum grouping in this context means that each cluster of stations has a maximum degree of homogeneity. This implies that the distances within a group lying in the taxonomic space are at a minimum. If a new and ungrouped station can only join a cluster of stations located next to it on the earth, then the grouping procedure yields *specific climatic regions*. The observations in each taxonomic cluster are also contiguous, or touching one another on the earth's surface. If the stations are allowed to group whether or not they are neighbors on the earth then *general climatic types* are defined.

Classification and Regionalization

The delimitation of regions is basically a problem in classification. Whether we are grouping land parcels, farms, census tracts, neighborhoods, or larger areas, our individual observations (operational taxonomic units) must be combined on the basis of similarities to form a smaller number of clusters. Each small place on the surface of the earth is unique, yet it has combinations of attributes that resemble combinations at other places. Although every place is unique, we can know about places only according to their resemblance or lack of resemblance to other areas.

We do not arrive at an understanding of the agricultural landscape of Iowa by examining farms one at a time. Nor do we attempt to make sense out of the physical and cultural diversities of St. Louis by a house to house search. Too much detail is just about as useless as too much generality. Yet statistics about the average Iowa farm are as meaningless as a case study of one farm if our objective is an understanding of the areal patterning of agricultural land uses across the state. In the city, average figures describing households and housing conditions fail to capture the essentials of the patterning of housing usage areas in the city. We avoid too much uniqueness and too much generality by assembling information into manageable areal aggregates termed *regional systems*. Regional systems are areal classifications lying somewhere between maximum uniqueness and maximum generality.

There are two types of regional systems: general and specific. In *general* (or *generic*) *regional systems*, types of places resemble one another according to a certain set of attributes like climate, language, cultural heritage, human use of the earth, and so forth. But all classifications are designed with a purpose in mind, and general regional systems are no exception. Attributes are therefore selected according to the purpose of the classification. The important feature of general regional systems is that places of each *type* in the system may be located at widely separated absolute locations. For example, one general regional system may come about from the classification of all places in the world according to the major mother tongue spoken in the place. In such a system the English language *type* includes such far-flung places as Australia, Jamaica, the United States, part of Canada, England, and others.

Specific regional systems are defined not only by combinations of intrinsic attributes, but by location as well. Whereas in general regional systems a type of place can occur at widely separated locations, in specific regional systems

all the parts of a homogeneous region must be spatially contiguous. This distinction is important when we begin to cluster similar places into larger homogeneous regions.

Let us examine the regionalization problem as a classification problem and pay special attention to the way a contiguity constraint affects the classification of places into homogeneous regional systems. Consider the forty-two counties of England in terms of their employment and industrial structures and let us suppose that it is necessary to group them into a smaller number of classes on the basis of their similarities. Nigel Spence studied this problem and grouped the counties first without regard to their contiguity, and again allowing only contiguous counties to cluster.

The raw data matrix consisted of sixty industrial employment categories and forty-two counties. Each county was an observation and employees in each county were assigned to the various employment categories. Then each employment value was expressed as a percentage of total employment within the specific county to suppress the effects of the vastly different employment totals in the different counties. After some transformations of the data, the z-scores were computed, and these were used as the original observations for the classification of the counties.

The 42 by 60 matrix of z-scores was submitted to a correlation analysis and the 60 by 60 matrix of correlation coefficients revealed that there were numerous redundancies in the patterns of variation of England's county employment values. The principal axis procedure was used to identify eight independent components of variation upon which each of the sixty variables had loadings of greater or less magnitude. The eight components accounted for 72 percent of the variation in the original data matrix. When the data matrix was multiplied by the matrix of factor loadings the product was a matrix of factor scores—that is, forty-two counties located in an eight-dimensional taxonomic space and ready for grouping.

The first grouping was accomplished without any requirement of actual contiguity in England. The second time, contiguity in England's absolute space was required before points in the

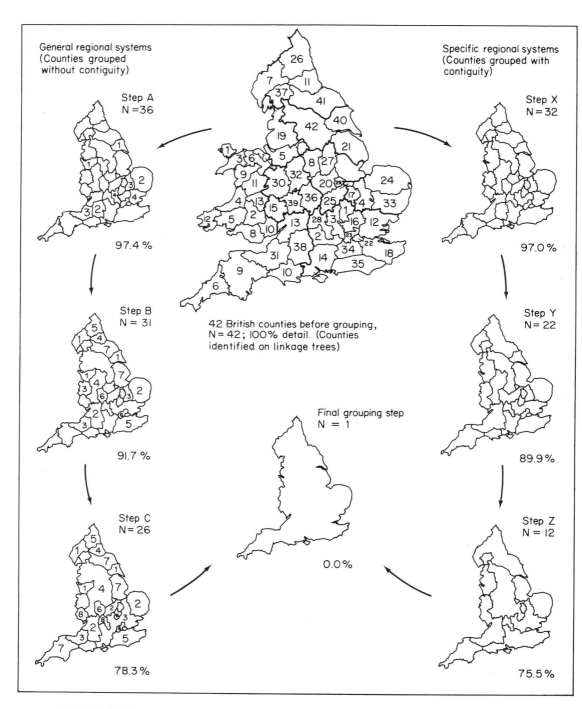

FIGURE 6–34

County classification without contiguity and with contiguity (adapted from N. A. Spence, Figs. 7 and 8).

taxonomic space could join together. The following procedure was used. A 42 by 42 matrix of distances between the counties (operational taxonomic units) in the eight-dimensional space was computed. Since close points are considered similar and distant points are unlike, the shortest distance in the distance matrix identified the pair of counties that clustered first. For purposes of identifying the location of the cluster in eight-space the two-county cluster is replaced by a new point whose location with respect to each dimension equals the mean value of the OTU's that the point represents. Next, a 41 by 41 distance matrix is computed and the shortest distance in it identifies the two points which cluster next. The points are replaced by a single point and a 40 by 40 distance matrix is computed, and so on. Each linkage step reduces by one the number of points remaining in the eight-space, yet at each clustering step the distance matrix gets smaller. The sum of the distances squared (ΣD^2) is a measure of the amount of detail present in the OTU's before the first pair is joined. As grouping proceeds, the ΣD^2 *between* groups declines as the number of groups declines and as detail is lost. Prior to the first grouping the *within*-group ΣD^2 is zero because there are no groups. As grouping proceeds the ΣD^2 *within* groups increases as the number of groups declines and as generality increases. Before linkage begins, 100 percent detail is retained; after all forty-two counties have been grouped into one class, 100 percent of the detail has been lost.

Any classification effort tries to produce the smallest number of classes possible while keeping the within-group ΣD^2 as small as possible and the between-group ΣD^2 as large as possible. If we are grouping the counties without regard to contiguity (Fig. 6–34), then at step A four pairs of places that have joined in the eight-dimensional taxonomic space because of their similarities are separated on the map of England. In the attempt to reduce the forty-two counties into a smaller number of relatively homogeneous areas the grouping has not been too successful at this stage. Thirty-six separate parcels remain, although eight of them are of four types. Even at step C twenty-six parcels still remain, forming a confused and ill-defined pat-

tern (Figs. 6–34 and 6–35). Eight of the clusters at step C have two or more noncontiguous segments. If the objective of the taxonomy is to produce a smaller number of relatively homogeneous regions, little is accomplished unless the clustering procedure reduces the areas to a manageable number before the within-group ΣD^2 mounts to high levels.

Linkage under a contiguity constraint is the practical alternative (Fig. 6–36). The linkage procedure when clustering under a contiguity constraint has one additional step. The 42 by 42 distance matrix is supplemented by another 42 by 42 contiguity matrix. If a pair of places is contiguous, then a one is entered in the respective cells. A zero entry in a cell indicates that two places are not contiguous. At step one of the linkage procedure the 42 by 42 distance matrix is searched for the smallest entry. When it is located the same position is checked in the contiguity matrix. If a one appears, the two places join; if a zero appears the places cannot join and the next smallest distance in the distance matrix must be checked, and so forth. After the linkage process is underway, and clusters have formed, a cluster may link with its nearest neighbor in the taxonomic space if any of the cluster's members is contiguous to the nearest neighbor (compare steps X, Y, and Z with steps A, B and C in Fig. 6–34).

When clustering proceeds under a contiguity constraint, each clustering step reduces the number of remaining parcels by one. At the end of step X there are thirty-two parcels; at the end of step Z there are twelve remaining. If we were interested in setting up twelve regional employment service offices to serve our forty-two counties and we wanted the labor market areas to be relatively homogeneous in employment patterns, it would make more sense administratively to cluster counties under a contiguity constraint than to cluster without the constraint. Moreover, we can usually achieve contiguity in our regionalization without much cost in higher within-group ΣD^2. Without contiguity at step C we have lost 21.7 percent of the detail and have twelve types of places, but they are located in twenty-six separate parcels (Figs. 6–34 and 6–37). Clustering under the contiguity constraint we

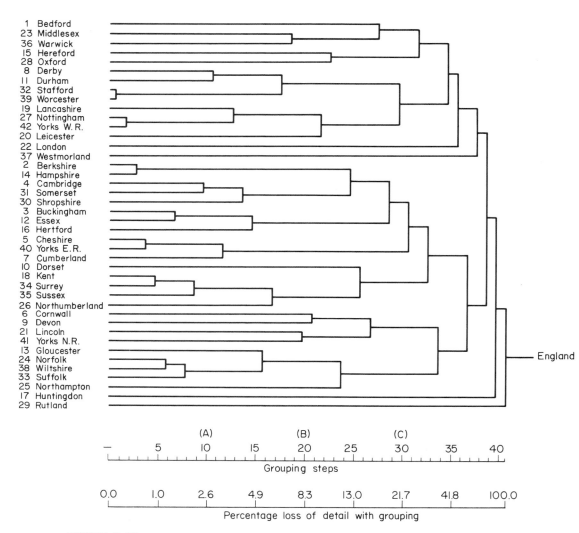

FIGURE 6-35

Linkage tree grouping of England's counties without contiguity (adapted from N. A. Spence, Fig. 4).

have lost only slightly more detail (24.5 percent), but we have just twelve compact and contiguous regions at step Z. The 2.7 percent detail that is paid for using the contiguity constraint is easily repaid through the elimination of fourteen of the twenty-six separate parcels produced without the contiguity constraint. As a general rule, since places on the earth are usually more like their neighbors than they are like distant places, classification of places under a

contiguity constraint usually carries a low cost and returns large benefits in the form of uncluttered regional systems.

The idea of multifactor uniform regions can be extended to every manner of geographical individual. The procedures used here have been used in one form or another to classify towns in Britain, Canada, and India. The same procedures have been applied to urban subareas, especially census tracts and blocks. When

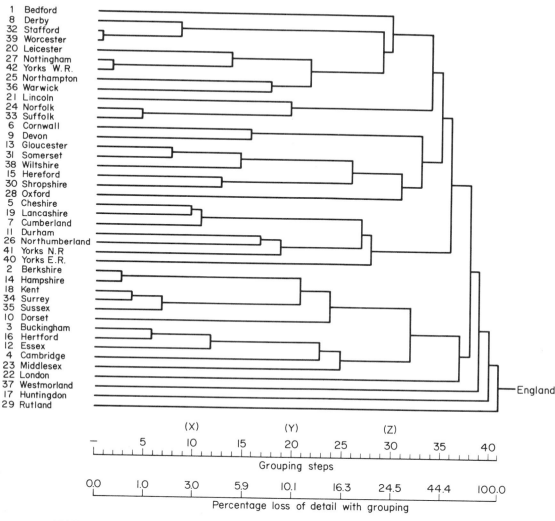

FIGURE 6–36

Linkage tree grouping of England's counties with contiguity (adapted from N. A. Spence, Fig. 5).

data comes in the form of county or township or farm crop information, the ingredients are available for the definition of agricultural regions of any desired detail or generality. Counties in New York State have been classified according to their economic health, and electoral districts have been evaluated to determine the extent to which gerrymandering has created excessively heterogeneous voting areas. The techniques described here have obvious appli-cation in regional economic planning, metropolitan planning, and the planning of public service facilities of every kind. Although the general ideas are quite logical and straightforward, the computations for any large problem must be performed by computer. Fortunately, virtually every well-managed computer center has, in its library, programs which will perform these taxonomic operations if you can supply the problem and the data.

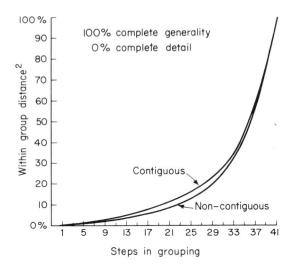

FIGURE 6–37
Within-group variance at each step of the grouping of England's counties with/without contiguity (adapted from N. A. Spence, Fig. 6).

Suggestions for Further Reading

AHMAD, Q. *Indian Cities: Characteristics and Correlates.* Department of Geography Research Paper No. 102. Chicago: The University of Chicago, 1965.

BERRY, BRIAN J. L. "Basic Patterns of Economic Development," in Norton Ginsberg, ed., *Atlas of Economic Development.* Chicago: The University of Chicago Press, 1961.

———. "A Method for Deriving Multifactor Uniform Regions," *Przeglad Geograficzny*, XXXIII (1961), 263–82.

———, and P. H. REES. "The Factorial Ecology of Calcutta," *American Journal of Sociology*, LXXIV: 5 (March 1969), 445–91.

BUNGE, W. "Delimitation of Uniform Regions," in *Theoretical Geography.* Lund: C. W. K. Gleerup, 1966, pp. 95–100.

———. "Gerrymandering, Geography and Grouping," *Geographical Review*, LVI (1966), 256–63.

FAIR, T. J. D. "A Regional Approach to Economic Development in Kenya," *The South African Geographical Journal*, XLV (1963), 55–77.

GRIGG, D. "The Logic of Regional Systems,"

Annals of the Association of American Geographers, LV (1965), 465–91.

HARMON, H. *Modern Factor Analysis.* Chicago: The University of Chicago Press, 1960.

JOHNSTON, R. J. "Choice in Classification: The Subjectivity of Objective Methods," *Annals of the Association of American Geographers*, LVIII (1968), 575–89.

KEMENY, J. G., et al. "Sets and Functions," in *Finite Mathematical Structures*, Chap. 2. Englewood Cliffs, N. J.: Prentice-Hall, Inc., 1959, pp. 51–111.

KING, L. J. "Cross-Sectional Analysis of Canadian Urban Dimensions: 1951 and 1961," *Canadian Geographer*, X:4 (December 1966), 205–24. Reprinted in Brian J. L. Berry and Frank E. Horton. *Geographic Perspectives on Urban Systems, with Integrated Readings.* Englewood Cliffs, N. J.: Prentice-Hall, Inc., 1970, pp. 154–67.

LANKFORD, P. M. "Regionalization: Theory and Alternative Algorithms," *Geographical Analysis*, I (1969), 196–212.

MOSER, C. A., and W. SCOTT. *British Towns: A Statistical Study of Their Social and Economic Differences.* Edinburgh: Oliver and Boyd, 1961.

RAY, D. M., and B. J. L. BERRY. "Multivariate Socio-economic Regionalization: A Pilot Study in Central Canada," in T. Rymes and S. Ostry, eds., *Regional Statistical Studies.* Toronto: University of Toronto Press, 1965, pp. 1–48.

SMITH, R. H. T. "The Functions of Australian Towns," *Tijdschrift voor Economische en Sociale Geografie*, LVI (1965), 81–92.

SOKAL, R. R. "Numerical Taxonomy," *Scientific American*, CCXV:6 (December 1966), 106–16.

———, and P. H. A. SNEATH. *Principles of Numerical Taxonomy.* San Francisco: W. H. Freeman, 1963.

SWEETSER, F. L. "Ecological Factors in Metropolitan Zones and Sectors," in M. Dogan and S. Rokkan, eds., *Quantitative Ecological Analysis in the Social Sciences*, Chap. 16. Cambridge, Mass.: The M. I. T. Press, 1969, pp. 413–56.

TAYLOR, P. J. "The Location Variable in Taxonomy," *Geographical Analysis*, I (1969), 181–95.

THOMPSON, J. H., et al. "Toward a Geography of Economic Health: The Case of New York State," *Annals of the Association of American Geographers*, LII (1962), 1–21.

U.S. BUREAU OF THE CENSUS. *Statistical Abstract of the United States: 1968*, 89th ed. Washington, D.C.: U.S. Government Printing Office, 1968.

Works Cited or Mentioned

BERRY, B. J. L., and P. H. REES. "The Factorial Ecology of Calcutta," *American Journal of Sociology*, LXXIV: 5 (March 1969), 445–91.

FAIR, T. J. D. "A Regional Approach to Economic Development in Kenya," *The South African Geographical Journal*, XLV (1963), 55–77.

GRIGG, D. "The Logic of Regional Systems," *Annals of the Association of American Geographers*, LV (1965), 465–91.

LESSING, L. "Systems Engineering Invades the City," *Fortune*, LXXVII (January 1968), 154–57.

SOKAL, R. R., and P. H. A. SNEATH. *Principles of Numerical Taxonomy*. San Francisco: W. H. Freeman, 1963.

SPENCE, N. A. "A Multifactor Uniform Regionalization of British Counties on the Basis of Employment Data for 1961," *Regional Studies*, II (1968), 87–104.

STEINER, D. "A Multivariate Statistical Approach to Climatic Regionalization and Classification," *Tijdschrift van het Koninklijk Nederlansch Aardrijkskundig Gerootschap Tweede Reeks*, LXXXII:4 (October 1965), 329–47.

U.S. BUREAU OF THE CENSUS. *Statistical Abstract of the United States: 1968*, 89th ed. Washington, D.C.: U.S. Government Printing Office, 1968.

PART III

Location
and
Spatial Interaction

CHAPTER 7

THE BASES
FOR SPATIAL INTERACTION

*The shortest distance between two points is approximately
seven inches.*

—*Ephraim Ketchall*

TRANSPORTATION DEVELOPMENT

Transportation and communications systems are created or improved only after there is a demand which the new systems will satisfy. For example, it was not until many settlers had already come to the United States that regularly scheduled steamship service was established between America and Europe. Similarly, pioneers pushed west of the Appalachians and started farming long before the Erie Canal connected the Great Lakes with the Hudson River and the Atlantic Coast, or before railroads were sent westward from the great east coast port cities. Transportation improvements fostered the increased interactions, but they were not solely responsible for them. The geographer Edward L. Ullman postulated three conditions affecting transportation development: complementarity, intervening opportunity, and transferability. Let us examine each condition in turn.

Complementarity

Traditionally it was assumed that interactions between places developed because of areal differentiation—the fact that places differed from one another. This is true to some extent of course, but mere differentiation never produces interaction by itself. For counterexamples we have only to consider the many different parts of the world which have no interaction of any kind with one another.

For two places to interact there must be a demand in one place and a supply at another, and the demand and the supply must be specifically complementary. To take an absurd example as an illustration, if there is a shortage of beef in New York, it makes little difference that there is a marketable surplus of pulpwood in Georgia. The two commodities are not substitutes, and a flow of wood from Georgia to New York will not come about because of New York's meat requirements.

193

The two places lack specific complementarity.

Complementarity is so important as a basis for spatial interaction that many very low-value bulk commodities can move many thousands of kilometers if complementarity conditions are properly met. Grain from continental interiors moves to sea ports, and from the ports moves by inexpensive ocean transport to markets on the other side of the world. Crude and refined petroleum products—commodities of relatively low value per unit of bulk—move enormous distances, from oil fields in the Middle East, the Canadian Prairies, and Venezuela to industrialized areas of Europe, Japan, and North America. When steel mills were built in the Chicago region, the need for coking coal was great enough to draw supplies from West Virginia despite the relatively low value of the coal and a distance overland of more than eight hundred kilometers. Without the specific complementarity of such supply and demand regions, the movements and interactions would never have taken place.

Intervening Opportunity

Complementarity between places can generate interchange only in the absence of intervening opportunities. If we are considering the potential for a movement of goods from place A to place B we have to consider any place C between them which might act as an intervening origin or an alternative destination. For example, New York is closer to Florida than to California so for New Yorkers anxious to take a winter holiday, Florida constitutes an intervening opportunity between New York and California. When young men and women are graduated from high school in eastern North Dakota, many of them migrate eastward for jobs or higher education. Some stop at the state universities at Fargo or Grand Forks, and others go on to the Twin Cities of Minneapolis and St. Paul. The Twin Cities would attract more young people than they do already if the intervening opportunities offered in the towns of North Dakota were absent. At a wider scale, Chicago would attract more migrants from western Minnesota and the Dakotas were it not for the intervening opportunities offered by the Twin Cities. Movements of commodities can be discussed at a very general level in the same terms as migration of people. Many fewer goods and services move from Philadelphia to Connecticut than otherwise would if New York City did not lie between them. In a sense, intervening opportunities are spatial sponges soaking up potential interaction between complementary places.

Intervening opportunities do not always curtail long-distance interactions. It is entirely possible that a sequence of such opportunities can help to create interactions between widely separated areas by making intermediate transport links profitable and thereby paying part of the costs of the link between distant places. This is what happened in the construction of the transcontinental railroads. Originally the lines were laid to connect nearby places with lake and river ports. Later, inland centers were connected with their nearest neighbors as larger trade and service centers tried to capture hinterlands for their central place activities. Finally, the various lines were adjusted to the same gauge (width between the rails) and linked together so that east and west coasts were connected by a continuous route.

Transferability

Besides complementarity and intervening opportunity, the third condition under which spatial interaction occurs is transferability—the friction of distance. Transferability is measured in real time and money costs. If the time and money costs of traversing a distance are too large, the movement will not take place despite perfect complementarity and the absence of intervening opportunities. Instead of reaping the benefits of interaction people will stay where they are and continue unchanged the life that they know. If goods cannot move because of the high cost of movement, other goods will be substituted if possible, or people will just go without. Transferability differs between places, between classes of movements, and between modes of movement. As we shall see in the next chapter, transferability changes dramatically through time.

As we consider interaction potential at the

worldwide scale, we recognize that all places are not equally linked. Distances, localized areal differences, and intervening opportunities may drastically reduce the number of long-distance interactions. As examples of two extreme situations, Great Britain and the United States contrast sharply in this respect. Britain has had to trade with many parts of the world to reach enough complementary resources and markets. The United States, on the other hand, has so many large and rich complementary areas within its borders that it spends the vast majority of its interaction efforts trading with itself.

The changing pattern of America's international trade reflects all three bases for interaction. Consider American imports in 1955 and again in 1967 (Table 7–1). The most important North American imports have been movements of pulp and paper, and machinery and transport equipment from Canada. The demand in the United States matched the complementary supplies available in Canada so the first of the conditions for interaction was fulfilled. The United States and Canada share a common border of several thousand kilometers. Intervening opportunities which might have intercepted the flows southward from Canada are absent. Transferability has been no obstacle because inexpensive rail, highway, and water shipments allow almost all classes of goods to move as easily between the two countries as they move between the states of the United States. Tariffs are the only real trade barrier. At the other extreme, Australia, New

Zealand, and Oceania are half a world away, and the United States can usually buy from closer sources. There are also intervening opportunities in the Pacific between these countries and the United States.

South America's exports to the United States were virtually stagnant for the twelve-year period from 1955 to 1967 (Table 7–1). During those years the American market for coffee, sugar, fruit, and meat expanded at about the same rate as the American population. American purchasing power rose much faster than its population, but the South American economies were not able to supply America's newly created needs and wants. Asia, on the other hand, and mainly Japan, was able to supply America with electronic equipment and components, precision tools, optical goods, motor vehicles, and machinery, which is exactly what Americans wanted to buy. When the Japanese industrialist thinks of markets, he thinks first of the American market which is the largest and richest and closest at the present time. China has a larger population and is closer, but it lacks the effective complementary demand for Japan's exports. Total African exports to the United States were small because much of what Africa had to sell the United States could buy from closer sources, and many African products went to intervening opportunities in Europe. Many of the riches of Africa, which could find markets in the United States and the rest of the world, cannot move easily to the sea coast. In many economically underdeveloped areas the transferability condition cannot easily be met and potential interactions occur only at severely reduced levels if at all.

Thus, spatial interaction systems are influenced by three factors: (1) complementarity, depending on areal differentiation, which results in a supply at one place meeting a specific demand at another place; (2) the intervening opportunities between places; and (3) transferability measured in time and money costs. When spatial interactions occur it is because each of these conditions has been satisfied. If two places have no interaction with each other, we can usually point to one or more of these factors as the reason.

TABLE 7–1

U.S. Imports by Continent of Origin: 1955 and 1967 (in Millions of Dollars)

	1955	1967
North America	$ 4,038	$ 9,073
South America	2,224	2,663
Europe	2,453	8,232
Asia	1,876	5,352
Australia and Oceania	174	581
Africa	619	905
Total	$11,384	$26,816*

Source: U.S. Bureau of the Census, *Statistical Abstract of the United States: 1968* (89th Edition), Washington, D. C.: U.S. Government Printing Office, 1968, pp. 814–17.
* Includes imports from unidentified countries.

WHY THINGS MOVE OVER THE EARTH'S SURFACE

Why do things move? Why do people move? Our knowledge about spatial interactions can be general or highly particularistic. Normally, it seems easy to state why a particular move takes place. The head of a household goes downtown to work. A housewife goes to the grocery store. University students go to Florida during winter holidays to lie on the beach. People's movements seem to be so obviously *for* some purpose that it seems a bit silly to ask: "Why do people move?" Give me an example, you say, and I will explain why a person moves from one place to another.

Why are oranges shipped from California to New York, or automobiles transported from Detroit to Montana? Why are television programs transmitted from New York to Omaha? Answers seem naively obvious. Anyone familiar with these examples is able to account to his satisfaction for each movement. But if questions about the bases for spatial interaction are to be answered in truly useful ways, allowing us ultimately to control events around us, we cannot ask the questions in particularistic terms. We must transcend specific examples and search for principles and concepts that make sense out of seemingly different facts. Newton's famous observation of the falling apple followed from the way he phrased his question. He did not ask: "Why did this particular apple fall from this tree at this time and place and land on my head?" Instead, he speculated: "Why do things fall?"

Any event can prompt general questions in addition to specific questions. "Why did Jones go downtown to work?" Rephrased and generalized, this question becomes: "Why do people make a journey to work?" By answering the second question satisfactorily, we discover a principle applying not only to Jones, but also to anyone else who travels daily to a job. Particularistic knowledge of a unique event is practically useless by itself as a contribution to wider understanding. An explanation for Jones' specific trip might not apply exactly to anyone else. On the other hand, efficient interaction models can provide useful explanation. They allow prediction as well as description of past events. Explanation of a region's interaction patterns permits their social control, and suitable controls can advance the general welfare. In contrast, particularistic explanations are scientific dead-end streets.

What are the main kinds of spatial interaction? Movements of *people*; movements of *goods and services*; and movements of *information*. Some places are well endowed with skilled people, physical resources, mainsprings of knowledge, and new ideas. Elsewhere supplies are scarce or demands are great or both. Severe inequalities from place to place can result in interaction if the three conditions for spatial interaction are satisfied: complementarity, lack of intervening opportunities, and transferability.

Why Do People Move?

The world's population is concentrated in a very few places and shows little tendency to spread into the empty areas. During the nineteenth-century explosion of European agricultural populations, the new farmlands of North America, South America, South Africa, Australia, and New Zealand attracted millions of migrants. After the 1840s, industrialization was underway in the United States. Industry expanded and agriculture started a relative decline. By the 1890s, much of the best American farmland had been settled. New employment opportunities were found only in urban areas. Cities absorbed successive waves of European immigrants together with incipient off-farm migration of homesteaders' children. In the United States the idea of a "surplus" of population at one place is based on the existence of better opportunities elsewhere. The New World was attractive to the colonist, and the frontier pulled him into the Middle West, the South, and the Far West. With the settlement of California, Oregon, and Alaska, cities began to pull harder than the open land. Recently, the possibility of a pleasant life in the suburbs has pulled down the population densities of the inner cities. In every age and at every spatial scale, populations are on the move in a continuing effort to reach the peaks of an ever-changing opportunity surface.

Before migration begins, population is distributed among a set of places. Alternative destinations usually differ significantly in their attractiveness to potential migrants. Migration potential depends on the strength of the specific complementarities from place to place in people's needs and available opportunities. Migration takes place between a set of origins and destinations. Once started, migration proceeds until the complementarities have been eliminated and can no longer stimulate further redistribution. Migration continues and complementarities dissolve until opportunities are exhausted, origins depleted, or until new complementarities develop between places.

Every population movement, from continent to continent or from room to room, ultimately depends on human needs and desires, and on place to place complementarities. In A.D. 1600 prior to the European explosion over the earth; in 1750 before the push westward of the American frontier; in 1910 before the peak of migrations from farms to cities in the United States; in 1945 before the suburban explosion; on June 15 prior to an annual vacation trip; at 6:00 A.M. each morning before the journey to work; or indeed, at 10:00 A.M. when it is time for a coffee break, people at one place prepare to move elsewhere. If the object of their need or want is not available, then either people must move to it or else the supplies must move to the demands.

Some migrations are forced by war, famine, disease, and other causes, but the majority of moves are voluntary. When human movements among places are voluntary, they depend on the information people have about those places. Sometimes the first migrants to a place find that they were fooled by incorrect information. Feedback of accurate information to the origin modifies images of destinations, and thereby adjusts the volume of the next wave. In time, ratings of places change, partly because places themselves improve or deteriorate absolutely but also because places change with respect to one another. Alaskan gold camps and Appalachian coal towns shared a common fate. Both declined as minerals were depleted and as new opportunities sprang up elsewhere. At the end of the nineteenth century, frontiersmen and displaced farmers of the upland South moved on to mining jobs. Such jobs seemed to be a superior alternative to impoverished life on the farm. Yet with the passage of time, the mining centers declined in turn.

Economic change is never ubiquitous in kind or intensity. New opportunities crop up in places different from those settled earlier. Migrations during a period depend on the location of opportunity at the beginning of that period. Then as populations are redistributed a new geographic pattern of opportunity is created by the end of the period. People move to a place because perceived opportunities at that place attract them more than their old home hangs on to them and more than any other destination entices them. Things may be bad around the home area, or perhaps they improved faster at some other place. It is the *difference* between present and perceived alternative opportunities that triggers the decision to move. When the perceived difference becomes great *and* no obstacles stand in the way, the migration takes place.

We can think of each migrant assigning one value to his present location and other values to places where he could be. He compares his present status with potential status elsewhere. Then he weights the different alternatives according to their distances and how risky he thinks each of them is. Finally he picks a strategy he thinks will be best for him. Such decisions, in the aggregate, determine the fraction of a population which moves during a given time period. Some people always opt for low risk and reap the accompanying low rewards. They value the certainty of a meager existence much more than they value a boom-or-bust proposition. Given a choice they prefer "a bird in the hand" over "two in the bush." The people who fled Ireland during the great famines between 1845 and 1850 were people who anticipated a better existence in the New World and were able to arrange the journey overseas. Another group that wanted to come could not arrange to pay the passage and many of them stayed behind and starved. A third group feared the famine but feared even more a new life in a strange land. As time passed, the size of the third group was reduced as ocean transportation was revolutionized. The arrival of the steam ship to replace the sailing ship cut the one to three month crossing time down to ten days and reduced the perils of trans-Atlantic travel, especially the risk of fatal disease. As information flows improved, and as migration costs dropped, the number of migrants soared because of the improved transferability.

Most of the European migrants came to the United States and Canada. Migrants preferred to settle in places that were substantially similar to the Old Country they left behind. They brought with them their European agriculture with its specific climate and soil requirements. They avoided the polar regions and tropical lowland regions because such places were too different from the places they left. They settled in midlatitude prairies and woodlands, and North America was the closest place to Europe which offered what they sought. South America was farther away, and so proportionately fewer Europeans settled there. South Africa, Australia, and New Zealand were even farther away and attracted still fewer migrants because of the intervening opportunities and the difficulties in transportation.

Movement of Goods and Services

The movement of goods from production points to consumption points resembles the movement of people in a migration pattern. Migrations are the way people smooth out the peaks of per capita opportunity which differ from place to place. As more and more people crowd into an attractive area, its desirability soon diminishes relative to other destinations. The late arrivals soon discover that they will be better off elsewhere and so they tend to move on. In the movement of goods a product moves from factory to consumer so as to maximize its net delivered price. For this reason the factory serves nearby markets first, to save transport costs, then those farther away. The penalties of trying to sell too much of the same thing in one place are similar to the problems which develop if too many people try to live in one place, such as Manhattan. If Detroit tried to dump in Wayne County its total output of automobiles by cutting prices until they were all sold, the local market would be glutted, prices would plummet, and severe losses would be incurred by the manufacturers. Instead of selling all the cars in one local market, manufacturers send cars all over the country, taking care to avoid shortages in some areas (shortages promote sales by competitors) as well as oversupplies (which lead to price cutting). By careful allocation of cars to market, Detroit manufacturers sell the greatest number of cars at the highest prices.

All commodity and service movements can be evaluated in terms of complementary supply and demand areas, presence or absence of intervening opportunities, and the transferability of the goods and services. Automobiles are shipped from place to place because production is localized at one set of places, while consumers are spread across the country and in foreign lands. Other commodities—Canadian wheat, Central American bananas, Cuban sugar, or Brazilian coffee—also illustrate how production capability in one place complements requirements of other places. The same principle applies in the production and distribution of personal services. Sometimes when supply occurs in one place and complementary demand is someplace else, we can choose to have the service delivered (home television repair); sometimes we must deliver ourselves to use the service (a hospital, a hairdresser).

If the same good or service commands a high price in one place and a low price in another, what could prevent sellers in the low price area from exporting to the high price market? Tariffs and import restrictions can prevent such movements, as can ignorance on the part of the seller in the low price market, or perhaps his reluctance to get involved with the risks connected with long-distance trade. The most common reason, however, is the transferability of the good or service. The difference in price between the two markets must be great enough to cover the cost of movement between the two places, otherwise long-distance trade would never come about naturally. In general, though, large price differences between nations, low transport costs, and free trade policies combine to support active trade among the regions of the world.

When the conditions are adverse, trade is negligible or nonexistent. Trade between Latin American republics and the United States is based on high prices and high demand levels in the United States combined with low prices and marketable surpluses in Latin America. Transport costs are low, many markets are guaranteed by U.S. government-backed purchase agreements, and trade barriers are no problem. The Latin American republics carry on little trade with one another. Interregional price differences are minor or nonexistent. Besides, Latin American countries seldom have what their neighbors want, or if they do the neighbors cannot pay for it.

Information Flows

Information consists of new facts, data, ideas, and routine communication. Like people, goods, and services, information flows from place to place, moving from places of production to demand areas. When we think of demand areas we should not confuse places of *need* with places of *effective demand*. In the flows of people, movements among big and vigorous centers are much larger than migrations among the depopulated backwaters of depressed areas. In the shipment of goods, the volume of trade among the developed countries far exceeds their trade contacts with the underdeveloped world. Similarly, information flows primarily among the vigorous idea centers in a system. Islands of lethargy are bypassed because they are literally out of the system.

Rich nations trade primarily with each other. Migrations often occur between clusters of mankind. Those who send the most messages get most of the return contacts. A tendency exists for flows to run from *places of abundance* to *areas of effective demand* according to the effective *pulls* exerted at each destination. As flows continue, a tendency toward equilibrium conditions seems common. In the physical world things move through space so that potential can equalize. Water tries to move under the force of gravity to the lowest point, and if free to move to sea level it will do so.

Similarly, air will move from a high pressure to a low pressure area. High pressure as a causative agent cannot move air very far in one direction. We know this from common experience when we try to blow air across a room with a fan or try to blow out a match at a distance. More air will be moved by putting the fan in one window blowing outward and opening another window across the room. Instead of emphasizing push factors, it makes more sense to think of air being pulled in a certain direction by a zone of low pressure, in the same way that water is pulled from a mountain top. Similarly in human affairs a *pull* exerts a *directed* force with respect to excess supply somewhere else. Therefore, to explain why something moves from A to B it helps to emphasize the pull at B rather than the push at A. The push at A often is unspecified directionally, whereas a pull at B is significant not only for A, but for any other potential supply source as well.

In locational analysis, demand and supply are complementary and regulate movement potential, yet the pull of demand is primary. We see this in the fact that during the Great Depression of the 1930s when pigs were buried in the midst of hunger the demand in existence was ineffective. People sometimes refuse to watch network television programs for the same reason; there is nothing on they want to watch. The push of supply alone cannot support a movement or an interaction. Effective pull in the form of demand will. Esoteric food wants are satisfied in every city, and national educational television was created to serve a previously unmet demand.

Information Production and Consumption

Production, collection, processing, and distribution of information are the activities that permit modern complex societies to operate. The city is a device for maximizing transactions, and message flows are the transactions which hold the fabric of the city together. A useful way to discuss this class of intangible activities is to consider information in general terms—as bundles of decisions, data, orders, instructions, visual entertainment, research findings, advertising, broadcasting, education, publishing, and so forth. Information activities, which Jean Gottmann called the quaternary activities, provide the foundation of modern urban society. J. K. Galbraith in *The New Industrial State* described the modern corporation as an institution with power based on its capacity to command technical knowledge, skill, and information. Control capacity in the giant firms of today depends on a coordinated team of human brains called a technostructure. The white collar group in each organization comprises the technostructure. Its function is continual information production, collection, processing, exchange, and dissemination. Head offices of modern corporations are decision factories where group decisions are made and where top management supervises.

A branch plant or a sales office supplies information to headquarters and receives instructions and other intelligence in return. News reporters blanket the world to collect the news, then ship it to headquarters so that it can be processed and printed in newspapers or broadcast to radio and television audiences. Students go to school to learn; teachers read, think, and teach. A massive communications system provides formal (telephone, postal system, classrooms) and informal (business lunches) channels to support the information enterprise. Information supply centers are highly localized in any national or subnational area. Of the five hundred largest industrial corporations listed by *Fortune* magazine in 1963, 163 maintained their head offices in New York City, the preeminent corporate management center in the United States. Chicago, Pittsburgh, and Philadelphia followed with fifty-one, twenty-one, and sixteen headquarters, respectively. The five hundred firms typically had domestic operations scattered over the United States, together with far-flung foreign operations, yet regular communications tied each corporate control center with its empire.

Governments at every level are information and control enterprises, too. Lower order centers such as the city hall or the county seat fit into a nested spatial hierarchy of government systems extending upward to include state capitals, national capitals, and even the operations of a supranational entity like the United Nations. At every level a control center manages the govenmental affairs within its territory just as a corporate headquarters directs operations under its control.

Part of the information sector of a nation's space economy is neither product-oriented (like most industrial corporations) or service-oriented (like most governments). Instead, its job is to produce new knowledge, new application of ideas, or to transmit formalized bodies of knowledge and to train people in the knowledge business. Thus, a nation's educational system together with its research and development effort forms a cornerstone for the entire society.

In varying degrees all parts of the nation consume the output of colleges and universities which in a sense are localized knowledge factories. In the United States, scientists are usually trained in one place, then live and work in another. Some states therefore export scientific expertise to other places (Fig. 7–1). States with strong universities, like New York, Massachusetts, and Wisconsin, are net exporters of trained scientists; places such as the District of Columbia, Delaware, and Alabama must import the scientific talents they need beyond what they can produce themselves. Scientists are trained in graduate schools and then find employment in colleges and universities as faculty members, or in industry and government at places where industrial and government research is located.

The Federal Government finances the majority of all scientific research and development. The geographic distribution of federal funds for scientific research and indeed the geographic allocation of the entire federal budget creates an opportunity surface which redistributes scientists to all parts of the country after they have been trained. Just as some states are better endowed with good graduate schools than others and can thereby produce more Ph.D.'s per capita, so also some places have better connections in Washington and receive more research money per capita and can thereby lure migrating scientists—at least on a temporary basis. The spatial distribution of research funds creates an opportunity surface which fluctuates as scientists redistribute themselves within the

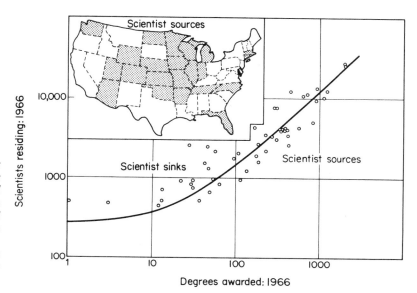

FIGURE 7–1

Resident scientists and doctoral degrees awarded, 1966. The least squares regression line defines an expected number of resident scientists for each level of doctoral degree output. Shaded states have fewer scientists than degree output suggests (scientist sources); states above regression line attract more scientists than local universities can produce.

nation's economy. Since scientists are suppliers as well as consumers of new ideas and information, and since they are also on the move from place to place, it is more difficult to describe the movements of ideas then of people or goods. Nevertheless, interaction principles for information flows—based on complementarity, transferability, and intervening opportunity—remain the same as those describing population and commodity movements.

VARIETIES OF URBAN COMPLEMENTARITY

Every economic era has its towns of different sizes, its characteristic growth industries, specific patterns of localized demand and supply, and a corresponding distribution of population. In the American experience, fertile forest and prairie soils attracted homesteaders during the nineteenth-century agrarian boom; steel mills and mining towns in coal and iron regions attracted additional immigrants at the turn of the century. Automobile, metal, and machinery factories monopolized growth through the first third of the twentieth century in the United States. Now, government, education, entertainment, decision-making, corporate control, and and research and development activities characterize the expanding edges of the new industrial state.

In any period the dominant human activity, the surplus resources, and the output are localized at one set of places while complementary market demands exist elsewhere. The stage is set for the two kinds of places to interact to eliminate surpluses and deficits. The bases for spatial interaction are interregional complementarities created by localized supplies and effective demands for people, goods, and information.

Most modern economic, social, and political activities are centered and controlled in urban centers. In any national territory, cities and towns appear in a bewildering variety of sizes, and range in importance from the national metropolis to the smallest hamlet. The very largest centers provide specialized goods and services for the whole nation. Smaller centers serve limited local markets with common needs and wants. Small centers themselves also provide part of the market for large regional and national metropolises.

The fact that urban centers come in different sizes implies that they perform widely varying roles. Towns specialize just as people do, and for the same reasons. Towns must earn a living. Some jobs cities perform for themselves. The rest of their efforts are devoted to the needs of other places. Towns are paid for the goods, services, and information that they sell. They use the cash received to buy what they need but do not produce themselves.

An analogy between humans and cities is apt. Each person engages in home-oriented as well as market-oriented or export production. Mowing the lawn, redecorating a piece of furniture, washing dishes, or building a basement amusement room are subsistence activities when done by the householder at his house by himself. The producer and consumer are one and the same. Cities can be classified by principal occupation just as individuals are grouped as teachers, engineers, and so forth. Sometimes people moonlight, selling their labors to more than one employer. Cities also can double up by pursuing several lines of activity instead of a single specialty. Prosperous and vigorous cities expand, attracting people and activities, just as talented and energetic individuals increase their personal income and their ability to serve others. Thus, city classification schemes are sometimes inadequate for the same reason that tags on individuals are unsatisfactory. San Francisco is more than a wholesale trade center, and Dwight D. Eisenhower had an identity beyond that of a military officer.

Specialized Places and Economic Base Studies

Complementarity in urban economic systems has been studied in three ways. The first approach emphasizes the *places* themselves (such as counties, states, or nations) and their import-export relations. The second approach focuses upon different industries and *sectors* of the economy (such as mining, manufacturing, trade, transport, and so forth) and describes the flows of goods and dollars in and out of each sector. The third approach is a combination of the first two. It considers flows in and out of *places* as well as flows among the *sectors*.

Any specialized region within a large and diversified economy must import to survive. To pay for its imports it must export goods and services to other regions. Regional specialization provides the basis for interregional complementarity and is responsible for the flows we observe among regions with different specialties. Export industries are called the *basic* activities in a region. When a regional economy is in its infancy, exports provide the initial base on which the total economy begins to grow.

Which activities in a region form the export sectors? In one region they might be agricultural; in another the specialty might be durable goods manufacturing. To estimate the export portion of an industry's output we need to know what portion of a region's activities comprises the minimum production needed to meet local consumption requirements. Production in excess of minimum requirements is presumed to be an exportable surplus. If a region's production, say of agricultural products, falls below minimum requirements, the difference has to be made up by imports.

Statistics of the working labor force can be used to measure the way a region divides its production effort among the different industrial sectors. Business establishments in the United States are classified into industries on the basis of their principal product or activity, in accordance with the *1957 Standard Industrial Classification (S. I. C.) Manual* which was amended by a 1963 Supplement. The Bureau of Labor Statistics collects from each employer a monthly report on employment, payroll, and hours worked. An establishment's classification by industry is determined from information submitted on an annual supplement to the monthly reports. If an establishment makes more than one product or engages in several different activities, its entire employment is included under the industry indicated by the most important product or activity.

The Bureau of Labor Statistics of the Department of Labor assembles the employers' statistical reports and estimates employment by industry each year for each Standard Metropolitan Statistical Area in the United States. Let us consider Capital City, a metropolitan area whose employment is distributed among various S. I. C. employment divisions as shown in Table 7–2.

What can employment statistics like these tell us about exportable surpluses and the require-

TABLE 7-2

Capital City Employment (in Thousands)

	Number Employed	% of Total
Contract construction	7.4	4.6
Durable goods manufacturing (primary metals, fabricated metals, transport equipment)	17.5	11.0
Nondurable goods manufacturing (food and kindred products, apparel, printing and publishing, leather and leather products, etc.)	21.2	13.3
Transportation and public utilities	12.1	7.6
Wholesale trade	7.2	4.5
Retail trade	22.2	13.9
Finance, insurance, and real estate	6.9	4.3
Service and miscellaneous	21.9	13.7
Government	43.2	27.1
Total	159.6	100.0

ments of the local market? When employment statistics are evaluated according to the *minimum requirements* method they can tell us a great deal. If Capital City's population is 390,000, first we must know how much employment is needed in each industry to satisfy local requirements, and how much employment is devoted to export markets to bring in the cash needed to buy import items that cannot be produced locally. Before we can answer this we must determine the minimum employment we would expect in each industry for a city this size. If we examined all American cities which have populations between 250,000 and 500,000 we would discover that each of them has a different percentage of its employed labor force engaged in each industrial class. Within an industrial class, say wholesale trade, the smallest percentage among the cities might be 2.2 percent. This means that when all cities between 250,000 and 500,000 population are considered, none of them has less than 2.2 percent of its labor force engaged in wholesale trade. Thus, the 2.2 percent can be considered the minimum requirement for wholesale trade for cities of that size class. Any employment beyond 2.2 percent is considered to be gross export or basic employment, serving markets outside the city and bringing cash into the local urban economy to pay for imports. Capital City has 4.5 percent of its work force in wholesale trade.

According to the minimum requirements method, therefore, 2.2 percent of the wholesale trade workers are nonbasic or local market employment; and the other 2.3 percent are busy serving the needs of people in places outside the local market.

The wholesale trade minimum requirement for smaller sized cities is lower than 2.2 percent; larger size classes have bigger minimum requirements. Using employment statistics we can identify the city in each size class which has the smallest percent of its labor force in wholesale trade. By plotting these minima (one for each city size class) we can estimate a best-fitting, least squares line to describe how the minimum requirement for wholesale trade increases gradually with increasing city size (Fig. 7-2). Using the same procedure a minimum requirements line can be defined for retail trade, for durable goods manufacture, and for each of the remaining S. I. C. industrial classes. If the minimum requirement in retail trade increases from 12 percent for cities of about 10,000 population, up to 15 percent for cities in the 10,000,000 class (Fig. 7-2), Capital City with 13.9 percent of its labor force in retail trade has practically no export employment in its retail sector. But this is what we would expect. Most retailing clientele is local except for visitors and tourists and the occasional rural resident who comes to the city to shop.

Capital City's main business is government. For a city in its size range we would expect only about 5 percent of its employees to be engaged in government service as a minimum requirement. Instead, the city has 27.1 percent of its workers in government employment at all levels—federal, state, and local. The difference between the minimum requirement percentage (what is needed for local needs inside the city) and the actual percentage is the export or basic employment. Capital City provides some federal government services for areas even beyond the state; it provides state government services for the whole state because it is the state capital; it serves the surrounding region as a county seat and center of metropolitan government. For all these exported services Capital City is paid in the form of taxes and by grants-in-aid from the federal government. Like other cities,

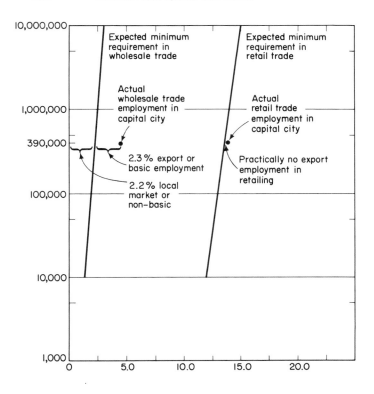

FIGURE 7–2

*Minimum employment needed in wholesale
trade and in retail trade is defined by the
upward sloping curves. In each sector the
minimum rises with city size.*

then, Capital City has its specialties and is
thereby provided with a basis for interregional
interaction.

Specialized Activities and Input-Output Analysis

When the basic industries in one place are
specifically complementary to needs elsewhere,
spatial interaction can occur if these conditions
are known in both places and transferability
conditions are met without any intervening
opportunities getting in the way. The minimum
requirements approach to the urban economic
base provides one kind of insight about why
specialized *places* interact. Input-output analysis
sets the foundation for an additional perspective
based on how the specialized industrial *sectors*
of the economy interact.

Input-output analysis is especially helpful
if we want to investigate why the economies of
some places expand as others languish. The
regional economic development process is
usually described as a sequence whereby one

event triggers the next. For example, during
Britain's industrial revolution declining death
rates meant higher rates of natural increase at
the same time that new lands were opening up
in the New World. European agricultural
technology could be applied directly to fertile
forests and prairie lands with high per capita
yields just as food needs were rising in in-
dustrializing Northwestern Europe. Immigrants
found new opportunities in both industry and
agriculture. Accordingly, both mill and farm
competed for immigrants. Urban factory labor
was scarce and wages were high in New England
factory towns because land was free in the
American West. Manufacturers began to substitute machinery for high priced labor, thereby
promoting ever-increasing per capita productivity and wages of factory labor. Urban demand
for food stimulated farm productivity; profits
from farming created demands for factory
goods. Productivity advances in one sector
stimulated production and improved benefits
in other sectors.

Besides the historical explanations, early attempts to describe the regional economic growth process used an analogy with human growth. It was argued that when a regional economy was in its infancy it emphasized primary extractive and raw material industries. Secondary activities, especially food processing, textiles, and clothing, predominate in the youthful stage. During adolescence, basic durable goods industries develop near power and transportation to serve both producer and consumer markets. When maturity is reached interrelated service and control enterprises dominate the economy. Regional economies which are overdependent on one narrow industrial base are referred to as immature. Ghost mining towns died in infancy. Sometimes technical revolutions or political change bring on premature death or decline, as when an industry is eliminated or a political capital is removed. Similarly, planned new towns are created artificially, just as stagnant towns are rejuvenated by transfusions or transplants of new investment of capital.

Analogies between support of an individual and support of the town break down if we assume that towns cannot exist largely by taking in their own washing. In fact, the larger and more diversified metropolitan areas consume a larger fraction of locally produced goods and services than do smaller, less diversified cities. Like economic base arguments, these analogies to individual growth are not incorrect or useless, but they are imprecise and often incomplete. Analogies are poor substitutes for analysis. They emphasize one part of a system and disregard the importance of others. Arguments from history to shed light on the present can also lead to pitfalls. For example, in the case of railroading, a boom industry in one period became a blighted industry in the next. Usually, when chambers of commerce initiate campaigns to bring new industry to their towns they are acting according to historical experience. They know that a new factory means new jobs, more local retail sales, more exports, more taxable property, and more economic activity all around town. But sometimes they are right for the wrong reason. Sometimes their efforts would have borne more fruit if they had paid attention to the important industries

of tomorrow rather than the growth industries of yesterday.

What kind of activity and *how much* change can be expected from a new factory, hospital, or insurance company can be predicted with an input-output model of a regional economy. In an input-output model, all economic factors in a region, whether companies, households, governments, or whatever, are first assigned to sectors. Members of any sector are relatively homogeneous in their production and consumption behavior. Sectors may be as generalized or specialized as is desirable.

A Two-sector Regional Economy. Let us assume an economy that has only two sectors: agriculture and manufacturing. The output of the agricultural sector each year is purchased partly by the manufacturing sector to be used as inputs (e.g., hides, cotton, oil seeds, etc.). The other part of the agricultural output is purchased for final consumption by the people in the economy. The sum of the shipments by the agricultural sector equals the total annual output of the agricultural sector (Table 7–3).

TABLE 7–3

Annual Input-Output Flows for a Two-sector Economy

	Output Sectors to			
	Agriculture	Manu- facturing	Final Demand	Total Output
Input Sectors from:				
Agriculture	0	5	15	20
Manufacturing	8	0	2	10

The manufacturing sector ships a total of ten units of output. Eight are absorbed by the agricultural sector as inputs to agriculture (e.g., tractors, plows, etc.); the other two units are purchased by consumers as final demand. If we look at the agriculture column in the input-output flow matrix we see a summary of the inputs purchased by that sector. The entries under manufacturing indicate the inputs purchased by manufacturers.

From the flow matrix a second matrix of input-output coefficients can be derived (Table 7–4). Each coefficient indicates the number of units from an input sector which must be used

TABLE 7–4

Input-Output Coefficients for a
Two-sector Economy

	Output Coefficients to:	
	Agriculture	Manufacturing
Input Coefficients from:		
Agriculture	0.0	5/10 = 0.5
Manufacturing	8/20 = 0.4	0.0

to produce one unit in the output sector. If the flow entries are in dollars, then in our two-sector economy $5 worth of agricultural output is purchased by the manufacturing sector in the year of operations described by the flow matrix. That same year the manufacturing sector had a total output of $10. To determine the amount of agricultural output which the manufacturing sector must buy per dollar of manufacturing output we divide the agricultural input ($5) by the corresponding manufacturing output ($10) yielding an input-output coefficient of 0.5 (Table 7-4). The coefficient means that for every dollar of manufacturing output, the manufacturing sector must buy half a dollar's worth of agricultural produce. If we do not know the exact output of the manufacturing sector and call it simply M, then in any given year the flows from agriculture to manufacturing equal 0.5 M. In the same way the other input-output coefficient is derived. Agriculture buys $8 worth of inputs from the manufacturing sector and uses them to produce an agricultural output of $20. Each dollar's worth of agricultural output requires 8/20 or 0.4 dollar's worth of inputs from manufacturing. If the total and unspecified agricultural output equals A, the flow of materials from manufacturing to agriculture equals 0.4A.

The levels of final demand, $15 worth of agricultural output and $2 of manufacturing output, are set independently of the input-output coefficients which are determined by the technologies of the two industries. Once we know the coefficients, and the levels of final demand, we can specify all the relationships in two equations. The input-output coefficients plus the levels of final demand determine the total annual output of each sector. The total

output of the agricultural sector A equals the shipments to manufacturing plus the level of final demand for agricultural commodities, that is: $0.5M + 15 = A$. The total output of the manufacturing sector M equals the shipments to agriculture plus the level of final demand for manufactured goods, that is: $0.4A + 2 = M$. So our two-sector economy can be represented by two equations and two unknowns taken directly from the matrix of input-output coefficients and the final demand column of the input-output flow matrix:

$$0.5M + 15 = A$$
$$0.4A \quad + 2 = M$$

which when solved yield $M = 10$, and $A = 20$.

Measuring the Impact of Change. Once the structure of the economy is described by a sequence of equations, we can ask questions about the impact of change. What happens, for example, if final demand for manufactured products rises from 2 to a new level of 6? If technology remains unchanged the input-output coefficients remain unchanged. But the higher final demand for manufactures causes reverberations in the economy and produces new higher levels of total output for both agriculture and manufacturing:

$$0.5M + 15 = A$$
$$0.4A \quad + 6 = M$$

which yield the solution: $M = 15$ and $A = 22.5$. The increase by 4.0 in the final demand for manufactures raises the total output of the economy from 30 to 37.5, so there is an important multiplier effect. An increase in demand of 4.0 brings an overall expansion in output of 7.5. The response is almost twice the size of the stimulus.

Final demands are important in keeping the whole regional economy on an even keel and keeping the various sectors smoothly interacting with one another. If the American government guarantees demand for domestic agricultural output, then the producers who sell inputs to farmers are protected. If the American government agrees to purchase agricultural output from Latin American countries, the effects of these purchases are multiplied inside those countries. The effect in those countries would be the same, however, if local entrepreneurs in those countries decided to borrow funds and invest in new plant and equipment, thereby raising the final demand for the output of manufacturers. Such a move would have the side effect of increasing markets for agricultural commodities because the different sectors depend on one another. If households borrowed money to raise final demand from either sector the effects on the economy are similar. Since the economic sectors are mutually dependent on one another as suppliers and customers, we can see that anything that increases final demand for the urban manufacturer helps farmers in the country, and anything that expands the final demand for agricultural output has the effect of stimulating the manufacturing sector as well. As long as the input-output coefficients can be specified, and the levels of final demand are independently set, the equilibrium values for total output of each sector are determined.

A Five-sector Economy. Let us assume that the firms, households, and other producing units in a region comprise five sectors which supply the inputs needed to run the region's economy. With careful measurement the flows among the five sectors can be defined (Table 7–5). The flows are of several different kinds. Firms sell goods and services and are paid in return. Governments provide services of many kinds and are paid when they collect taxes and fees. Households sell their services to firms, governments, and other households and receive payment in the form of wages and salaries. In this hypothetical example each input sector sells goods or services to every other sector (Table 7–5). For example, the household sector sells services to the primary sector when a man goes to work for a corporate farm. If another man owns and operates a grain farm, the farm is a firm in the primary sector which sells its produce perhaps to a milling company in the secondary sector. In the two-sector economy which we discussed earlier all households were assigned to either the agriculture or the manufacturing sector.

TABLE 7–5

Annual Input-Output Flows in a Five-sector Regional Economy

	Output Sectors to:				Consumption by Households and Other	
	Primary	Secondary	Tertiary	Quaternary	Final Demand	Total Output
Input Sectors from:						
Primary Sector (P): Agriculture, forestry, fishing	50	35	16	6	293	400
Secondary Sector (S): Durable goods manufacture, nondurables, mining, and construction	18	180	85	60	457	800
Tertiary Sector (T): Business and personal services	20	65	30	70	265	450
Quaternary Sector (Q): Information production, headquarters activities, educational systems, control enterprises	15	80	30	35	240	400
Household Sector (H):	125	300	150	125	150	850
Totals	228	660	311	296	1405	2900

The input-output coefficients are computed for the first column by dividing each flow by the total output of the primary sector. In the second column each flow entry is divided by the total output of the secondary sector, and so forth (Table 7–6).

TABLE 7–6

Input-Output Coefficients for a Five-sector Regional Economy

Sector	P	S	T	Q	Final Demand
Primary	.12	.04	.04	.02	293
Secondary	.04	.22	.19	.15	457
Tertiary	.05	.08	.07	.18	265
Quaternary	.04	.10	.07	.09	240
Household	.31	.38	.33	.31	150

Once the coefficients are established we can see how they determine each sector's total output (which we shall designate as P, S, T, Q, and H, respectively) and the region's total output. For example, the total output of the primary sector equals sales to itself ($.12P$), plus sales to the secondary sector ($.04S$), plus sales to the tertiary sector ($.04T$), plus sales to the quaternary sector ($.02Q$), plus sales to private consumers and other forms of final demand including exports (293). This summation can be expressed in an equation, and the same operation can be performed to define the total output of each of the other sectors:

$$P = .12P + .04S + .04T + .02Q + 293$$
$$S = .04P + .22S + .19T + .15Q + 457$$
$$T = .05P + .08S + .07T + .18Q + 265$$
$$Q = .04P + .10S + .07T + .09Q + 240$$
$$H = .31P + .38S + .33T + .31Q + 150$$

Any time we have five equations with five unknowns we can solve them to yield the values which satisfy all equations simultaneously. In this case the total outputs for each sector are the same as those in Table 7—5. The more detailed the input-output description of a region's economy, the closer we can come to predicting the consequences of changes in levels of final demand or of adjustments in the input-output coefficients due to technological change. Understanding spatial interaction requires an understanding of an economy's intersectoral

relations. Geographers study sectoral interrelations because such interrelations help them make sense out of why things move from place to place.

Specialized Activity Places: Interregional Input-Output Analysis

The easiest way to see the spatial implications of intersectoral transactions is to broaden the input-output model to include more than one regional economy. The input-output model focuses attention on the *sectoral* basis of a region's interaction patterns. Economic base studies focus on the import-export or *spatial* basis of

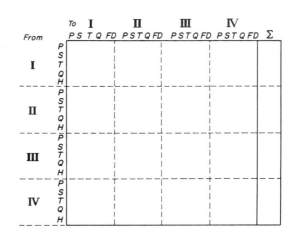

FIGURE 7–3

A regional economy with four subregions (I–IV). Each subregion has five producing sectors selling their outputs to themselves, to other sectors in the same subregion, and to sectors in other subregions. P = primary; S = secondary; T = tertiary; Q = quaternary; H = households; FD = personal consumption and other final demands.

regional economies. The two approaches can be combined to form an intersectoral, interregional input-output model. Such a model provides still deeper insights into the complementarity of both the *sectors* and *places* that comprise a space economy.

Consider a regional economy with four subregions (Fig. 7–3). Two of the regions are mainly rural. Region I is an important cattle raising, mining, and vacation spot; region II is a rich farming area. The other two regions are metropolitan centers. Region III is largely a manufacturing and transportation center, and region IV is a management and service center with corporate headquarters, government, finance, and trade being the major functions.

If within each subregion there are five producing sectors, the whole regional system can be thought of as comprising twenty sectors. If each of the producing sectors purchases inputs from every other sector in addition to purchasing inputs from itself, then the basis for spatial interaction is revealed in sharp relief. Moreover, it becomes a simple matter, once annual flows have been measured, to compute input-output coefficients and then to describe how an increase in final demand in one sector of one subregion causes changes in the production levels and flows between all other sectors in the four subregions. In this way functional dependencies are translated into spatial transactions, and the bases for spatial interaction are better understood.

THE METROPOLIS: AN EXAMPLE OF SPECIALIZED ACTIVITY PLACES AND INTERACTION POTENTIAL

As a city becomes larger and more self-contained, it requires relatively less trade with other areas. The export share of its employment drops as metropolitan size rises—that is, the minimum requirement percentage in each producing sector rises with metropolitan size. The bigger the place, the more it lives by taking in its own washing. Moreover, the larger the city, the more efficient its large-scale enterprises can become, up to some optimum size beyond which

congestion costs outweigh scale economies.

Inside the city, interaction potential increases with metropolitan size. Individuals and organizational units progressively specialize as the city generalizes. Specialized units occupy different places within the metropolis, and their specific complementarities provide the potential for spatial interaction within the metropolis. When the specialized units are geographically separated, interaction occurs only after distance barriers have been overcome. This means that people, goods, and information must move not only from one functional unit to another, but from one place to another as well.

Urban traffic patterns depend primarily on trips which start or end at home. The basis for trips by people rests on the locational structure of different but complementary activities; their type, location and intensity; and the geographic patterns of various population units and their social and economic characteristics from one residential area to the next. Transportation facilities of different types tie specialized residential and other activity subareas together. Resulting travel flows comprise the city's traffic pattern (Figs. 7–4a and b).

The complementarity of urban subareas is the basic factor in the generation of urban trips. Changes in complementarity alter the level and pattern of interaction. Such changes are critically important in the transportation planning process at regional and urban levels. Technical elements in the transportation planning sequence include: population and economic studies to determine the magnitude of each activity for future dates at every location; land use studies to determine where activities will be located at the target date; trip generation studies to determine level of demand for interactions among the tracts in the urban area; and trip assignments to determine the allocation of trips among alternative destinations. Study of future *interaction bases* is followed by an appropriate transportation plan to handle projected travel. A fraction of projected travel volume must be assigned to each transport mode, and routes must be specified for every trip. The transport plan concludes with an evaluation of money and nonmoney costs vis-à-vis benefits returned so that the best system can be devised.

FIGURE 7–4a

Person trip destinations on nonresidential land in the Chicago transportation study area. Over 4.6 million person trips are destined for nonresidential land on the average weekday. The highest value shown on the model represents 133,000 destinations. The lowest vertical value shown is 5,000. Shaded areas represent values between 2,500 and 5,000 destinations per quarter square mile tract (per .65 quarter square kilometer) (from C.A.-T.S., Fig. 35, p. 65).

FIGURE 7–4b

Model of all person trip destinations. On an average weekday over 10.2 million persons have destinations in the study area (from C.A.T.S., Fig. 31, p. 59).

210

Trips and Travel

Consider the trips made within a city. Trips are made for profitable purposes. Sometimes the profits are intangible and difficult to measure but they are nevertheless real. A man leaves his home in the morning for the office and his salary justifies the journey's costs. A woman drives to the library to borrow a book and her intangible reward comes from reading. About 10 percent of consumer income is spent for travel; the rewards from travel justify this expenditure. To understand trip generation and travel behavior we must turn to the activities which underlie travel.

Land use analysis is a convenient way to study the activities that provide the basis for trip generation because travel patterns (routes and flows) are dictated by network structure and land use arrangements. At this point, it is important to distinguish between a *trip* as an event linking an origin and a destination, and *travel* as a journey defined by route and length and time. When trips are counted or classified by origin or destination, each trip is one unit. When considering travel in an area, trip length must be incorporated into the measure. Amount of travel is then recorded in units like vehicle kilometers or person kilometers. Travel is also assigned to the appropriate media within an area.

The travel volume within an area depends on the trips starting and ending there, as well as the relative location of the area with respect to

other origins and destinations beyond it. If an area lies between a downtown and a suburb, much of its travel is through traffic. The expected levels of through and local traffic in an area can be depicted with a desire line map. A desire line map depicts the straight line connections between origins and destinations for one class of trips. The desire line is thus the shortest line between origin and destination (Figs. 7–5a and b). On a map a desire line presents visually the basis for travel behavior by expressing how a person would like to go if such a way were available. A desire line map provides a strong impression of the geographic pattern of travel demand.

Neither the trip patterns nor travel behavior of each person can be reliably predicted, but it does not matter; transport systems are built to serve large groups of people. Trip patterns and travel behavior of large groups can be reliably described and predicted. In *trip* generation studies attention turns to origin and destination characteristics of urban trips. In *travel* surveys, vehicles en route are stopped and occupants queried as to the establishment or land use at each end of the trip.

On an average weekday in 1956 there were 10.2 million person trips made in the study area of the famous Chicago Area Transportation Study (Table 7–7). Residential, manufacturing, and public lands generated trips at about the same average rates. Commercial land generated trips at a rate nearly four times greater. Other lands, such as those devoted to railroad yards or public open spaces and parks, generated few

TABLE 7–7

Person Trip Destinations and Generation Rates by Type of Land Use

Type of Land Use at the Destination	Person Trip Destinations	Area in		Person Trip Destinations per	
		Square Miles	*Square Kilometers*	*Square Mile*	*Square Kilometer*
Residential	5,606,527	180.6	467.8	31,000	12,000
Manufacturing	779,340	24.7	64.0	31,600	12,200
Transportation	280,270	50.7	131.3	5,500	2,100
Commercial	2,449,468	21.1	54.6	116,000	44,800
Public buildings	781,960	23.1	59.8	33,800	13,000
Public open space	314,833	114.9	297.6	2,700	1,000
Total	10,212,398	415.1	1,075.1	24,600*	9,500†

Source: Chicago Area Transportation Study, *Final Report, I, Survey Findings* (Chicago: Chicago Area Transportation Study, 1959). Data apply to Chicago, 1956.
* 10,212,398 ÷ 415.1 = 24,600 Person Trip Destinations per square mile.
† 10,212,398 ÷ 1,075.1 = 9,500 Person Trip Destinations per square kilometer.

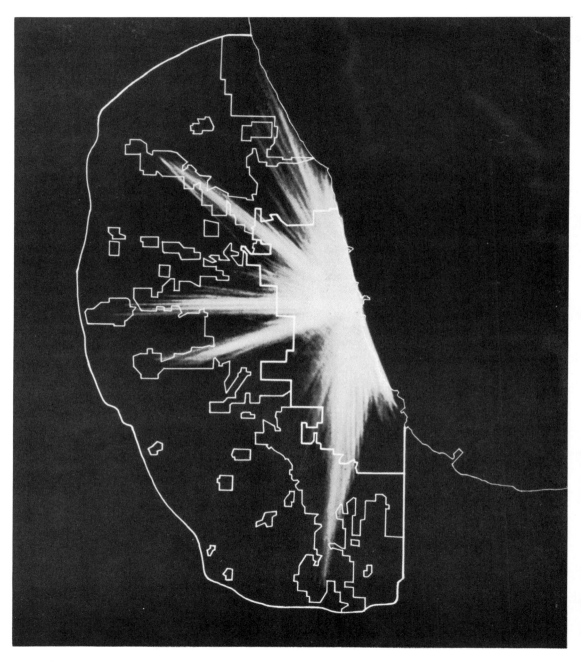

FIGURE 7–5a

Desire lines of internal person trips using rapid transit. Each desire line is traced from a transit rider's trip origin directly to his destination (from C.A.T.S., Fig. 18, p. 40).

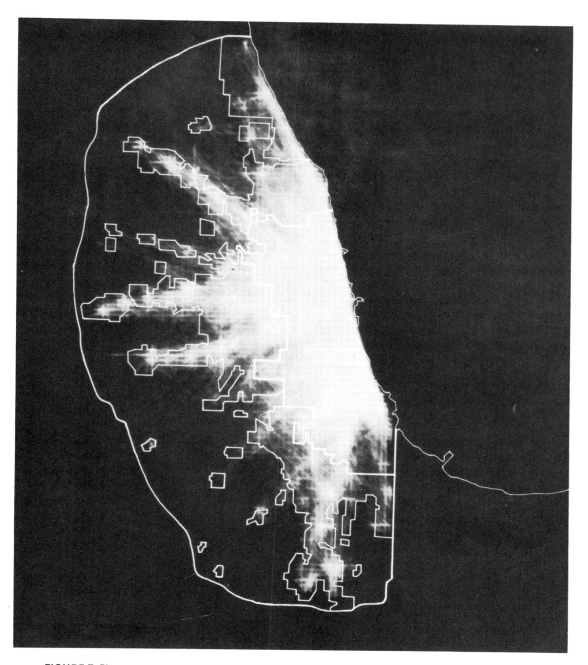

FIGURE 7–5b

Desire lines of internal automobile driver trips. By comparison with desire lines of transit users, trip ends for automobile trips are more scattered. Transit lines reflect a severe limitation in trip options (from C.A.T.S., Fig. 22, p. 44).

213

trips per unit area. The geographic pattern of land uses combined with trip generation behavior underlies spatial patterns of travel behavior. In planning new transportation systems, therefore, land use changes must be predicted along with variations in trip generation rates. Conversely, by carefully planning the geographical pattern of land uses, urban areas can drastically reduce future investments in transport facilities and at the same time reduce the time and money costs per capita of intra-metropolitan travel.

Trip Generation by Tract Density

A trip generation analysis attempts to identify and quantify trips beginning and ending in the tracts within a transport study area. Trip volumes are difficult to forecast, depending as they do on individual decisions to be made in a future context. The characteristics of the tracts and household units are more easily discovered. Trip generation analysis, therefore, establishes

the functional relationship between trip volumes at the ends of the trips—the origins and destinations—and the land use and socioeconomic character of the trip ends.

In each tract of a study area, trip generation levels vary systematically with (1) land use intensity, (2) socioeconomic character of population units in a tract, and (3) the location of activities in the urban area. Land use intensity describes the functional size or importance of a tract. Intensity is usually measured in terms of density variables like dwellings per unit area or workers per unit area. Variations in intensity cause a direct impact on the volume and kinds of trips that are generated within a tract. The number of person trips and automobile trips per family, for example, usually shows a sharp decrease as the number of dwellings per unit of residential area rises (Fig. 7–6). Even so, the relationship between trip generation rates and land use intensity accounts for only part of the tract-to-tract variation in trips generated. If all variations were accounted for, the plotted data points would line up along the best-fitting trend line.

Trip Generation by Population Characteristics

The socioeconomic characteristics of population units in a tract often vary independently from land use and density attributes of tracts. For example, car ownership rates and family income levels are closely related to trip-making behavior. As the number of cars owned by the household rises, daily trips per family rise as well (Fig. 7–7). As family income rises, car ownership rates also rise; and not only are more trips made per household, but the mode of travel changes too (Fig. 7–8a). Prosperous families living at the edge of the city make the most vehicular trips per dwelling unit, make many trips downtown, and are thereby responsible for generating an enormous number of very long trips. A different and more rational spatial organization of the metropolis could radically curtail the average length of trip for all purposes.

Low income families often own no car and rely heavily on public transportation, thus generating fewer auto trips. With higher

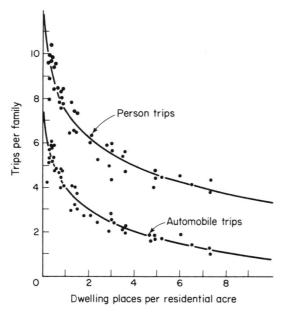

FIGURE 7–6

Effects of residential density on trip production. With higher density (closer to downtown) fewer trips are made per family per day (from C.A.T.S., Fig. 37, p. 68).

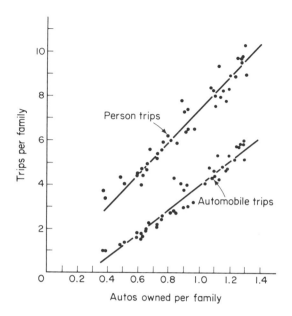

FIGURE 7–7

Effects of automobile ownership on trip production. The higher the automobile ownership rate, the greater the number of daily trips made per family. Automobile ownership rates depend, in turn, on average annual family income (from C.A.T.S., Fig. 38, p. 68).

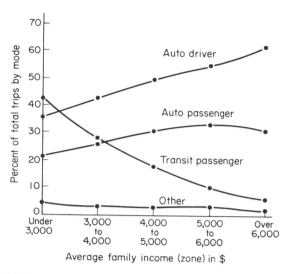

FIGURE 7–8a

The mode of transportation used depends on average family income. The poor are tied to transit; the wealthy travel primarily by car (from Guidelines for Trip Generation Analysis, p. 12).

income, families display increased travel from home for nonwork purposes, especially for shopping, social, and recreational purposes. Number of work trips per person is fairly stable as income rises or residence changes (Fig. 7–8b).

Trip Generation by Activity

The geographic pattern of activities within urban areas has profound effects on urban travel behavior. Interaction between an origin and destination varies with (1) the distance between them as well as with (2) the number of other possible origins and (3) the variety of alternative destinations.

In cities, most trips start or end at home. The journey to work and the return home account for a large fraction of the travel within an urban area. Work places are concentrated, although far from exclusively, at the urban core. Residential land lies in concentric rings around the

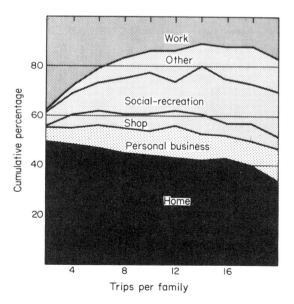

FIGURE 7–8b

A family's daily trip mix changes as it makes more and more trips. The largest fraction of trips end at home (from C.A.T.S., Fig. 39, p. 69).

central business district. The daily ebb and flow of radial journeys to and from work has been relatively easy to describe in model form up to the present time. Land uses were segregated and most cities had well defined annular forms.

Recently, new jobs have been created on the edge of town next to new residential developments. As a result, a complex fabric of land uses and traffic patterns has developed. Conceptually, however, identical interaction principles underlie both the old and new patterns. All that is lost is the visual elegance of old-style radial flow maps.

The metropolitan transportation study, then, applies both the idea of specialized places and specialized functional units in the urban system. When specialized places are complementary, interaction potential is created. For any place or group of specialized units, potential interaction levels can be estimated from past flow analysis. Especially helpful in estimating interaction potential are the potential model and the gravity model. The potential model is a way of summarizing the geographical complementarities in any spatial system; the gravity model describes flows through which places take advantage of the complementarities.

THE POTENTIAL MODEL

Let us assume that a town i is located in a region with several other cities. In a modern economy of specialized people and places we can reasonably expect that town i has some potential for interacting with each of the other towns in its region. But how much interaction should we expect? The concept of potential suggests a measure.

Borrowing from ideas of Newtonian physics, we might expect the same kinds of regularities in the attraction between social units as we observe among physical units. Any two physical objects in the universe attract each other with a gravitational force that varies directly with the product of the masses of the objects; the larger they are, the greater the attractive force. The attractive force or gravitational potential between two objects diminishes as the (square of the) distance between them increases.

We can apply these gravitational ideas to the towns in a region and derive a measure of interaction potential at each town location. Potential at a point may be thought of as a measure of the proximity of that point to all other places in the system, or as a measure of aggregate accessibility of the point to all the other points in a region. Potential at a point is simply an aggregate measure of the influence of all distant places on that point.

Within a bounded region containing n points, *total* potential at one point i is computed as the sum of the *separate* potentials created by the existence of every point including point i. The potential V created at i by each point j is equal to the mass at j divided by its distance from i. In symbols:

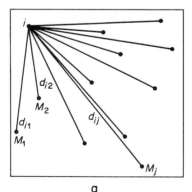

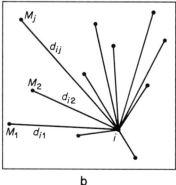

a   b

FIGURE 7–9

Total potential at any place i *is equal to the sum of the mass of every other place divided by its respective distance. If a total potential value is computed for each point* i, *a map of a potential surface can be constructed by drawing isolines connecting points of equal potential.*

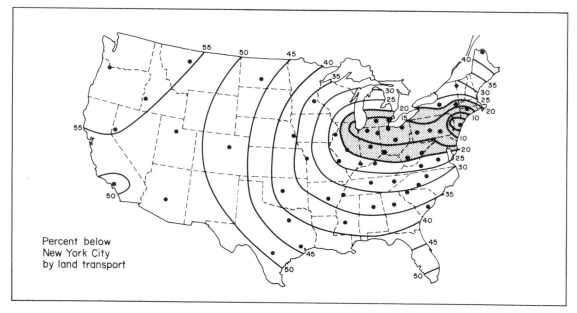

FIGURE 7–10

Distribution of market potential for the United States, using retail sales per county in 1948 as the measure of each county's market mass, and land transportation cost between production point and market as the measure of distance. The production points for which calculations were made are indicated by dots (adapted from C. D. Harris, Fig. 4).

$$V_i = \frac{M_1}{d_{i1}} + \frac{M_2}{d_{i2}} + \cdots + \frac{M_j}{d_{ij}} + \cdots + \frac{M_n}{d_{in}} + \frac{M_i}{d_{ii}}$$

that is: $V_i = \sum_{j=1}^{n} \frac{M_i}{d_{ij}}$

where: V_i = the representation of the total potential at place i,

M_j = the size of another place in the bounded region, and

d_{ij} = the distance separating i and j.

Thus, V_i is a summation of the effects of all n places on place i including the effects of i itself on itself (Fig. 7–9).

The mass at i itself is included in the summation, but clearly d_{ii} cannot be taken as zero because $M_i/0$ would drive V_i to infinity. Sometimes half the distance between point i and its nearest neighbor is used for d_{ii}. The total potential at each point is calculated accordingly. Each time the potential is computed for a point the masses remain the same but distances are recomputed. (Compare Figs. 7–9*a* and *b*.)

Constructing a Potential Surface

Interaction potential is a general concept and can be applied to marketing, migration, commuting, communication, and other kinds of problems. For example, market potential is an index of the intensity of possible spatial interaction between producers and markets. Thus, market potential at place i:

$$V_i = \sum_{j=1}^{n} \frac{M_j}{d_{ij}}$$

is the summation of all markets (M_j) in a region accessible to the place i with mass in each case divided by some distance measure (d_{ij}). When potential is computed for a large number of points in a region, a map of a potential surface can be constructed (Fig. 7–10).

Chauncy D. Harris studied market potentials using the United States as a region, retail sales per county as the measure of mass, and transportation costs as a measure of distance. With these measures he computed market potential at each major metropolitan center and constructed an isarithmic map of market potentials for the United States. The market potential was highest along a belt between Massachusetts and Illinois. Potential dropped sharply to the north and south. The peak of the potential surface was at New York City, at the center of the coastal ridge of high potential between Boston and Washington. As buying power shifts and as transport costs decline, the market potential surface fluctuates, settling at some places and rising elsewhere.

Potential is a state or a stock idea. When applied to populations rather than market, potential is a measure of the nearness of people to one another in the aggregate. Each individual in a country contributes to the total potential at any place i by an amount equal to the reciprocal of his distance away from i. Population potential is an abstract, macroscopic (large scale) variable measuring the relative position of each place with respect to all other places in a region. Potential represents a force underlying interaction among places.

Alternative Definitions of Mass and Distance

In the potential model the appropriate measure of mass depends on the class of interaction anticipated. The market potential model considers retail sales in a county to be a useful measure of county importance. In other applications, population or purchasing power are useful measures of mass, but their appropriateness depends on the problem. Alternatives to population and purchasing power include: total invested capital at a place, number of families, car registrations, hospital beds, investment in tractors and farm equipment, commodity output, value added in manufacturing, gross regional products, newspaper circulation, church attendance, school enrollment, and so forth. We can also weight population by income per capita, education, or employment. The class of interaction studied will dictate the appropriate population or nonpopulation measure of the mass at a place.

The potential for population movement and

other flows between places diminishes regularly as distance increases. Thus, every category of movement, whether telephone calls, business contacts, newspaper circulation, retail trade contacts, passenger traffic, or commuter flow, displays a lapse or fall off of movement with distance. How can we measure the effect of distance on interaction potential? Kilometers will be inadequate in certain cases. For the average person, the psychological and economic view of distance might really be a logarithmic function of kilometers or a function of time or of stimulation along the way. Nearby areas might be strongly differentiated while distant areas are perceived uniformly. If the friction of distance varies in logarithmic fashion away from the home origin, then the daily commuting distance would be perceived as a greater obstacle to movement than an extra three hundred kilometers when an extended vacation trip is planned (Fig. 7–11).

Other problems in distance measurement and space twisting arise when distances are not commutative. Almost everyone has had the experience that the journey to work often takes less time than the journey home from work. This is but one of a class of situations wherein the friction of moving from A to B is less than the friction of moving from B to A. How can

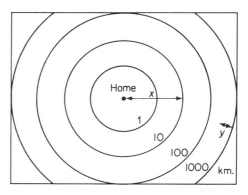

FIGURE 7–11

If the psychological and economic view of distance is really represented by a logarithmic function of kilometers, then daily commuting distance X *is perceived as a greater obstacle to movement than an extra three hundred kilometers,* Y, *at the margin of a planned vacation trip.*

this be? In the case of the journey to work, people must get to work at widely varying times in the morning, yet they all seem to leave for home at once. In another case *A* and *B* might be at the two ends of a one-way street. Or, when walking, *A* might be at the top of the hill when *B* is at the bottom. In this case the lineal distance is the same regardless of uphill or downhill direction but for someone walking or bicycling the direction of movement makes a difference in the distance friction.

The rent charged for private rental of trucks and trailers is normally based on a flat charge plus a charge per kilometer. Sometimes when a surplus of vehicles accumulates in one part of the country with a corresponding shortage somewhere else, penalties are temporarily assessed on certain movements to help redistribute the trucks and trailers. With a penalty in effect Illinois might be moved farther away from California for migrants to California, but for movers to Illinois from California, the dollar costs of distance would remain unchanged. Air fares between Europe and the United States across the Atlantic River are adjusted upward during peak seasonal demands. Then when discounts are granted in May and September when business is lagging, the effect is to bring the two places closer together and thereby increase the interaction potential by reducing the friction of distance.

Applications of the Potential Model

The map of potential describes the amount and location of interaction energy at each point in a region. In a regional system of equally spaced points of equal mass, the potential surface comes to a peak at the center of the region. In and around the center, the potential for interaction with other places is greatest. At the edges of the region potential falls to a minimum. In applications to follow, the potential model is shown to be the major building block of the gravity model of geographic interaction. Whereas the potential model considers places one at a time with respect to *interaction potential* with all other places in the system, the gravity model describes the *interactions themselves* among all points in the system.

The attenuating effect of distance on movement has been studied for a long time. Ravenstein's observations on the connection between distance and the flow of migrants proved so striking that many attempts were made to express the relationship concisely in general mathematical terms. As a first approximation of the general form of the relationship:

Flows per unit time = f (distance)

a scatter diagram can be constructed and a trend line estimated to depict the average *lapse rate*—that is, the rate at which the volume of flow falls off as a function of distance.

The volume of oceangoing freight moving between several pairs of places of equal size declines steadily as distance between origin and destination increases (Fig. 7–12). Over shorter distances the number of deliveries from a central city to nearby towns diminishes as distances increase. Inside a city a different number of trips per unit time are made to a store according to the distances people live from the store.

We must note two things about the effect of distance in regulating flows. First, lapse rates reflect the attenuating effect of distance and they differ for different kinds of movements; and second, the rate is different at different

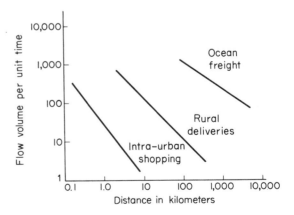

FIGURE 7–12

With increasing distance the flow volume per unit time diminishes for each class of movement. Time and money costs of movement determine the slope (distance decay rate or lapse rate) of each curve.

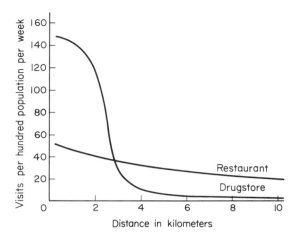

FIGURE 7–13

Distance curtails visits to two different establishments in different ways. Substitute establishments also curtail visits. In the cases shown, the drugstore seems to have more effective competitors than the restaurant.

distances. Consider how visits to a drugstore and visits to a restaurant might decline with distance (Fig. 7–13). On the vertical arithmetic scale are plotted visits per 100 people per week. Distance is plotted on the horizontal scale. Assume that trend line R refers to the restaurant and D is the rate at which customers visit a particular drugstore. Because the lines do not coincide we can infer that distance affects patronage differently. Trend line D drops faster than R as distance increases. This means that most people living close to the drugstore shop there, but between two and four kilometers away the average patronage drops off to almost zero, in part because there are competing drugstores nearby. Thus, a distance unit close to the drugstore is a more important barrier than one farther away. The restaurant's interaction curve declines in a different way. Here the first kilometer is more important than succeeding kilometers, but farther out a kilometer is less critical than in the drugstore case. Clearly, competing restaurants are less suitable substitutes for one another, either because there are fewer of them or because they are able to offer something special.

The effect of distance in regulating a class

of interaction (say messages) can be measured by calculating the *rate* at which the number of interactions per unit time declines with distance between origin and destination. In symbolic terms:

$$M = a \cdot \frac{1}{D^b}$$

which is the same as $M = a \cdot D^{-b}$

where: M = messages,
a = a constant, and
D = distance weighted by an exponent b.

Compare the ways this relationship can be described on a graph (Fig. 7–14). R is a hypothetical trend line based on a sample of long-distance telephone messages made by rich people; trend line P represents the relationship between distance and number of calls made by poor people.

Because each of the lines is described by an equation of the form:

$$M = a \cdot D^{-b}$$

we can take the logarithm of each side to produce the expression:

$$\log M = \log a - b \log D$$

where: $\log a$ and b = the parameters or constants which describe the location of the trend line.

Log a is where the trend line intersects the vertical axis; and b describes the slope of the line; that is, it corresponds to the change in the log of telephone calls with each unit change in the log of distance (Fig. 7–14a). When the relationship is expressed in logarithmic form rather than in its original form, it is computationally easier to estimate the effect of distance on telephone calling behavior. If the relationship is expressed in its original form it is difficult with sample pairs of M's and D's to estimate the a and b which define the trend (Fig. 7–14b). If we convert the equation to logarithmic form, then solve for log a and b using simple linear regression techniques, we can get the estimates of a and b with a minimum of fuss.

Once b has been estimated for the rich and again for the poor long-distance callers, we have an index or a measure for each group of the way the friction of distance curtails its telephone

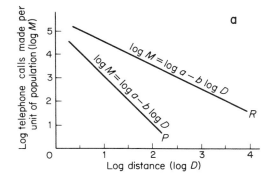

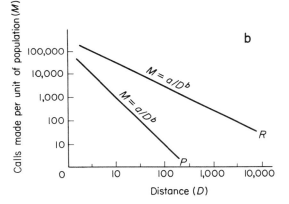

FIGURE 7–14

Two ways to describe how, as distance increases, the number of telephone calls made per person per unit time declines rapidly for poor persons (P) and much less rapidly for the rich (R). Either way, the slope of the curve is an index of distance friction. The friction is higher for the poor.

usage. The negative exponent of distance in the formula

$$M = a \cdot D^{-b}$$

gives the equation its distinctive character. Functions of this negative exponential type are usually described as Pareto curves after the Italian social statistician who first described their properties and their importance in social research. In movement studies, low values for b indicate wide and easy fields of movement. Conversely, when b values are high, lapse rates are steep and movements are sharply curtailed with increases in distance from the origin. This is

the case with long-distance calls by poor people. The cost of a call rises with distance, and when cost goes up, number of calls drops off sharply.

THE GRAVITY MODEL

In studies of spatial interaction the ideas of specific geographical complementarities and the friction of distance are brought together in the gravity model. The gravity model has captured a great deal of attention because of its pervasive simplicity. It says so much and says it so well. Simplicity or elegance has always been a virtue of good scientific theories. For example, in the sixteenth century the Copernican Heliocentric theory—that the earth and planets revolve around the sun—could not be verified any better than could the prevalent Ptolemaic doctrine. With minor exceptions both theories accounted for existing data. What distinguished the Copernican theory was its simplicity.

The theory was also logically fertile. It led to postulated explanations for data discrepancies which led to new discoveries. In social science the gravity model has been used to explain much of the variation in data describing movement of people, goods, and ideas, and to suggest insights about geographic structures formerly hidden from view. The gravity model is thus a valid representation of (P-plane) facts about spatial interactions.

What is the logical basis of the gravity model? Simply that two places interact with each other in proportion to the product of their masses and inversely according to some function of the distance between them. That is:

$$I_{ij} = f\left(\frac{M_i M_j}{d_{ij}}\right)$$

where: $I_{ij} =$ the number of interactions between i and j during some time period,
$d_{ij} =$ the distance between i and j, and
$M =$ some measure of the size or mass of the interacting pair of places.

Consider, for example, a set of six towns (Fig. 7–15). During a given period of time telephone calls are made within and between the towns (Table 7–8).

TABLE 7–8

**Telephone Calls, Distances, and the Interaction Energy Factor
for Six Neighboring Towns**

From Towns (Populations):	To Towns (Populations):					
	A (30)	B (80)	C (60)	D (30)	E (50)	F (40)
A (30)	21* 125‡ 43§	42† 193 57	45 95 40	58 75 16	70 54 21	73 162 16
B (80)	42 142 57	4 3860 1600	8 1325 600	19 340 126	29 350 138	40 190 80
C (60)	45 103 40	8 1560 600	4 2450 900	14 295 129	28 310 107	45 135 53
D (30)	58 49 16	19 271 126	14 470 129	5 455 180	16 160 94	43 35 28
E (50)	70 60 21	29 410 138	28 195 107	16 230 94	8 740 313	33 203 61
F (40)	73 38 16	40 240 80	45 110 53	43 72 28	33 160 61	17 240 94

* By convention, the distance of a place from itself is taken as half the distance to its nearest neighbor.
† Distance
‡ Telephone calls
§ $\dfrac{M_i M_j}{d_{ij}}$ = the interaction energy factor

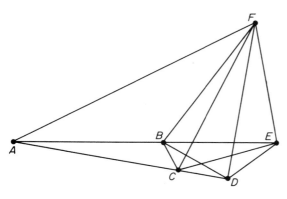

FIGURE 7–15

Six towns in a telephone system.

Using the population as the measure of mass at the origin and the destination, an interaction energy factor (IEF) equal to $(M_i M_j / d_{ij})$ can be calculated for each cell in the flow matrix. The IEF can be interpreted as an independent variable accounting for the level of calls reported in that cell. The larger is M_i or M_j, the greater will be the value for the IEF and the greater the number of calls. The bigger the distance between M_i and M_j, the smaller the value and the fewer the calls.

The relationship between the interaction energy factor as the independent variable and the number of calls reported in each cell as the dependent variable can be illustrated with a scatter diagram (Fig. 7–16). A best-fitting trend line summarizes the relationship between

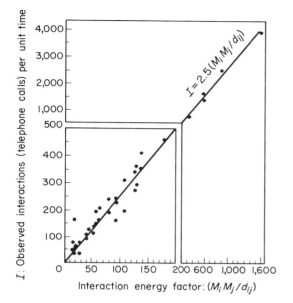

FIGURE 7–16

Thirty-six telephone flows among six towns during a sample time period.

interaction and the IEF. The trend line has an upward slope of 2.5. This means that for every unit increase in (M_iM_j/d_{ij}), two and a half more calls, on the average, are made from M_i to M_j in each time period.

Although the interaction energy factor accounts for most of the observed variations in the telephone calls, we might want to take a still closer look at the pattern. The IEF lumps all our information together into a single index. With little additional effort we can separate its parts and examine the influence of each component variable (M_i, M_j, d_{ij}) on the observed interaction pattern. In pulling apart the interaction energy factor we make use of some basic arithmetic. A generalized form of the interaction relationship is:

$$I_{ij} = a \cdot \frac{M_iM_j}{d_{ij}^b}$$

which is the same as $I_{ij} = aM_iM_jd_{ij}^{-b}$

Unless otherwise indicated, the exponent of each variable is unity. If we decide to estimate precisely the role of each variable in generating

telephone calls, we do not force or *constrain* the exponents of the M's to be unities (or ones); instead we leave them as parameters to be estimated. After we replace the exponents with a symbol representing a number of unknown size, the general form of the gravity model becomes:

$$I_{ij} = a \cdot M_i^{b_1} \cdot M_j^{b_2} \cdot d_{ij}^{-b_3}$$

By taking the logarithm of each side of the equation we produce the expression:

$$\log I_{ij} = \log a + b_1 \log M_i + b_2 \log M_j - b_3 \log d_{ij}$$

which has several *variables*, which we know for each sample observation in the interaction matrix $(I_{ij}, M_i, M_j, d_{ij})$, and several parameters to be estimated by simple multiple regression procedures (a, b_1, b_2, b_3). Parameters are the numbers which describe the statistical relationship between the independent variables (M_i, M_j, d_{ij}) and the dependent variable (I_{ij}).

The size of log a depends on a group's inherent tendency to make phone calls. A poor, backward, antisocial group will make far less use of the telephone for business and social purposes than a vigorous, prosperous, socially active group. With every unit increase in log M_i, log I_{ij} rises by b_1; when log M_j increases one unit, log I_{ij} rises by b_2. Interaction thus varies directly with the masses of the interacting bodies. As log d_{ij} rises by a unit, however, the value of log I_{ij} *drops* by b_3. This too is expected. In situations where interaction is sharply curtailed by distance, b_3 is relatively large. When distance is only a minor barrier to interaction, b_3 is small. Clearly b_3 can vary from place to place in the United States at any one time, and it has certainly dropped sharply through time for all classes of geographic interaction.

The gravity model assumes that the effect of distance varies smoothly and continuously over geographic space. Normally this is so. Political boundaries, however, create discontinuities in patterns of interaction among places in different political territories. To demonstrate this fact, Ross Mackay studied telephone traffic between Montreal and: (1) Quebec cities; (2) other Canadian cities; and (3) United States cities (Fig. 7–17). Canadian cities outside

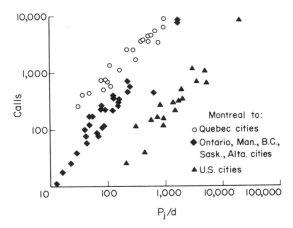

FIGURE 7–17

Telephone calls from Montreal to three sets of cities in a ten-day period. The boundary between French- (Quebec) and English- (other Canadian) speaking cities can be equated to an increased distance factor of 5 to 10. The international boundary acts like a fifty-fold increase in distance (adapted from J. R. Mackay, Fig. 2).

Quebec province interacted with Montreal at levels far lower than expected given the values of the interaction energy factor, and other Canadian cities behaved as though they were five to ten times as far from Montreal. United States cities interacted with Montreal as though they were fifty times as distant. The displacements can be interpreted as due to the discontinuous friction of the boundaries as they retarded interaction.

Fitting Commodity Shipment Data into the Gravity Model

California is the most populous state in the United States. It is also the leading agricultural state and second only to New York in the annual value of its manufactures. California's huge fruit and vegetable output depends on a warm, sunny climate and a vast irrigation system. Its livestock output is second only to Iowa. It is second to Florida in oranges, and the third-ranking state in cotton. California's commodity output is marketed largely within the state itself, containing as it does about a tenth of the national market. Yet every state imports com-

modities from California. The larger, richer, and closer a state is to California, the more it buys. Small, poor, and remote states buy proportionately less.

California sells a substantial portion of her manufacturing output (including processed agricultural commodities) by rail shipments in carload lots. Each carload of a commodity is accompanied by a waybill, which is a document prepared by the railroad company and which contains a description of what is being shipped in the rail car; the names and addresses of the sender and receiver; the origin, route, destination of the shipment; and the shipping charges. The Interstate Commerce Commission collects a 1 percent sample of these waybills every year to keep track of the interstate flows of commodities in the United States. The flow statistics are then published for analysis.

From the waybill sample and the decennial population census we can accumulate the following data pertaining to commodity shipments within California and from California to the forty-seven coterminous states and the District of Columbia: (1) carloads of commodities shipped from California to each state in 1957; (2) total personal income received in each state in 1957; and (3) average rail haul in miles between California and each state in 1957. Let us fit this flow data into the gravity model. Interaction between California and state j consists of the number of carloads shipped from California to that state: I_{cj}. The mass of each destination state j is its market demand which can be represented by personal income in 1957: M_j. The distance between California and state j equals the average haul to that state by rail: d_{cj}.

The gravity model tells us that in a system of flows, the flow level between an origin and a destination varies directly according to the product of their masses and inversely with the distance between them. In the California commodity flow problem, one of the masses is always California's. For every observation it is a constant, so if we incorporate California's mass into the constant of proportionality, or a, we are left with the simplified expression:

$$I_{cj} = a \cdot \frac{M_j^{b_1}}{D_{cj}^{b_2}}$$

which can be rewritten:

$$I_{cj} = a \cdot M_j^{b_1} \cdot D_{cj}^{-b_2}$$

From our earlier discussion we know that it is easier to estimate the parameters in a gravity model if we take the logarithm of each side of the equation:

$$\log I_{cj} = \log a + b_1 \log M_j - b_2 \log D_{cj}$$

By taking the logarithms we are left with an expression of the form:

$$Y = a + b_1 X_1 - b_2 X_2$$

in which the parameters a, b_1, and b_2 control the relationship between variables Y, X_1, and X_2. In the California commodity flow example the parameters ($\log a$, b_1, and b_2) control the relationship between the transformed variables ($\log I_{cj}$, $\log M_j$, and $\log D_{cj}$) in exactly the same way, and using 1957 waybill data they are estimated according to the multiple regression technique introduced earlier, producing:

$$\log I_{cj} = 4.56 + 0.86 \log M_j - 1.88 \log D_{cj}$$

This expression describes the relationship between the variables very well. The coefficient of multiple determination, R^2, equals 0.88; the gravity formulation accounts for 88 percent of the variation in the commodity flow data. The constant of proportionality, 4.56, is an index of the average propensity of places to import commodities from California in 1957. If $\log M_j$ rises by one, then $\log I_{cj}$ rises by 0.86. The effect of distance is the opposite. If $\log D_{cj}$ increases by one, $\log I_{cj}$ drops 1.88.

Consider the Montana observation as an example:

$$\log I_{c \to mont} = 1.1139$$
$$\log M_{mont} = 3.0645$$
$$\log d_{c \to mont} = 3.1186$$

If we plug log mass and log distance values into our gravity model we find that instead of 1.1139 (rounding to 1.11), the model estimates:

$$\log I_{c \to mont} = 4.56 + 0.86\,(3.0645)$$
$$- 1.88\,(3.1186) = 1.34$$

Thus, if Montana conformed precisely with the average relationship defined by the gravity model we would expect that $\log I_{c \to mont}$ would equal 1.34. When we turn to our actual data we observe that $\log I_{c \to mont} = 1.11$. The difference (observed $I_{c \to mont}$ — expected $I_{c \to mont}$) is called the residual. In the Montana case the residual equals —0.23. We interpret the residual by pointing out that Montana imported somewhat fewer commodities from California than we would have expected, given its income and distance from California, and based on the gravity model. Using the $\log M_j$ and $\log D_{cj}$ data for each of the other states, we can compute an expected $\log I_{cj}$ value for each of them and derive a residual for each state. With these residuals, some positive and some negative, we can make a map of residuals to show how the variation in the transformed data that was not explained by the gravity model is distributed among the various states and regions of the country. Recall that the gravity model explained 88 percent of the variation. This means that only 12 percent remains unaccounted for. The 88 percent is due to a combination of personal income and distance from California. What is the other 12 percent due to? By making the map of residuals and studying their spatial patterns we can begin to get some new ideas and form additional hypotheses to explain the 12 percent residual variation (Fig. 7–18). For example, if we look just at the negative residuals we are faced with a question: why did certain states import less than expected? Perhaps the Upper Midwest states of Minnesota, the Dakotas, Montana, and Wyoming diverged slightly from the gravity model because they imported disproportionate amounts of merchandise from sources in the northeastern manufacturing belt. Perhaps the Kansas-New Mexico-Oklahoma cluster was well supplied with certain commodities from other sources as well and had relatively less need for California products. The cluster of southeastern states have Florida sources competing with California in many processed food lines. Moreover, the southeastern states generally have very low incomes per capita. Now it is true that the gravity model takes differences in state income into account, but perhaps when people's incomes drop below a certain level, their consumption of the sorts of things California sells changes in unusual

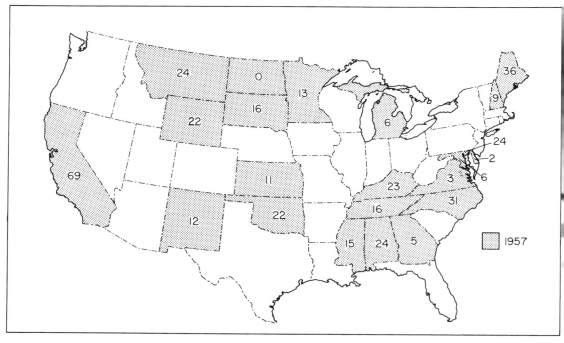

FIGURE 7–18

States receiving fewer commodity rail shipments from California than would be expected, based on destination states' personal incomes and rail distances from California. Numbers refer to relative sizes of negative residuals.

ways that the gravity model does not account for. Elsewhere, Michigan is an exceptionally large producer of manufactured goods itself, and there are many intervening opportunities accounting for the large negative residual in Maine.

The other side of the story is the pattern of positive residuals formed by states that imported California commodities in 1957 in excess of what the gravity model predicts. One thing that seems to stand out is the corridor of states between California and New York which are the places best served by the transcontinental railroads as trains streak back and forth between Megalopolis on the east coast and the sprawling valley farms and factories in sunny California in the West. The more we study a map of residuals the greater the number of new hypotheses we can generate to be tested. Until we test them, they seem perfectly reasonable. Imagine our surprise therefore when we repeat

the analysis for commodity flows from California in 1963. The old gravity model (1957) is:

$$\log I_{cj} = 4.56 + 0.86 \log M_j - 1.88 \log d_{cj}$$

Compare its parameters with those in the 1963 model:

$$\log I_{cj} = 3.81 + 0.91 \log M_j - 1.76 \log d_{cj}$$

The income coefficient rose from 0.86 to 0.91, indicating that in the latter year income had relatively more explanatory power in accounting for the flows. At the same time, the distance coefficient dropped slightly from 1.88 to 1.76, revealing a general decline in the importance of distance in explaining flow patterns. The R^2 dropped from 0.88 to 0.84 as well. Whereas it is possible that these small changes could have arisen from slight fluctuations in the 1 percent waybill sample, the changes seem to be in the right direction. Possibly other factors became more important. What could they have been?

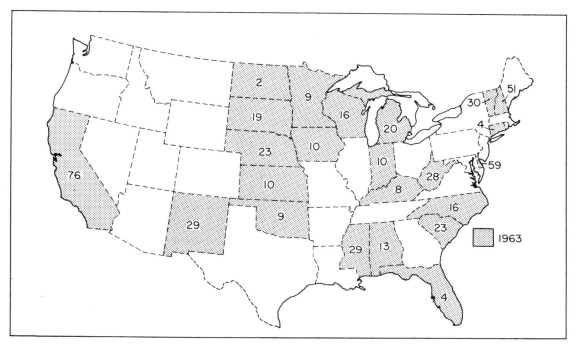

FIGURE 7–19

States receiving fewer rail shipments from California in 1963 than expected, again using a gravity model with 1963 personal income as the mass of a destination and rail distances.

If we construct a map of residuals to get a few ideas about the source of the 16 percent residual variation, many of our earlier guesses based on the 1957 map fly out the window (Fig. 7–19). The pattern of negative residuals is quite different, perhaps due to a crop failure or killing frosts in Florida which damaged fruit output and allowed California producers of manufactured agricultural goods to make major inroads in traditional Florida markets, but at the same time creating shortages in states where California traditionally sold more than expected.

Fitting Migration Data into the Gravity Model

Between 1955 and 1960 thousands of black Americans moved out of the southeastern United States, to settle in larger metropolitan areas of the North and West. During that five year period 23,397 blacks moved into the twenty-three largest American metropolitan areas after leaving the state of Alabama. A different fraction of this group went to each destination. How can the allocation by destination be accounted for? In terms of the potential model, we would expect that each of the twenty-three cities would exert a pull on the Alabama migrants proportional to its mass and inversely proportional to its distance from Alabama.

If the distance between the center of Alabama and destination is equal to d_{aj} and the mass of the destination is taken as its population P_j, then the twenty-three cities create a total potential on Alabama equal to the sum of the potentials created by each of the cities:

$$V_{\text{alabama}} = \sum_{j=1}^{23} \frac{P_j}{d_{aj}}$$

If the total potential created at the center of

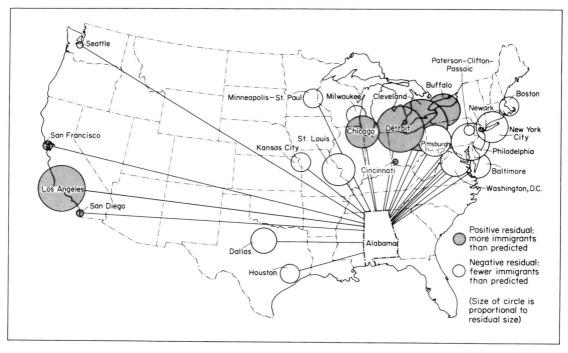

FIGURE 7–20

Positive and negative residuals when black migrants from Alabama to 23 cities between 1955 and 1960 are allocated according to population size of destination and its distance in miles from Alabama.

Alabama equals V, the fraction of that total accounted for by the jth place equals:

$$\frac{\text{The potential contributed}}{\text{by the }j\text{th place}} = \frac{P_j/d_{aj}}{V_{\text{alabama}}}$$

New York's 1950 population was 9,556,000 and its distance from the center of Alabama was about 1,125 miles. Population (in thousands) divided by distance yields 8.494. Repeating the computation for each of the other twenty-two metropolitan destinations and adding the results produces $V_{\text{alabama}} = 48.971$. Thus, the fraction contributed by New York is 8.494/48.971, or 17.3 percent. If we expected the Alabama outmigrants to distribute themselves among the twenty-three centers according to their respective potentials, then New York would receive 17.3 percent of the 23,397 mi-

grants. The census records show that only 13.1 percent actually went to New York—4.2 percent less than expected. If the other residuals are computed for the other cities a map of residuals can be constructed to help us generate additional ideas about why black Alabama outmigrants moved and resettled in the ways they did (Fig. 7–20).

In several cases the expected percentages diverge widely from the observed values (Fig. 7–21). The logic of the model is, of course, very simple and can be improved. Perhaps destination population in 1950 was not a good predictor of its attractive mass between 1955 and 1960. Perhaps a superior measure of a city's drawing power would be manufacturing jobs, or black population already at the destination and sending information back to the South; or all jobs weighted by annual pay levels; or new factory jobs created between 1950 and 1955. There are many possibilities which can be used with greater theoretical justification than population alone.

With total population used as the mass of a destination the positive residuals occur at places with large numbers of factory jobs and relatively less union discrimination than elsewhere (Fig. 7–20). Moreover, the positive residuals are closer to the Alabama origins of the migrants. Out-migrants from states east of Alabama move northward to the metropolitan centers of Megalopolis. Migrants from Alabama are relatively closer and more aware of opportunities directly north; migrants from Mississippi and Texas are inclined to move to the west coast. The migrants from Alabama somewhat avoid southern cities like those in Texas and Missouri. Perhaps these places are just too similar to home to suit people who have decided to leave the South. California cities are well known through television, movies, and the experiences of military service. The trip from the South to California is perceived to be much shorter than kilometers alone would suggest. Alabama is closer

in kilometers to Maine than to California, but that fact is of little consequence to the typical Alabama migrant looking for a better life. Some cities like Milwaukee and Minneapolis-St. Paul are a long way from Alabama in another sense. Important intervening opportunities lie between the South and these destinations. Boston, at the northern end of Megalopolis, occupies a similar position in terms of intervening opportunities and receives fewer southern migrants than expected. The positive residual at Newark is interesting considering the fact that most Americans know very little about this New Jersey city. Yet people moving into New York learn about Newark very quickly and many southern migrants who come first to New York find Newark a preferable destination. Such stepwise relocation is important in migration analysis. In the 1960 census many black Newark residents reported that in 1955 they lived in Alabama. Perhaps between the time they left Alabama and the time they were interviewed by the census taker many of them spent some time in New York City.

FIGURE 7–21

Observed and expected shares of the black Alabama out-migrants to 23 cities. Expectation is based on a gravity model with mass equal to urban population at the destination, and distance equal to miles from Alabama.

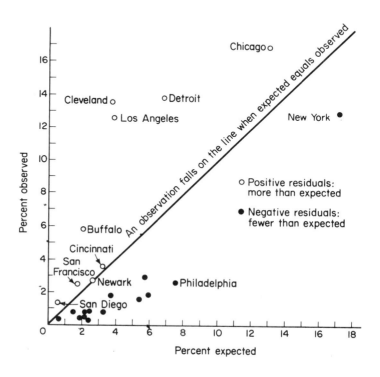

A group of people moves because they think it will pay them to move. Another group stays because perceived benefits from moving do not exceed moving costs. Economic opportunity is a major cause of interstate migration. Opportunity, as perceived by Alabama residents, underlies the desire lines which direct travel behavior. Initial errors in perception are eliminated as pioneers feed back information from new homes to old relatives and friends. Changes

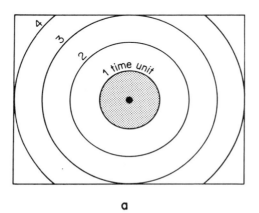

a

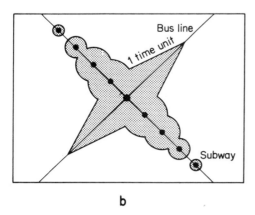

b

FIGURE 7–22

When everyone walks, an isotropic plane is created (A). Movement ease is equal in every direction and isochrones are concentric around the city center. If a bus line is introduced on one main street, and subway service with limited stops is begun along a second arterial route through the city center (B), the area which can be reached within one unit of time (shaded) expands in an irregular fashion.

in employment, housing, or social conditions produce regular revisions in the weight assigned to alternative destinations by people about to migrate. Differential changes in time and money costs of moving bring about adjustments in the new attractiveness of destinations as well. If the costs of a trip to Los Angeles are reduced relative to the costs of a trip to Detroit, the added attractiveness of California will divert extra migrants west instead of north.

Movements at the interstate scale reflect the distance distorting effects of transport and communications systems in the same way that urban geographic space is twisted by expressways and mass transit (Fig. 7–22). If everyone in a town walks to and from the downtown center to work and shop, the isochrones (lines of equal travel time from downtown) form an annular pattern around the center. If a mass transit system is constructed with limited stops, some outlying places are brought closer to downtown in a time sense than the inner-city neighborhoods. When studying interaction between the downtown and various neighborhoods, defining distances in terms of absolute space (kilometers) has less explanatory power in a gravity model than measuring movement costs in relative space (time). This model of movement in the city is directly applicable to questions of population movements across the country. For an Alabaman about to migrate, a place in south-central Illinois might, in an accessibility sense, seem farther away from Alabama than the city of Chicago, just as in the city a transit stop at the edge of town might be easier to reach from downtown than other places closer to downtown in absolute space.

Barrier Effects of Social Distance

If the gravity and potential concepts are useful in explaining interactions which do occur, they may also shed light on the absence of interactions where they might otherwise be expected to occur. Some differences within cities seem to be noncomplementary, or if complementary and contributing to interaction potential, the interactions are thwarted. Black ghetto residents, for example, have little interaction with residents of white ghettos. When the ghettos are side by side, distance in simple absolute terms loses its explanatory power as a control on interneighborhood interactions.

In 1968 the National Advisory Commission on Civil Disorders (the Kerner Commission) reported that:

Our nation is moving toward two societies, one black one white—separate and unequal.... Discrimination and segregation have long permeated much of American life,... The deepening racial division is not inevitable. The movement apart can be reversed.... [but] ... To pursue our present course will involve the continuing polarization of the American community and, ultimately, the destruction of basic democratic values.

This is a situation so serious that everyone in a new generation will have to contribute to the effort to solve it. It demands imaginative interpretation of what constitutes distance or separation in a city's social geographic space. If the gravity and potential models predict interaction among complementary areal units when it is to their mutual advantage, we must think imaginatively about what must change for the interactions to proceed unencumbered. Presently, social distances between black and white areas are great and movements are thwarted. Perceptions of opportunity are clouded on both sides of the ghetto boundary; perceptions must be sharpened and broadened.

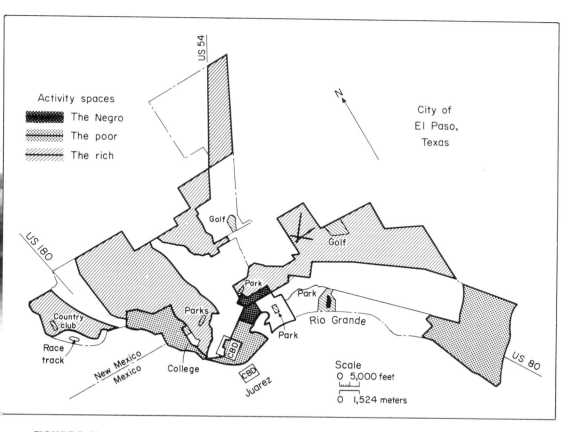

FIGURE 7–23

Each social group occupies a separate part of the city and can carry on much of its life without crossing over into the activity spaces of others (based on a map by Eugene L. Ziegler).

Not all social barriers are racial. Race just aggravates separations based on cultural or ethnic differences. The spatial separations of socioeconomic groups inside American metropolitan areas are severe, and the result in the long run is certainly detrimental to the national well-being. Inside metropolitan areas, ethnic separation shows up as residential segregation. Metropolitan El Paso-Juarez straddles the U.S.-Mexican border and offers an excellent example of how socioeconomic neighborhood boundaries within the city raise greater obstacles to spatial interaction than does the closely patrolled international border which separates the metropolitan area into two approximately equal parts.

El Paso is located in the extreme western tip of Texas and shares its city boundaries with both Mexico and New Mexico. The city is surrounded almost entirely by desert. Juarez, Mexico, comprises the other half of the urbanized area. The combined population is approximately 700,000, of which El Paso accounts for less than half.

Inside El Paso the rich live in one part of town, the poor in another, and the blacks (a scant 2 percent of the city's total) in a third (Fig. 7–23). The significant ethnic minority in El Paso is the portion of the population which is of Spanish origin, presumably Mexican. For the city as a whole, 45.4 percent of the 1960 population had Spanish surnames, and one

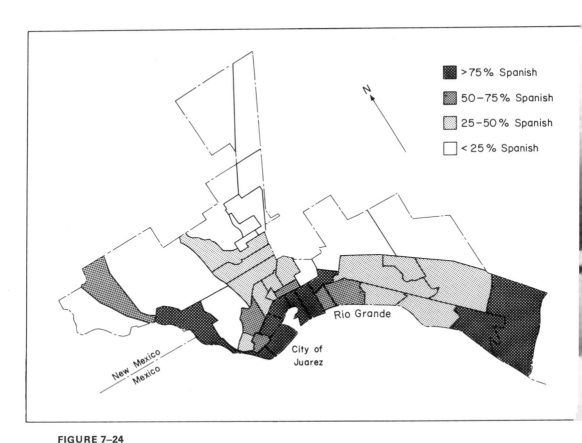

FIGURE 7–24

Distribution of Spanish surnames in El Paso (based on tract data from the 1960 Census of Population and Housing, and a map by Eugene L. Ziegler).

fourth of these were born in Mexico. The Spanish Americans or Mexican Americans in El Paso occupy a second-class position in terms of affluence and residential segregation. Each of the socioeconomic groups in El Paso has its own residential areas and its private activity spaces. What is distinctive about the residential areas, however, is the clustering of the people with Spanish surnames along the Mexican border (Fig. 7–24). It appears that the social and ethnic barriers inside El Paso are more important obstacles to human interaction over space than is the international boundary separating the United States and Mexico.

SUMMARY AND CONCLUSIONS

Mere areal differences do not necessarily provide a basis for interaction. Numerous differentiated areas are and have been unconnected by trade, migration, or message flows. To develop a basis for interaction there must be a supply in one place and a specific complementarity elsewhere. But complementarity alone, although necessary, is an insufficient basis for interaction. Interchange occurs only if intervening opportunities do not interfere. The final factor is transferability. Interaction is possible only if the friction of distance measured in real terms (transfer and time costs) is not prohibitive.

The laws of geographic interaction among places describe *aggregate* human behavior, not individual behavior. Ian Burton has suggested that a social science which recognizes random behavior at the microcosmic level and predictable order at the macrocosmic level is a logical outgrowth of the quantitative revolution in social science. This distinction is one way to differentiate macro-spatial from micro-spatial behavior in geographical inquiry. Areal patterns and flows over the earth are at once the cause and the effect of one another. An idealized movement pattern generates an areal pattern with predictable regularity. An idealized, homogeneous landscape will generate interaction patterns which are perfectly predictable. But why discuss idealizations when the real world is such a messy place? Because by studying ideal-

ized models and their implications we can sharpen our analytical tools and broaden our range of insights for real world applications.

Distributions imply movements; and movements produce and modify distributions. Models of geographic relationships and interaction bases are the intellectual frameworks providing insights about the events we observe. The competitive positions of the east coast cities during the frontier push westward after the American Revolution are good examples of flows regulating geographic pattern and geographic patterns producing flow patterns. With each advance in transportation and communication, the east coast cities sent out a new set of routes, setting off new development patterns in the Midwest, and culminating in a new flow pattern of people, goods, and messages between east coast cities and the interior. The cities created the conditions for the interaction pattern and were subsequently modified by them.

Similarly, in the Upper Midwest, the Twin Cities of Minneapolis and St. Paul built a railroad network serving western Minnesota and the Dakotas. The cities aided in the settlement of the area served, then tapped the timber, agricultural, and mineral wealth of the region. Resource distribution patterns regulated the flow potential, and once the resources were tapped, the flows created a new set of complementarities, one of which was a large consumer market which was promptly served by Twin Cities merchants. By examining the bases for spatial interaction we move one step closer toward an answer to geography's central question: why are spatial distributions structured the way they are?

Suggestions for Further Reading

BERRY, BRIAN J. L., and DUANE F. MARBLE, eds. *Spatial Analysis: A Reader in Statistical Geography.* Englewood Cliffs, N. J.: Prentice-Hall, Inc., 1968.

BUNGE, W. "Toward a General Theory of Movement," in *Theoretical Geography*, Chap. 5. Lund Studies in Geography, Series C, General and Mathematical Geography No. 1. Lund: C. W. K. Gleerup, 1966, pp. 112–33.

BURTON, IAN. "The Quantitative Revolution and Theoretical Geography," *Canadian Geographer*, VII (1963), 151–62. Reprinted in Brian J. L. Berry and Duane F. Marble, eds. *Spatial Analysis: A Reader in Statistical Geography*. Englewood Cliffs, N. J.: Prentice-Hall, Inc., 1968, pp. 13–23.

GARNER, BARRY. "Models of Urban Geography and Settlement Location," in R. J. Chorley and P. Haggett, eds. *Models in Geography*, Chap. 9. London: Methuen and Co., Ltd., 1967, pp. 303–60.

HAGGETT, P. "Movement" and "Surfaces," in *Locational Analysis in Human Geography*, Chaps. 2 and 6. London: Edward Arnold, 1965, pp. 31–60, 153–82.

HAMILTON, F. E. IAN. "Models of Industrial Location," in R. J. Chorley and P. Haggett, eds. *Models in Geography*, Chap. 10. London: Methuen and Co., Ltd., 1967, pp. 361–424.

HARVEY, D. "Models of the Evolution of Spatial Patterns in Human Geography," in R. J. Chorley and P. Haggett, eds. *Models in Geography*, Chap. 14. London: Methuen and Co., Ltd., 1967, pp. 549–608.

ISARD, WALTER, et al. "Gravity, Potential, and Spatial Interaction Models," in *Methods of Regional Analysis*, Chap. 2. New York: The M. I. T. Press and John Wiley & Sons, Inc., 1960, pp. 493–568.

LEONTIEFF, WASSILY. "The Structure of Development," *Scientific American*, CCIX: 3 (September 1963).

OLSSON, GUNNAR. *Distance and Human Interaction: A Review and Bibliography*, Regional Science Research Institute Bibliography Series No. 2. Philadelphia, 1965.

PRED, ALLAN R. "American Metropolitan Growth: 1860–1914; Industrialization, Initial Advantage," in *The Spatial Dynamics of U.S. Urban-Industrial Growth, 1800–1914*, Chap. 2. Cambridge, Mass.: The M. I. T. Press, 1966, pp. 12–85.

SMITH, R. H. T. "Method and Purpose in Functional Town Classification," *Annals of the Association of American Geographers*, LV: 3 (September 1965), 539–48. Reprinted in Brian J. L. Berry and Frank E. Horton, *Geographic Perspectives on Urban Systems, with Integrated Readings*. Englewood Cliffs, N. J.: Prentice-Hall, Inc., 1970, pp. 106–12.

THOMAN, RICHARD S., and EDGAR C. CONKLING. *Geography of International Trade*. Foundations of Economic Geography Series. Englewood Cliffs, N. J.: Prentice-Hall, Inc., 1967.

THOMPSON, WILBUR R. "Economic Growth and Development: Processes, Stages, and Determinants," in

A Preface to Urban Economics. Baltimore: Johns Hopkins University Press, 1965, pp. 11–60.

ULLMAN, EDWARD L. "Minimum Requirements after a Decade: A Critique and an Appraisal," *Economic Geography*, XLIV (1968), 364–69.

———. "The Role of Transportation and the Bases for Interaction," in William L. Thomas, ed. *Man's Role in Changing the Face of the Earth*. Chicago: The University of Chicago Press, 1956, pp. 862–80.

———, and MICHAEL F. DACEY. "The Minimum Requirements Approach to the Urban Economic Base," *Papers and Proceedings of the Regional Science Association*, VI (1960), 175–94.

VANCE, JAMES E., JR. *The Merchant's World: The Geography of Wholesaling*. Foundations of Economic Geography Series. Englewood Cliffs, N. J.: Prentice-Hall, Inc., 1970.

WARNTZ, WILLIAM. *Macrogeography and Income Fronts*. Regional Science Research Institute, Monograph Series No. 3. Philadelphia, 1965.

Works Cited or Mentioned

BURTON, IAN. "The Quantitative Revolution and Theoretical Geography," *Canadian Geographer*, VII (1963), 151–62. Reprinted in Brian J. L. Berry and Duane F. Marble, eds. *Spatial Analysis: A Reader in Statistical Geography*. Englewood Cliffs. N. J.: Prentice-Hall, Inc., 1968, pp. 13–23.

CHICAGO AREA TRANSPORTATION STUDY. *Final Report, I: Survey Findings*. Chicago: Chicago Area Transportation Study, 1959.

GALBRAITH, JOHN KENNETH. *The New Industrial State*. Boston: Houghton Mifflin Company, 1967.

GOTTMANN, JEAN. *Megalopolis: The Urbanized Northeastern Seaboard of the United States*. New York: The Twentieth Century Fund, 1961.

HARRIS, CHAUNCY D. "The Market as a Factor in the Localization of Industry in the United States," *Annals of the Association of American Geographers*, XLIV (1954), 315–48.

MACKAY, J. ROSS. "The Interactance Hypothesis and Boundaries in Canada: A Preliminary Study," *Canadian Geographer*, XI (1958), 1–8. Reprinted in Brian J. L. Berry and Duane F. Marble, eds. *Spatial Analysis: A Reader in Statistical Geography*. Englewood Cliffs, N. J.: Prentice-Hall, Inc., 1968.

RAVENSTEIN, ERNEST G. "The Laws of Migration," *Journal of the Royal Statistical Society*, LII (1889), 241–305.

ULLMAN, EDWARD L. "The Role of Transportation and the Bases for Interaction," in *Man's Role in*

Changing the Face of the Earth, ed. William L. Thomas. Chicago: The University of Chicago Press, 1956, pp. 862–80.

U.S. DEPARTMENT OF TRANSPORTATION/FEDERAL HIGHWAY ADMINISTRATION, BUREAU OF PUBLIC ROADS. *Guidelines for Trip Generation Analysis.*

Washington, D. C.: U.S. Government Printing Office, June 1967.

ZIEGLER, EUGENE L. "The Social Areas of El Paso, Texas, and Juarez, Mexico," Unpublished paper, Department of Geography, The Pennsylvania State University, University Park, May, 1969.

CHAPTER 8

MOVEMENT AND
TRANSPORT SYSTEMS

*The aim of transportation is to set men free in respect to place relations,
to make these relations more plastic to social needs.*

—Charles H. Cooley

WHY THINGS MOVE

Why Things Move versus How They Move

A trip is an event, whereas travel is a process. Businessmen take trips; vacationers travel. When we talk about trips the emphasis is on linking origins and destinations. When we study travel we examine route, mode, and speed; we are primarily interested in what happens en route. When we take a trip in a subway, we want to get from one station to the next with a minimum of time and discomfort. Seldom are we interested in what happens along the way. Airplane trips carry us along "subways in the sky" from one airport to the next while the stewardesses perform elaborate food and drink rituals to break the monotony. Travel, on the other hand, is done for its own sake. When we take a walk or a drive with someone whose company we enjoy, what happens en route is the important thing. Trips and travel

are even reported differently. Trips are counted; travel is described in terms of cost, time, distance, route, and events and stimuli encountered along the way.

To explain *why* things move from an origin to a destination we refer to ideas of complementarity, the relative attractiveness of alternative destinations, the technology needed to overcome distance friction, and intervening obstacles to interaction. Questions of *how* things move are resolved in terms of modes of transport, rates of speed, and efficiency. A trip is a fact, an event. Movements over the earth are processes extended over time and space.

Movements of every sort create spatial structures, and once established such spatial arrangements influence subsequent movement. Movements and actions of ice in glaciated areas created a variety of landforms, which in turn influenced later water movements, which· in their turn slowly modified the landforms in a continuous process of mutual cause and effect.

Processes of erosion and degradation created new structures, then the structures channeled subsequent processes. We do not have to look far for parallel examples from the human world. Migrations redistribute populations in one period and thereby influence movements in the next. The edge of the built-up city moves into the countryside and produces a new pattern of commuter movements. The technology of the industrial revolution spread south and east over Europe producing marked interregional differences in wealth and opportunity and generating a need for new communication links to handle the flows of people, ideas, and material.

Spatial arrangements are reciprocally tied to movement processes. A structural remnant like a terminal moraine is the outcome of glaciation, a spatial process in the physical world. A contemporary group of native-born Americans in Frenchville in central Pensylvania who learned French as a first language in the twentieth century is the outcome of a spatial process in the human world. Regular outcomes imply regular processes, and when we observe regular or symmetrical patterns which seem to have occurred naturally, they command our attention. Symmetry and repetitive patterns in the P-plane have a surprise value that prompts us to speculate, leading to constructs and concepts in the C-field. We say: I wonder why the bee's cell in a honeycomb forms such a nice hexagonal shape? The reason is that things moving in nature try to reach their goals by the least costly route. At first this idea was hypothesized in social science as the *principle of least effort:* that natural events reach their goal by the easiest route. At an earlier stage in the development of theory social scientists thought that a direct analogy would be useful in social science where it had long been held that rational economic men, if given a choice, preferred more pleasure and less pain. With the principle of least effort in mind, social scientists began to hunt for structural regularities as outcomes of human affairs, to gain insight into the processes at work which make us tick.

This makes sense. We see regularities and symmetrical patterns all around us. When we

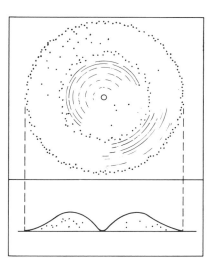

FIGURE 8–1

When ants build an ant hill they carry soil particles out and drop them, carrying them only so far as is necessary to prevent them from rolling back in. The principle of minimum net effort constrains a process and thereby promotes a symmetrical form.

see an undisturbed ant hill straddling a crack in the sidewalk in the summertime, we are struck by its symmetry (Fig. 8–1). The ants carry grains of sand out and drop them, carrying them no farther than necessary to prevent them from rolling backwards into the nest. The grains are distributed over an area that is almost perfectly circular. The depth profile resembles that of the density of litter deposited away from each edge of a highway. It is reliably reported that the railway line from Adelaide to Perth, Australia, is lined with beer cans for 1,600 kilometers and on a sunny day the route is visible from the air by the light reflected from the ribbon of shiny cans. Because of the principle of least effort a beer can, cigarette wrapper, or junked car is seldom more than a few feet beyond the edge of the road. In each case the process is waste disposal away from a point or a line source. In each case a symmetrical density profile is the result because structure follows process and the process operates under a simple rule or principle called least effort.

The Reason We Study Why Things Move

We describe how things move to understand movement laws, and because we want to predict and control social and natural events. If we discover a principle that governs many movements, such a principle stores up a wealth of information, and allows us to consider all sorts of implications of the principle.

What kinds of questions can we ask about transportation and communication networks? A major concern of society is whether such structures allow people, goods, and messages to flow efficiently. We demand low cost, accuracy, speed, safety, and comfort. Consider the first requirement. How can we tell whether a particular set of movement costs is as low as it could be? Do we know, for instance, that the Bell Telephone System and the Post Office are collecting messages and distributing them at the lowest cost? Usually we do not unless we have some *standard* or *yardstick* with which to measure their performance. But even if we did, what would we measure? The *flows* through the network? Or the *network* used by the flows? That depends on whether we want to minimize flow costs (e.g., cents per ton kilometer) or network cost (e.g., dollars per kilometer of highway constructed). Flows must move through their proper channels, but if the network is clumsy and inefficient, then flows cannot move at very low cost. Yet to bring flow cost per kilometer down to a minimum might drive the network costs out of sight.

There really are several different problems here involving flow or user costs and network or builder costs:

1. Given a required set of flows which must be accommodated, what network has the lowest construction costs for the builder (cost of structure)?
2. Given a required set of flows, what network design has the lowest operating cost for the users (cost of flow)?
3. If the fixed investment cost of the network structure and the variable operating costs based on flow processes are borne by the builders who are also the users, what design is best?
4. If a network is already established, what route selections minimize user costs?

To answer such questions we devise *normative models* and use them as standards against which observed networks and flow patterns are compared. The degree to which observed patterns match our model is a measure of the efficiency of those patterns. In the following sections we first deal with movements, next with networks and flows, and then with network design and performance. The last section applies the ideas to several common geographical problems.

MOVEMENT GEOMETRY

The building blocks for movement models are the points, lines, areas, and volumes among which movements occur. Movements or flows in earth space begin with a static distribution at an instant in time and end with a static distribution at a later point in time. Sometimes the distributions are identical but often they differ. We have several familiar techniques for describing spatial distribution patterns, such as population dot maps, isopleth maps of per capita income, or choropleth maps of consumer buying power. But a map describing the spatial distribution of a population, a ratio of values, or the level of some variable represents only one state of the system at an instant in time.

Movements have attributes aside from their static origin and destination end points. Yet it is difficult to map flows and it is harder still to portray changes in flows through time. We can, however, classify flows or movements according to their basic spatial dimensions. Let us look at a classification of movements and ignore for the moment whether they refer to the human world, the physical world, or a combination of the two. By ignoring actors and activities, the basic spatial dimensions of moves are displayed in sharp relief and different kinds of moves are seen as basically identical in their spatial structure.

When we consider the origins and destinations of movements, it is obvious that all movements are basically point-to-point movements. Even when areas are invaded by waves, as in the case of an army invading a foreign nation, the areal movement can be broken down into the movements of individual men and pieces of equipment. Similarly, when it appears to us that things are moving from a point to an area, as when workers in the central business district of a city make their evening migration to the suburbs, it is also possible to resolve this mass movement into individual movements from point to point. Thus point-to-point moves are the foundation for all kinds of movements, and the channels by which people and phenomena move from point to point comprise our various transportation and communications media. The spatial structures of the origins and destination points are important determinants of the structure of movement systems and deserve further attention.

Sometimes origins and destinations are single points, but more often we wish to consider group movements from several points to one or many points. Before we consider the way in which these various kinds of movements influence structure, let us specify the range of movement possibilities under alternative spatial arrangements of origins and destinations. Points which are origins or destinations may be located in four basic way. Origins and destinations may be located at points, along lines, in areas, or in volumes (Fig. 8–2). A set of origins with a volumetric spatial structure from which things are moving to a set of destinations with a

linear structure will require movement systems different from those needed for a set of origins with an areal spatial structure from which things are moving to a single point destination. Classifying point-to-point movements on the basis of the spatial structures of the origin and destination points is a useful way of breaking down the complexity which exists in the almost infinite number of movements which we observe around us. And because the four spatial structures we mentioned exhaust all possibilities, a matrix obtained by considering the permutations of these four spatial structures includes all possible forms of movement (Table 8–1). Let us turn to some examples of these movements.

TABLE 8–1

Sixteen Different Classes of Movements and Spatial Interactions

From (source or origin):	*To (destination):* Point	Line	Area	Volume
Point	1	2	3	4
Line	5	6	7	8
Area	9	10	11	12
Volume	13	14	15	16

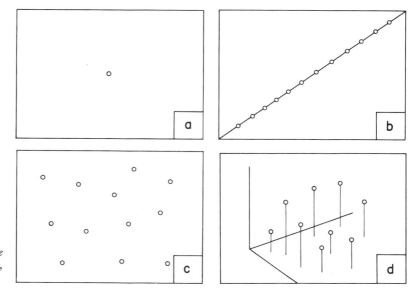

FIGURE 8–2

Origin and destination points can be located singly (a), *along lines* (b), *in an area* (c), *or in a volume* (d).

Ideal point to point movement.

Small boy returning home from school.

Path of a person attempting to walk directly across a football field.

Riding a horse to the top of a mountain.

The most direct path in the journey to work: driveway–local street–freeway –one way streets–parking lot– sidewalk.

Path of a migrating waterfowl.

A bridge minimizing bridge building costs. The shortest bridge from the builder's viewpoint.

A bridge built to minimize bridge using costs. The shortest bridge from the user's viewpoint.

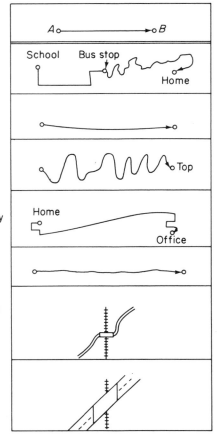

FIGURE 8–3

Some least net effort paths for various point-to-point movements. Each path provides a minimum of net effort, that is, costs minus benefits are at a minimum.

Movements from Single Points to Single Points (Class 1)

Point-to-point moves are basic because all moves can be reduced ultimately to this scale. Consider point-to-point moves in the light of the modified principle that events reach their goal by the route incurring the *least net effort* and returning maximum net benefits (that is, benefits minus costs are at a maximum). In an ideal, undistracted, unencumbered movement, a person wastes no time going directly from *A* to *B* (Fig. 8–3). In other movements, the benefits gained en route repay diversion, such as when a small boy travels home from school, or when a farmer travels from home to field by way of the pub and repeats the pattern on the way home. In other cases, faulty orientation encumbers direct movement (walking across a football field), or the presence of a grade means that a longer route will demand the least effort. The shortest (least net effort) journey to work when driving in a city is fundamentally similar to that of a migrating bird journeying to wintering grounds in the South. In every case a point-to-point movement generates a line, typically a least net effort line, and in that sense a shortest line. In network design the question "shortest for whom?" is settled in favor of builders, or users, or by some compromise.

We should not forget that a time interval is

implied when generalized movement patterns are described in model form. At the daily scale we consider the journey to and from work important; whereas viewed in an annual or larger context, daily work and shopping travel becomes local vibration framed by interstate changes of residence or extended vacation and business travel. Similarly, the migrating duck, en route from Canada to subtropical areas, follows a direct international route. By resting and feeding in a New Jersey marsh it may range over a wide water area. The marsh is a *point* or an *area* depending on the temporal and spatial frames of reference.

Movements from Single Points to Points Located along Lines (Class 2)

When water springs from a hillside source it immediately trickles downward toward a lower point. Eventually it joins another stream of equal or larger size. In simple geometric terms a point (the spring) has been joined to a line (a stream) by another line (a rivulet, Fig. 8–4). The spatial arrangements are similar to sewer and trash collection from a house. Waste moves along a path of least net effort to a larger collection channel (trash truck route, sanitary sewer). The object is to move something (a

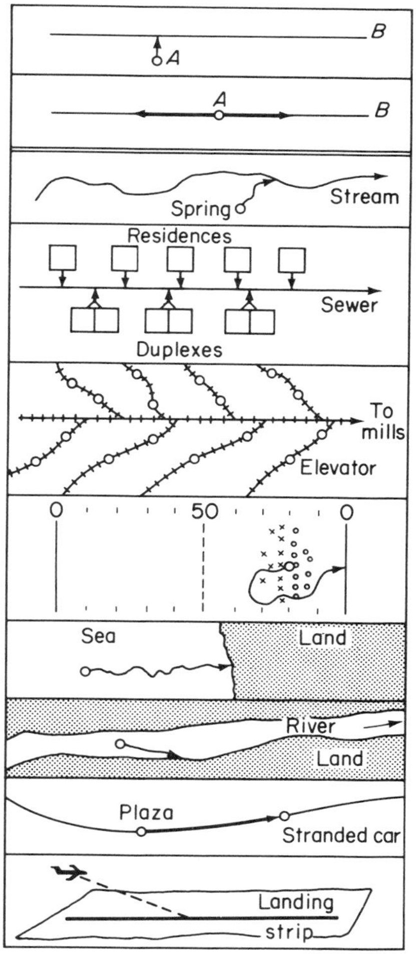

FIGURE 8–4

Least net effort paths defined by various point-to-line moves.

Ideal point to line movements:
  From any point off the line to any point on the line.
  From a point on the line to another point on the same line.

A spring (point) is joined to a stream (line) by a rivulet (another line).

Sewage lines collect from residential point sources.

Grain is assembled at elevators in rural areas. Branch rail lines link elevators with main rail lines.

Path of a football as it moves by a least net effort path to the goal line.

Path of a person who fell from a boat and swam to shore in choppy water.

Swimming directly to shore from a point in the middle of a fast flowing river.

One of many point to line movements defined by service vehicles on a toll road.

As it lands, an airplane moves from a point aloft to a line of points on the ground.

carload of wheat, a football, a person) from one point to another point, a place on the destination line. The path followed usually defines a line of least net effort.

Movements from Single Points to Points Located in Areas (Class 3)

When something at a point source is to be spread over an area, the least net effort principle is again at work although sometimes it is hard to detect (Fig. 8–5). On a still day, an irrigation sprinkler will distribute water to a circular area. If winds are predominately from one direction grains of soot from a smokestack will still take a path of least net effort in falling

to the ground, but the pattern formed appears distorted because the regular principle of least net effort was acting in an environment of irregular forces.

Sometimes the environment includes forces of competition or interference which modify point-to-area movement patterns. The manager of an isolated radio station serving a market area of similar listeners might expect that the proportion of listeners should decline regularly and identically in each direction. But atmospheric disturbances due to storms, or mountain barriers might interfere. If the station is not isolated, competitive disturbance due to another radio station can cut down the audience. On a map, loss of part of the radio

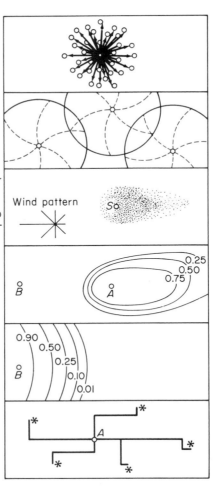

An ideal point to area movement pattern.

Irrigating crops with three sprinklers; no wind blowing.

Soot grains falling through a moving air mass define a least net effort path downward and form a pattern on the ground that is eccentric with respect to the fixed smoke stack(S), but symmetrical with respect to the moving air.

Proportion of radio listeners who were tuned to the radio station at city A when sample was taken.

Proportion of families who read the daily paper published in city B. Compare the pattern with the radio pattern shown above.

Path of fire trucks rushing directly from station A to fires in area protected by the fire department.

FIGURE 8–5

Paths and distribution patterns based on point-to-area moves.

FIGURE 8–6

Variations over space in the legal (a) *and effective* (b) *jurisdictions of a national capital. Modern states with good communications can exercise effective jurisdiction over most residents and most places.*

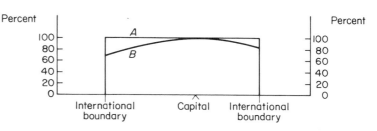

audience means that the isolines describing the market area are pushed back toward the station. Similarly, the presence of a large nearby city will curtail newspaper penetration into the tributary area on that side of a metropolis but leave the other side free for monopoly control.

It has been traditional in geography to speak of point-to-area (and area-to-point) movements in terms of *fields* such as a city's sphere of influence, a port's hinterland, or in the reverse case, the source area of the city's inmigrants. When a point interacts with an area a field is established. Some fields are theoretically continuous, with interaction rates declining regularly with distance from the center. Most readers of *The New York Times* live in the New York area. Moving away from the five boroughs, copies sold per thousand population drop steadily. Yet copies are read daily as far away as Peking and Sydney so the *Times* field covers most if not all of the world.

Legal political hegemony extends from a national capital to the boundaries of a state. Beyond a state's frontiers its legal jurisidiction drops to zero. Whatever the *legal* jurisdiction, *effective* jurisdiction varies from one part of a state to another. Modern states with good communications can provide effective government control to all areas; but the influence of the national government might temporarily be weak in peripheral areas, such as on the American frontier in 1794 or in northeastern Brazil today (Fig. 8–6).

Sometimes arrangements are made to extend by agreement the boundaries of effective jurisdiction, as when a city annexes unincorporated rural land. Sometimes there is no agreement, as when one nation invades another and annexes new territory. When a colony acquires independence, the opposite occurs. The jurisdiction of the former mother country shrinks. In the

United States, effective legal jurisdiction is a major problem in metropolitan areas, especially when they overlap state boundaries. In Washington, D.C., Maryland, and adjacent parts of Virginia, traffic tags received by residents of any of these three jurisdictions are accepted for payment in courts of any of the other participating jurisdictions. This is a recent solution to a problem raised by motorists who were tagged when commuting to work. Before the participating governments came to an agreement, tagged motorists were treated as out-of-state residents and forced to put up bail before leaving the state. State boundaries raised problems in this case that normally would not have arisen if the metropolitan area were contained within the same state.

For economic and other societal interactions, a center's range of influence is often described in terms of its *mean field*: the area nearest the center accounting for half the interactions. If the closest 50 percent of a supermarket's customers live within the first three kilometers of the store, the mean field is three kilometers. This procedure is troublesome when a center lies outside its field, as it sometimes does in the case of *fragmented* migration fields. A more explicit procedure describes point-area relations by measuring the proportion of each subarea which interacts with an internal or external center. Around a metropolis a different proportion of the people in each place commute to the central city *A* (Fig. 8–7). After the proportion is determined for each subarea, isolines are drawn to describe the interactions of a central city with its commuter shed.

In commercial applications such as supermarket studies this procedure is quite useful. If we operate a supermarket and we construct a map describing the fraction of people in each city block that is attracted to our store we can

A supermarket's trade area is described by a field map. Concentric circles around the store (S) enclose a larger and larger fraction of the store's customers. The mean field encloses half the customers. Mean field size varies by function and by location.

A migration field is described by a map indicating the fraction of the population in each place at time t_1 which moved to a destination, say town T, during the time interval $t_0 - t_1$. A fragmented field may result as in the example shown here.

A commuter shed map describes the proportion of a resident labor force at each point in a region that commutes daily to a downtown employment center c. Dashed line denotes city limits.

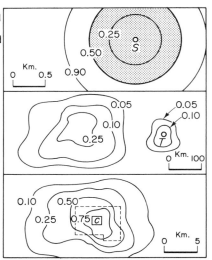

FIGURE 8–7

Fields created by the movement of people and goods.

construct an isoline map describing our market area. At the 50 percent isoline half the potential customers are shopping at our store and half are shopping elsewhere. At that line our competitor's pull equals our own store's pull. A series of such maps for successive dates indicates how well we are holding our own against the competition.

Distorted fields are often the consequence of spatial competition created by intervening opportunities. Without competition, fields can sprawl out great distances, but spatial competition abruptly curtails them. In the nineteenth-century city the field around the downtown center was supplied from one central business district. In a typical city today, outlying commercial centers compete with downtown. The outlying centers were established at streetcar intersections as the city grew. A shopping center at y (Fig. 8–8) between the downtown and the city's edge is an effective intervening opportunity for residents at z on the suburban fringe. People living on the suburban margin will not travel all the way downtown for what they can buy at center y. So the locus of points at which people are indifferent between the downtown and the shopping center defines the limits of the fields of each. Sometimes customers are located midway between shopping centers

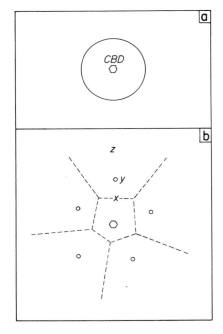

FIGURE 8–8

Retail distribution fields around trade centers. Originally the city is served only from the central business district (a). Later, as the city expands, outlying centers are added to compete with the original center and with one another (b).

and are indifferent as to where they shop. Again, the zone of indifference establishes field boundaries.

Movements from Single Points to Many Points in a Volume (Class 4)

Because mankind effectively occupies such a thin slice of the interface between the air above and the bedrock below, we fall into a habit of thinking our world is flat, or at least two-dimensional. But when someone lights a cigar in a small and crowded room, the sickening odor quickly spreads from that point source to pervade the volume of enclosed air which we occupy. It is for this reason that many feel smoking should take place only in private among consenting adults. At a wider scale, the smokestacks on lime kilns, paper mills, and power plants act in the same way with costly and troublesome

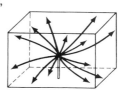

Ideal movement of a gas, liquid, solid, sound or electromagnetic waves from a point source throughout a volume.		
	Through a gas volume	Through a liquid volume	Through a solid volume

	Through a gas volume	Through a liquid volume	Through a solid volume
From a gas point	Hydrogen sulfide gas from a smokestack to the city's air A female butterfly emits an odor to attract a mate A woman uses perfume	In the basic oxygen process of steel manufacture, pure oxygen is pumped through molten iron	Gas storage in porous underground rock formations
From a liquid	Gasoline escapes from automobile gas tanks into the air A room deodorant covers up one smell with another	Treated sewage is poured into a lake or stream A river mouth empties into the ocean Clorine and fluorine are added to a water supply	Atomic wastes are pumped into under-ground rock formations
From a solid	Moth crystals protect clothes in a closet Methane gas escapes from a coal mine during a low pressure atmospheric condition	Mine drainage into fresh water streams Garbage and sewage are dumped into the sea	DDT is applied to crops and ends up in the soil
	City air space fills with noise	Sonar	Monitoring earthquakes
	Short and long wave radiation: TV, heat	Sunlight on shallow lakes and ponds	x–rays; radar ovens

FIGURE 8–9

The variety of point-to-volume movements.

consequences when our air sewer fails to flush itself. In the Great Lakes or in the ocean off New York City, garbage and sewage is removed to a point source and then dumped with the pious hope that the sewage, diluted by tides and currents, will leave the water almost as good as new (Fig. 8–9).

Radio transmitters send electromagnetic signals through the air even though most receivers are on the ground. Noise sources at a point also fill the air spaces as well as any solids than can transmit sound waves. In Colorado attempts to dispose of atomic wastes in holes drilled at the base of the Rockies seem to have increased the incidence of severe earthquakes as folded and fractured rock strata were lubricated by the wastes spreading through volumes of stratified rock. The rate and degree of saturation of a volume by a point emission depends on receptivity and irregularities within the volume. The principle of least net effort implies a

somewhat steady rate of spread through the volume. But winds and currents introduce irregularities in gas and water volumes, and rocks vary in their densities and permeabilities. Although the principle of least net effort remains constantly in force, the medium within which it acts may not be homogeneous.

Movements from Points on Lines to Single Points (Class 5)

Line-to-point movements are the last stage in many delivery systems providing gas, oil, mail, and other channeled flows to users located at dispersed points, whether individual people, households, or cities (Fig. 8–10). In the ideal pattern, something in or on a line must move to or through a point located on or off the main line. In all cases the principle of least net cost operates.

But cost to whom? The delivery problem

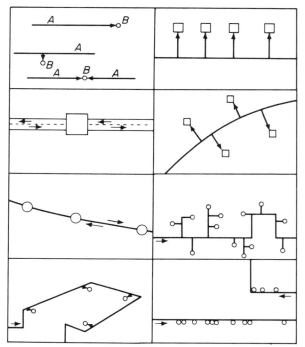

FIGURE 8–10
Typical line-to-point delivery patterns.

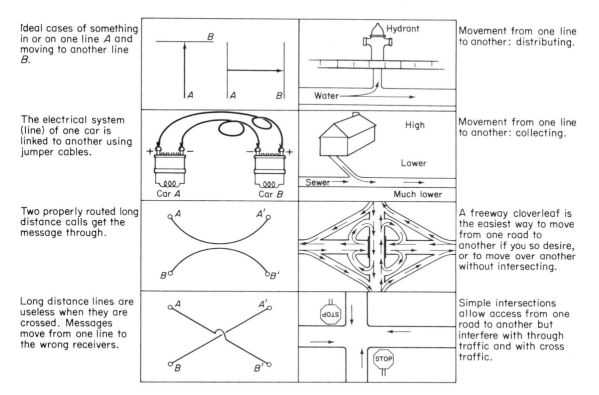

Ideal cases of something in or on one line *A* and moving to another line *B*.

Movement from one line to another: distributing.

The electrical system (line) of one car is linked to another using jumper cables.

Movement from one line to another: collecting.

Two properly routed long distance calls get the message through.

A freeway cloverleaf is the easiest way to move from one road to another if you so desire, or to move over another without intersecting.

Long distance lines are useless when they are crossed. Messages move from one line to the wrong receivers.

Simple intersections allow access from one road to another but interfere with through traffic and with cross traffic.

FIGURE 8–11

Line-to-line movements. Some raise problems; some are solutions to problems.

can be solved in favor of customer convenience or carrier convenience, but seldom both. Sometimes customers need electricity, piped gas, or bottled gas delivered right to the room where it is to be used. Sometimes such delivery is too expensive for the carrier and not required by the customer, as in rural mail delivery, so once a day people are forced to visit the mail box. More severe cases of delivery organized for carrier convenience are elevator and subway systems, neither of which can be easily shifted around the way an electric wire or iron pipe can be rerouted.

Movements from Points along One Line to Points on Another Line (Class 6)

Sometimes there is a need to move from one line to another (Fig. 8–11). If the lines intersect the transfer is direct; If not, a connecting path must be used. If water is being shunted from a water main to a hydrant, the orientation of the main line is left undisturbed. If a sewer pipe is collecting sewage from a household, care must be taken so that a flow will never slow down or reverse. Some flows, unrelated to gravity, are undisturbed by the orientation of lines and the relative elevations in the system, such as when jumper cables are attached to a car to start another one with a dead battery.

Sometimes during long-distance telephone conversations lines get crossed and messages go to the wrong receiver. Whereas telephone message channels are useless *if* they cross, highway systems are useless *unless* they cross. Users must have access and egress to roads at low net cost or else they cannot be used. At the same time, too many intersections impede the smooth flow of road traffic raising user costs *and* construction costs. By controlling access from one rural road to another with a simple stop sign, building costs are minimized, but users might never get on or across a busy main road. A stop-and-go signal raises construction costs but improves access. It lowers user costs for cars on the cross street, but raises them for users on the main street when queues start to form. The highway department assumes, however, that savings to cross street users more than offset costs to main street traffic. A grade separation and cloverleaf access ramps at an intersection impose outrageously high construction costs but reduce net user costs practically to zero.

Movements from Points along Lines to Points in Areas (Class 7)

Think of the many ways man and nature use a line or a set of connected lines to distribute something to an area. A leaf is an excellent example of an efficient distribution system using a set of lines to provide water and nutrients to the leaf area (Fig. 8–12). The fact that leaves come in so many shapes and sizes suggests that there is a wide variety of efficient solutions to line-to-area movement problems. Lines are used to cover an area in plowing, seeding, and fertilizing a field. A farmer plowing and a house-painter painting are in a sense doing the same

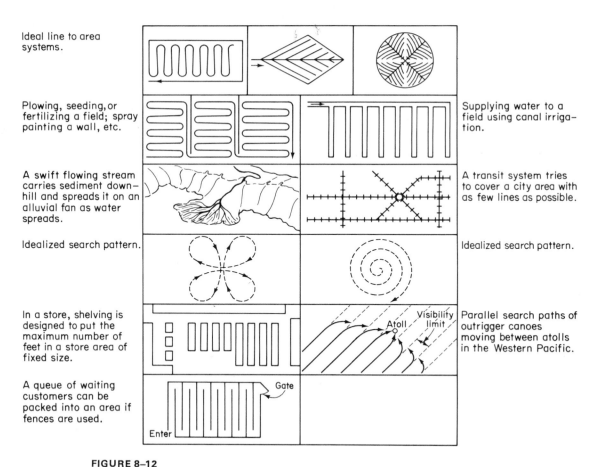

FIGURE 8–12

Typical line-to-area moves: arranging a line to serve an area or packing an area with a line.

thing and following the principle of least net effort. Both attempt to cover an area with a continuous line. If three adjacent fields are fenced, plowing in one must be completed before work on the next may begin. Similarly, a housepainter may work in an area one or two meters wide without moving his ladder, scaffolding, or feet.

An irrigation canal waters an area from a line source, but, as with the leaf, there is no need for the line to return to the origin. Yet a highway system, which also is attempting to serve an area with a set of connected lines, cannot tolerate too many dead ends. The layout of a supermarket resembles a road system. Store designers want to put a long line of shelving into an enclosed area, arranged so that there are no dead ends and so that customers can move easily about the entire area of the store. The old-fashioned general store with a pot-bellied wood stove in the center and merchandise to the ceiling around the walls was an inefficient system. It kept the proprietor running and the customers waiting. Merchandise was arrayed on wall areas instead of along shelf lines. Efficient search patterns resemble flowers opening because each tries to reach effectively, with the least net effort, the largest area possible. Because the jobs are the same, nature and man do the jobs in the same way.

Movements from Points along Lines to Points in Volumes (Class 8)

A radiator is surrounded by a volume of air. Inside the radiator a hot liquid or gas courses through thin pipe lines so that heat will be dissipated into an adjacent volume of air. When used for heating, a radiator delivers heat from a line to a volume (Fig 8–13). Sometimes the purpose of the radiator is to heat the surrounding air (as in home heating); sometimes the radiator's function is to cool the material flowing through it (as in a car radiator). Regardless of purpose, the movement process is identical.

When a student browses through the stacks in the library, access to the volume of books is provided through low and narrow passageways designed to allow maximum book storage in the library while access to all books is made as

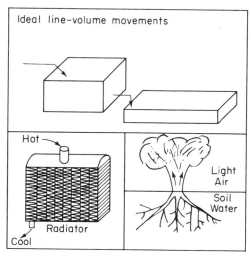

Ideal line-volume movements

Hot —

Radiator

Cool

Light Air

Soil Water

Other lines which fill or serve volumes:

1. A book
2. A library
3. A multi-story department store
4. A coil of rope
5. A coal mine
6. The blood circulation system:
   systems of arteries and veins link the heart with the body
7. The central nervous system:
   a system of nerves linking the brain with the rest of the body

FIGURE 8–13

Common patterns of line-volume interactions.

easy as possible. When books are returned this process is in turn reversed. In this way many thousands of meters of book shelving can be located within a few dozen meters of the circulation desk, an excellent example of the principle of least net effort. A department store is designed along similar lines to allow easy stocking of shelves and convenient customer access to all parts of the store from the front door.

Other examples are structurally identical. Underground coal mine tunnels are systems of lines designed to allow access to a volume of coal. The coal is removed from a deposit (volume-to-line movement), carried along passageways to the mine mouth for processing (line-to-point), and transferred to rail cars or barges (point-to-point), from whence it is distributed to market points and market areas. The root system of a tree does the same thing, extracting water and nutrients from a volume of soil to supply demands above ground. The delivery system transfers these materials to the branches and leaves which fill a volume of light and air above the ground. In the human body, the spatial structure of the blood circulation and central nervous systems are least net effort solutions to the movement problem between lines and volumes.

Movements to a Single Point from Scattered Points in an Area (Class 9)

Movements from an area to a point can take a direct one-step course or use a two-step procedure (Fig. 8–14). In the one-step case the movement is direct, as when wild animals converge in the evening to a water hole from their daytime habitats. College students from a wide tributary area gather at a campus and, whether the campus is central or peripheral to its student supply area, students are observed to follow a principle of least net effort in their movements to and from school.

Least net effort routes are used when the people in a neighborhood mail letters. If they walk *directly* to the mailbox, however, they usually follow paths comprised of sidewalks. The movements involved in getting the letters mailed can best be evaluated by breaking them into two steps: area-to-line (sidewalk) moves, followed by line-to-point (mailbox) moves, and then the journeys home. This is what is done when a rubber tree is tapped to collect the sap which flows continuously under the bark. The Mississippi-Missouri drainage network is another least net effort system whereby the water over a wide *area* is collected in successively larger channels (lines) for delivery to the river mouth (point).

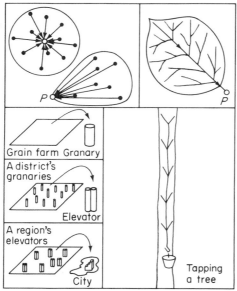

Ideal one step area to point (*P*) moves

Grain farm Granary

A district's granaries

Elevator

A region's elevators

City

Ideal two step moves:
area — line
line — point (*P*)

Tapping a tree

Draw some sample sketches of the following events:

1. Students go from home to grade school; high school; college.
2. Wild animals converge in the evening at a water hole.
3. A mail collection box serves a residential neighborhood.
4. A river basin drainage system collects water from a wide area.
5. Farmers from a rural area visiting town on Saturday.

FIGURE 8–14
Area-to-point movements.

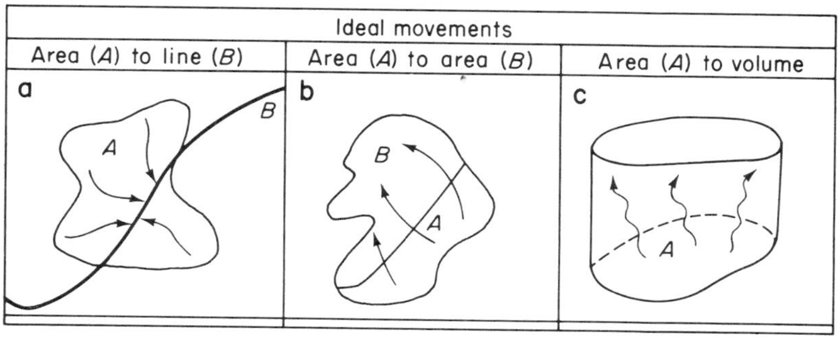

Ideal movements		
Area (*A*) to line (*B*)	Area (*A*) to area (*B*)	Area (*A*) to volume

a

b

c

1. Water from a roof to a gutter
2. Water from a road to a ditch
3. Commuters from a suburb to a highway
4. Animals from a forest to a river
5. Territorial species moving to common boundary when it is attacked or threatened
6. Sheet erosion of soil into ditch

1. Agression: one country attacks another
2. Advance of an ethnic ghetto
3. Expansion of a trade area into the territory of a competitor
4. Urban sprawl into hitherto rural area
5. Urban renewal project cuts into blighted neighborhood
6. An epidemic moves from one area to another
7. An innovation spreads from one region to another

1. Evaporation of water from a lake or sea
2. Radiant energy released from the earth shortly after sundown
3. Perspiration evaporates from the skin

FIGURE 8–15

Some typical movements from areas: (a) points in areas to points along lines; (b) points in areas to points in areas; and (c) points in areas to points in volumes.

In agriculture, farmers gather grain from their fields directly into a granary, grain merchants assemble grain from the local supply area into their elevators, and distant cities drain the assembled grain from a broad region. The different stages in the collection process illustrate how an area-to-point movement, viewed at one scale, becomes a collection of point-to-line and line-to-point moves when examined from a detailed perspective.

Other Movements from Areas
(Classes 10 to 12)

By now the point has been emphasized that all spatial processes have attributes in common, whether they occur solely in nature, or whether human activity plays a part. Moreover, a principle of least net effort regulates movements and helps us account for observed patterns. When water runs off a roof and into a gutter during a rainstorm the dimensions of the moves are the same as those when animals come out of the forest to drink at the river; or when commuters leave their garages for the street; or when soil is wasted through sheet erosion into ditches, gullies, and stream beds. In all cases something moves from an area to a line with the least net effort expended (Fig. 8–15).

Area-to-area moves suggest invasion or aggression. An expanding urban area cuts into the countryside, a black ghetto advances into a previously white neighborhood, or an aggressive advertising campaign expands the trade area of a supermarket into a new territory. The same thing happens when an idea or an epidemic spreads from one area into another.

Because most of our activities take place in our two-dimensional world, we have fewer examples of area-to-volume movement. In the physical world evaporation of water from lakes and oceans fits here, as does the release of radiant energy by the earth into the atmosphere after sunset. Perspiration evaporating from the skin is a movement process of a similar dimensional order.

Movements from Points in Volumes (Classes 13 to 16)

Before discussing models of networks and flows, let us complete our taxonomy by considering movements from points in volumes to points elsewhere (Fig. 8–16). Something is moved from a three-dimensional zone to a point, as when groundwater leaks out in the form of a hillside spring, or when a deep sea fishing boat returns to port with a catch, or when ore from an open-pit mine is processed at the edge of the pit or at another point elsewhere. The movement can be to a single point, or to a point or set of points along a line. In all cases the movement is accomplished so as to minimize the net effort expended.

The most common volume-to-area movement is falling rain or snow (Fig. 8–16). The area of deposit may be adjacent to the source volume or located elsewhere, as when glaciers advance and melt or as in the mass wasting of rock and the disintegration and dispersal of litter. When solid three-dimensional structures (ice, rock) disintegrate, gravity draws them downward over a two-dimensional area.

When an air mass is replaced by another at a place we have an example of a volume-to-volume movement (Fig. 8–16). Fog or smog which accumulates in one valley and then spills into an adjacent valley furnishes another example of a movement occurring with a minimum of net effort expended.

Movement Dimensions: A Basic Spatial Viewpoint

To describe and understand how things move in earth space we need a useful spatial or geographic viewpoint. The geometric dimensions of moves provide us with one such viewpoint, allowing us to group an apparently endless hodgepodge into classes defined simply in terms of points, lines, areas, and volumes.

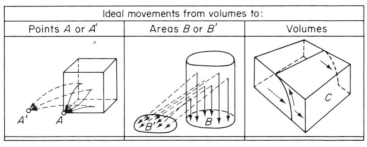

Ideal movements from volumes to:		
Points A or A'	Areas B or B'	Volumes

1. Ground water escapes the earth at a spring
2. Fishing: fish are taken from the sea and hauled to port
3. Production of liquid oxygen, nitrogen, hydrogen, etc., from the atmosphere
4. Open-pit mining of iron ore

1. Mass wasting of rock
2. Glacial advance
3. Rain
4. Snow

1. A cold front advances and replaces warm air
2. A warm front moves ahead and over cold air, gently pushing the cold air mass ahead
3. A fog bank rolls on land from the sea
4. Smog invades a valley

FIGURE 8–16

Movements from volumes and movements of volumes.

At a sufficiently detailed scale all movements become point-to-point moves, but this viewpoint is too general to be really useful. By considering movements at different scales we produce the sixteen movement classes shown in Table 8–1. All point-to-point moves produce lines and on an isotropic plane a point-to-point move would coincide with its desire line and produce a direct path between origin and destination. But the real world is not an isotropic plane, smooth like a billiard ball. Instead, it has a highly irregular surface. Communication and transportation require ways or channels along which movement is easier than over valley or hill or through the bush. Ways are expensive to build, maintain, and use, but they have been far cheaper than the alternative of going without them. Because of their expense, ways are limited in number and their use is restricted. Traffic of each type must use the way built for it. Tractors are not permitted on the turnpike; motorcycles are prohibited on the sidewalk; trains must stick to the rails or disaster will occur. Clusters of ways comprise our transport systems: sidewalks, elevators, radio stations, telephone lines, railroads, airlines routes, ocean shipping lanes, rivers, mail routes, alleys, city streets, subway lines, and every other kind of path or channel. Each is created in response to desire lines for point-to-point movement.

Each transport system that is built represents a trade-off between conflicting objectives: (1) the desire for direct point-to-point movements; and (2) the need to keep the cost of the network of ways and channels as low as possible. Innovations in transportation and communication are attempts to reduce the cost of this trade-off. Radio-telephones for automobiles, or helicopters and hovercraft for passenger movement come close to providing interaction from any point to any other point at costs that are tolerable for many users. Some media that are theoretically point-to-point, like jet air transport or ocean shipping, require elaborate terminal facilities and are thereby prevented from operating as flexibly as they might. The terminals are part of the way, and because they are so expensive there can be only a few of them. The transistor radio is a case at the opposite extreme.

It is an extremely cheap receiving terminal and can be placed practically anywhere.

Throughout our review of movement types, we stress the principle which seems to describe best one facet of how things move. They tend to move according to a principle of least net effort. We stress *net* effort to emphasize that the very movement process itself may carry benefits at the same time that costs are incurred. A commuter sitting in a traffic jam inhaling gasoline and carbon monoxide fumes pays a high cost for his trip, but he also has relative peace and quiet twice a day, a radio to listen to, and the feeling that for a while at least he is boss. If his job were to move next door to his home he would probably move.

Effects of Movement Time and Other Movement Costs

We speak of *least effort* solutions to movement problems because we cannot usually minimize distance and time and other costs simultaneously. The most direct path in a distance sense between a chair behind a desk and the hallway might be a straight line through the desk and the wall (Fig. 8–17), yet the least effort (least time and cost) path goes around the desk and through the door.

Terrestrial movements link points, lines, and areas in the many combinations reviewed so far. Movements produce fields, and we can speak of a field as part of a *functional nodal region*:

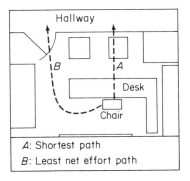

A: Shortest path
B: Least net effort path

FIGURE 8–17

Alternative paths out of an office.

a center and its dealings with a tributary area. Let us look now at the effect of movement costs on the shape of such fields.

On a homogeneous frictionless plane, a center would extend its influence to some degree all the way to the end of the plane. For example, *The New York Times* and *The Washington Post* are two outstanding morning newspapers. If they could be dropped on doorsteps anywhere in the United States at 6:00 A.M. for 15 cents a copy, each of them would be read nationwide to a far greater extent than they are already. New York, Chicago, and Los Angeles television programs are watched almost everywhere in the United States because people can receive them without paying a price differential due to distance from the studio. Transfer costs for people, goods, and messages are significant obstacles to frictionless movement. Movement costs prevent fields of undiminished intensity. In the case of metropolitan morning newspapers, if movement costs are too high for a city to import its morning papers, it substitutes local production for national distribution.

Field shapes reflect *all* obstacles to interactions over earth space. In the shopping center example, interaction depended on the level of consumer demands, the supply of available goods and services at a center, the distance between store and customer, and the competition of intervening opportunities. If a center (a city, a downtown, a grocery store, an amusement park, etc.) serves the population living in an adjacent homogeneous area, and movement friction is equal in every direction away from the center, the mean field served by the center will be a perfect circle around the center. If we then introduce an important political boundary, the shape of the center's mean field will be distorted. If a political boundary can be crossed at any point yet raises some obstacle to free movement, (perhaps a tax is levied at p on people who work at p but live east of the boundary) the boundary friction can be expressed as an equivalent distance (Fig. 8–18a). Without the boundary the field would have extended from center p to q. The boundary causes the field to contract a distance of y to r. The distance y is thus the distance equivalent of boundary friction when the boundary can be crossed

anywhere. If points s and t are on the east side of the boundary, distances ps and pt are equal to pr.

If the border may be crossed only at a customs point s, the mean field gets distorted in a different way (Fig. 8–18b). The mean field in the border's absence would extend from center p to the field's limit at q. The border obstacle reduces the field to r, illustrating again that border friction is equivalent to additional travel of y miles. But since anyone who wants to travel from p across the border must pass through customs point s, that point acts as a subsidiary center, creating a semicircular subfield around it with radius sr.

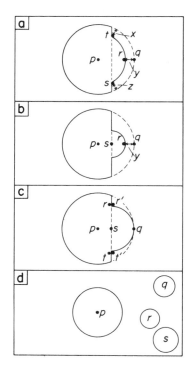

FIGURE 8–18

Mean field distortions. Some fields are distorted by boundaries. In a the boundary may be crossed anywhere but it creates a friction. In b the border may be crossed only at a checkpoint (s) and a friction is created. In c the checkpoint (s) raises no friction but it is the only crossing point. In d a fragmented field indicates that movement is easier between the center C and remote areas than between the center and some closer areas (adapted from Lösch, Fig. 60).

If the barrier is a river which can be crossed only at bridges and tunnels, a third type of situation may exist. If the bridge or tunnel also restricts travel because of congestion or a toll charge, we have the second case repeated (Fig. 8–18b). If, however, there is only one bridge but traffic moves freely and unimpeded across it, a third situation arises (Fig. 8–18c). Travel between p and s is unhindered as before. Travel to r on the west side of the river proceeds in a direct line as well. Travel to r' east of the river is possible only after crossing at s and heading north. Once a traveler is at s, q is as easy to reach as r' or t', so the shape of the new subfield centered on s is controlled by movement costs away from s east of the river.

The three cases remind us of many spatial patterns in human geography, from telephone calls across international boundaries to the commuter sheds of cities like Detroit (across the river from Windsor, Ontario), St. Louis (across from East St. Louis, Illinois), or Philadelphia (across the river from Camden, New Jersey). Field shapes *appear* irregular, but in terms of travel time, travel costs, convenience, interest, or influence, regularities and symmetries still exist. The fields are distorted in terms of absolute distance and compass directions, but they are regular when considered in relative spaces.

Fragmented fields (Fig. 8–18d), suggest that in some sense other than kilometers q, r and s are closer to center p than to other apparently intermediate locations. Such a field pattern is not uncommon in migration when a center competes with no intervening opportunities and intervening areas can generate no interaction. A colonial empire of scattered possessions has such a structure.

In manufacturing cities in the northeastern United States, it frequently happens that the majority of the workers in a factory comes from the same European town or from the same county in the American South. Because of historical reasons a set of widely separated places may have unusually high information and interaction levels. Unequal feedbacks may have reduced the barrier effects of distance friction between the destination center and far-flung places more than for nearby areas. If ignorance

forms a major obstacle to migration interaction, whereas information dispels ignorance, there is no reason why information levels *must* decline with distance from the center. When they do not, fragmented fields can develop. If people at q are looking for employment and housing at p, movement from q to p will depend on the vacancies at p, and the amount of information at q about the situation in p (Fig. 8–18d). As either or both the vacancies or information level rise, moves from q to p will increase. But if the population around p rises, the vacancies will soon be used up and we would expect moves from q to p to decline. It seems pretty clear that information levels can be high in far-off places and low close to the city, and thereby contribute to the formation of fragmented fields. Returning once again to Fig. 8–18d, if early migrants to p came from q, r, s, and other places, but only the migrants from q, r, and s found success and wrote enthusiastic letters back "home," only those three places will continue to send large numbers of additional migrants to region p.

ANALYSIS OF NETWORKS AND FLOWS

Networks are structures designed to tie together various points which are located in one-, two-, and three-dimensional space. They produce regional geographical systems by carrying flows of people, goods, messages, and anything else that is moved from one place to another. Three-dimensional systems are exceptionally hard to describe satisfactorily on a single map or diagram, but a two-dimensional network such as a pipeline system can be represented by a map. Lines of different widths can represent various capacities. Broken or different colored lines can indicate whether the pipelines carry gas, oil, water, or something else. Sources, destinations, and direction of flows are conveniently represented as well.

Although maps can summarize much, there are aspects of networks and flows they cannot effectively cope with. The major limitation is in their computational possibilities. A map is far too concrete and detailed to permit much in the way of direct manipulation and analysis. Maps frequently are general purpose so they may be adapted to a wide range of specific problems. A topographical map, for example, contains an enormous store of information, presented in an efficient format. But such a map is not usually designed for ready quantitative analysis, although it may provide all the raw materials for such study.

Maps by their very format are static and two-dimensional. They describe spatial structure. All spatial processes involve a flow quantity with respect to some time interval such as gallons per day or vehicles per hour or long-distance calls per minute. We can always select an appropriate time slice (a day, a minute, etc.) and construct a map describing a network and a set of flows with respect to that time interval, but it is difficult to analyze either the network or the flows when they are presented in this way.

To meet the need for a convenient representation of a network and its flows some crude general measures of network structure and its flows possibilities have been developed. Once we have measured network structure and the relative locations of places in a network, several doors are opened for us. One regional network can be evaluated and compared to another one in the same or in a different region. If the structure of some ideal or perfect theoretical system is known, an existing network can be compared to such a normative model to see how far reality falls short of perfection.

Certain cities and towns now are being served relatively better by the Interstate Highway System than they were in the rail era, and other places are falling behind. Our socioeconomic spatial system is expanding at locations different today from those in the past century. At the same time, due to changing resource requirements, ever-changing mixes in production and consumption patterns, and shifts in population locations and life styles, some of yesterday's growth poles are today's depressed areas.

Graphs to Represent Networks

Graphs and associated matrices permit us to look at entire transport networks and their parts in terms of the whole. A graph is made up of *edges* representing routes, and *vertices* representing nodes or places. Ordinary graphs have many properties in common with simple transportation networks:

1. Each network has a finite number of places.
2. Each route joins two different places.
3. A pair of places is joined by no more than one route.
4. Routes allow two-way movement.

Probably the most familiar example of a simple transportation network represented in graph terms is a subway or transit map (Fig. 8–19). Such a graph shows *topological* position only; the location of a node is reckoned in terms of position *on the graph*, regardless of its position in the real world. Distance is determined in terms of intervals between stations.

On a graph a *route* connects two places and a *path* is a collection of routes linking a series of different places. The *length* of a path is, in topological terms, the number of routes in it. The *topological distance* between two places is the length of the shortest path joining them. Using these concepts and definitions, a number of measures have been devised to measure (1) how well connected the places in a regional network are, and (2) the relative locations or accessibility of different places on the network. One of these measures, the *associated number* of a place, is the topological distance from that place to the most remote place in the network (Fig. 8–20). In Fig. 8–20 the associated number of place P_4 is three, based on its distance from P_1. Every place in a network has an associated number.

In topological terms the most central place in a network is the place with the lowest associated number. P_3 with an associated number of 2 is the central place in Fig. 8–20. The highest associated number in a network establishes the network's *diameter*. In Fig. 8–20

257

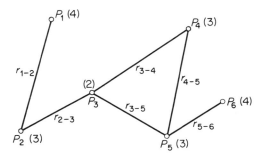

FIGURE 8–20

A simple graph consisting of six places (P_1–P_6) *and six routes. A path is a collection of routes. Places* P_1, P_2, *and* P_3 *are linked by path* r_{1-2}–r_{2-3}. *The path's topological length is 2. The topological distance between* P_1 *and* P_6 *is 4. The associated number of each place appears in parentheses.*

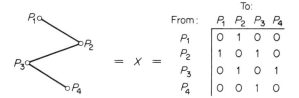

FIGURE 8–21

A graph can be represented by a matrix of 0's and 1's.

both P_6 and P_1 have associated numbers of 4; hence 4 is the diameter of this simple graph. Since the diameter of the graph is the number of routes in the shortest path between the two vertices or nodes that are farthest apart, and the *central place* has the lowest associated number, these two values are limits which establish the range of associated numbers within the graph. The centrality or remoteness of any other place in the graph can be evaluated within the distribution of associated numbers.

A Matrix to Represent a Graph

We can construct a matrix, that is, an array of numbers ordered in columns and rows, to describe a network. Anyone who has consulted a mileage chart while driving cross country has been introduced to the way this is done. On a mileage chart the matrix rows and columns are identified by origins and destination places. Each entry at the intersection of a row and a column tells the distance between two places in a highway system.

The linkages in a simple graph can be expressed in matrix form in a similar way (Fig. 8–21). The places in the graph identify the rows and columns (origins and destinations) in the matrix. We usually give the matrix a name, like X, to keep track of it. If a pair of places is directly linked in graph X we put a 1 in the corresponding cell in matrix X, otherwise we enter a zero. For example, P_1 and P_2 are directly linked. We enter a zero for pairs of places not *directly* linked, such as pair P_2–P_4. Since a route in this graph can only link two *different* places, the diagonal in the matrix, representing points connected with themselves, contains only zeros.

Matrix X is only partially connected. In matrix Y (Fig. 8–22), every place is connected directly to every other place. A 1 is placed in the appropriate matrix cell if and only if a route exists between some origin place i and a destination place j. If no route exists, then y_{ij} (the entry at the intersection of origin row i and destination column j) equals zero.

The connectiveness of the graph can be studied by examining a connectivity matrix such as X or Y. Because two-way traffic is permitted on any route, the matrix is symmetrical. In a symmetrical matrix such as X, x_{ij} equals x_{ji}. Once a route is established between P_i and P_j it is just as easy to go from P_i to P_j as it is to move from P_j to P_i. If some of the routes were one-way, the matrix would not be symmetrical.

Rows and columns can be totaled. A glance at a row total indicates the number of destina-

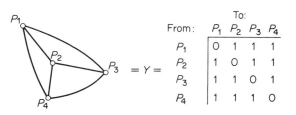

FIGURE 8–22

A completely connected graph represented by a matrix.

tions which can be reached directly from the origin in that row. A column total tells the number of origins linked directly with the destination at the head of the column. Each row and column total reveals the number of routes serving each place.

Matrix and Graph Connectivity

If there are m places in a network, what is the maximum possible number of routes it could have? If there are m origins, there are as many possible destinations (Fig. 8–23). Thus, there are m times m cells in the matrix, or m^2. From this total we deduct the m cells in the diagonal because a route cannot connect a place with itself.

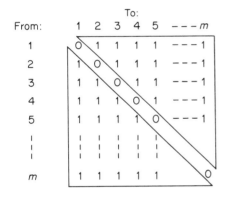

FIGURE 8–23

The matrix of a completely connected graph of m *points. The number of routes involved equals half the number of 1's in the matrix.*

After we deduct the zeros we have $(m^2 - m)$ cells each with a 1 in it. But is this the maximum possible number of routes needed to connect m places with one another? No, because there are redundant routes. Once a route is built between place i and place j, there is no need to duplicate the route by building another parallel route from j to i in a symmetrical network. So once the diagonal is removed, a symmetrical matrix remains and half of it (either half) describes the maximum possible number of routes: $\frac{1}{2}(m^2 - m)$. If there were fifty places, each place could be linked to forty-nine other places. With two-way flows the maximum number of separ-

ate routes $R_{max} = \frac{1}{2}(50^2 - 50)$ which is 1,225.

We can measure the degree of connectivity of a network by comparing the *existing* number of routes with the *maximum* number of routes possible. The observed number of routes divided by the maximum number of routes is the connectivity index. The measure R_{max}, the maximum number of routes, has a counterpart in R_{min}, the minimum number of routes needed to link up m places. The minimum is simply one less than m. For two places one route is needed; for three places, two routes; and so forth. If no place is connected to another, of course, the number of routes is zero. Thus the connectivity index can vary from zero to unity. Such an index is expressed as the ratio of observed routes to the maximum number of routes in the system of m places:

1. absolute nonconnectivity $= \dfrac{0}{\frac{1}{2}(m^2 - m)} = 0$

2. minimum connectivity
$$= \frac{(m - 1)}{\frac{1}{2}(m^2 - m)} = \frac{(m - 1)}{\frac{1}{2}(m)(m - 1)} = \frac{2}{m}$$

3. intermediate connectivity
$$= \frac{\text{(observed number of routes)}}{\frac{1}{2}(m^2 - m)}$$

4. maximum connectivity $= \dfrac{\frac{1}{2}(m^2 - m)}{\frac{1}{2}(m^2 - m)} = 1.0$

where: $m =$ the number of places or points in the system.

Consider, for example, the fifteen Standard Metropolitan Statistical Areas wholly or partly in Ohio in 1960 ($m = 15$), and their connectivity according to the Interstate Highway System. A path with minimum connectivity would require no fewer than fourteen routes ($m - 1$). Maximum connectivity would require 105 routes. It turns out that there are only seventeen routes, and two centers are not connected at all (Fig. 8–24a). Which system is best? Minimum connectivity? Maximum connectivity? Some intermediate scheme? What is the impact of each scheme on the accessibility of each place to every other place? Do we want to favor one place with a position on the network that is superior to the positions held by all the other places? If so, which position should be superior?

If Ohio were isolated from the rest of the

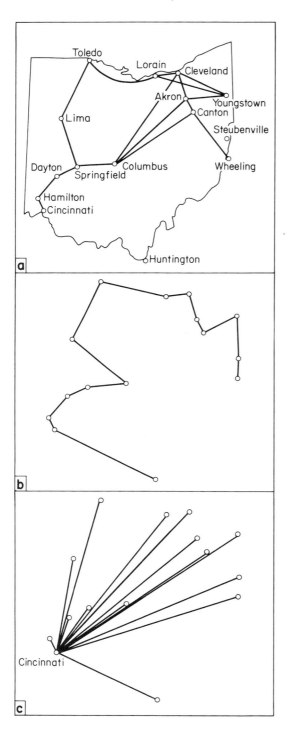

world and the fifteen places were linked by a minimum connectivity network, Columbus, the center with the lowest associated number, could dominate the system (Fig. 8–25a). If, on the other hand, Ohio were merely an appendage of a larger system, and the fifteen places were still linked by a minimum connectivity network, Columbus could dominate the system only if it were at the end where the path linked up with the rest of the world (Fig. 8–25b). This is the kind of position enjoyed by New York City after the Erie Canal opened in 1825 before other canals and canal-rail systems were built to link other coastal cities with the agricultural interior. Evaluating locations on an existing network is different from planning the location and dimensions of a network to be built. In a developing area, where capital construction costs are high compared with over-the-road operating costs, minimum connectivity might be the goal, but in such a case the relative positions of places on the minimal network would be very different from the positions occupied on a more complete network.

In highly developed societies, capital construction (builder) costs are often rather low compared to the time and money (user) costs of daily operations. That is why the railroad, road, and highway systems in places like the United States, England, and Japan are more complete networks than those found in the developing countries. In modern economies the more completely connected networks result in lower overall (builder plus user) costs than either minimum networks, with unusually high time and money user costs, or maximum networks, with impossibly high builder costs. So far in our discussion we have discussed only places and routes. In a graph we consider only a topological attribute called connectiveness, not route

FIGURE 8–24

*Alternative connections for metropolitan areas in Ohio:
(a) routes of the Interstate Highway System in Ohio;
(b) a minimum connectivity path linking all metropolitan areas yet keeping construction costs down to a minimum;
and (c) Cincinnati with maximum connectivity with all the other metropolitan areas at the lowest user cost.*

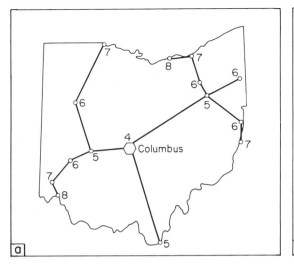

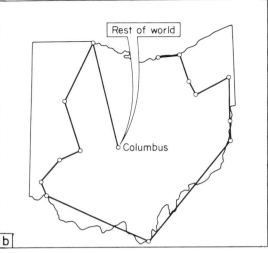

FIGURE 8–25

Optimal location of Columbus, Ohio, on alternative minimum connectivity networks: (a) if Ohio is isolated from the rest of the world, Columbus with the lowest associated number occupies the position that is topologically the centermost; (b) if the metropolitan areas are linked to the outside world through Columbus, a position at the head of the line insures dominance.

length. Understanding the simple topological properties of networks as graphs is an important building block in our geographical understanding of network designs and the impact of network structure on the growth and decline of places in regional systems.

How to Analyze Linkages and Moves in a Network

One matrix can be multiplied by another according to simple rules of matrix multiplication. Consider matrix A multiplied by itself:

$$A = \begin{pmatrix} 2 & 3 \\ 4 & 5 \end{pmatrix}$$

$$A^2 = \begin{pmatrix} 2 & 3 \\ 4 & 5 \end{pmatrix} \times \begin{pmatrix} 2 & 3 \\ 4 & 5 \end{pmatrix} = \begin{pmatrix} (2 \times 2) + (3 \times 4) & (2 \times 3) + (3 \times 5) \\ (4 \times 2) + (5 \times 4) & (4 \times 3) + (5 \times 5) \end{pmatrix} = \begin{pmatrix} 16 & 21 \\ 28 & 37 \end{pmatrix}$$

The entry in each cell of the product matrix A^2 is computed by summing a pair of subproducts.

Larger matrices are handled in the same way. Consider matrix B:

$$B \times B = \begin{pmatrix} a & b & c \\ d & e & f \\ g & h & i \end{pmatrix} \times \begin{pmatrix} a & b & c \\ d & e & f \\ g & h & i \end{pmatrix} = \begin{pmatrix} (aa + bd + cg) & (ab + be + ch) & (ac + bf + ci) \\ (da + ed + fg) & (db + ee + fh) & (dc + ef + fi) \\ (ga + hd + ig) & (gb + he + ih) & (gc + hf + ii) \end{pmatrix} = B^2$$

If a matrix is multiplied by itself, the matrix is raised to the power 2. The powered matrix contains interesting and valuable information in concise and usable form, especially when the matrix represents a graph of a transport network (Fig. 8–26). In matrix C, an entry c_{ij} equal to zero means that there is no direct route running between origin i and destination j. If the entry is a 1 this indicates that the two places are directly linked.

Matrix C^2 is nothing more than C (understood as C^1) multiplied by itself or squared: $C^1 \times C^1 = C^2$. When a matrix is raised to a power above unity, it is a powered matrix. A matrix can be raised to any power n by multiplying the matrix by itself n times:

$$C^1 \times C^1 = C^2;\ C^1 \times C^2 = C^3;\ C^1 \times C^3 = C^4;$$

and so forth:

$$
C^1 = \begin{array}{c|cccc}
 & P_1 & P_2 & P_3 & P_4 \\
\hline
P_1 & 0 & 1 & 0 & 0 \\
P_2 & 1 & 0 & 1 & 1 \\
P_3 & 0 & 1 & 0 & 1 \\
P_4 & 0 & 1 & 1 & 0
\end{array}
\qquad
C^2 = \begin{pmatrix}
1 & 0 & 1 & 1 \\
0 & 3 & 1 & 1 \\
1 & 1 & 2 & 1 \\
1 & 1 & 1 & 0
\end{pmatrix}
$$

$$
C^3 = \begin{pmatrix}
0 & 3 & 1 & 1 \\
3 & 2 & 4 & 4 \\
1 & 4 & 2 & 3 \\
1 & 4 & 3 & 2
\end{pmatrix}
\qquad
C^4 = \begin{pmatrix}
3 & 2 & 4 & 4 \\
2 & 11 & 6 & 6 \\
4 & 4 & 7 & 6 \\
4 & 6 & 6 & 7
\end{pmatrix}
$$

In a powered matrix such as C^2, the exponent 2 means that each matrix entry c_{ij} tells how many different ways one can go from place i to place j in exactly two steps. When the original connectivity matrix C^1 is squared the exponent is 2 as in C^2 above. In matrix C^2, entry c_{22}

(from P_2 to P_2) is three. This means that there are exactly three different paths that can be used to make two-step round trips from P_2. At the same time, since c_{12} (from P_1 to P_2) is 0, we are told that it is impossible to move from P_1 to P_2 in a two-step trip. In matrix C^1 which describes one-step moves, entry c_{12} is 1, meaning that a one-step trip is possible between P_1 and P_2. In matrix C^3, referring to three-step moves, we see that a three-step move from P_1 to P_2 can be made in three different ways. There are two different ways to make four-step moves between the same two points.

If a destination column in matrix C^2 is summed, the total represents the number of different ways the place at the head of the column can be reached from all origins using two-step moves. Because the matrix is symmetrical, the same information is presented in a different light by adding across an origin row. A row total gives the number of ways that two-step moves can be made from that origin. By either standard, row totals or column totals, place P_1 is most remote. Place P_1 can reach other places or be reached by them with only three different two-step paths (Fig. 8–27).

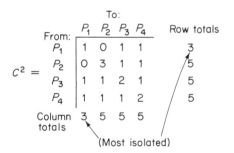

$$
c^2 =
$$

From:	P_1	P_2	P_3	P_4	Row totals
P_1	1	0	1	1	3
P_2	0	3	1	1	5
P_3	1	1	2	1	5
P_4	1	1	1	2	5
Column totals	3	5	5	5	

To: (above column headers)

(Most isolated)

FIGURE 8–27

The number of two-step moves which can be made from each origin to each destination.

Matrices can be added by simply summing the corresponding elements in each matrix:

$$
A + A = \begin{pmatrix} 2 & 3 \\ 4 & 5 \end{pmatrix} + \begin{pmatrix} 2 & 3 \\ 4 & 5 \end{pmatrix}
$$
$$
= \begin{pmatrix} (2+2) & (3+3) \\ (4+4) & (5+5) \end{pmatrix} = \begin{pmatrix} 4 & 6 \\ 8 & 10 \end{pmatrix}
$$

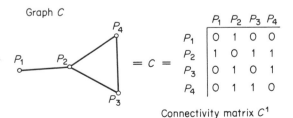

Graph C

$$
= C = \begin{array}{c|cccc}
 & P_1 & P_2 & P_3 & P_4 \\
\hline
P_1 & 0 & 1 & 0 & 0 \\
P_2 & 1 & 0 & 1 & 1 \\
P_3 & 0 & 1 & 0 & 1 \\
P_4 & 0 & 1 & 1 & 0
\end{array}
$$

Connectivity matrix C^1

FIGURE 8–26

Graph C and its connectivity matrix.

and,

$$B + B = \begin{pmatrix} a & b & c \\ d & e & f \\ g & h & i \end{pmatrix} + \begin{pmatrix} a & b & c \\ d & e & f \\ g & h & i \end{pmatrix}$$

$$= \begin{pmatrix} (a+a) & (b+b) & (c+c) \\ (d+d) & (e+e) & (f+f) \\ (g+g) & (h+h) & (i+i) \end{pmatrix}$$

If matrix C^1 is powered once to yield matrix C^2 and the powered matrix is added to C^1, a new matrix T is generated:

$$C^1 + C^2 = T = \begin{pmatrix} 1 & 1 & 1 & 1 \\ 1 & 3 & 2 & 2 \\ 1 & 2 & 2 & 2 \\ 1 & 2 & 2 & 2 \end{pmatrix} \begin{matrix} \rightarrow & 4\ (P_1) \\ \rightarrow & 8\ (P_2) \\ \rightarrow & 7\ (P_3) \\ \rightarrow & 7\ (P_4) \end{matrix}$$

Row Totals

Powering and summing continues until all zeros are eliminated from matrix T. When that point is reached, matrix T is termed the solution matrix. In the example here, $T = C^1 + C^2$ and all zeros have been eliminated.

Because no element in T equals zero we know that every point can be reached by every other point in at least one one-step or one two-step move. The graph's diameter is the power to which C must be raised to eliminate all zeros. The diameter of graph C is 2 (no place has a higher associated number) so our solution matrix T contains no matrix beyond power 2. Graph

C is evaluated by summing rows or columns of the solution matrix T. If rows are summed, a column vector of accessibility scores is generated, revealing that P_1 is topologically the least accessible (most remote or isolated) place in the graph, and place P_2 has the highest score and is therefore the most accessible.

A Hierarchy of Places in a Network

Specialized activities of every kind—business, governmental, social, educational, and so forth —concentrate in towns and cities. Urban places form nodes in regional transportation and communication networks. Each activity in an urban place maintains a regular set of contacts with related activities in other places. Contact can be measured by monitoring mail flows or telephone calls, or passenger and freight movements. These data yield indices of the direction and volume of a city's functional contacts with other places.

When a region is just beginning to develop, interactions between centers are sporadic and it is uncertain which of the early patterns will persist. In early development phases a week may pass with no contact. With the passage of time, however, towns begin to contact each other more frequently in a typical day. In a modern setting, messages, goods, and people flow readily among places, prompting the query: what is the basic *structure* underlying the flows among the components (Fig. 8–28)?

FIGURE 8–28

Interaction patterns during regional development and the emergence of a dominant flow structure.

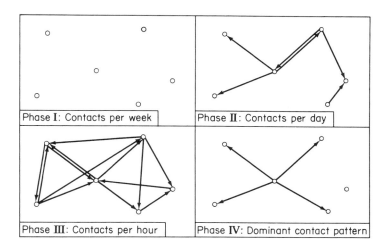

Phase I: Contacts per week

Phase II: Contacts per day

Phase III: Contacts per hour

Phase IV: Dominant contact pattern

One type of nodal region, as we said in Chapter 3, includes an urban center and a tributary area. Tributary areas might include a group of smaller nodal regions. Geographical relationships of this sort are profitably analyzed as hierarchical structures. Each center dominates an area around it, and, in turn, a small town and its adjacent rural area is assigned to that larger urban place with which it has the greatest number of functional linkages. Usually this "dominant association," as it is called, is with a nearby city. A nodal region, therefore, comprises a *central place* and the *surrounding areas* predominately subordinated to it. Nodal regions exist at several scales. The major hinterland of Chicago, for example, is defined by its dominant

association with smaller cities. Each of the smaller places is the dominant center for a set of still smaller towns. In regular fashion cities and towns in rural areas are nested within a hierarchy of nodal regions (Fig. 8–29).

In a nodal hierarchy, described by a network of points and lines, the points represent central places and lines represent functional associations between dominant and subordinate centers. Within the hierarchy, the links to subordinate points may be direct or indirect. For example, when connections are indirect (Fig. 8–30a) deliveries from center *C* are handled in two steps, such as when Sunday newspapers are delivered to small towns, then to individuals. Nested hierarchies can have more than two levels. National political parties are organized by block, precinct, ward, county, state, and national committee. Another kind of arrangement often functions parallel to the two-step case (Fig. 8–30b). For example, a wholesaler at center *C* could receive mail orders from retailers at outlying points and ship directly to them without relying on intermediate jobbers. If this happens, the hierarchy has only two levels, yet the hierarchical idea and dominance of small outlying points by larger urban centers are easy to see.

Let us consider an example prepared by Nystuen and Dacey to illustrate how we analyze flows in order to discover underlying hierarchical structures. Consider twelve cities interacting with each other to various degrees. Within the network of all intercity flows of goods, services, information, and people there exists a pattern of dominant flows. The network of dominant flows defines the skeleton of the nodal organization of the entire region. The network of dominant flows of each type, such as wholesale goods or metropolitan newspaper circulation, usually reveals a nodal structure similar if not identical to the nodal structures delineated in terms of other sets of flows. If a farmer in northeastern Iowa reads *The Chicago Tribune* on Sunday instead of *The Minneapolis Sunday Tribune* he probably follows the Cubs or the White Sox rather than the Minnesota Twins.

Telephone calls between cities are especially convenient and meaningful when studying interurban linkages and dominance. In a city firms

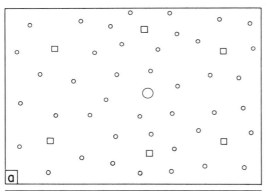

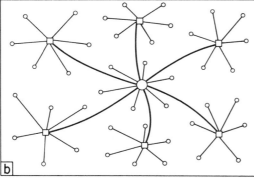

FIGURE 8–29

(a) *An apparently random spatial distribution of towns and cities; and* (b) *a hierarchy of nodal regions. That there is more geographical order on the earth than is generally supposed is not apparent until that order is looked for.*

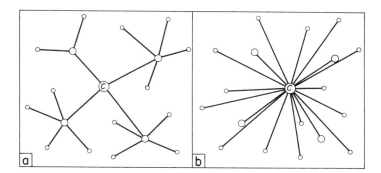

FIGURE 8–30

Within an urban hierarchy, when a good or service is delivered, the interactions between the dominant center and subordinate centers may be indirect (a) or direct (b).

engaged in each specialized activity, such as pharmaceuticals or automobile wholesaling, must maintain functional contacts with other places. With few exceptions each firm decides which other cities it will contact and which ones it will ignore. The choices are usually two-sided. A small town drugstore in central Illinois may choose a drug supplier in St. Louis instead of one in Chicago. Similarly a Chicago wholesaler may decide to expand his territory into Wisconsin instead of into Indiana. Despite freedom on both sides, competition and the principle of least net effort prompt sellers to create contacts in easy to service markets, and buyers shop as close to home as they can if they can get satisfactory service.

Whatever arrangements are settled upon, telephone calls reflect the prevailing structure of linkages and dominance. Returning now to our twelve city example, the distribution of calls from each city to each of the eleven others reveals the degree to which the city is oriented to each of those places. If we examine the outflow of messages from the first city (city *a*) to every other city in a study area, we can discover its dominant association by finding the largest flow. The largest flow defines a single outdirected line to one other point. We can do the same for each of the other eleven cities in our sample study region (Table 8–2). The interaction matrix discloses the number of telephone messages between each pair of cities during a

TABLE 8–2

Telephone Message Flows among 12 Cities per Unit Time

		a	b	c	d	e	f	g	h	i	j	k	l
							To city*						
	a	0	**75**	15	20	28	2	3	2	1	20	1	0
	b†	**69**	0	45	50	58	12	20	3	6	35	4	2
	c	5	**51**	0	12	40	0	6	1	3	15	0	1
	d	19	**67**	14	0	30	7	6	2	11	18	5	1
	e†	7	40	**48**	26	0	7	10	2	37	39	12	6
From city	f	1	6	1	1	10	0	**27**	1	3	4	2	0
	g†	2	16	3	3	13	**31**	0	3	18	8	3	1
	h	0	4	0	1	3	3	6	0	12	**38**	4	0
	i	2	28	3	6	43	4	16	12	0	**98**	13	1
	j†	7	40	10	8	40	5	17	34	**98**	0	35	12
	k	1	8	2	1	18	0	6	5	12	**30**	0	15
	l	0	2	0	0	7	0	1	0	1	6	**12**	0
Column totals ‡:		113	337	141	128	290	71	118	65	202	311	91	39

*Largest flow in each row in boldface.
†Largest flow from these cities is to a smaller city; thus they are dominant centers.
‡A city's size is determined by the number of messages it receives (column totals).

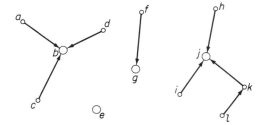

FIGURE 8–31

Nodal structures defined through the analysis of message flows among twelve cities during a time interval (adapted from J. D. Nystuen and M. F. Dacey, Fig. 1).

given time period. The largest flow from a city is the largest entry in its row. The functional size of a city is measured by the number of messages it receives. Column totals in the interaction matrix are the number of messages received by each city. A city is *independent* or *central* if its largest flow is to a smaller city. A *subordinate* or *satellite* city sends its largest flow of outgoing messages to a larger city. These relations are transitive. If *a* is subordinate to *b*, and *b* is subordinate to *c*, then *a* is subordinate to *c*. This is what we expect in a nested hierarchy of urban centers and nodal regions.

The nodal structure of dominant and subordinate cities in a region is represented by a *directed* graph (Fig. 8–31). Sometimes an independent city, like *e*, has no satellites. City *e* certainly has dealings with many smaller nearby centers, but in every case the smaller places send their largest message flows to centers other than city *e*.

The Growth of Cities on Transport Networks: An Application of Matrix and Graph Ideas

A communications network graph can be represented by a matrix with 0 and 1 entries, and the graph can be used to study a network's topology. But whereas a graph can retain geographical information as it reveals the topology lying submerged in a map, graphs and maps have serious computational shortcomings. If we represent a network in matrix form, we can analyze network properties that otherwise would remain hidden or obscure. Consider a network

of places and one-way routes. Up to now we have considered only two-way routes such as roads, telephone lines, or two-way radio. A one-way link might be a messenger, a signal light, or a fast flowing river in pre-steamboat days. By convention we continue to assume that a place has no links to itself. Every network of one-way and two-way routes can be represented by a *directed graph*. Such a graph has a corresponding communications matrix of 0's and 1's (Fig. 8–32). The graph and the matrix contain the same information but in different forms. We could reconstruct the graph from the matrix just as we filled in the matrix with information from the graph.

If a place (say P_1) can communicate with another place (P_3), we enter a 1 at the intersection of the first row and the third column. But in a directed graph, the fact that we can go in one step from P_1 to P_3 does not mean that we can return in one step. Yet we can analyze matrix D^1 and the relative accessibility of different nodes in the network (at the top of p. 267) in the same way that we analyzed graph C.

Matrix D^1 describes the one-step links between places. Note that the matrix is not symmetrical. Adding across rows we determine that P_2 can reach two other places in one step whereas the other places can reach only one. When the matrix D^1 is powered, D^2 indicates the two-step linkages possible within the network. When D^1 and D^2 are added, four zero entries remain. The zeroes represent four one-way trips which cannot be made in either one-step or two-step moves (P_1 to P_1; P_1 to P_4; P_3 to P_3).

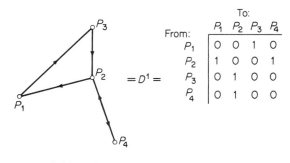

FIGURE 8–32

A directed graph and its connectivity matrix.

Row Totals

$$D^1 = \begin{bmatrix} 0 & 0 & 1 & 0 \\ 1 & 0 & 0 & 1 \\ 0 & 1 & 0 & 0 \\ 0 & 1 & 0 & 0 \end{bmatrix} \quad \begin{matrix} 1 \\ 2 \\ 1 \\ 1 \end{matrix}$$

$$D^2 = \begin{bmatrix} 0 & 1 & 0 & 0 \\ 0 & 1 & 1 & 0 \\ 1 & 0 & 0 & 1 \\ 1 & 0 & 0 & 1 \end{bmatrix} \quad D^1 + D^2 = \begin{bmatrix} 0 & 1 & 1 & 0 \\ 1 & 1 & 1 & 1 \\ 1 & 1 & 0 & 1 \\ 1 & 1 & 0 & 1 \end{bmatrix} \quad \begin{matrix} 2 \\ 4 \\ 3 \\ 3 \end{matrix}$$

$$D^3 = \begin{bmatrix} 1 & 0 & 0 & 1 \\ 1 & 1 & 0 & 1 \\ 0 & 1 & 1 & 0 \\ 0 & 1 & 1 & 0 \end{bmatrix} \quad D^1 + D^2 + D^3 = \begin{bmatrix} 1 & 1 & 1 & 1 \\ 2 & 2 & 1 & 2 \\ 1 & 2 & 1 & 1 \\ 1 & 2 & 1 & 1 \end{bmatrix} \quad \begin{matrix} 4 \\ 7 \\ 5 \\ 5 \end{matrix}$$

and P_4 to P_3). Powering the matrix once again and adding matrix D^3 to $D^1 + D^2$ yields a solution matrix and no zero entries. The three sets of row totals reveal that in every case P_2 is topologically the best connected place in the network. With the addition of D^2 to D^1 it becomes clear that P_1 is the place on the network that is hardest to reach.

Consider now a region settled before the steamboat and railroad era. The principal means of interior transportation through the wilderness among eleven places is by river flatboat powered by river currents. Obviously, only one-way traffic is possible (Fig. 8–33). Which place is most easily reached from the others? The centers are linked by three separate and unconnected subnetworks. Centers in one subnetwork cannot have contacts with centers of other subnetworks. Moreover, an upstream center can send goods, people, and messages downstream, but there can be no return traffic. The centers that dominate others in such an arrangement are those which can collect better than the other centers on the networks.

Consider one-step moves. Column totals in matrix E^1 disclose the number of one-step moves that terminate in each place. Five places are at the head of river navigation and each can draw from no other place ($P_1, P_3, P_6, P_8, P_{10}$). Four centers can draw from one other place.

Two centers (P_9 and P_{11}) can act as entrepots in the way that Louisville and New Orleans dominated Ohio and Mississippi river navigation before the introduction of the river steamboat in inland water transportation.

If the water channel is without falls and rapids, two- and three-step moves are also possible but four-step moves are not. If matrix E is powered up to 3 (the diameter of the largest subnetwork), column totals from the solution matrix $T = E^1 + E^2 + E^3$ permit us to rank the places according to their access to other places. P_{11} has direct access by one-, two-, and three-step moves to five places in addition to itself. P_9, in an intermediate location, can tap three places. P_5 dominates its own little world, as does P_2. In the early nineteenth century, Great Lakes ports such as Cleveland and Detroit resembled P_5 and P_2 in their small local monopolies just as New Orleans emerged as the giant for a time monopolizing the Mississippi-Ohio system.

Two-way traffic shattered the old order along with its locational monopolies. In United States river transport, the steamboat's arrival was greeted with a spate of canal building in a frantic effort by each city to gain a superior position on the newly transformed network (Fig. 8–34). Consider the hypothetical region of eleven places transformed by the addition

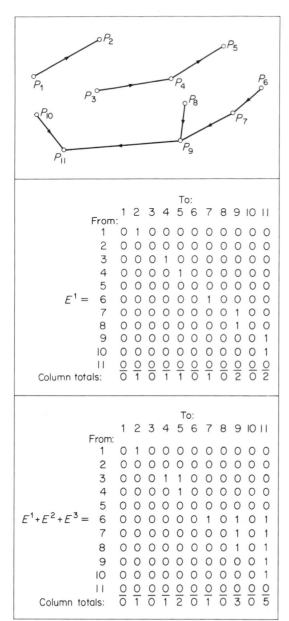

			To:								
		1	2	3	4	5	6	7	8	9	10	11

From:

	1	2	3	4	5	6	7	8	9	10	11
1	0	1	0	0	0	0	0	0	0	0	0
2	0	0	0	0	0	0	0	0	0	0	0
3	0	0	0	1	0	0	0	0	0	0	0
4	0	0	0	0	1	0	0	0	0	0	0
5	0	0	0	0	0	0	0	0	0	0	0
6	0	0	0	0	0	0	1	0	0	0	0
7	0	0	0	0	0	0	0	0	1	0	0
8	0	0	0	0	0	0	0	0	1	0	0
9	0	0	0	0	0	0	0	0	0	0	1
10	0	0	0	0	0	0	0	0	0	0	1
11	0	0	0	0	0	0	0	0	0	0	0
Column totals:	0	1	0	1	1	0	1	0	2	0	2

$E^1 =$ (at left of matrix)

			To:								
		1	2	3	4	5	6	7	8	9	10	11

From:

	1	2	3	4	5	6	7	8	9	10	11
1	0	1	0	0	0	0	0	0	0	0	0
2	0	0	0	0	0	0	0	0	0	0	0
3	0	0	0	1	1	0	0	0	0	0	0
4	0	0	0	0	1	0	0	0	0	0	0
5	0	0	0	0	0	0	0	0	0	0	0
6	0	0	0	0	0	0	1	0	1	0	1
7	0	0	0	0	0	0	0	0	1	0	1
8	0	0	0	0	0	0	0	0	1	0	1
9	0	0	0	0	0	0	0	0	0	0	1
10	0	0	0	0	0	0	0	0	0	0	1
11	0	0	0	0	0	0	0	0	0	0	0
Column totals:	0	1	0	1	2	0	1	0	3	0	5

$E^1 + E^2 + E^3 =$ (at left of matrix)

FIGURE 8–33

Eleven towns linked by three river networks. Matrix E^1 describes connectivity of places in terms of one-step moves. Matrix $E^1 + E^2 + E^3$ describes how destinations may be reached in terms of one-, two-, and three-step moves. Place P_{11} can be reached by five different towns in one-, two-, or three-step moves and has the highest connectivity score.

of two canals and the start of two-way traffic on the entire network of water routes. Column totals from the connectivity matrix F^1 disclose that places P_4 and P_9 are each directly tied to three places in one-step moves. When the matrix F^1 is powered for two-step, three-step, and four-step moves we begin to see the larger picture of places' general topological positions on the network. Place P_9 is clearly the most advantageously located. P_8 and P_4 are close behind but never take the lead. Without the canals and two-way traffic, P_{11} was the leader. Then P_9 takes the lead just as Cincinnati dominated the American Midwest for awhile because of its temporary locational advantage over Louisville, New Orleans, and other interior centers. But then came railroads. Water transport was pushed into a secondary position. The more completely connected network redefined once more the centermost place in the regional system. In the American Midwest's urban competition, Chicago emerged as the clear winner.

Boundaries can interfere with flow patterns and dominance arrangements if there is a limit imposed on the number of crossing points. If a boundary exists or is imposed, whether political or physical (such as a river with a single bridge), a hierarchical system usually develops differently from what might have happened without the barrier. Seldom does a transport or communications network become less complete. But if it does, the relative locational advantages of places served by the network are adjusted and dominance patterns are rearranged (Fig. 8–35). These principles also find application in commercial planning. When selecting a store site or shopping center location in a suburban area, for example, attention must be paid not to the transportation system at the moment, but rather to location within the system once the area is built up and served by an extended and fully linked network (Fig. 8–36). As networks evolve, the relative positions of nodes on the network change.

Flow Rates, Service Capacities, and the Problem of Queues

Every network has a limit to the amount of flow it can handle in a given time period. If the

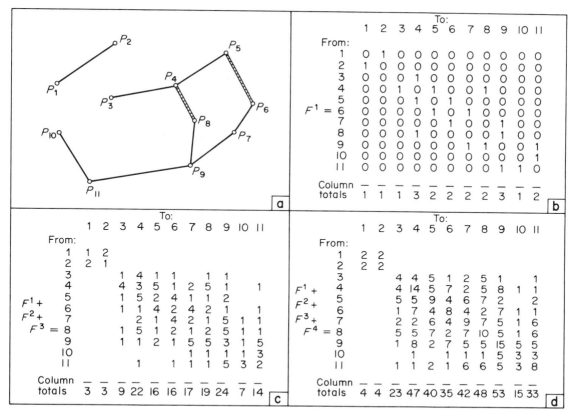

FIGURE 8–34

Dominance patterns shift with two-way traffic (e.g., steamboats) and additional links between cities (e.g., canals in a). Matrix F¹ in b presents the accessibility scores for each city in terms of one-step moves. Connectivity matrices in c and d describe possible moves and positions on the network in terms of multiple-step moves and one-step moves combined. City P₉ is clearly the best located city.

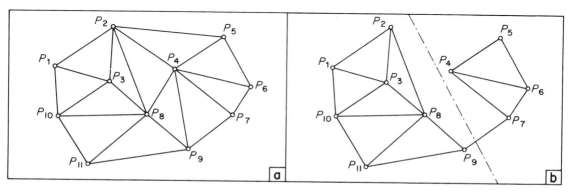

FIGURE 8–35

Relative locations on a network are adjusted when a boundary prevents linkages. Which center has the highest accessibility score in terms of one- and two-step moves in network a? In network b?

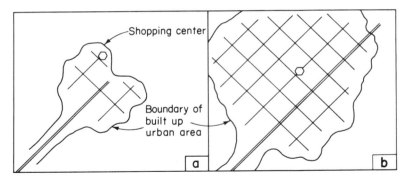

FIGURE 8–36

Relative locations are seldom permanent. Here, a location changes as the built-up urban area expands and as the transport system moves outward. A shopping center constructed at the edge of the urban area in one time period (a) *is centrally situated with respect to its trade area at a later date* (b).

network is overtaxed, queues of people, merchandise, messages, and other things start to form. Let us investigate this problem by examining some problems connected with coin tossing, climatic change and glaciation, and floods and traffic jams. Each is a time series process and much can be gained by examining them as a group.

If we have a storage system of very large or perhaps infinite capacity, what are the long run consequences if there is a fifty-fifty chance of the storage system's gaining or losing one unit in a time period? Intuitively we might regard such a system as in equilibrium, but our suspicions are unreliable. Common sense tells us one thing but science tells us something else— and science is right. No matter how long we watch the system, once started there is a greater than 50 percent chance that *storage will never return to its original size or will return only once,* than that any other course of events will occur.

Such a process resembles two persons with unlimited funds tossing a coin. Such a game can be simulated with a table of random numbers (Fig. 8–37). The coin is tossed and *A* wins one cent when heads shows; *B* wins when tails turns up. Once the game is begun, the probability is very high that the lead will remain on one side. In fact, in an extremely long game, say one million tosses, one side or the other will build up some impressive gains just tossing a penny. Yet on any two tosses there is a 50 percent chance that players will break even. The possible outcomes from two tosses are: *HH, HT, TH, TT.* In two of the four possible outcomes (half the time) the players will come out even (*HT, TH*). The other half of the time they do not.

Think of what this means in nature. If, in the Arctic, ice accumulation in winter equals ice melt in the summer, then *on the average* net ice gain or loss is zero. But regardless of what

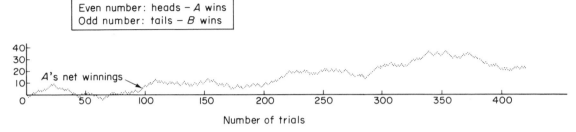

FIGURE 8–37

Simulation of a coin tossing game of 420 tosses using a table of random numbers. A fifty-fifty chance of winning still permits large net winnings at an average point in time. Even though average winnings equal average losses over a short period of time, in a long time series enormous deviations from parity are probable.

happens on the average, continental glaciers developed and moved out at their margins as a result of massive accumulations poleward. In other words, it is possible to account for glacial accumulations of ice under the simple assumption that ice accumulation equals ice melt, on the average, each year.

If two competitors build identical gas stations across from a freeway exit, located so that cars have a fifty-fifty chance of stopping at one or the other, a car that stops at one must bypass the other. Ignoring the possibility that bandwagon effects operate to draw more business to busy places, we can predict that one station will do better than the other, and that over time one will do substantially better than the other, even though each has a fifty-fifty chance of attracting the next car.

From some starting point, say a joint grand opening day, the most likely course of events is that the system will show a major imbalance over the long run, with one station gaining substantially over the other. As the time period is increased, provided that operating conditions are unchanged, the number of returns either to equality or to their initial states, if they were unequal at the beginning, will not increase proportionately. Instead, for a given probability level, the chances of a return to the initial value are proportional to the square root of the time period under study.

This idea of the cumulative results of chance events can be applied in the analysis of lines or queues in transportation and communication systems. What if automobiles *arrive* at a toll booth *on the average* of one every fifteen seconds. Let us say that it takes *on the average* fifteen seconds to *service* each car. It might appear that if average arrival time equals average service time things would move along like clockwork. But some arrivals are more frequent than once every fifteen seconds and others come less frequently. Service times also vary. Occasionally a driver will need change for a $100 bill whereas the next driver will have his arm extended with exact change. If people arrive at the toll booth at intervals shorter than fifteen seconds, arrivals are stored up in a queue. Yet if service times are unusually speedy, there can be no storage of the time saved. Thus, with only one toll window

per traffic lane, in theory, as average arrival time *approaches* average service time (it does not even have to equal it) a line of infinite length will form. Broadening two lanes to four or more at toll stations means that average service time can be kept substantially below average arrival time, thereby preventing the formation of long queues.

A reservoir behind a dam is another example of a solution to a movement problem of this sort. Over a year average monthly water arrival in a region is equal to, more than, or less than the water removal capacity of the regional stream system. But averages mean nothing if all the water comes at one time of the year. The broadening of a river bed into a wide flood plain is an attempt by nature to prevent a queue from forming. As such it is identical to the broadening of the toll road into six to twelve toll booths to prevent a waiting line. A dam across a river retards the flow and induces a queue to form. In a river valley we call this a reservoir; on the Northwest Tollway from downtown Chicago at 5:00 P.M., we call it an intolerable traffic jam.

If we designate the rate at which units arrive for service as A, and designate the potential service rate as S, then the ratio of input rate to output rate A/S may be thought of as a load factor or system utilization factor. When the load factor approaches unity, the expected length of the waiting line lengthens explosively (Fig. 8–38). As already suggested, this idea has many applications in human geography and elsewhere. Consider farmers bringing livestock to market at an average rate of so many head per day, and the slaughterhouse purchasing animals according to its capacity to handle them. If the marketing of the animals is not coordinated through a producers' cooperative or some other arrangement, some days the farmers will glut the market and will either sell their stock for very low prices or be forced to take them back home again. Other days there will be a shortage of animals and prices will rise.

When it is raining, the rain arrives on the ground at one rate and the soil soaks it up at another rate. As the load factor approaches unity, the soaking in cannot keep up with newly arriving rain and runoff occurs. We experience

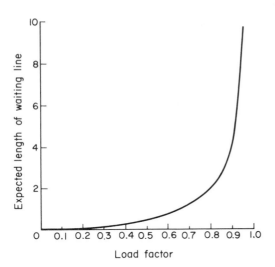

FIGURE 8–38

How the expected length of a waiting line or queue increases explosively as the load factor approaches unity. The load factor is the ratio of the average arrival rate to the average service time (adapted from C. D. Flagle, Fig. 4, p. 410).

the same thing in stores and shops when customers' arrival rates approach the store's service rates. Dissatisfied customers "run off" to other stores where service is better.

At a wider scale we can consider the various residential subareas of a metropolis as places where people are served with housing. Newcomers are always arriving and others are regularly departing. Because of information gaps and special arrangements that have to be made before people can move from one place to another in the same city, arrival and service are not instantaneous. They are time series processes and queues can develop, which take the form of severe overcrowding in some parts of the city while vacant housing is going begging elsewhere. Thus, a good supply of vacancies is usually needed to keep the load factor comfortably below unity.

In the same vein but at a worldwide scale, we can think of the birth rate as arrivals on earth and the death rate as an index of the average service time alloted to each person. The queue is the number of people living on earth at any one time and waiting for death. Even if

average arrival time *equaled* average service time, the queue (world population) could still increase rapidly in a short period of time. In fact, in recent decades birth rates have greatly *exceeded* death rates, producing a load factor substantially *in excess of 1.0*. It is understandable, then, why Fig. 8–38 resembles the world population growth curve we see so frequently. To bring the queue under control, service rates have to be increased relative to arrival rates. The most acceptable way to do this is to reduce arrival rates (birth control) because the alternative of increasing the world death rate by shortening life spans is abhorrent.

NETWORK DESIGN AND NETWORK PERFORMANCE

Despite the diverse origins and functions of the various geographical networks, all of these networks have properties in common. Common properties are discovered through a study of network geometries. If we restricted ourselves to investigating a network's development history or the character of flows through it, important network characteristics would escape our notice. The basic building block of even the most complex of networks is the line or route. We might expect that the location of lines in a network would be easy to explain in terms of the shortest distance between pairs of points. Sometimes this is true when routes are direct, but many routes and paths (a string of many connected routes) wander instead of forming direct connections.

On an isotropic plane where movement friction is equal in every direction, the most direct way to connect two points is with a direct path. But what is the location of the *shortest* route network connecting *three* points, perhaps three villages? If all three lie along a straight line then the solution path is a straight line. If not, the three places form the vertices of a triangle. If any one of the triangle's interior angles equals or exceeds 120 degrees, the shortest route network is a pair of direct lines from that vertex to the two other points (Fig. 8–39). If all three angles are less than 120 degrees, three routes radiate from a new point *W* within

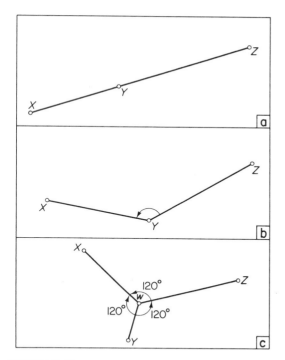

FIGURE 8–39

Shortest route networks linking three villages: (a) when all three villages lie on a straight line; (b) when interior angle Y exceeds 120°; and (c) when all interior angles are less than 120°. Interior angles are those formed when the three villages are the vertices of a triangle. In case c, the shortest route network radiates from a new point W located such that three 120° angles are formed by the roads to the villages.

tance between X and Z declines faster than the distance between W and Y increases. If W were relocated a short distance in any direction, two places would be brought closer together, but the savings in road length between them would be lost, and more, tying the third place into the network.

We should emphasize here that minimizing the *road length* differs from minimizing *movement* effort along a route network. If a small college's three major activity nodes consist of a dormitory X, a library Y, and a classroom building Z, located on a flat tract of land without hedges, walls, or fences, we might expect unfettered students to establish direct paths between buildings (Fig. 8–40*a*). A system of direct paths minimizes user movements. Campus planners have other things, such as capital and maintenance budgets, on their minds, however. Often they try to cut route construction and

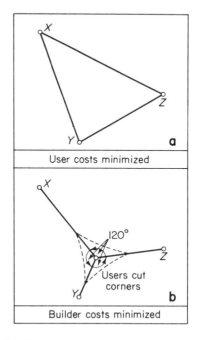

FIGURE 8–40

A minimum user cost network (a) linking three centers, compared with a minimum path length network (b), linking the same three centers. In the latter case, users cut corners to save time and effort.

the triangle, and form three 120 degree angles as routes link that point to each of the other three original points.

What is the reason for these solutions? Figure 8–39*a* is pretty obvious. It is hard to find a more direct path on an isotropic plane than a straight line. In Fig. 8–39*b*, a more direct connection between X and Z plus a siding to serve Y would have meant an overly long total network length. The opposite is true in Fig. 8–39*c* where an XYZ path (linking X directly to Y and Y directly to Z) would have produced a longer route than the network centered on W. In Fig. 8–39*c*, as the network center moves from Y toward an optimal location at W, the road dis-

maintenance costs by minimizing aggregate path length, but thereby invite optimizing students—who sleep late or fail to watch the time—to cut across lawns or through hedges (Fig. 8–40b).

On most campuses two sets of paths develop as solutions to the two minimization strategies. An official minimum path length solution is enshrined in concrete and asphalt. The unofficial path, minimizing movement, is easiest to see after a snowfall but before official paths are shoveled. In summer movement minimization paths never have any grass growing in them. Occasionally when a new campus is built or an old one expanded, planners recognize that it does not pay to try nudging students away from their preferred least effort paths. Instead, when designing foot paths, planners plant grass everywhere and postpone planning the official routes until travel patterns are beaten into the grass. This way they can find out not only where paths should be installed but the appropriate widths as well. In the long run this procedure may be a least cost solution for the builder because there will be little need to install and maintain fences, hedges, and other obstacles to minimum movement.

If several points must be connected by the shortest path, the 120 degree rule is still in effect after the optimal solutions are found by trial and error. Given five scattered points $(P_1–P_5)$ as in Fig. 8–41, vertices U, V, and W must be located simultaneously for the optimal solution. This is no easy matter. How to do it in the general case has not yet been solved.

Studying networks such as the one shown in Fig. 8–41 suggests that displacing any one of the intersections U, V, or W, would reduce path length for certain movements but would increase it by larger amounts for movement in other directions. The fact that displacement in any direction cannot improve things, and may make matters worse depending on the criterion—in this case overall path length—is an important feature of an *optimum solution*. Each set of locations or places carries with it two implied sets of minimum costs, one for builders and one for users. Other sets of places imply other minimum cost arrangements.

Costs and Benefits in Path Design

The problem faced by a railroad company of how to locate a railroad to serve a set of towns usually means considering how to maximize the difference between costs and revenues, and not how to minimize one or another set of costs alone. The railroad company knows it should make the route as short as possible to save construction and maintenance costs, yet the rail line must be long enough to ensure that traffic levels and revenues will be maintained at high levels. Put another way, the problem is to figure out the best way to locate a path between major places in order to pass most profitably through minor intermediate places (Fig. 8–42).

Let us assume that if only the terminal cities X and Y are joined, the path length equals costs which equal 10.0, and traffic revenues equal 20.0. Connecting one or more intermediate towns with the terminal cities means added network costs but also more freight revenue. Which towns should be connected and which ones should the railroad network bypass?

The first extreme solution (Fig. 8–42a) linking only the terminal cities yields a net return of 10.0 units (20.0 revenue units less 10.0 cost units). The other extreme solution is to connect all six intermediate towns with the railroad and thereby maximize the traffic and revenues (Fig. 8–42d). But connecting all the towns also maximizes construction and maintenance costs. Yet even though costs rise, net returns rise to 19.1 units (34.0 revenue units less 14.9 cost

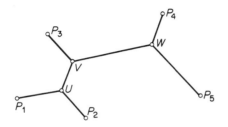

FIGURE 8–41

A shortest path network linking five scattered places, $P_1–P_5$. Vertices U, V, and W must be located simultaneously so that routes intersect at 120° angles.

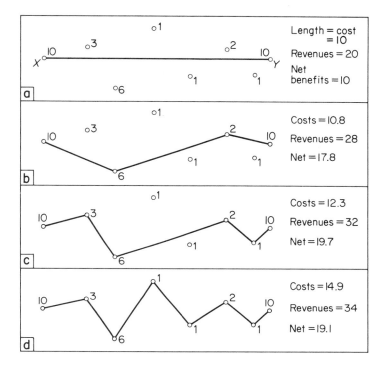

Length = cost = 10
Revenues = 20
Net benefits = 10

Costs = 10.8
Revenues = 28
Net = 17.8

Costs = 12.3
Revenues = 32
Net = 19.7

Costs = 14.9
Revenues = 34
Net = 19.1

FIGURE 8–42

Locating a railroad to maximize net revenues (revenues — costs) per unit time. The examples shown are only four of sixty-four possible ways in which the eight cities can be linked or bypassed with a single rail line. Each of the sixty-four arrangements carries with it a set of costs and a set of revenues. Turned on its side, the sketch resembles the line of old port cities along the Atlantic coast of the United States— the BosWash Corridor.

units) so that connecting all intermediate centers is more profitable than skipping them all.

If our planning strategy is to design a network with a net return exceeding some threshold level of acceptable net profit, we may stop with alternative *d*. If we want maximum profits we should look further. With six intermediate centers there are sixty-four ways to route a railroad between terminal cities *X* and *Y* (Table 8–3). Each of the sixty-four options carries a set of costs, produces a certain level of revenue, and yields a net profit. The *best* strategy yields the highest net profit and lies somewhere among the sixty-four. Each option could be computed separately and the best one selected. This simple hypothetical example of how a railroad company should select an optimal route pattern has counterparts in any situation where routes must be specified to handle a system of flows or deliveries. Sometimes all possible routing strategies are defined and evaluated individually. Other times a more gross decision procedure is employed. When the American railroad network was laid down, some towns and cities were well

TABLE 8–3

Rail Routes between Two Terminal Cities X and Y, with Six Intervening Towns and Cities*

Number of Intermediate Towns Served	Number of Shortest Paths Linking Terminal Cities X and Y with Indicated Number of Intervening Centers
0	1 (Fig. 8–42*a*)
1	6
2	15 (Fig. 8–42*b*)
3	20
4	15 (Fig. 8–42*c*)
5	6
6	1 (Fig. 8–42*d*)
Total	64

* Any set of *r* objects (say four towns) selected from a group of *n* objects (say the six intervening towns) without regard to order within the set (i.e., the set containing the second and fourth towns is the same as the set containing the fourth and second towns) is said to be a combination of the *n* objects taken *r* at a time. If $_nC_r$ represents the number of combinations of *n* objects taken *r* at a time, then:

$$_nC_r = \frac{n!}{r!(n-r)!}$$

If we want to serve two intermediate towns (*r* = 2), there are fifteen different combinations we could choose ($_6C_2 = 15$).

served and others were passed by. Recently this process was repeated somewhat when the Interstate Highway System bypassed small towns in favor of larger centers. Similarly, before the days of the Rural Electrification Administration in the United States, many American farmers went without electricity and telephone service because utility companies decided that by serving towns and cities, and bypassing farmers, the companies' net profits could be maximized.

The Traveling Salesman Problem

A network connecting a set of points can assume the form of a treelike structure of connected lines without any complete loops, or it can be a single continuous path connecting many places. The path can also be closed in a circuit or a loop. The former two are typical of collection and delivery systems; the last is called the traveling salesman problem because a salesman's itinerary takes him away from home to make his calls but brings him home at the end of his trip.

What is the shortest path through a set of points, touching each of them just once and ending at the starting point? This problem is important in many modern business and transportation areas. It might seem that the traveling salesman problem could be solved by measuring all the possible paths and then picking the shortest one, but even with a computer this is a staggering job. Comparing a million paths per second, for instance, it would take several billion years to compute and compare all the paths in just a twenty-five point problem.

Short-cut methods exist for finding efficient routings around many points but they are slow when over sixty points are involved and yield only approximate solutions. Frequently expensive routing problems are solved by inspection of maps and graphs, and occasionally special computer programs are used. Mathematician Shen Lin of the Bell Telephone Laboratories tackled this important problem and developed a way of getting correct solutions to problems of up to 145 points. Because the method is fast, it is possible to find many such approximations and then to pick the shortest one. The path chosen is almost always the shortest possible one, but

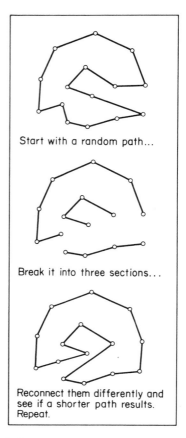

Start with a random path...

Break it into three sections...

Reconnect them differently and see if a shorter path results. Repeat.

FIGURE 8–43

Shen Lin's rapid route to the shortest traveling salesman path (adapted from a method developed by Shen Lin of Bell Telephone Laboratories).

even if it is not the approximate solution is good enough.

To make one approximation of the shortest traveling salesman's path the computer chooses a starting path at random (Fig. 8–43). Then it removes three links from the path, breaking the path into three sections. The three sections are connected differently to see if a shorter path results. If not, the computer starts again systematically removing other combinations of three links from the random starting path until all combinations have been tried.

Whenever a reconnection does produce a shorter path, this becomes the new starting path and a new series of breaks, reconnections, measurements, and comparisons begins. An *approximation* is completed when no further improvement results from additional breaks and reconnections. In the same way, beginning each time with a new and different random starting path, additional approximations are found. Usually they have path sections in common. Probably these identical sections are part of the absolute minimum path so they are incorporated into every new starting path and are no longer broken up.

The solution to the traveling salesman problem depends on the kind of surface over which movement between centers must take place.

If the centers are located on a uniform transport surface in Euclidian space, one solution will obtain. If the centers occupy positions in a relative cost or non-Euclidian space, another solution is in order. Consider a set of twenty-three towns in southern Ghana. Fishing towns and villages on the coast sell fresh fish to truckers who carry the fish to interior towns and sell them. Since fresh fish spoil rapidly in warm weather, the truckers require the shortest path to follow as they pick up fish on the coast, carry them to interior towns, and return to the coast again to repeat the process.

If a fish merchant used a helicopter to visit each of the twenty-three towns and villages on

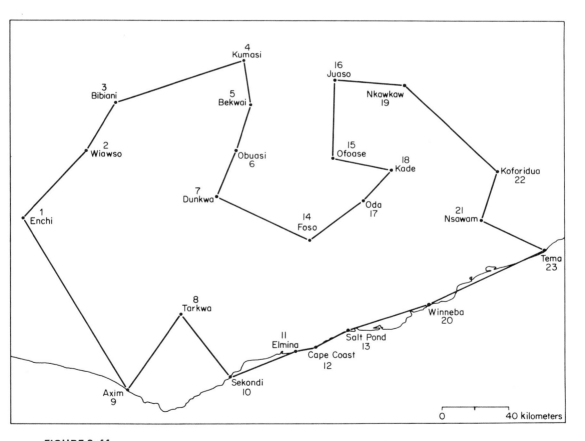

FIGURE 8–44

Shortest traveling salesman path among twenty-three towns in southern Ghana assuming that movement is equally easy in any direction (computations courtesy of Shen Lin, Bell Telephone Laboratories).

the fish circuit, he could move through the air on a uniform transportation surface and his shortest traveling salesman path would be that illustrated in Fig. 8–44. Instead of helicopters or ground effect machines, Ghanaian fish merchants ply their trade using trucks over roads of widely varying quality. When the road routes and differences in road quality have been incorporated into the traveling salesman problem, another solution path different from the first emerges (Fig. 8–45).

Other Minimum Distance Networks

We have seen a variety of ways in which points are linked by networks of different types.

We have also spoken of optimal networks, their objectives and their structures. In one case, we may want to connect separated points with the shortest path and thereby minimize builder costs. In other cases we may want to minimize user costs by designing a completely connected network. In a hierarchical arrangement we may wish to connect a capital city or a leading center directly with its satellites without connecting the satellites to one another. The traveling salesman problem seeks the shortest round trip path touching all places. A fifth problem, dubbed the Paul Revere Case by William Bunge, considers a path starting at one point and linking that point to all the others without returning. It applies, for example, to car pools when the

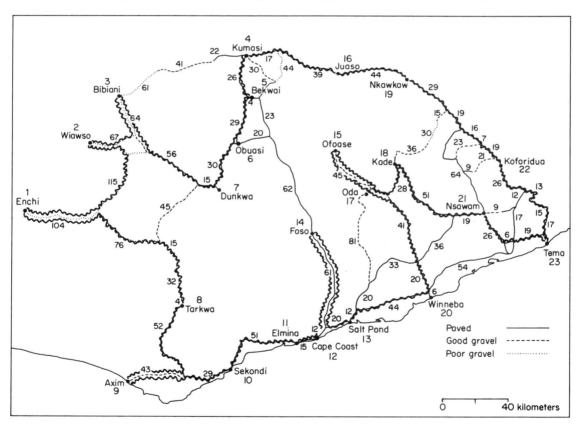

FIGURE 8–45

Shortest traveling salesman path among twenty-three towns in southern Ghana considering the varying road qualities and actual route lengths between towns (computations courtesy of Shen Lin, Bell Telephone Laboratories).

in the United States was installed in the corridor between Boston and Washington, D.C. Marginal areas often are inconveniently served, if served at all, because they offer low potential revenue for high service costs.

Defining and Identifying the Best Path in a Network

Rather than asking "What is the shortest path connecting a set of points sequentially?" or "What is the shortest road serving a set of points?" or questions about other problems of path design on an isotropic plane, we might ask "What is the best path when resistance to movement varies from one path to another?" There are two separate groups of network problems. In the first group, only the nodes are given. The job is to establish efficient or best paths connecting the nodes. Best is defined in different ways: best for users, best for builders, or best for operators. In a second group of problems, routes are given and we must *pick* from available paths the least effort paths or maximum flow paths if flow capacities are introduced into the network.

When routes in a network are constrained, they have a limit on their carrying capacity—for example, multi-lane roads versus one-lane roads; or twenty-centimeter pipe versus ten-centimeter pipe. Perhaps a channel permits flows only one way at a time, such as a street or a pipeline. With such constraints we might ask: "How can a flow be *routed* through a network containing directional and capacity constraints so as to move a maximum amount of oil or automobiles or telephone calls or mail?" The first step is to find out the maximum flows the network can handle. The *maximum flow* from a source to a destination through a network equals the sum of the route capacities in the *minimum cut*. A cut is a collection of routes in a network which, if removed, would separate a source from a destination. For example, consider the network connecting W and Z (Fig. 8–47). What is the maximum possible flow from W to Z? If only one-way flows are allowed, four different cuts (*a–d*) could separate W from Z. Each cut comprises two or three routes. The sum of each cut is derived by adding the capac-

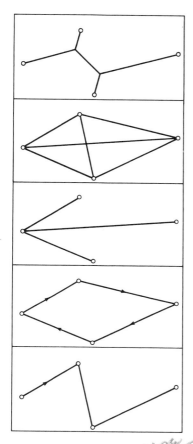

FIGURE 8–46

Five minimum paths. What is the minimization principle in each case?

starting point (the driver's house) and the end point (the office) are given (Fig. 8–46). Finally, and realistically, many networks are designed to maximize net economic benefits (revenues minus costs) to the agency in control. Often the path chosen in these cases (railroad, highway, electricity, telephone, and so forth) lies between a minimum construction cost network and a maximum user convenience system.

The most expensive solutions to problems of inadequate network capacity are usually installed first between nodes of great interaction. For example, almost every major innovation in the movement of mail and telephone messages

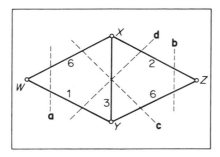

FIGURE 8–47

Any one of several cuts (a–d) in a network will separate origin W from destination Z. A cut is a collection of routes which, if removed, would separate the origin from the destination (from Haggett, "Network Models in Geography," Fig. 15–4, adapted from Akers, p. 312).

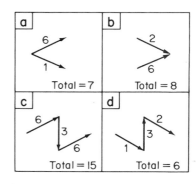

FIGURE 8–48

Each cut equals the sum of the capacities of the routes it contains. The minimum cut controls the maximum flow in the network from origin W to destination Z. Cut d is the smallest, so the maximum flow between W and Z is 6 (from Haggett, "Network Models in Geography," Fig. 15–4, adapted from Akers, p. 312).

ities of each route included in the cut (Fig. 8–48). The *minimum cut, d,* has a value of six, and thus the *maximum flow* between *W* and *Z* is six. Minimum cut *d* controls the whole system of flows between *W* and *Z*. Traffic analysts and city planners use these principles to locate potential congestion spots in urban transport planning. It is also possible to locate the critical links in a network and this is often done in traffic studies and in military planning.

Establishing the Best Path on Irregular Surfaces

Transportation arrangements in North America illustrate how paths have been located across areas of variable resistance to movement. Pioneer farmers of French origin often built houses next to navigable streams and partitioned farmland into *long-lots* so that a maximum number of farms could be created, with each farm having water access. Roads were too difficult to build away from the river bank, so farmsteads were erected on the river bank instead of scattering the farms and then connecting them by expensive roads (Fig. 8–49). The land surface was irregular because the friction of distance was high for overland movement, but relatively low for movements on water.

Most inland colonial towns in America were located at the centers of good farming areas. Roads linked them as directly as possible with their nearest neighboring towns (Fig. 8–50). Good locations at route intersections, through a

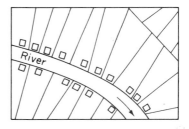

FIGURE 8–49

The best path on an irregular surface. It was easier for French pioneer settlers in North America to move by water than to travel overland. They built their farmsteads near riverbanks and divided farmland into long lots.

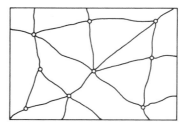

FIGURE 8–50

In colonial times, inland movement was difficult in every direction. Settlement was scattered because there was no incentive to cluster or line up along a superior transport route. Towns were established at the centers of good farming areas. Crude roads linked each town with its near neighbors.

circular cause-effect process, reinforced those superior positions at network intersections. Existing towns prospered and new towns never came into existence unless a transport revolution redefined the critical nodes in the geographical system.

The nineteenth-century homestead acts permitted farmland to be parceled out to settlers who agreed to improve it. The various land disposal schemes conformed to the township and range system of land survey adopted by the United States in 1785 and applied to most of the country except the southeast and eastern seaboard states. The township and range system's principle feature was the definition of east-west strips of land six miles (9.7 km.) wide called townships; and north-south six-mile-wide strips called ranges. Various latitudinal base lines and longitudinal principal meridians served as reference lines in identifying townships and ranges. Most homesteads were about 160 acres in size (65 hectares). Some in the East were forty and eighty acres; in the dry West homesteads were granted that were 640 acres and larger (Fig. 8–51).

Because American farmers lived on their farms, the job of linking *farmstead* to *towns* and to *other farms* without cutting across farm lands

meant that the only adequate yet acceptable road scheme was an almost completely connected rectangular network of section roads, so named because they followed section boundaries. This rural grid of main roads was largely adopted by cities as they spread into rural areas. The major streets and avenues in most Midwestern cities, for example, are a half-mile (0.8 km.) or a mile (1.6 km.) apart. Downtown areas settled ahead of the survey do not conform to the survey scheme.

Water routes, road systems, and the surveys were not the only surface distorting systems in United States settlement history. The railroads played an important role as well. In the eastern half of the United States, towns and cities were already in existence when the railroad came along. Rail lines were laid to connect centers within the existing urban hierarchy. The basic rail network in that part of the country, therefore, reflects a prior urban pattern. In parts of the Great Plains, however—and North Dakota is an excellent example—the rails were laid *ahead* of settlement. Points (cities and towns) were located with respect to rail lines (Fig. 8–52).

The three federal highway systems built in the United States in the twentieth century

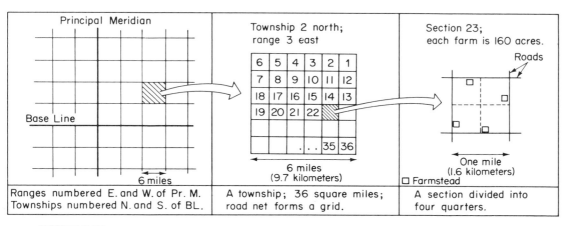

FIGURE 8–51

The Township and Range Survey as a surface distorting system. Because farmers lived on their farms, the job of linking farmsteads to towns and other farms without cutting across farmland meant that the only adequate yet acceptable road scheme was an almost completely connected rectangular network of section roads.

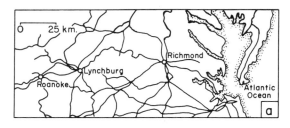

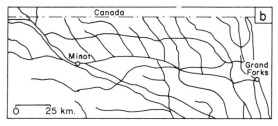

FIGURE 8–52

Locating rail lines. In Eastern Virginia (a), lines were laid to connect existing towns and cities; in eastern North Dakota (b), lines were laid to serve areas, not points. Towns and cities were founded with respect to the location of nodes on the railroad.

state highway systems linking rural America to urban America. These F.A.S. roads were planned, built, and maintained by state highway departments. Next up the hierarchy is the 400,000-kilometer Federal Aid Primary (F.A.P.) system known as the U.S. highway system. Each state unit in the system was cooperatively planned by state and federal highway bureaus. Financing came from state and federal sources. The objective of the U.S. highway system was to tie together the nation's cities. The points in this network were the centers of the downtowns of all the leading cities in the country. Small towns were bypassed unless they happened to lie on the path between two important centers (Fig. 8–53).

By the time World War II had begun it had become clear that the F.A.P. system of U.S. highways could not possibly meet the highway transportation demands of an urbanized nation's great metropolitan areas. A fourth system of roads was begun in 1956, the 69,000-kilometer Interstate Highway System, designed to link

again represent network solutions to problems of servicing points, lines, and areas. The Federal Aid Secondary (F.A.S.) system, also known as the farm-to-market system of federally assisted roads, connected farming areas to urban market points. These roads, fanning out from each city and enabling a flow of farm output cityward and a return flow of urban goods and services to the farms, became the basic framework of the various state highway systems. They were paved between 1920 and 1940 and their use by automobile driving American farmers had social implications of enormous consequences. Other nations in the world which have invested heavily in railroad development in rural areas have lagged considerably behind the United States in rural social development.

The network of over 4.8 million kilometers of rural roads in the United States can be broken down in hierarchical terms. In the lowest position are the rural section roads linking farm to farm and farm to village. These roads were locally built and are maintained by the local counties. Next up the road hierarchy are the

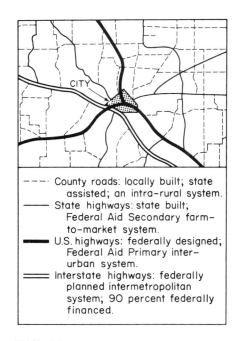

- - - - County roads: locally built; state assisted; an intra-rural system.
——— State highways: state built; Federal Aid Secondary farm-to-market system.
▬▬▬ U.S. highways: federally designed; Federal Aid Primary inter-urban system.
═══ Interstate highways: federally planned intermetropolitan system; 90 percent federally financed.

FIGURE 8–53

A hierarchy of United States rural road systems.

metropolitan areas with a limited access, multiple-lane, high-speed highway network without grade crossings. Every effort is made in the Interstate System to reduce user costs to a minimum. Deep cuts are made even in the lowest hills to eliminate grades and to prevent the need for roundabout routes. Links in the Interstate System follow fairly direct paths in a geographical sense, removing obstacles rather than going over or around them, thereby rendering an irregular surface regular.

Other Irregular Surfaces and Isochrones

On a flat desert, an ice cap, a plowed field, or an empty parking lot, ease of movement might be equal in any direction. On such an isotropic plane the least cost route between two places is a straight line. Around a point on such a plane isochrones are concentric circles. If resistance to movement is spread unevenly over the plane, the least effort path is still the shortest route in a time or cost space, but it usually will not be a straight line. Such minimum effort lines will appear twisted in absolute geographical space even though they are direct in the appropriate relative space.

August Lösch presented an example to illustrate least effort paths in irregular surfaces. What is the best path from an island town X to an inland town Y (Fig. 8–54)? There are several possible paths and the one chosen depends on the movement costs on water S compared to the movement costs on land L. If S is free and L is not free, then the sensible course is to make the relatively expensive distance over land as short as possible (Fig. 8–54a). If S and L are equal, then the shortest or least effort trip between X and Y is a straight line route (Fig. 8–54b). If land travel is free and water travel is not, the situation is reversed. Now it pays to make the X-to-shore trip as low-cost as feasible (Fig. 8–54c). Finally, if both S and L are positive but unequal, the minimum cost for the completed trip from X to Y will obtain at that crossing point where (S/L) equals the ratio of the sines of the angles formed (sin y/sin x). Thus, if S drops, the fraction S/L drops, and in order to make the ratio (sin y/sin x) drop, the crossing must be made further north so that

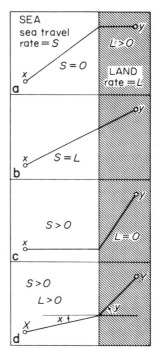

FIGURE 8–54

Lowest cost user paths over regular (b) *and irregular* (a, c, d) *movement cost surfaces* (adapted from Haggett, Locational Analysis, Fig. 3–2, p. 63, based on A. Lösch [1954], p. 184).

angle y along with its sine will drop also (Fig. 8–54d).

A similar problem arises when a decision must be made on whether to build a canal or to go around. If water travel were free ($S = 0$), then overland costs dominate the route decision and they are minimized, perhaps to zero, by ships going completely around (Fig. 8–55, route D). If land route travel is costless but sea passage is expensive, then water transport is minimized to minimize total trip costs (Fig. 8–55, route A). Ordinarily, though, neither transport rate is zero so a compromise route between extremes (all water route vs. all land route) is reached on the basis of the ratio between the two transport rates (S/L) In the Americas at one time the cheapest way to get from New York to San Francisco was around Cape Horn, an all water route.

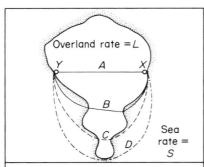

Route A: if S is costly but L is zero.
Route B: if S/L is less than in case A but not as low as in case C.
Route C: if S/L is less than in case B but not as low as in case D.
Route D: if S is zero but L is costly.

FIGURE 8–55

To build or not to build a canal or road to shorten a sea journey depends on the ratio of the sea transport rate S *to the overland rate* L, *part of which is assigned to the amortization of the overland construction costs (adapted from Haggett,* Locational Analysis, *Fig. 3–3, based on A. Lösch [1954], p. 184).*

Later, with the completion of the railroad, an overland route was best for passenger travel. Still later, with new digging technology, a canal was dug across Panama, similar to route C in Fig. 8–55. Had water passage been more expensive and digging cheaper at the turn of the century (i.e., had S/L been higher), the canal might have been put through Nicaragua, shortening the sea route. Similar analysis is used in planning whether to build a tunnel through a mountain or to find a path through which a road can be diverted.

Consider an island over which the maximum travel rate is one unit of distance per day, surrounded by the ocean on which travel at two units per day is possible. The questions are: how far can a person travel from A per time period (Fig 8–56)? And what does an isochrone map look like which describes accessibility of A to all land and sea points in the area? From the viewpoint of a person at A the most remote place in the area mapped is B, not because it is farthest away, but because it is hardest to reach.

William Bunge has evaluated the early circumnavigation of the earth in these terms. In isochronic or effort terms, polar visits such as those by Perry or Byrd are spectacular because such places were the most remote places in the world at the time of their first explorations. They resemble trips from A to B. A voyage such as Magellan's, on the other hand, is similar to path C in Fig. 8–56. Marco Polo's travels were more noteworthy. At the turn of the twentieth century, isochrones were widely spaced for sea travel from an origin in northwestern Europe. Inland the ease of travel depended on railroads, vehicles drawn by horses and oxen, the backs of horses or camels, or native porters. Travel speed varied from several hundred kilometers a day in steamships, to only a few kilometers a day through the rainforest (Fig. 8–57).

Within each city neighborhood a specified maximum rate of travel is possible. What does a rate of travel map look like and how is it constructed? A rate of travel or isotachic map is derived from isochrone maps. Each isochrone map is drawn for a specific point. By measuring rates of travel speed at different places on the isochrone map, we can assemble observations for each neighborhood and create the isotachic map. The procedure can be reversed and an isochrone map for any *one* point can be drawn

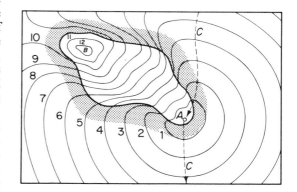

FIGURE 8–56

An isochronic map showing number of time units needed to reach each point by the fastest means, overland or over water, moving away from A. *In terms of accessibility, point B is most remote. A circumnavigation of the globe resembles a trip of type* C *(based on a suggestion by Bunge).*

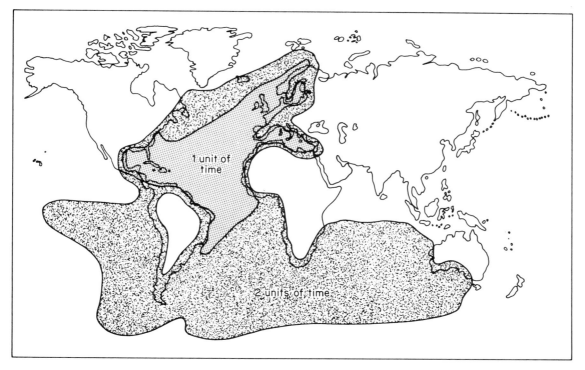

FIGURE 8–57

An isochrone map based on England about A.D. *1800. Sailing ships depended on wind and currents. Overland travel depended on foot travel and horses or other beasts of burden. The most remote corners of the world were the poles and the interiors of the large land masses. They were remote, of course, only from England's viewpoint (suggested by Bunge).*

by integrating about the desired point. In fact, one isotachic or rate of travel map underlies an infinite number of isochrone maps of an area (Fig. 8–58).

In the city of Seattle, sample automobile runs were made from the downtown to the suburbs during rush hour so that an isochronic map of the city could be constructed. Another set of runs was made from the University of Washington to the city edge to develop another isochrone map. The two maps permitted traffic analysts to estimate the maximum rush hour rates of travel at different points in the city, and from these point estimates a rate of travel choropleth map was drawn. The rate of travel or isotachic map underlies each isochrone map. It reveals how irregular the transportation surface in the city of Seattle becomes at rush hour.

Trees

In network analysis, arrangements of connected routes having no complete loops (circuits) are called *trees*. A stream system is a perfect example of a least net effort transport system with a treelike structure. A tree's properties teach us principles about collection and distribution as they apply to lines which are required to serve points scattered over areas. The largest or highest order rivers, such as the Amazon and the Mississippi, considered as networks extending down to their smallest tributaries, are all parts of treelike or dendritic systems in which several regularities have been observed. For example, the number of streams of a given order varies inversely with stream order, that is, there are only a few streams of the highest order and progressively more in the

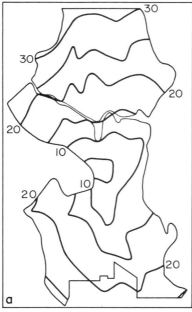

Peak hour travel time from central Seattle, in five-minute intervals.

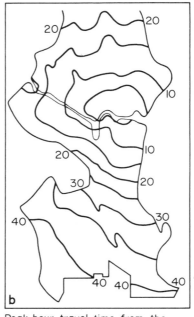

Peak hour travel time from the University of Washington, in five-minute intervals.

Steps in the construction of a rate of travel or isotachic map:

1. Construct one or more isochrone maps based on different points (e.g., *a* and *b*)
2. Estimate the average rate of travel for a large sample of points throughout the area
3. Construct a rate of travel choropleth map based on the set of point estimates (e.g., *c*).

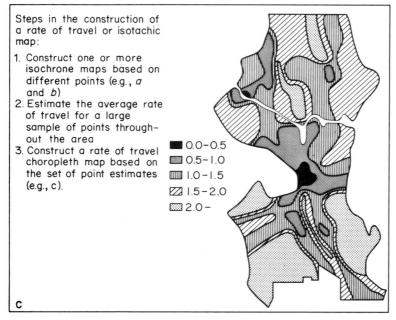

0.0–0.5
0.5–1.0
1.0–1.5
1.5–2.0
2.0–

FIGURE 8–58

A rate of travel or isotachic map (c) can be derived from one or more isochrone maps (a, b) of the same area. An isotachic map underlies each isochrone map (courtesy, Traffic Engineering Division, City of Seattle; adapted from Bunge, "Metacartography," Figs. 2–13, 2–17, and 2–18, pp. 55 and 58).

lower orders. Stream length varies directly with stream order. The largest streams are also the longest, and the low order streams are shorter in proportion to their order. Flow volume also varies with stream order as we might expect, as does the area drained by a stream and its tributaries. Each of these regularities observed in nature seems to be repeated in networks created by man to handle flows of people and materials.

As we saw earlier, urban water supply systems, sewer collection networks, gas delivery pipelines, and the systems of drainage tile that farmers bury in their fields are all trees in the network sense. In many recently glaciated portions of states around the Great Lakes in the American Midwest, land is flat and undissected. Stream development is so immature that in the spring and after heavy rains water lies on the ground, delaying planting or damaging crop roots. To prevent these harmful effects, a network of clay tile pipe, each pipe being about ten centimeters in diameter and thirty centimeters long, is buried at depths of from one to three meters. The pipes are laid end to end but not sealed. As excessive water percolates down to the water table, it is drained away from the field through the network of tiles to roadside ditches which are then linked to countywide drainage networks which ultimately empty into natural lakes and rivers (Fig. 8–59).

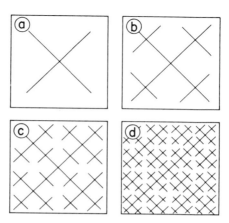

FIGURE 8–60

The optimum location of lines with respect to areas. Whether we are considering a system of roads, drainage tiles, stream systems, etc., the lines, of limited total length, try to pack an area as best they can (adapted from Haggett, "Network Models in Geography," Fig. 15–13).

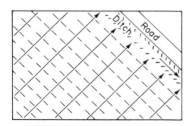

FIGURE 8–59

A network of drainage tiles is laid under a field to speed up removal of excess water in flat, poorly drained farmland. Nature has not had time to create an efficient drainage system of closely packed streams. Once the tiles are installed water is removed as quickly and efficiently as it would be eons in the future after nature would have had time to create a drainage system. Water flows into the ditch beside the road.

The tree-shaped tile network is trying to accomplish what nature has not yet had time to do. A tree is an attempt to fill an area as completely as possible with a set of connected lines, while keeping the total length as short as possible—like a hand spread to catch a ball or signaling for help, or a tree trying to capture ever more light and air. The older the system, whether it includes streams, oak trees, or roads, the greater the density of routes (Fig. 8–60). Compare old stream-dissected regions of southern Illinois with the bogs in northern Wisconsin, or the forked-stick appearance of young suburban trees with the mature trees in central city parks, or, indeed, downtown road and alley systems with their suburban counterparts (Fig. 8–61).

If a tree is created when a network expands under spatial constraints, what happens when the constraints are relaxed? Consider the problem of city expansion and the orderly provision of utilities (transportation systems, water, sewer, electricity, gas, etc.) to the newly built-up areas. With the introduction of the automobile, passenger movements are practically frictionless so the transport constraint has been relaxed.

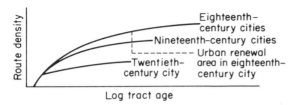

Log tract age

FIGURE 8-61

Route density per unit area rises as the route network in an urban tract ages. If an old tract is renewed, density of routes is reduced. Renewal in this sense can be compared to the pruning of a tree.

When everyone used public transportation, new urban areas grew up right next to older built-up areas. Today with the automobile, a new housing development can be miles away from any other built-up area and people can still get to work, to shops, and to school. It might appear, therefore, that public utilities are the last tie binding the urban area together. Yet with proper management and in suitable physical environments, even these last bonds can slip away. For example, water supplies and sewage disposal are possible on the same suburban lot if housing densities are low enough. In fact, complete city water and sewer systems are necessary only in high-density urban areas. This is not to say that all suburban developments should be low-density or that all low-density development can get by with onsite water supply and sewage disposal, but only that tree networks are a response to constraints and will not exist when constraints are absent. It is an instructive exercise to list the tree networks around us and to define the maximization (of scarce areal or volume resources) or minimization (of network length) problems that they respond to.

Land transportation systems grow like trees throughout successive periods in developing nations. Starting at coastal market towns where a territory links itself to the rest of the world, fingers of the incipient transportation tree creep inward, trying to grasp as much territory as they can with each route increment (Fig. 8–62). It is no surprise that rail nets resemble stream networks while they grow. Once established, closed circuits are created (unlike stream sys-

tems) because two-way traffic and trade among points in the region are desired. In certain strictly exploitative networks draining a region's outputs, circuits are redundant, as they are in pipelines. If a railroad is built for exploitation, however, locomotives must take empty cars *in* and then haul them *out*, so two-way traffic inevitably develops and once it is underway, circuits allow more flexible use of the network.

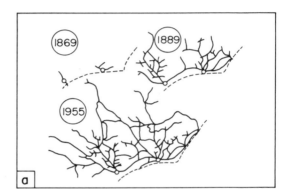

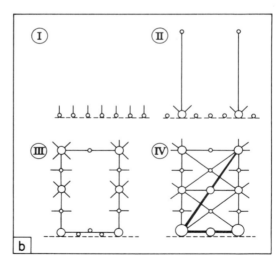

FIGURE 8-62

Treelike transport networks grow to serve developing areas: (a) stages in the growth of the railway network in southeastern Brazil (adapted from Haggett, Locational Analysis, Fig. 3–14); (b) four stages in the development of a transport network in an underdeveloped country (adapted from Taaffe et al.).

MOVEMENT, TRANSPORT SYSTEMS, AND TRANSPORT GAPS

New Transport Systems for New Transport Problems

Everyone knows that activities succeed partially as a function of relative location. Nowhere is this principle illustrated more dramatically than in instances where transportation and communications improvements have transformed isolated, self-sufficient, subsistence farmers into prosperous commercial producers. Although two-thirds of the world's people are economically bogged down as they were two thousand years ago, some have escaped into a new day. For the fortunate third, absolute space has been transformed into relative by space adjusting systems.

Wilfred Owen, a world renowned transportation-development authority, has put the issue succinctly. According to Owen, the farmers of the world know that immobility perpetuates poverty. Most of them earn their bread and pass their lives in isolation. They have little contact with the outside world. Yet widespread hunger in the world is a problem not only for the poor nations; it threatens everyone by undermining world peace. Transportation is a key to solving this problem. As the supply of new land disappears, the principal way to get more food is to provide water, fertilizer, improved seed, and other inputs needed to grow more on the land already in use. These inputs involve extensive transport from urban factory to farm, and then the surplus that is produced has to be moved from farm to city. As the population of the world increases, there will be a compelling need for more transport in both directions.

Where roads are impassable, where transport costs are high, and where marketing is uncertain, progress in agriculture will be limited. This is the way things are in much of the world today. In central India, over 30 percent of the rural villages have no roads at all. Another third have only the roughest cart tracks to connect them with the nearest town. Without a road, no one can sell anything at a market. Fertilizer never gets delivered, technicians rarely visit a village, children are unable to get to schools, the doctor never comes to see the sick, no one will show the villages how to dig a well.

Without a road there is little chance of getting poles and wires installed for electricity, so information and knowledge are unable to move. People communicate by carrying messages, and if new ideas move at all, they have to make their way through the sand and the mud. Information about markets and prices and new techniques may never reach the rural hinterland. The very concept of economic progress is foreign to those who live beyond information barriers created by impassable distances.

The possibilities of overcoming these problems through transport and communications have been demonstrated in an Indian village. The villagers of Wazipur in central India were asked what had happened after their new road was built. They said that they used to receive cash only once or twice a year, at harvest time. Now cash is entering the village daily because the new road permits vegetables to be grown and marketed every day. This shift in the cropping pattern from wheat to vegetables has tripled the returns per hectare. It has also made it possible to sell milk in neighboring towns at good prices, instead of disposing of it locally. The milk collector, who once shunned this village, is now able to carry twice as much milk on his bicycle over the improved road. What is perhaps most important, teachers can now reach the village. Examples from other parts of the developing world show similarly dramatic transformations.

In an Indian village or in the United States, everyone knows that activities succeed partially as a function of their relative location with respect to other places. The Interstate Highway System induces changes in the relative location of urban centers, and of relative location within urban centers. Activities in the cities are affected accordingly. Favored places prosper and relatively isolated centers fall behind. When the Interstate System is completed, the location of cities relative to each other will have changed again as they have changed in the past (Fig. 8–63). Areas tributary to cities will be twisted and modified according to their proximity to the Interstate route. Within metropolitan areas some sites will be endowed with superior access. Others will be shut out.

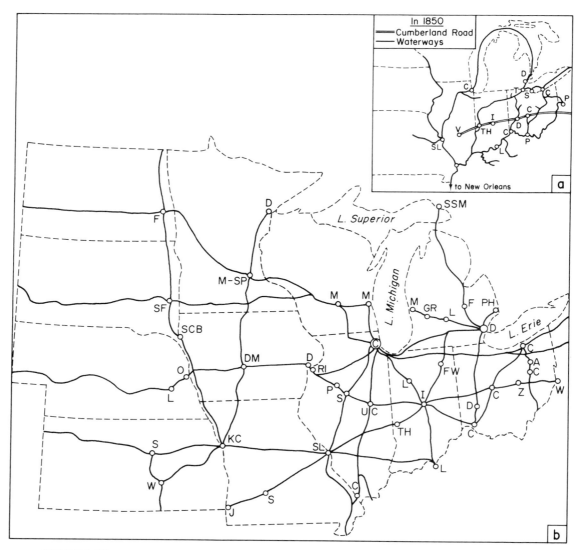

FIGURE 8–63

Three phases in Midwestern transport development: (a) *the Cumberland Road, canals and canalized rivers in 1850; and* (b) *Interstate Highways in 1965, completed or under construction.*

Let us consider the Midwestern portion of this network and evaluate its structure and impact as a space adjusting system. The Interstate System lies on top of the two previous federally assisted highway systems (Fig. 8–64) and performs a different task, tying together as it does the nation's major metropolitan areas and bypassing unimportant spots. Within metropolitan areas, radial and circumferential Interstate highways have their primary impact on commuter patterns, and guide residential, commercial, and industrial land development.

Previous highway systems catered primarily to rural areas and the needs of small urban centers. In the first decades of the twentieth century the nation's highway needs were those of small towns and rural America. Today, an ever-growing fraction of the American people live and work in the nation's two hundred largest metropolitan areas. The Interstate System is addressed to the high speed highway requirements of this group.

The Interstate System is rapidly reducing unit transport costs between and within metropolitan areas. Time space and money cost space convergences are occurring at every scale. When transportation costs drop low enough, places become more and more alike in the ease with which they can be reached. No place, therefore, retains a locational monopoly in the way that downtown centers once did. In the days of the streetcar, the central business district was easier to reach from all points than any other place in the city, and it could thereby monopolize a city's retailing business.

The Interstate highways undermine the locational advantages once enjoyed by places well served by public transportation. With the automobile there is no need for activities to cluster at a few transport nodes. Kilometer after kilometer of metropolitan land is served by the new highways. In the past, unit transport costs were successively reduced by the steamboat, canal, and railroad eras, and resources which were localized at places too remote for profitable exploitation gradually became useful when networks grew in extent and efficiency. In the twentieth century, a parallel development in urban transportation has meant that suburban land which once had value only for agriculture is now available for profitable exploitation by hordes of suburban developers and home builders. Everything has been brought closer together in relative space.

Today, when we assess the impact of the Interstate System on locational and human affairs we should think of our national area, not in terms of the traditional absolute space which is based on our rural and agricultural heritage, but in terms of the relative spaces of the future. An overland kilometer in 1840 meant

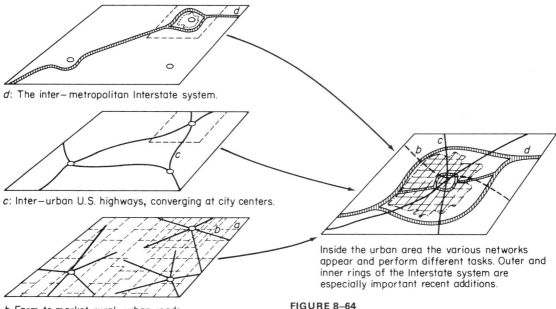

d: The inter−metropolitan Interstate system.

c: Inter−urban U.S. highways, converging at city centers.

b. Farm to market rural−urban roads.

a. Country roads, section roads, old intra−rural paths.

Inside the urban area the various networks appear and perform different tasks. Outer and inner rings of the Interstate system are especially important recent additions.

FIGURE 8–64

Strata in a highway network hierarchy and their appearance within a metropolitan area.

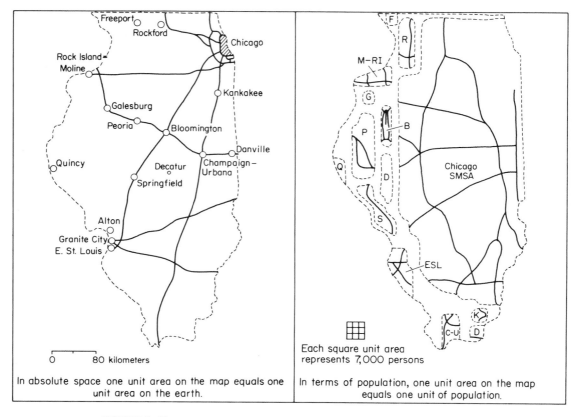

In absolute space one unit area on the map equals one unit area on the earth.

Each square unit area represents 7,000 persons

In terms of population, one unit area on the map equals one unit of population.

FIGURE 8–65

Illinois Interstate highways in geographic space and in relative (population) space.

something quite different from a kilometer in 1940. Even more profound is the difference between an urban kilometer today and a rural kilometer today. They differ so in time, money, reliability, convenience, and comfort that it hardly pays to retain the concept of linear kilometer.

Late twentieth-century America is metropolitan America and our new land transportation system is metropoltan in character. If we continue to use the traditional and outworn version of our nation's map, we will be prevented from devising the most sensible perspectives on the geographical problems which vex us. We must transform our maps of static distributions into areal cartograms, and begin to view spatial processes in terms of relative spaces. Consider, for example, Illinois' Interstate

highways mapped in conventional absolute space, and again in a relative space where map area appears proportional to population (Fig. 8–65). Roads are supplied in proportion to people's needs, not in proportion to hectares of land. Accordingly, the density of roads appears highest in densely settled areas. On a map where each unit area represents a given number of people the design of the Interstate System is easier to evaluate.

Nineteenth-century transportation systems moved the products and the resource input requirements of those days. They continue to do so today. An imposing share of our wood, coal, oil, iron ore, grain, meat, and other physical resource-tied commodities still move by boat and rail. But as the years pass a smaller and smaller fraction of the country's work force has been assigned to the production of food, fuel, fiber, clothing, and building materials, and a diminishing fraction of the consumer dollar is spent in these directions. Because the national economy continues to expand, proportional declines are offset so that absolute activity levels in primary (extractive) and secondary (goods processing) sectors are stable, if not advancing slowly.

We think too often in terms of past economic organizations and emphases instead of in terms of the patterns of the present and future. The principal expansions in the two decades after World War II were in business services and personal services—the tertiary sector of the economy. From now on, the emphasis will be on the quaternary or information and management sector. Growth will be greatest in education, research, corporate management, government regulation and control, recreation of all kinds, finance, insurance, publishing, property brokerage, and the telecommunications systems which permit such activities to flourish.

Collection, processing, and distribution of information will be the activities characteristic of the late twentieth century. Geographers will continue to describe and explain locational structure and spatial behavior, together with the communications and transportation systems that will be needed. The Interstate Highway System is but one cog of a larger machine including network television, direct distance dialing for little if any more cost than local telephone service, next-day mail service, high speed air travel, and so forth. Each network helps get a movement job done, and in the end the whole system is not only served but also transformed by the way the network is set up and how it works.

By responding in traditional ways to a need for more and better intraurban transport, we create a system that is in tune with requirements of a bygone day. A city is a human-physical system needing transportation, but network increments decisively influence the very activity systems they are designed to serve. For example, at any moment the need for additional highway capacity seems unlimited, yet when new roads are built to ease congestion they are immediately choked with traffic. Sometimes additional highway demand is elicited by the creation of new roads, leaving the system on balance worse off than it was before.

Some observers cynically suggest that one solution to traffic problems is to close all the highways built in the last twenty-five years. Perhaps. Such an action would force an explicit recognition that highway interaction needs are a function of activity location patterns and that highway changes induce activity location changes. The relationship is circular and explosive, and when improperly understood leads to public transport policies that are needlessly disruptive.

The most obvious example is the journey to work. If zoning practices and residential development policies put jobs at one location and homes at another, the journey to work is the costly outcome. Commuting may return therapeutic psychic benefits of privacy twice a day, away from family and boss, but aggregate social costs are far higher than a least net effort solution. Any policy or lack of policy which permits location of jobs and homes ever farther apart exacerbates already crushing rush hour jams, no matter how optimal the network design may be. This is one principle yet to be incorporated into the metropolitan plans of the Interstate System (Fig. 8–66). System designers, builders, and operators need performance criteria along with external checks so that when communications systems fail to do their jobs they can be scrapped, not expanded.

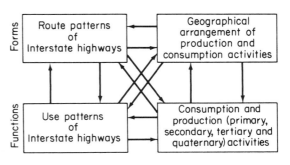

FIGURE 8–66

Metropolitan activity locations and new highway location: mutual cause and effect. One type of influence is exerted in the direction of each arrow.

Obsolete Systems, Transport Psychology, and Transport Gaps

Following the Industrial Revolution, the specialization which led to modernization caused two things to happen at the same time. First, the very specialization demanded by the modernization process forced people to come together into cities for the purpose of assembling specialized inputs, processing them, and exchanging their specialized outputs. Secondly, new technology was developed and applied to the problem of moving people and goods around the city, horizontally and vertically.

The specialization which causes growth and modernization begins before the full fruits of the process are shared by a large fraction of the society. People gather in growing cities to specialize in factories, for example, before prevailing technology is applied to make the city more livable personally or socially. In every period of American urban growth, density of urban population has been higher than existing technology would have required. But there has always been a lag between what was possible and what was done, partly because people have brought to the cities an agrarian heritage and have assigned higher priorities to production problems than to the creation of more habitable cities.

In intra-urban transportation, for example, cross country transport problems have always been solved ahead of parallel problems inside the city. The technology was available but not applied promptly to ease human effort and to help accomplish more useful work. The plan of every American city discloses the transport constraints of each era of city building and the lag in technological application. From the city center to the city edge, the transport technology applied in each successive era has had to cope with movement channels established in an earlier day. Old residential parts of cities are retained to minimize short run replacement outlays, but in keeping the residences we also hold on to obsolete transport systems.

The juxtaposition of new vehicles in old channels, which were inappropriate channels even when they were developed, is at the root of the problem. Accordingly, the best way to get around downtown is by walking. It was designed for walking in the horse and buggy era. The best way to get around in the suburbs is by car. Suburbs were designed and built in a different transport era. Disaster can accompany a switch in transportation technology, as testified by downtown traffic jams or a commuter whose car refuses to start in the morning. In the former case, cars are being driven in channels built for pedestrians; in the latter case, things are so far apart that a pedestrian is out of luck.

Transport lags are compounded at every spatial scale by transport gaps in the complete transport system which moves people, goods, and messages. According to Gabriel Bouladon, an overall transport hierarchy exists which complements and includes the network hierarchies and urban hierarchies already discussed. Gaps are identified first by asking: what is the psychological law relating time to distance for each individual? This law is clearly not uniform for people of different levels of economic development, and it also changes from era to era. Subconsciously most travelers accept a given amount of time as necessary to cover a given distance. They attach less importance to distances which are remote than to those which are close at hand. They also accept a relatively longer period of time to cover short distances than long ones. For example, we might willingly take six minutes to walk about five hundred meters but would not agree to devote an hour to covering five thousand meters. Instead, we would look for some other means of transportation which would allow us to cover the five thousand meters in a quarter of an hour at the most. Even this speed would not satisfy us for a longer journey, for in the United States we would not devote twenty-five hours to travel a mere five hundred kilometers.

How can this subconscious law be translated into practical terms? According to town planners, people are indeed prepared to walk five hundred meters. But as the years pass this limit declines. For any mode of travel the same holds true. The distances and high speeds of last year are always too slow today (Fig. 8–67). But there is a limit. In a dynamic and unified theory of transportation, in which each system is put into its rightful place from the point of

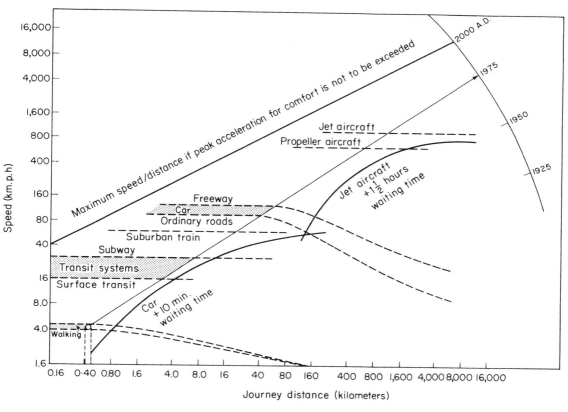

FIGURE 8–67

Demand for speed depends on distance traveled. Long distances mean greater speed. With time, all forms of transport theoretically move at higher speeds, as indicated by the pivot arrow. All systems falling above the pivot at a point in time are regarded as fast enough; those below the line are too slow. Dashed lines (horizontal) describe theoretical performance of each system. Continuous curves (cars; jet aircraft) show practical performance; performance lower than theoretical is due largely to the waiting times indicated (adapted from Bouladon, cover fig.).

view of requirements to be met, we can identify areas where there are competitive systems of transport, and also those where there are transport gaps which must be filled by new systems (Fig. 8–68).

The greatest transport demand is for very short distance transport. The top line of Fig. 8–68 represents optimum use as defined by maximum satisfaction to users. The line below represents a point at which half of the users would declare themselves satisfied, and thus it marks a lower limit for reasonable use. In each of the first five areas into which the figure

is divided there should be an optimum means of transportation. In practice, such optima are available in only three areas, in which pedestrian, car, and air transportation dominate. In fact, these three methods dominate the whole hierarchy of transport. Between these three regions many other modes of transport are currently in use, but these give less than optimal satisfaction. There are two significant gaps, one in the second area for trips from about 0.5 to 5.0 kilometers and the other for trips of from about 50 to 500 kilometers. The time and speed scales on the abscissa of Fig. 8–68 are complementary.

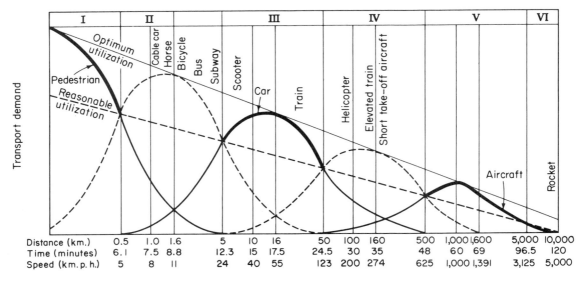

FIGURE 8–68

Transport gaps. When demand for transport (vertical axis) is plotted against the speed and optimum range of existing transport systems, we see that the transport range has three areas (I, III, and V) which are well taken care of by pedestrian, car, and air transport. Major gaps occur in areas II and IV (adapted from Bouladon, Fig. 1).

For any journey it is possible to establish the optimum speed for a car. The greater the speed, the greater must be the separation distance maintained between successive vehicles. As the separation increases rapidly between fast cars, the flow actually decreases. Conversely, the flow is also decreased at too low a speed.

How can the gaps best be filled? There is no doubt that supersonic aircraft satisfy a need for journeys longer than 1100 kilometers. With three times the speed of present jets, they provide six times the range of action, so that in practice all the great world links can be flown satisfactorily. The need for such high speed is real, even though it pertains to only a small number of people who regularly make transcontinental or transoceanic journeys.

Of the two major gaps, it would be more worthwhile to fill first the gap between the car and the airplane. The train might have satisfied this requirement, but for a variety of reasons it was not understood until too late that the train's real value in passenger movements is over distances of from 160 to 500 kilometers at speeds

between 250 and 360 kilometers per hour. Instead, commuter railroads have been struggling to compete with the automobile, offering comparable overall average speed but much less flexibility, or else with the airlines for long-distance travel. In both areas the railroad does poorly. The only country which seems to have moved in the right direction is Japan, with its 250 kilometer-per-hour Tokaido Line. Aircraft firms are working on convertible aircraft capable of 500 kilometer-per-hour speeds between the hearts of cities. In other systems, the train rides on air but follows a track. In all systems a speed of 400 kilometers per hour is the aim.

But what is being done about the third transport gap between pedestrians and car users, the area of interest to by far the greatest number of users and, in particular, town dwellers? If we do not adjust transportation to housing, people, and cities of the present and future, clearly we will be forced by default to adjust society's spatial organization and spatial behavior patterns to prevailing transport modes and transport networks.

Suggestions for Further Reading

BLACK, W. R., and F. E. HORTON. *A Bibliography of Selected Research on Networks and Urban Transportation Relevant to Current Transportation Geography Research.* Evanston, Ill.: Northwestern University, Research Report No. 28, 1968.

CURRY, L. "Climatic Change as a Random Series," *Annals of the Association of American Geographers,* LII (1962), 21–31. Reprinted in B. J. L. Berry and D. F. Marble, eds., *Spatial Analysis: A Reader in Statistical Geography.* Englewood Cliffs, N. J.: Prentice-Hall, Inc., 1968, pp. 184–194.

GARRISON, W. L. "Connectivity of the Interstate Highway System," *Papers and Proceedings of the Regional Science Association,* VI (1960), 121–37.

GAUTHIER, H. L. "Transportation and the Growth of the São Paulo Economy," *Journal of Regional Science,* VIII (1968), 77–94.

HAGGETT, PETER. "Network Models in Geography," in *Models in Geography,* eds. R. J. Chorley and Peter Haggett. London: Methuen & Co., Ltd., 1967, pp. 609–68.

HORTON, F., ed. *Geographic Studies of Urban Transportation and Network Analysis.* Evanston, Ill.: Northwestern University Studies in Geography, No. 16, 1968.

KANSKY, K. *The Structure of Transportation Networks.* University of Chicago, Department of Geography Research Paper No. 84 (1963).

MEIER, RICHARD L. *A Communications Theory of Urban Growth.* Cambridge, Mass.: The M.I.T. Press, 1962.

THOMPSON, WILBUR R. "Traffic Congestion: Price Rationing and Capital Planning" and "Interactions among Problems: The Problems of 'Solutions,'" in *A Preface to Urban Economics.* Baltimore: Johns Hopkins University Press, 1965, pp. 333–83.

ULLMAN, E. L. "The Nature of Cities Reconsidered," *Papers and Proceedings of the Regional Science Association,* IX (1962), 7–23.

Works Cited or Mentioned

AKERS, S. B., JR. "The Use of the Wye-Delta Transformations in Network Simplification," *Operations Research,* VIII (1960), 311–23.

BOULADON, G. "The Transport Gaps," *Ekistics,* XXV (1968), 6–10.

BUNGE, W. "Metacartography," in *Theoretical Geography,* Chap. 2. Lund: C. W. K. Gleerup, 1966, pp. 39–72.

FLAGLE, C. D. "Queueing Theory," in *Operations Research and Systems Engineering,* Chap. 14, eds. C. Flagle, W. Huggins, and R. Roy. Baltimore: Johns Hopkins University Press, 1960, pp. 400–24.

HAGGETT, P. "Networks," "Nodes," and "Hierarchies," in *Locational Analysis in Human Geography,* Chaps. 3–5. London: Edward Arnold, Publishers, Ltd., 1965, pp. 61–152.

———, and R. J. CHORLEY. *Network Analysis in Geography.* London: Edward Arnold, 1969, Chap. 15.

LÖSCH, A. *The Economics of Location,* trans. William H. Woglom with the assistance of W. F. Stolper. New Haven: Yale University Press, 1954.

———. "Regulation from Without," in *The Economics of Location,* Chap. 20. New York: John Wiley & Sons, Inc., 1967, pp. 315–59.

NYSTUEN, J. D., and M. DACEY. "A Graph Theory Interpretation of Nodal Regions," *Papers and Proceedings of the Regional Science Association,* VII (1961), 29–42. Reprinted in B. J. L. Berry and D. F. Marble, eds., *Spatial Analysis: A Reader in Statistical Geography.* Englewood Cliffs, N. J.: Prentice-Hall, Inc., 1968, pp. 407–18.

OWEN, W. *The Metropolitan Transportation Problem.* Washington, D.C.: The Brookings Institution, 1956.

———. "Transportation, Communication and the Future," *Ekistics,* XXV (1968), 3–5.

TAAFFE, E. J., R. L. MORRILL, and P. R. GOULD. "Transport Expansion in Underdeveloped Countries: A Comparative Analysis," *Geographical Review,* LIII (1963), 503–29.

CHAPTER 9

LOCATING HUMAN ACTIVITIES

A place for everything . . .

THE LOCATION PROBLEM AND
THE PUBLIC INTEREST

How do public and private decision-makers decide where to locate something, like a mailbox, a tavern, a library, a power plant, a home? What criteria are important? What kinds of distribution patterns are created when location decisions are made rationally and deliberately? What do the location patterns look like in the long run if decision-makers are poorly informed or act haphazardly? In the geographical analysis of human activities we must describe and analyze decision-making processes in order to explain the areal distributions which are created. Sometimes decision-makers are trying to locate facilities so as to maximize a value, such as rent per hectare, family enjoyment, or net profits per year. In other location decisions, attention focuses on how to minimize something, perhaps unemployment, disease, transport expenses, or a firm's operating costs. In every instance the decision to put an activity in one place rather

than another makes a difference. Deciding what to do about location is often termed "the location problem."

The location problem has several varieties, depending on what happens to be permanently fixed and what can be moved around. If we want to plan an optimal locational arrangement of activities and roads, for example, it makes a difference whether the activities are fixed and the roads movable, or whether the roads are permanently located and the activities are movable. In Chapter 8 we examined movements and the ways networks are arranged to permit movements among existing points along lines, in areas, and in volumes. Let us now consider the opposite case. In this chapter movement planes and networks are given. We want to know how activities will be located when transportation facilities are already installed. This chapter describes how people arrange activities on the earth and what the distribution patterns which are created look like.

We may study the location of human activities

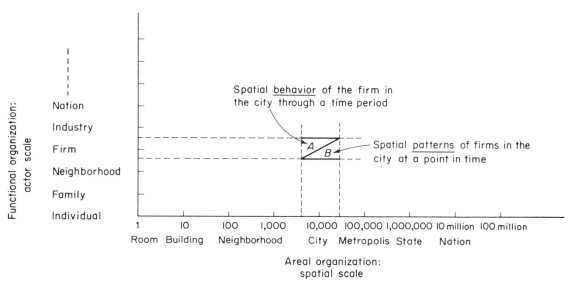

FIGURE 9–1

Approaches to studying the location of human activity. Zones along the actor scale and spatial scale define classes of problems. An investigation may emphasize spatial behavior, or spatial patterns, or both.

and emphasize the spatial or locational behavior of decision-making units such as individuals, firms, industries, or governments. Alternatively, we can examine the location patterns they create. Both kinds of investigations may be undertaken at any spatial scale: neighborhood, city, region, nation, or world (Fig. 9–1). Locational studies often have an economic flavor because most human activities involve the allocation and use of scarce resources. If something at one location is needed somewhere else, different locations imply spatial interaction costs which must be paid out of limited time, money, and effort budgets. The resources used in overcoming distance cannot be used for other purposes. Such opportunities foregone are the opportunity costs paid by society to overcome locational separation.

Costs, Revenues, and the Economics of the Firm

Many of the geographer's ideas about the location problem were first elaborated by the German economists Johann Heinrich von Thünen, Alfred Weber, and August Lösch. They were uneasy about the haste with which mainstream economists dismissed location questions. All economists readily acknowledged that places traded more with near neighbors than with far-off lands, but analysis never went much further.

Sciences are frameworks or habits of thought by which we make sense out of human events. Back in Chapter 1, we discussed the P-plane and C-field, and spoke of the networks of theory and law which each discipline invents to structure its conceptual interpretations of perceived reality. Despite the importance of location in resolving questions of what shall be produced, which inputs are to be used, and who will receive the proceeds from production, economic theory has seriously underemphasized the concept of location. Economists habitually think of places as points, not as areas.

Let us review some basic ideas about the economics of the firm to illustrate the importance of location in economic events. Consider the way in which a firm's average production cost per unit of output varies for different production

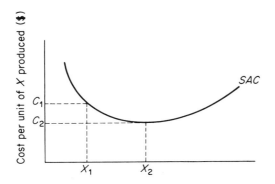

FIGURE 9-2

In the short run, when the plant cannot be moved or modified, the average cost of producing a unit of X *varies with the total level of output. At output level* X_2, *the short-run average cost* (SAC) *is a minimum for the plant.*

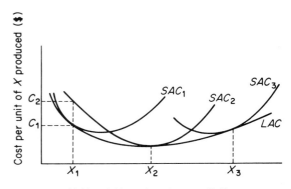

FIGURE 9-3

The long-run average cost curve (LAC) *describes the minimum average production cost for every level of output if any scale of plant can be built. Each short-run average cost curve* (SAC) *describes a different scale of plant.*

levels (Fig. 9–2). The short run average cost curve describes how costs can fluctuate for different output levels using *existing* plant and equipment. In the short run, investment in plant and equipment is fixed. We must wait until the long run to adjust the plant size and its furnishing. If X_1 is the current output level, the average production cost per unit produced is C_1. By expanding production to X_2, average cost drops to C_2 per unit.

Most plants have some optimum rate of output at which short-run average costs are lowest. At that optimum level, the cost of resource inputs per product output unit are a minimum. The optimum level of output varies for each different plant size. If a plant twice as large were erected, average costs might be much lower than C_2, but the optimum production level might be higher than in the smaller plant described in Fig. 9–2. We call such cost reductions "economies of scale." Big firms can cut costs by ordering inputs in large lots, workers can specialize further, and more efficient management techniques can be used that would be too cumbersome in a small place. Each scale of plant has a different short-run average cost curve and a corresponding optimum rate of output (Fig. 9–3). In the long run a firm can

build a plant of any desired scale. Nothing is fixed in the long run; everything is variable.

Assume that a firm can build any scale of plant and that SAC_1, SAC_2, and SAC_3 in Fig. 9–3 represent the short-run average cost curves for three of the possibilities. Each SAC curve summarizes the range of production possibilities once that specific plant scale is chosen. The long-run average cost curve (LAC) defines the rock-bottom minimum long-run average costs possible for each output level. The LAC curve is an envelope curve tangent to each of the SAC curves. In its long-run plan a firm may decide on a specific output level and then construct the scale of plant best suited for that output level. If it selects output level X_1, it should build a plant (SAC_1) for which average short- and long-run costs are C_1, the lowest possible for that output level. If the firm built a larger scale plant (say SAC_2) but held production down to X_1, average production costs per unit would rise to C_2.

But costs are only one side of the economics of the firm. Every firm also faces an average revenue curve called a demand curve. If a firm produces X_1 units (Fig. 9–4) at an average cost of AC_1 and sells each of them for a price of AR_1, the firm makes an average profit equal to

$AR_1 - AC_1$ per marketed unit. We usually assume that a firm adjusts its output so that per unit profit (the difference between average revenue and average cost) is maximized.

Do firms really behave this way? At a time and a place in earth space they try to. Firms try in the short run and in the long run to maximize the difference between average revenue and average cost. According to the model just presented, firms try to maximize by adjusting output in the short run and by adjusting output and scale of plant in the long run.

Now assuming that the revenue and cost situations behave as they do in Figs. 9–2, 9–3, and 9–4, the model prescribes correct behavior for the firm according to economists. What geographers quibble about is the assumption that average costs and average revenues behave in the same way at the same magnitudes at all places. Of course they do not. Some inputs are cheap in one place and expensive elsewhere. Prices are high at one place and low in others. Moreover, in the long run, not only is the plant scale variable, but plant location is variable too. The economist's general model is correct as far as it goes, but it does not go far enough to suit geographers. It excludes the location variable. For every firm, each different location implies a different set of revenues and cost curves. Part of the location problem as seen by the geographer is to describe cost and revenue patterns which prevail over areas and to evaluate a firm's behavior when it picks a location and operates at that spot.

Public Welfare Considerations about the Location Problem Today

We usually talk about firms but the argument applies with equal validity to a family or any other decision-making unit. Location decisions made by a firm or a family affect the public interest as well as the private interest of the decision-makers. If a factory decides to locate downtown rather than in an industrial suburb, it adds to rush hour traffic, slowing everyone else down. The factory may preempt downtown land which might have been better used by an activity that really needs to be located downtown. If the factory makes a noise or a smell it may disturb many more people in cramped downtown than out on a large suburban tract. If average costs, average revenues, and average profits are the same in both downtown and suburban sites, a firm is generally indifferent as to where to go. If a firm is indifferent in its preference for alternative locations, the choice should be made in the public interest because the public cannot afford to be indifferent about location patterns. The community at large suffers when unnecessary congestion and pollution are concentrated at a point (Fig. 9–5).

If a firm finds a downtown site somewhat preferable on a cost-revenue basis, whereas the community would find the downtown factory somewhat objectionable when social costs are deducted from social benefits, then a conflict arises between the firm's private interests and the community's public interests. Private activity locations are a matter of public interest and location or zoning laws are passed to coerce compliance with locational arrangements that are in the public interest. In some countries location decisions are made to maximize the public interest, and firms are reimbursed so that the difference between revenues and costs is acceptable.

Obviously a full range of possible philosophies about location exists. Sometimes the public

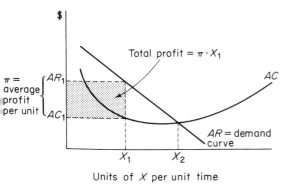

FIGURE 9–4

If a firm has a cost structure described by average cost curve AC, *and faces a demand described by average revenue curve* AR, *and if it produces at level* X_1, *average profit per unit is* π *and total profit for the operating period is* $(X_1 \cdot \pi)$. *What is the per unit profit if production is expanded to* X_2?

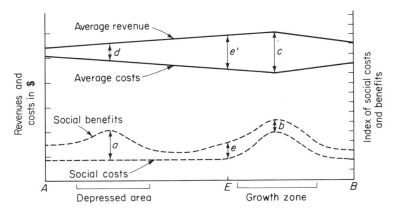

FIGURE 9–5

Hypothetical example of how private costs and benefits (measured on left vertical scale) and public or social costs and benefits (index on right scale) might vary at alternative factory sites along a traverse A–B. The public is interested in the difference between social benefits and social costs which a factory would create at each alternative site along the traverse. The firm typically wants to select a location maximizing the difference between average revenue and average cost. If the firm gets its choice, the society trades a social gain for a private benefit (a minus b for c minus d). A decision is ultimately made in the political arena—intentionally or by default—as to which outcome comes about. Perhaps a compromise solution is reached at E. Sometimes it is hard to say which is worth more: (a minus b) or (c minus d), or how much higher one should be to swing the decision one way or the other.

interest is completely submerged, as it is in many United States industrial suburbs. Such places were incorporated as municipalities and company towns to escape taxation and community controls on smoke, noise, health, and safety conditions, building codes, and waste disposal. This is *laissez faire*, or unfettered free-enterprise capitalism, with a vengeance. At the other extreme is state planning in which the ownership of investment resides in the public domain, and location choice and other decisions of the individual firm are completely dominated by wider scale interests of the community.

Location under free-enterprise capitalism is criticized by socialists because it inevitably produces inequalities among regions in economic opportunities. Industry increasingly agglomerates in growth regions, while the backwaters, declining, stranded, and ruined, are ignored by profit-minded capitalists who locate someplace else. If opportunities are created at one place when people who need these opportunities are

located somewhere else, one or both should be relocated so that the private *and* public welfare is advanced. Up to now government planning efforts toward remedying such situations have been too little, too late, and too ineffective.

In centrally planned economies, geographers argue that a state planner could compare many alternative locations with greater certainty than could a private capitalist. In such states a planning office has more variables under its control and can thereby make location decisions to maximize either total output or total public welfare. In the short run the two are not usually coincident. To produce one commodity, for example, means to divert resources away from other lines of production. Thus, to maximize production of *X* might mean cutting production of *Y* and *Z* to zero, which might seriously harm the public welfare. Such mistakes are made in totally planned states but this does not deny that locational arrangements are possible with state planning which would not be in the best private interests of capitalist entrepreneurs.

A state planning office might direct that a factory be built in a depressed area instead of a thriving growth center. But putting the factory in the depressed area might turn out to be a more costly solution than putting it elsewhere and then sharing higher benefits with residents of backward areas who will not or who psychologically cannot move. Just as alternative factory locations imply nonmoney benefits and costs, to move traditionally minded people away from ancestral homes carries high nonmonetary personal costs. Sometimes such costs are so high that to move people would literally kill them.

Subregional industrial complexes are a popular compromise in nations where employment location and residential location are planned in an effort to serve the public interest. The subregional industrial complex avoids the two costly extremes of overconcentration, which unregulated free enterprise capitalism leads to; and overdispersion which shortsighted welfare state planners might create by underemphasizing production goals.

Attention so far has focused on firms but the argument is general and can be applied to individuals and families just as well. If six million people want to live in Manhattan, should the rest of society let them? If hundreds of thousands choose to locate their homes in depressed coal mining regions in South Wales, central Pennsylvania, or eastern Kentucky, should the society allow them to pass on to future generations the high social costs that are thereby created? As we noted earlier, one of the great locational problems of our day is to contrive ways of measuring social costs and benefits which attach to alternative locational distributions of people and their activities. Semi-socialist states like the United Kingdom and semi-capitalist states like the United States and Canada urgently need practical solutions to this problem.

So why do activities locate where they do? In traditional agricultural society, the optimum social welfare location pattern for people and activities meant distributing families over arable land. As population grew, additional land was brought into cultivation, for example, by clearing European forests. Elsewhere, land was farmed more intensively—that is, on each unit of land inputs such as labor, fertilizer, and capital were increased as in irrigated riverine rice areas in China, Japan, and India.

In modern urbanized states the principal resource is not land, but human skill and its organization. Dispersal was necessary for agricultural societies; agglomeration was essential for many activities during the industrialization period; but somewhat less spatial agglomeration is needed today. Indeed, too much consolidation today is positively disastrous both from a private as well as from a public welfare viewpoint.

Today the location of the market is vitally important in making location decisions for industrial (factories, refineries), commercial (supermarkets, insurance offices), and public service facilities (schools, hospitals, fire stations). The market is important not because customers pull more than suppliers, but because both are at the same place—in the metropolis. This is especially true in tertiary and quaternary industries where input sources and the output markets are people, usually living in the same city.

In the urbanized commerical-industrial-service nations, all economic units, whether families, firms, or government decision and control agencies, are drawn together by: (1) a mutual interdependence of supplier and consumer roles; and (2) a rapid decline in the proportion of industrial activity of all kinds that uses localized material inputs not available in the metropolis. In other words, all the elements in the metropolis are interdependent on themselves and on other metropolises, and to an increasing degree are independent of rural areas. In dealing with other urban areas a metropolis may choose to be less independent than it could be in order to capitalize on its comparative advantages. The urban growth industries today, including finance, telecommunications, corporate and government intelligence and decision-making operations, education, insurance, property brokerage, and so forth, do not use much in the way of material resources. Trained human brains and their ability to think and communicate are the important resources today in the same way that coal, oil, iron ore, timber, and fertile soil were the cornerstones of growth in past eras. In fact, if forced to, cities could live off their own garbage by reprocessing water, sewage, trash, smoke, and other wastes. Although this is now technologically possible, it is not yet economically necessary. (It has, however, become ecologically mandatory. Chicago, for example, now ships daily a trainload of processed sewage to southern Illinois for use as liquid fertilizer. Increased crop yields more than pay the costs, and water free of inorganic and organic pollutants [including viruses] is returned to ground water supplies.)

Given the interdependencies of our current urbanized system, then, two questions arise: (1) how do decision-makers decide where to locate events? and (2) what kinds of distribution patterns are produced? Our discussion considers events of every kind, at different spatial scales, and at several economic development levels. Our approach is abstract. We feel that location laws and theories depend on general spatial properties, not on specific cases and case studies. The general approach is a more powerful approach because it can do so much so efficiently. In a sense it is easier too. We do not have to probe deeply into the details of an event to make sense out of an important fraction of its behavior. Additionally, man and society are at the center of our inquiry. We want to understand location patterns and human spatial behavior so that we can bring the spatial organization of man's world under his deliberate control so that social as well as individual welfare objectives can be realized.

LOCATION THEORY VERSUS LOCATION PRACTICE

Given the locations of all activities but one, what is the best location for that one? Consider the most general case in which different benefits accrue to every location in an area (Fig. 9–6a), and different costs exist at every point as well (Fig. 9–6b). If we compute for a set of sample points in the region the level of net benefits at each point, a third statistical surface can be

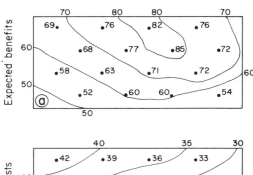

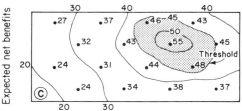

FIGURE 9–6

A net benefits surface (c) is estimated by sampling a gross benefits surface (a) and a costs surface (b). For each sample point some level of expected net benefits can be specified and the configuration of the net benefits surface estimated. If a firm needs net benefits of 45 or more to stay in business, points within the stippled area will probably be satisfactory on the average.

derived (Fig 9–6c). Assume now that the difference between benefits and costs must exceed some threshold value for a location to be acceptable to a decision-making unit (household, firm, etc.). If the threshold is 45 units, anticipated net benefits must be 45 or higher for a location to be acceptable. Any location within the shaded part of Fig. 9–6c could be chosen, and if expected costs and benefits prevail the outcome at the end of an operating period will be satisfactory. But in the real world nothing is certain—neither the gross benefits surface, nor the cost surface, nor the final *net benefits* surface.

Not only are things like benefits and costs inherently uncertain, but our information about events is incomplete as well. Uncertainty about final net benefits is due to the inherent uncertainty in the cost and benefit situations at each location. Moreover, even if things were "certain," our information might be so incomplete we would be forced to act as if they were uncertain. And finally, it might turn out that 45 units is not a realistic threshold. The threshold, after all, is the level of net benefits which the proprietor or investor feels must be earned per unit time for the enterprise to be operating within the bounds of acceptability. At the start of the operating period he might think that 45 units is reasonable as a threshold, but toward the end of the period, perhaps due to inflation or the discovery of more attractive alternative lines of business, he might be forced to raise the threshold to say 50 or 55.

Most decision-makers seek locations where risks are fairly low and net benefits are comfortably above acceptable levels. It is too hard to define the optimal location where risks are lowest and final net benefits will be highest, so why look? If a family looking for a house wanted the best buy, they would never make a selection. In the first place, they would never be able to specify what they mean by "best." Furthermore, in a big city there are more vacancies created each month than could be inspected. The same is true in locating an office,

a store, or a factory. So instead of hunting for an optimum location, we find a low risk location with an acceptable level of anticipated net benefits and then wait to see how things work out. If some firm, for example, locates too close to the margin of low risk operations, say at the 95 percent line, in Fig. 9–7, and he needs 45 to stay in business, a rise in costs or a decline in benefits can wipe him out, whereas another firm, located within the 99 percent zone, may weather the storm and survive a business recession or competitive threat, or an emergency which raises costs above normal levels. Frequently a sound location strategy consists in identifying the firms in the same industry which have been most successful and then imitating what they have done.

To summarize, net benefits per unit time must meet some threshold for a decision-making unit to survive. The exact threshold level of net benefits that will be needed may not be perfectly known ahead of time in this uncertain world. At the start of a period the decision-maker may think he will need, say, 45, and at the end he may well be happy with 43 or may find that 45 was too low.

The net benefits reaped at the end of the operating period are controlled by gross benefits and costs. These also are subject to ups and downs according to certain probabilities. If you buy a house at a certain location you might get a flooded basement (an unexpected cost), but then you might get exceptionally nice neighbors (a distinct, but perhaps remote extra benefit). We make decisions based on what we think is going to happen. Nature, our competitors, and customers decide what will happen. At the end of a period, families and firms decide if they have made satisfactory location decisions at the beginning of the period. Sometimes they are happy with the way things turned out. Sometimes they are disappointed. If they are happy, they may stay put and try the same thing during the following time period at the same location. If they are disappointed, one of two things happens. Either they give up and move, and start over again someplace else, or else they look ahead to the next time period, compare expected benefits and expected costs, define

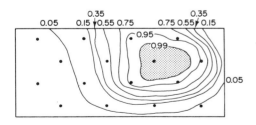

FIGURE 9–7

Probability at each location in a region that a factory will earn at least 45 in net benefits per unit time. Even if the factory locates in the zone where benefits of 45 are virtually assured (shaded area), one time out of a hundred they will fall below 45.

some threshold level of acceptability, and then wait to see what happens. If at any location final benefits less costs are the maximum possible anywhere, the locational choice is called an optimum. Yet we can see that even maximum net benefits might still fall below threshold net benefit levels if the threshold is high enough. In such a case, the activity is simply not economically viable.

Final net benefits at the end of the first time period might equal or exceed acceptable levels, and at the end of the next period they may fall short for any one of three reasons or for a combination of reasons: benefits may have declined, costs may have risen, or the net benefit threshold may have been revised upward. In any case, the zone of feasible locations will shrink, and firms, families, or other units on the margin of the zone will have to go out of business or relocate.

The foregoing is the general argument to be elaborated in different ways in this chapter. Some students feel more at home with specific examples like the steel industry, a supermarket, or a hula-hoop factory. Examples are helpful in elaborating an abstract case or argument, but geographical ideas are more general and more useful than one or several case studies might suggest. That is why we emphasize concepts and spatial dimensions. By doing so we are able to organize our experience in earth space in terms of constructs, concepts, laws, and theory for which we have the vocabulary and the ideas. Constructs are handles for experience. Experiences for which we have no constructs do not usually exist for us. In geography, as in other social-behavioral sciences, relationships among constructs seem significant after we identify them, name them, and discuss their importance in terms of some process or model.

The Location Problem at Each Stage of Economic Development: Some Definitions

Most productive activities require the assembly of inputs, input processing, and output distribution. If inputs are initially located at one set of places and outputs are to be sent elsewhere, the question arises: "Where should processing take place?" The same question arises whether we are locating a kindergarten, a post office, or a factory. As national economies pass through the modernization process, different inputs are emphasized during each development stage. Traditional societies and those undergoing early modernization stress *primary* (raw materials based) activities such as agriculture, forestry, fishing, and mining. Often the necessary inputs are available over wide areas, like farmland. Other resource inputs occur only sporadically, like soft-wood forests, mineral fuels, and metallic ores. They are abundant in some places and absent or scarce elsewhere. An activity needing a ubiquitous resource can locate anywhere. Activities requiring sporadically occurring inputs have less locational freedom.

As modernization advances in an economy, goods-producing activities such as manufacturing and construction become for a time the prominent growth sectors. Such *secondary* activities take raw material inputs, transform them by adding *form utility*, and then provide the output to consumer markets. Goods-producing activities require transportation industries, which increase the value of goods by moving them from one place to another, adding *place utility*. Factory mass production implies temporarily stored inventories of inputs and outputs to prevent supply gaps in the increasingly interdependent economy. Warehousing and storage services add *time utility* to goods. Processing points and storage points each must be located somewhere in economic space.

Tertiary industries are the basis of the mature economy where production is highly automated and where the fruits of production are widely shared. Tertiary industries include all commercial traders, but especially a group of essential business, personal service, and recreational enterprises. In prosperous societies, producers farm out many activities such as bookkeeping, advertising, business machine maintenance, cleaning, and other things which were taken care of "inside the store" in an earlier day. Individuals do the same thing by paying someone else for fixing the car, cutting their hair, and straightening their teeth. Tertiary outputs are often intangible, and the demand for them is usually income elastic—that is, as corporate and individual incomes rise by a certain percent, tertiary consumption rises by an even larger percent. Tertiary activities usually locate close to their clients.

Quaternary activities comprise the fourth set of economic activities which must be located by the modern society. Quaternary enterprises specialize in the assembly, processing, and transmission of information and the control of other enterprises. Sometimes entire firms and industries are devoted to quaternary pursuits, yet every firm and industry in the primary, secondary, and tertiary category devotes a substantial fraction of its activity and its personnel to handling quaternary functions.

It is helpful to classify occupations separately from the firms and industries in which they are carried on. Within a given firm, we can point to individual departments and employees and designate their separate functions as primary, secondary, tertiary, or quaternary. A university, for example, is definitely a quaternary organization, yet all four types of occupations are represented. The professor and the research worker are clearly quaternary by occupation along with the administrative structure including deans and various executive officers. Tertiary or service personnel do the typing, run the heating plant, mow the grass, and perform a wide range of services for the classrooms, offices, dorms, and other places on campus. Secondary occupations are relatively rare although colleges of science, for example, have elaborate shops where special experimental apparatus is constructed. If the university operates a dairy or a university press, the production personnel are secondary by occupation. In colleges of agriculture, experimental plots of grain are grown, orchards are maintained, and other agricultural pursuits are carried on. Students and maintenance crews who tend these activities are engaging in primary occupations whatever the scientific objective of the activity may be.

As a society modernizes, the fraction of the gross national production which originates in each industrial sector (primary, secondary, tertiary, quaternary) undergoes steady adjustment. The primary sector's share of production drops; the secondary sector's share expands then stabilizes; the tertiary expands rapidly for a while after the secondary stabilizes;

finally, the quaternary share expands explosively. The mix of occupations also undergoes adjustment as modernization proceeds. In the traditional society almost everyone works on the land or sea—in agriculture, forest gathering, or fishing. In advanced economies, many firms and industries are almost exclusively comprised of quaternary occupations—collecting, processing, and disseminating information and control. In nonquaternary industries, an ever larger share of the jobs are in management, research, advertising, public relations, and other quaternary operations (Fig. 9–8). The quaternary sector includes firms and organizations such as the following:

Census bureaus
Research institutes
Universities
Colleges
Think tanks
Insurance companies
Securities brokerage firms
Real estate brokerages
Movie studios
Communication Satellite Corporation (COMSAT)
U.S. Senate
U.S. House of Representatives
Executive Office of the President
All the independent regulatory agencies of the U.S. Government
The cabinet departments
State government offices
Local governments
Libraries
Planning offices
Advertising agencies
AT & T
Radio and TV networks and stations
Western Union
Associated Press
Newspaper companies
Magazine publishing companies
Accounting firms
Churches and synagogues
Corporate headquarters
The Joint Chiefs of Staff
Banks
Federal Reserve System
School boards
Book publishers
Stock, bond, and commodity exchanges
Consulting firms
Data processing centers

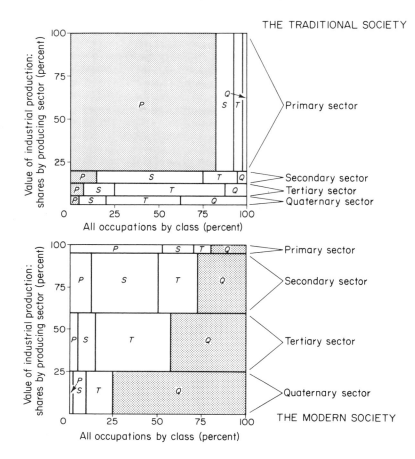

FIGURE 9–8

In the Traditional Society, most industrial production originates in the primary sector and most people work at primary occupations. In the Modern Society, the value of industrial production is distributed over all sectors, and the primary sector is least important. Among the four classes of occupations, the quaternary ultimately dominates in proportion and in importance.

The list is staggering. It makes us wonder who is minding the store. As it turns out, an impressive fraction of the quaternary workers are trying to figure out ways either to get rid of the store altogether, or to develop a machine to monitor it as it runs. In the post-industrial state, a continuing effort is made to render people redundant by task simplification and by the replacement of people with accurate and reliable machines.

What are the locational requirements of activities in the quaternary sector, and quaternary occupations in the other sectors? In the near future, information and control activities may locate almost anywhere and continue to function optimally. That is because they are concerned with information and information can be moved conveniently, cheaply, and instantaneously. If the President of the United States can communicate with much of the world from Air Force One (his private jet liner), a secretary certainly can take dictation five miles from her boss. Unless business is far more than business (which it is), there really is no compelling reason why many office employees and their superiors could not work in their respective homes in different states. Picture phones and facsimile transmission of letters and other documents could make the journey to work more of a chore than it would be worth in economic and social benefits. For the present, we cling to the old ways and traditional locational arrangements through force of habit ("Miss Jones, step into my office."), not because they are indispensable to satisfactory performance. But before we probe modern and possible future departures from traditional practice, let us examine some simpler cases and fix in our minds some general locational principles.

Point Sellers and Linear Markets

Consider the effect of distance on the location of a trade boundary between competitors. Assume that buyers are uniformly distributed along a line of length L (Fig. 9–9). The line may be thought of as the main street in a town or a transcontinental railroad. At distances a and b respectively from the ends of the line are stores A and B. Each buyer transports his purchases home at a cost of C per unit distance.

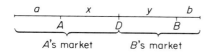

FIGURE 9–9

A linear market of length L, *served by two sellers at* A *and* B.

Assume also that assembly and processing costs are equal to zero everywhere; and that consumption along the line equals one per unit time at each unit of the line length. Demand is therefore perfectly inelastic—that is, completely unresponsive to selling prices. No buyer has any preference for either seller except on the basis of price and transportation costs. In general, of course, this is not true of most commodities and services, but we can assume here that differences are subsumed in transport costs.

If A and B charge identical prices, consumers are indifferent between the two sellers at the point of market division D along line L. In other words, for a customer at the market division point, the price charged at A plus the transport costs over distance x equals the price at B plus the transport cost over distance y: $P_A + Cx = P_B + Cy$. If the two prices are equal on both sides: $x = y$. The two sellers will split the market that lies between them, and each will monopolize one of the end portions, a or b, of the linear market.

Now let us look at a graphical solution to the

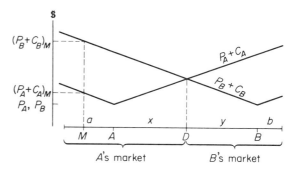

FIGURE 9–10

Two sellers divide the market between them when both charge identical prices, when demand is distributed evenly along a linear market, and when each buyer must transport his purchase home at a cost which increases linearly with distance.

same problem (Fig. 9–10). Consumers at A and B pay no transportation charges. They pay only the store price, P_A or P_B. At other points along line L delivered price is higher due to transport charges. Customers buy at the store closest to them. A customer at M, for instance, shops at A and pays the price at A plus the transport cost from A, or $(P_A + C_A)_M$, which is less than $(P_B + C_B)_M$, what he would have had to pay at his location if he shopped at B. The two delivered price curves $P_A + C_A$ and $P_B + C_B$ intersect where $x = y$. The two competitors divide the market between them and this solution prevails when the locations of seller A and seller B are fixed and $P_A = P_B$.

Consider the situation which develops if the store prices are set independently. P_A and P_B therefore are different (Fig. 9–11). At the market division point *delivered* prices from the two stores are still equal, that is: $P_A + Cx = P_B + Cy$. Customers at the division point therefore are indifferent about which store to shop at. But where is the division point? What is x and what is y? We know that the linear market is made up of four segments: $a + x + y + b = L$. When $P_A = P_B$ the length of x equals half of the distance between A and B (Fig. 9–10). When prices are unequal the length of x equals half of the distance between A and B plus some displacement that depends on the difference between the two prices and C, the transportation

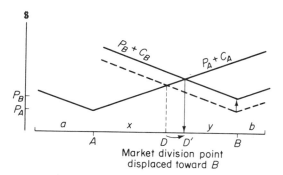

FIGURE 9–11

How two sellers split a linear market when store locations are fixed and seller B raises his price so that store prices differ.

rate (Fig. 9–11). If P_B exceeds P_A, then x increases as some of B's customers near the old trade division point swing over to store A (Fig. 9–11). If P_A is greater than P_B, the division point moves closer to A, A's market shrinks, and y increases.

Note in Fig. 9–11 that the amount of shift in the market division point depends on the difference in P_A and P_B *and* the size of C, the transport rate measuring the friction of distance. The higher the transport rate, the less will be the boundary point shift for a given difference in price. Alternatively, improvements in transportation that cut the transport rate C emphasize differences in prices from place to place, thereby promoting spatial competition and reduced prices. Poor roads or other transportation difficulties in urban or rural areas usually mean higher prices and lower public welfare. When it is expensive for customers and goods to move, isolated sellers can exploit inelastic demand as monopolists, charging very high prices and losing very few customers.

Neighborhood pharmacists who sold prescription drugs before the age of the high speed automobile provide a good case in point. In those days, once your doctor called your neighborhood druggist with a prescription, it was virtually impossible to transfer the prescription to another pharmacist when you discovered that the price charged on delivery seemed excessive, which it often did. Today, intra-city movement

costs are lower for most people and large well advertised cut rate drugstores have made massive inroads on the neighborhood pharmacies, who continue to survive by catering to the poor, the elderly, and the immobile in highly localized markets, or to customers trapped by their ignorance of lower cost alternatives.

Price Adjustments and Profit Levels

If we assume that business costs are zero along our linear market, then profits π (pronounced: pi) for a firm equal sales—that is, price times quantity: $P \times Q$. For the stores A and B: $\pi_A = P_A Q_A$ and $\pi_B = P_B Q_B$. In terms of our linear market, since there is one unit sold per unit length of L: $\pi_A = P_A(a + x)$ and $\pi_B = P_B(b + y)$. In graphical terms the shaded portion of Fig. 9–12a represents total profits and their division between seller A and B when prices are equal, perhaps because of some artificial condition in the market such as collusion or retail price maintenance. If B

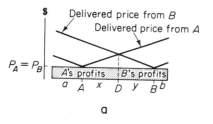

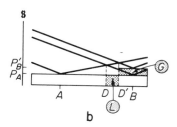

FIGURE 9–12

Profit levels and their division before and after seller B raises his price above that charged by seller A. After B's price is raised, the market division point moves toward B and B's profits are cut by stippled amount L which is added to A's profits. The higher price enables seller B to earn a higher profit per unit sold and B's profits rise by stippled amount G. B's net gain in profit is G minus L.

decides to increase his profits by raising his price (Fig. 9–12*b*), he has to decide if the gain from a higher price will yield more than the loss due to the diminished extent of his market penetration. In Fig. 9–12*b*, seller *B* loses part of his profits by raising his price to P'_B and thereby losing part of his market to *A* whose price remains P_A. But *B* earns more per sale in the market he retains and makes a big gain in his total profits. In the example illustrated, *B*'s net gain (*G* minus *L*) is substantial. A point is soon reached, however, where unilateral price increases by *B* will result in net losses rather than net gains.

If it is in *B*'s interest to raise his price by a small amount, why should *A* stand pat? Why should they not act together as duopolists and both raise their prices to very high levels and amicably exploit their customers? Remember we assumed *inelastic* demand—that is, the amount of demand does not diminish or expand with ups and downs of price. Consider Fig. 9–13, the situation which followed a doubling of prices by both *A* and *B*, from P_A and P_B to P'_A to P'_B. The market division point remains unchanged: $x = y$ and profits are twice as large for each firm. Price increases may not be agreed upon in advance but may proceed upward by alternate steps, each seller in turn increasing his price so that profits reach a maximum, given his competitor's price level, yet not so high that all his business will be driven away to his competitor.

Sometimes price escalations involve tacit understandings that become formalized when professional associations circulate lists of "ethical" price levels below which association members must not go or they will be charged with unprofessional or unethical (i.e., competitive) practices. Familiar examples include minimum prescription charges, minimum attorney's fees for handling an estate probate, minimum brokerage commissions for selling a house, or indeed the normal or customary consulting fees collected by professors who are not busy writing textbooks. If all members of the formal or informal price maintenance association cooperate, the group can act as a single monopolist. If they are cautious the law leaves them alone. Unless arrangements are

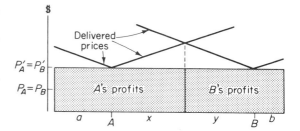

FIGURE 9–13

Profit levels and market division after price doubling by both sellers A and B.

highly formalized, however, tacit understandings between competitors are fragile. If *A* and *B* are cooperating to maintain a high price position and *B* runs short of cash, he can always cut his price a little and perhaps increase profits a little at *A*'s expense. *A* can retaliate and a downward spiral of prices may result until all excess profits are eliminated. This point is reached when further cuts begin to hurt both parties. So fragile are these kinds of arrangements that unless a group can impose sanctions such as licensing of members, maverick members sometimes cannot be kept in line.

Market Shares When Sellers Can Move

In the context of our linear market case consider three modifications or alternative situations: If *A*'s location is fixed, and *B*'s is not, where will *B* locate? *B* moves to the left so as to make *b* as large as possible by coming as close to *A* as he can yet still staying to the right of *A*. In such a location *B* can dominate the largest part of the market (Fig. 9–14, case 2).

If both *A* and *B* can relocate with ease—say they are selling ice cream bars on a beach—where will they end up? Assuming they sell an identical product at the same price and that customers always buy from the closest vendor, both *A* and *B* eventually locate beside one another, each serving half the market, $a = b$, and $P_A = P_B$ (Fig. 9–14, case 3). Prices necessarily will be equal. If all customers are coming to the middle of the beach anyway, a slight price reduction by one seller will leave the other with no customers.

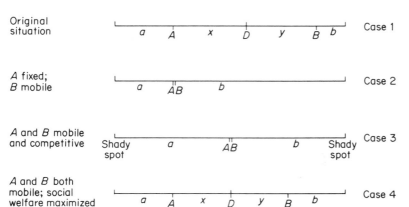

Original
situation — Case 1

A fixed;
B mobile — Case 2

A and B mobile
and competitive — Case 3

A and B both
mobile; social
welfare maximized — Case 4

FIGURE 9–14

Two sellers on a linear market. When one or both can move, where will they settle?

If a and b both can be adjusted, the competitors split the market but customers are put to considerable inconvenience because they all have to go to the middle of the beach. In fact, if there are shady spots at each end of the beach (Fig. 9–14, case 3), both A and B could get together and agree to relocate, one at each shady spot, and sell ice cream comfortably, with no decline in profits but with equal inconvenience to the customers. But if the Ministry of Beaches and Ice Cream Vendors is interested in maximizing sales and minimizing customer inconvenience, the wastefulness of such private profit-seeking competition is obvious. To minimize the social costs of customer movement to and from the vendors the following situation must obtain: $a = b = x = y$. A's sales equal B's sales in both the free location and the assigned location cases (Fig. 9–14, cases 3 and 4), but social costs created by customer inconvenience are much lower in the latter.

A similar phenomenon occurs in planned residential developments where subdividers and community planners impose zoning restrictions on commercial land to maximize the value of the land which is earmarked for business and commercial uses, rather than to minimize inconvenience and transport costs for shoppers by putting shopping centers in the most accessible spot. In Fig. 9–15 notice the location of the boundary between city and suburb. If a new shopping center is planned for site X, its trade area will extend inward toward the city to a

trade area division point W as it competes with existing shopping center C. The new center at X will monopolize all trade between W and Z (the edge of the built up area) because there is no intervening center. By putting the center at X, the commercial site will bring a maximum price within the suburb, yet it will cause grave inconvenience to residents between Y and Z. If the new shopping center were located so as to minimize traffic and maximize public welfare it would be put at Y.

Moreover, if the center is placed at Y instead of at X this would mean a longer life for the shopping center at C, a reduction in wasteful duplication of facilities at X, and less land devoted to commercial uses. If the shopping center at C is owned and operated by the developers of the new subdivision, the private interest and the public welfare solutions will

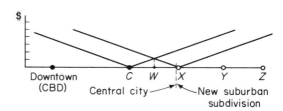

FIGURE 9–15

Locating a new shopping center along an arterial road radiating outward from the central business district (CBD) through a new suburban subdivision.

coincide because if the two shopping centers are too close together their mutual owner will suffer. Normally, however, the developer's profit interests are not consistent with the community's welfare and taxation aims.

This problem of the two shopping centers resembles the case of the linear market and the two sellers. If customers must add freight charges to the market price, if the amount consumers buy drops sharply with increases in the delivered price (i.e., if demand is elastic), and if both stores are owned by a single monopolist, he will put the stores at quartiles along the linear market which extends from the *CBD* to *Z* at the edge of the suburbs. This way he can maximize his sales by minimizing the movement of his customers. This is the same as a public welfare solution in Fig. 9–14, case 4. In Fig. 9–15 if the same company owns both shopping centers and can influence suburban zoning commissions, it may well put the second shopping center somewhat to the right of *Y* in anticipation of further suburban growth and maximum total sales. The main difference between two competitors and a two-store monopolist is that each competitor tries to occupy more than half of the market while the monopolist tries to maximize profits in both halves of the market.

Continuing the case where customers must add a freight bill to the store price and demand is elastic, if both competitors believe that there are no profits to be gained from invading a competitor's territory, they stay put. If one of two competitors along a linear market tries to encroach on the competitive region between the two sellers, once he moves inward from a quartile location he increases competition with his rival and loses part of his successful exploitation of his hinterland due to higher delivered costs to customers on the other end of his trade area. Neither store, therefore, has anything to gain by moving toward the center unless his competitor moves first. Since both feel this way neither may move toward the center. In such a case there really is nothing to be gained by invasion. This is the major consequence of elastic demand where the quantity purchased by a consumer depends on *delivered price*. Competitors repel each other.

SOME TYPICAL POINT AND LINE LOCATION PROBLEMS

Locating Mail Collection Boxes and Mail Storage Boxes

Every neighborhood in America is dotted with mailboxes. Some of the boxes are for mail collection. The others are painted an inconspicuous dark green and are for mail storage, the next to the final step in mail distribution. Collection and storage boxes are always located at streetside so that truck pick-up and delivery is convenient. The location of each set of boxes represents a solution to a point-line problem.

Consider first the number and location of collection boxes. Streets are lines and serve boxes at points. Each box serves a tributary area. Think of the constraints in this location problem. The Post Office wants to make it easy for people to mail letters (maximize the number of collection boxes) yet it also wants to pick up letters with as few drivers and truck kilometers as possible (minimize the number of boxes). Because the public ultimately picks up the tab for the Post Office anyway, a social optimum would exist if the sum of the costs to the Post Office plus other costs to the public is least (Fig. 9–16). The cost accountants at the Post Office know how much it costs to maintain an additional collection box, but they do not know precisely how to price the individual and the aggregate mailing effort that is borne by the public and which increases if the number of collection boxes is reduced. Thus, the number of collection boxes which will be installed is derived by trial and error methods until an apparently satisfactory solution is achieved.

The same kind of problem develops in specifying how many storage boxes will be needed. Again, the problem is settled according to the volume of mail received in each neighborhood and the amount the mail carrier is able to carry at one time. Additional storage boxes are one way mail carriers can avoid carrying extra-heavy mail loads over long distances, yet as the number of storage boxes rises, additional trucks, drivers, and stops are necessary to distribute bundles of sorted mail to the storage boxes.

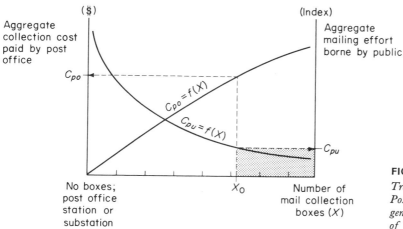

FIGURE 9–16

Trade-offs between costs borne by the Post Office (C_{po}) and costs borne by the general public directly (C_{pu}), as a function of the number of mail collection boxes in a city area.

After the question of the number of collection and storage boxes is settled, we have to decide where to put them. Mail generation does not equal mail receipt in each neighborhood. Commercial business areas send out more mail than they receive, whereas most residential neighborhoods receive more mail than they send out. Thus, it is easy to find a mail collection box in a business district. Among residential areas variation exists in the density of storage boxes—variation that is not due to differences in population densities.

Recently direct mail advertising firms have begun to analyze ZIP code areas in terms of their social, ethnic, and economic composition. Then, when they rent mailing lists to large publishing houses, book clubs, record clubs, and other advertisers, customer groups can be preselected. Areas offering good prospects for responding to a direct mail appeal will receive far more mail than other neighborhoods. Mail storage boxes will necessarily be located closer together in such neighborhoods than in other neighborhoods of different composition but of similar population density.

The mail collection problem faced by the Post Office then is (1) to estimate the threshold mail level needed to justify a collection box; (2) to establish the probable maximum range people will travel to mail a letter without complaining; (3) to estimate an approximate minimum number of collection boxes needed for an area; and (4) to install the collection boxes so that they are as close as possible to the neighborhoods served, while forming the shortest traveling salesman path for the mail pick-up truck. The assignment of collection boxes to a new neighborhood usually is made according to an iterative or trial-and-error procedure. Following a first approximation, some collection boxes are bulging at every pick-up, and others only have a few letters. Some additional collection boxes can be installed at nearby corners to take the pressure off the overflowing boxes, or perhaps two collection boxes, one for local mail and one for out-of-town, will be set up side by side. Seldom used boxes can be replaced with smaller boxes or eliminated altogether. Each adjustment in box size, box location, and pick-up route represents a closer approximation to an optimal strategy. Finally, an arrangement is settled upon and remains fixed until neighborhood change induces shifts in mail generation rates.

The location of storage boxes and the delivery of sorted mail ready for the letter carrier represents the same problem in reverse. Mail is at a point, the substation, and is delivered to storage boxes along a line. Each storage box is a point from which mail is delivered house to house over an area. Again the problem is to locate storage boxes as close as possible to the area served while keeping the delivery routes from substation to storage box as short as possible.

Locating Housing Units along City Streets
in Different Eras

Competing points can consume location on lines. We saw in Chapter 8 how French pioneers settled along river banks and arranged their farmland in long strips away from the river. The river was the scarce resource and strips narrowed as the number of competitors increased. Houses are arranged along city streets according to the same principle. They are brought together along the street because their occupants need access to other city areas, but at the same time they are pushed apart because people need privacy. If movement around the city is difficult, people will live at very high densities in order to retain access to jobs, stores, social life, entertainment, and other wants and needs available at other locations. But people, like many species of animals, seem to need a certain private territory in which to move or stress will develop. The need for space pushes people apart as other needs pull them together.

How tightly people are pulled together depends on where their needs are located and the ease of covering the intervening distances. Before 1890 people living on farms needed little that they could not supply themselves. In the city, the opposite prevailed. People depended on a specialized economy. Movement in the city was largely restricted to walking or riding slow and inconvenient horsecar lines. Difficult intra-urban travel had imposed a limit on the maximum diameter of cities and inchoate skyscraper and elevator technologies had not yet removed constraints on building height. Cities could not grow upward nor outward, so they accommodated continued population growth by squeezing ever more people into urban containers of fixed maximum dimensions. In Manhattan before 1920, for example, some neighborhoods had densities above 1,850 persons per residential hectare.

In the days when movement in the city was difficult, streets were narrow but city blocks were small so as to maximize the length of building frontage and the number of different buildings per square kilometer. Residential and commercial lots were very narrow and with blocks less than 100 meters long, a maximum number of activities could be located within easy walking distance of the greatest number of people. Wider lots would have meant fewer activities per block; longer blocks would have meant fewer front meters of street per square kilometer as well as fewer kilometers of street to carry pedestrian traffic. Elevated trains and subways expanded street capacities in major cities where the congestion was simply out of hand. Today what remains of the pre-1890 core in places like New York, Philadelphia, and Chicago is a legacy of the first era of American city building.

When locating houses along a city block, residents' movement requirements and transport constraints dictate the degree of crowding. In an analogous way, when spectators view a football game and occupy bleacher seats they also consume location along a line. Visual limits restrict the maximum size of a football stadium. If a stadium were twice as large as the biggest built now, the most distant spectators would see practically nothing. Compare the stadium situation with the seating arrangement in a living room when two or three people watch a football game on television. The game has been spread out (broadcast) over a wide area and crowding is thereby eliminated, unless of course the television set has only a twenty-centimeter screen, in which case the spectators must again crowd together. Normally, however, spectators can be spread out if the game is broadcast. During American urban development residents could spread out (1) when localized activities were dispersed throughout the city, or (2) when the movement constraint was eliminated by the streetcar and later the automobile.

Three revolutions in urban transportation occurred after 1890. Each transport era included two residential building cycles (Fig. 9–17). Each construction cycle created a ring of new and distinctive housing around the downtown core. Population density levels in each new residential ring depended on the urban transportation technology that prevailed during that construction cycle. Just as walking forced high densities in the city before 1890, electric streetcars permitted the city to spread out between 1890 and 1920. Movement ease controlled density levels (Fig. 9–18).

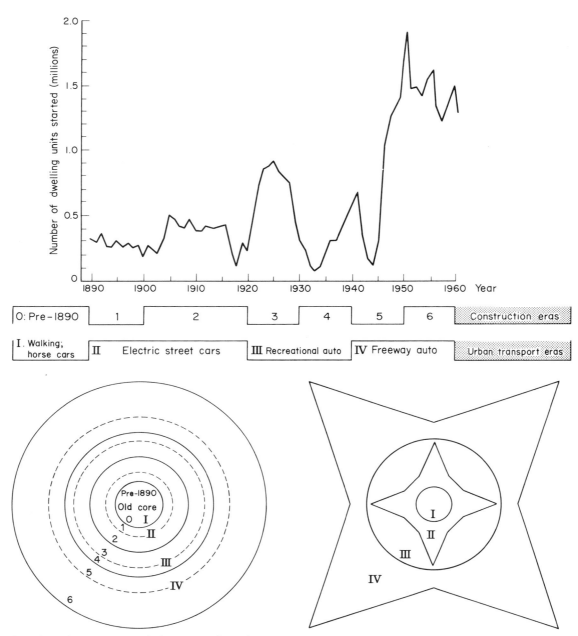

The size of each concentric increment depends on construction levels through time and housing densities.

City shape is modified by the predominant mode of transport during each construction era.

FIGURE 9–17

Building cycles, urban transport eras, and city shapes in the United States, pre-1890 to 1960 (reproduced from J. S. Adams, "Residential Structure of Midwestern Cities," Annals of the Association of American Geographers, *LX [1970], by permission of the publisher).*

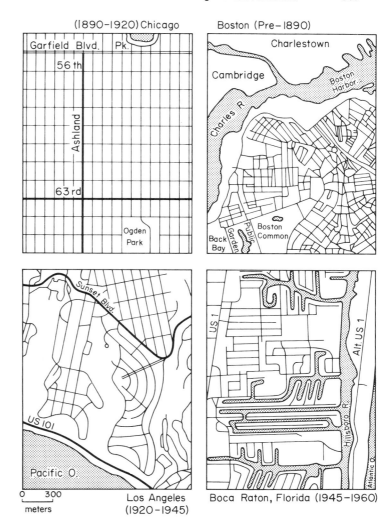

FIGURE 9–18

Street patterns reflecting four ages of American city building.

The automobile and motorbus permitted still lower densities for residential areas erected from 1920 to 1945. Since 1945 the freeway-automobile era, bringing with it exceptional ease of movement, encouraged urban sprawl into far-flung suburbs and exurban areas. Nowadays there is almost no limit to the length of new roads which can be built into suburban areas. Along such roads travel is virtually frictionless. Thus, competition among points such as houses or location along lines such as streets has been all but eliminated in the suburbs. The consequence has been very wide and deep residential lots in suburbia, many of them a half-hectare or more in size.

Locating Billboards along Highways

Advertising billboards are located along most American highways but their frequency varies from place to place. Whereas most of us would just as soon eliminate all roadside eyesores, the location of billboards is a useful example of events consuming point locations along lines.

Out on the open road, the signs we encounter advertise brand name products and services distributed to regional and national markets. The closer we get to a city the greater the frequency of signs advertising firms located within that city. Several constraints control sign placement. If the sign is too far out in the country its message may be forgotten when you reach town. Thus signs are pulled in toward town. Closer to town and within the city limits signs are more tightly regulated, permission to install them is much harder to obtain, and locations are more expensive to lease. Thus signs are pushed out into the country. The two forces, one centripetal, the other centrifugal, interact to cause most signs to accumulate within a few kilometers of the built-up city edge. Within this sign zone signs must be spaced or none can be read. One force pulls signs inward, another force pushes them outward, and a third force causes them to repel one another. A fourth, of course, keeps them all close to the road so they can be read, and still another force keeps them outside the highway right-of-way. The signs compete with each other for the driver's attention. The location and spacing pattern we observe is the outcome of all these location forces.

Out on the open road, signs are subject to few locational constraints. They are therefore located only sporadically. Some interesting exceptions, which have become bits of vanished Americana, are the Burma-Shave signs, sets of which dotted American highways for two generations (Fig. 9–19). Six signs were erected at intervals of about thirty meters. The first five presented a jingle and the sixth carried the name of the sponsor. When Burma-Shave signs were first erected in the 1920s, they were read by motorists passing at speeds of about 40 to 65 kilometers per hour. Accordingly, they were more closely spaced than those along high-speed roads of the 1960s. The signs consumed location along a line. If they were too close together part of the jingle would be missed. If they were too far apart the meter of the jingle would drag and lose its impact. Finally, if they were too far apart along a straight stretch of road, then lease negotiations would be necessary with two farmers instead of one. Since the signs were erected on farm land another constraint was soon recognized. Originally the signs were placed about 152 centimeters from the ground but grazing cattle began using them to scratch their

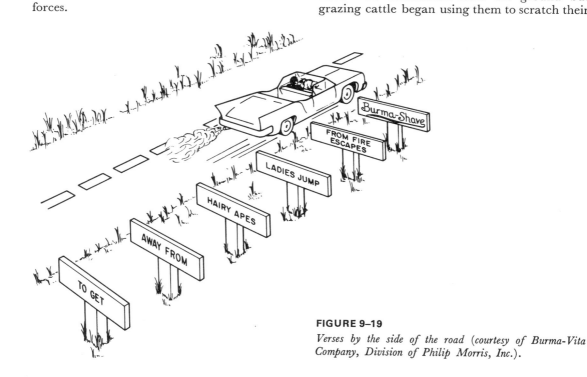

FIGURE 9–19

Verses by the side of the road (courtesy of Burma-Vita Company, Division of Philip Morris, Inc.).

backs. Thus, it was soon decided that the bottom of the sign had to be higher than the height of the back of the average cow and 183 centimeters was settled on as a standard. Such spatial pushes and pulls along a line, up and down, in and out, created a discreet series of "billboards" enjoyed by millions.

Service plazas along toll roads also consume location along a line. If the plazas are too frequent, they run at an inefficient scale and cannot provide either fresh coffee or a complete line of services. If there are too few plazas, people leave the toll roads for gas, food, and services in nearby towns. Turnpike commissions try to get by with as few service points as possible along the line without encouraging motorists to take their business elsewhere. One force spreads plazas out; another pulls them together. Location patterns which persist, whether billboards or service plazas, represent viable arrangements which are resolutions of these countervailing forces.

POINT-TO-POINT MOVES AND FACTORY LOCATION

The location of a factory, an office, a store, a shop, a home, or any other kind of a facility which processes inputs gathered from other points must be chosen by a decision-maker. Final choice of a location depends on objective, factual considerations as well as on subjective opinions and preferences.

Location choices are goal-oriented. Decision-makers recognize that places are different in their internal (site or intrinsic) character as well as in their external position relative to the location of other places (situation or relative location). Confronting alternative locations, decision-makers bear certain objective facts in mind but these alone cannot dictate location choices. Under apparently similar circumstances, two decision-makers engaged in the same business might make very different locational choices. For one thing, they may have imperfect or conflicting information about external circumstances now and how they might change in the future. It is also likely that their subjective criteria may differ in degree and kind.

Time and circumstances decide whether a venture will succeed. Subjective failure may mean unhappiness. Objective failure will be brought on by bankruptcy. The two go together exactly only when subjective success is a function of and only of objective success, but this is rarely the case.

Locating a Factory: One Input and One Market

Let us consider the problem of locating a facility in terms of deciding a factory's location. Assume that a firm manufactures one commodity. It uses one input available at a point source, and it sells the product produced at only one market located at another point. Assuming that movement ease is equal in every direction, where should the mill or factory be located to minimize the aggregate transport bill? Assembly costs, the cost of bringing each input unit to the mill, must be paid, as must marketing costs for each corresponding product unit shipped to the market. The total transport bill per unit of marketed output consists of assembly costs plus marketing costs (Fig. 9–20).

Raw material is supplied at the raw material source point R at an established price. If the raw material is delivered to points away from R a freight bill accumulates at a rate of two per kilometer for each raw material unit shipped. At any point one kilometer from R an assembly cost of two must be paid for each unit of raw material delivered. Additional concentric circles at one kilometer intervals represent lines of equal transport cost and are called *isotims*. A factory located two kilometers from R pays a freight bill of four per unit of raw material delivered. At three kilometers it pays six, and so on.

Once a unit of input is processed into a unit of output the product must be shipped to the market. In our example, marketing costs are two per kilometer on each unit of product, so that a factory six kilometers from the market must pay a delivery charge of twelve to move the product to market. For factory locations closer than six kilometers to the market, marketing transport charges per unit of product decline at a rate of two per kilometer. Accordingly, the market is surrounded by concentric circles indi-

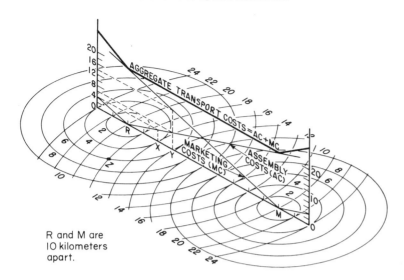

R and M are
10 kilometers
apart.

FIGURE 9–20

One input available at R and one market at M. Where should the factory be placed to minimize aggregate transport costs per unit product?

cating marketing costs per unit of marketed output for all possible factory locations.

If the factory is located at X (Fig. 9–20), assembly costs for one raw material unit equal four. If the factory is at Y, assembly costs per unit are equal to six. A finished product at Y must move seven kilometers to the market so marketing costs when the factory is put at Y are equal to fourteen. The total or aggregate transport bill for a factory located at Y equals $6 + 14$ or 20 per marketed unit. The per unit aggregate transport bill at X is 20 as well, because although X is closer to the raw materials source, it is farther from the market. Location Z is three kilometers from the raw material source R and nine kilometers from the market M. For a factory located at Z the total transport bill on each unit of product marketed is 24.

Lines of equal aggregate transport cost per marketed unit of output are called *isodapanes*. In our one-resource and one-market example, every location along the line between R and M carries an aggregate transport bill of twenty. Successive locations to the left of R and to the right of M move away from both the raw material source R and the market M and aggregate transport charges rise sharply at a rate of four per kilometer. An aggregate transport bill can be computed for any point in the region and the bill's magnitude will vary directly with the distance from R plus the distance from M.

Where is the optimum location in such a situation? At the place where the aggregate transport bill for each marketed unit is least. Any place along the line R-M therefore is an optimum.

It is unlikely that any enterprise would be located within the simple framework suggested by Fig. 9–20. In more typical but still oversimplified cases, assembly costs per distance unit on an input unit are greater or less than the marketing costs on the corresponding output unit. When the two transport rates differ, then the optimum location or point of minimum aggregate transport costs dissolves from a line as in Fig. 9–20 to a point, either at the raw materials origin or at the market (Fig. 9–21).

Terminal Costs and Transport Gradients

Up to this point we have assumed that transport charges on raw materials and on finished products increase from zero in a linear fashion with distance. Of course this is an oversimplified notion. If it costs $25 to truck a piano one kilometer across town, it does not cost $12.50 to ship it half a kilometer. For very short trips, in fact, the loading and unloading, the insurance, and the filling out of forms are the major cost components of the trip. Fixed costs which must be paid regardless of the length of trip are called *terminal costs*. Even if you want the piano moved

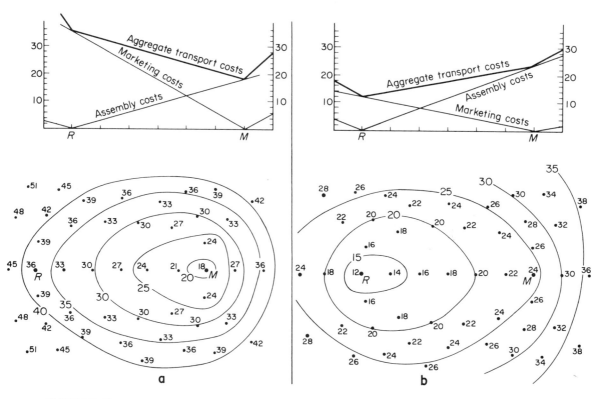

FIGURE 9–21

(a) *Aggregate transport costs at all potential factory sites for a bulk expanding enterprise. In such a market-oriented enterprise, it costs more per unit distance to ship a unit of product than it costs to ship the raw materials that go into it.* (b) *Aggregate transport costs for a bulk-reducing manufacturing process. This enterprise is raw materials oriented, because it costs much less per unit distance to ship a unit of the finished product than it costs to move the needed raw materials. Lines of equal aggregate transport costs are isodapanes.*

from the living room to the downstairs amusement room, terminal costs must be paid.

On top of terminal costs are added an increment depending on distance. We assumed in the examples so far that transport charges for raw materials and finished products increase in a *linear* fashion with distance—that is, with an equal cost increment per kilometer. But this is only one possibility. Consider the movement of raw materials from a source R to alternative mill sites X_i at different distances when the transport function is not linear (Fig. 9–22).

A mill located at R would pay neither terminal costs nor transport costs. Assembly costs

would be zero. For sites away from raw material source R, assembly costs depend on the level of terminal costs, and on the transport cost gradient. Case A (Fig. 9–22), the linear case, is unlikely. Terminal costs are zero and per kilometer charges vary directly with distance. Case B with terminal costs plus a linear transport cost function is less unlikely but still hard to imagine in the real world. Case C might obtain where the factory or mill owner had his own truckers picking up materials rather than hiring common carriers to do it. Terminal and other fixed charges include the cost of trucks and the annual wages of drivers. Over-the-road costs do

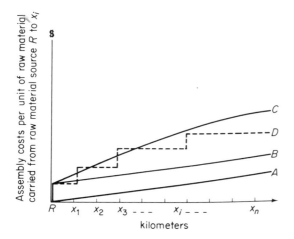

FIGURE 9–22

Terminal costs and alternative transport gradients. If there are no terminal costs, transport charges start at zero and increase as a function of distance (case A).

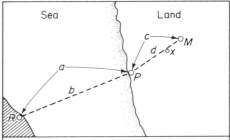

If the mill is located at the market *M* and raw materials must come from source *R*, the following must be paid:

  a: Terminal costs: loading and unloading the ship
  b: Transfer costs: moving the materials from *R* to *P* over the water
  c: Terminal costs: loading and unloading the land vehicle (train, truck, pipeline, etc.)
  d: Transfer costs: overland movement of materials from port to mill

FIGURE 9–23

When a break in bulk occurs, an extra set of terminal costs is incurred.

not rise linearly. On longer trips the per kilometer increment drops a bit.

Probably the most common situation is Case *D* where the total transport bill per unit of delivered material is the same over a whole rate zone. We experience such an arrangement during a metered taxi ride. Putting the flag down immediately adds a terminal charge of say 40 cents or more. Then every one-third of a kilometer or so the meter clicks, adding 10 cents. A graph of cab fare by trip distance defines a step function instead of a linear function or a curvilinear function. Railroad rate zones and truck freight schedules are step functions. Parcel post schedules and long-distance telephone rate structures also define step functions. Thus, to the extent that location decisions depend on aggregate transport charges, precise information about terminal costs and rate schedules is necessary.

Sometimes transport gaps or changes in transport mode interrupt the flow of raw material between its origin and the mill, or the flow of product between the mill and the market. Transport breaks occur naturally when ocean vessels reach port and goods are unloaded, or when a pipeline ends at the water's edge or at a rail line. Any combination of transport modes

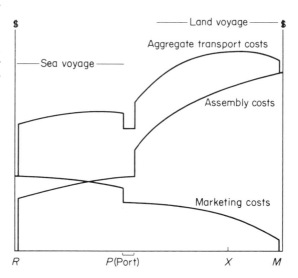

FIGURE 9–24

The cost of a break in bulk en route from raw material source R *to mill at market* M. *A mill located at* R, P, *or* M *pays two sets of terminal charges. At any other point—for example, potential mill site* X—*a mill would have to pay three sets of terminal charges.*

is possible, but where such interruptions occur, a *break-in-bulk* is created. Materials or products that have been packed for shipment by one mode must be disassembled and prepared for shipment on another mode. This means that an additional set of terminal costs must be paid (Fig. 9–23).

From Fig. 9–23 another figure can be constructed to show how assembly costs vary for mill sites anywhere between raw material source R and the market M (Fig. 9–24). The same may be done to illustrate how marketing costs vary with differing combinations of terminal and transfer costs. When the assembly costs are added to the marketing costs for each possible mill site between R and M, the aggregate transport cost profile shows sharp declines at R, at M, and at the break-in-bulk point. At each of these three points only *two* sets of terminal costs must be paid; if any intermediate points were selected, *three* sets of terminal costs would be incurred. A mill at X, for example (Figs. 9–23, 9–24), would pay for (1) loading raw materials at R and unloading them at the port P; (2) loading raw materials at P and unloading them at X; and (3) loading the product at X and unloading it at the market M. Moving the mill to the port P or to the market would permit elimination of a third set of terminal costs. The break-in-bulk phenomenon is one reason why many industries involved with foreign and domestic markets and material sources cluster in places like London, Boston, or New York.

Industrial Linkages

The single mill with one input, one output, and one market is meant to illustrate how an aggregate transport cost surface blankets the earth for every class of activity. Activity costs vary from one place to another, and aggregate transport charges are one component of activity costs. There are other costs, and other benefits as well, and so a statistical surface of net benefits blankets the earth for each activity. Decision-makers really try to locate their activities in places where risks are low enough and net benefits are high enough. Attention to transport charges is important but it is only a first step toward

defining the shape of the net benefits surface.

Benjamin Chinitz and Raymond Vernon studied the enormous concentration of manufacturing industries in the New York metropolitan region. They concluded that the chain of processes between raw materials and final product markets has been growing ever longer. Increasingly, the tendency for any one plant in the chain is to use materials which are already processed. As a result, and to an increasing degree, plants hold down their assembly and marketing costs by locating near other plants. When they all do this, raw materials entrepots, mills, factories, offices, homes, and markets are located nearer and nearer to each other. When every activity is movable all can move to a common location and eliminate much of the long-distance transport.

The extended New York area comprises a tenth of the national market, and thus much of the area's output is consumed locally. If two linked firms ultimately serve the same local market, they might as well be in the same city. Once there, they might just as well occupy adjacent tracts. The urban-manufacturing-commercial concentrations of the American manufacturing belt represent the same phenomenon as the New York-centered Boston to Washington Megalopolis, only at a wider scale.

Nonmonetary Costs and Benefits and the Location of a Public Facility

Logical, reasonable people know that the term "locational costs" implies more than terminal and transfer charges. Noneconomic criteria are especially evident in the location of public facilities, as an editorial in the *Minneapolis Star* implied:

Illinois has been picked for the Atomic Energy Commission's $375 million atom smasher. Two years ago Twin Cities' civic leaders thought they had a good case for locating the plant at the University of Minnesota Research Center at Rosemount. A committee was organized, and the Minnesota arguments were put into an impressive brochure.

The AEC had logical reasons for the Illinois choice—proximity to Chicago and the Argonne National Laboratories, six universities within an hour's drive, etc.—and probably nothing Minnesotans could have done would have changed the decision.

But there are some things the people and their officials here can do about bringing in other plants and the jobs they would provide. A number of companies have passed up Minnesota because they feel the tax climate is unfavorable, particularly in the matter of personal property levies. How about the tax reform which takes into account Minnesota's present competitive disadvantage with other states. . . .

Minnesota has lagged behind most other states in programs to attract industries and tourists. . . .

Huge atom smashers don't come along often, but an alert state could lure other installations to provide jobs for our bright young citizens.*

Such commentaries appear from time to time and represent a story far larger than a short editorial could convey. In the first place, several viewpoints exist and should be noted regarding the atom smasher's location and accompanying locational benefits and costs. These viewpoints include those of (1) the editorial writer; (2) the public; (3) the AEC decision-makers; and (4) the Minnesota business community. Did the state of Minnesota want the atom smasher? The man in the street probably did not know or care. The business community, on the other hand, thinks that all growth is good. It especially likes growth which it *thinks* carries high multiplier effects—that is, growth which it thinks will create further growth. Merchandising enterprises, especially the department stores which support metropolitan newspapers with their advertising, wanted excessive property taxes on inventories and equipment reduced or eliminated. Their spokesman, the editorial writer, related the loss of the atom smasher to the state tax structure. In an appeal to several groups—in a classic *non sequitur*—he implied that tax laws keep firms out of some places and attract them to others.

Why did the atom smasher go to Illinois rather than another place? Probably the decision was made in two stages. A list of acceptable sites was prepared using important objective criteria such as bedrock characteristics, availability of personnel, construction costs, and access by public transportation to all parts of the country.

*Minneapolis Star, December 1966, courtesy of the Minneapolis Star and Tribune Company.

At each such site expected net benefits (scientific, political, social, economic, etc.) would have been acceptable to all decision-makers and decision-influencers, lobbies, and power groups. One can only speculate on the designation procedure used in selecting the Illinois site from among the acceptable ones. Perhaps Minnesota winters, Mayor Daley of Chicago, or Senator Dirksen of Illinois played a part. From the final decision-maker's vantage point, expected aggregate net benefits, both economic and noneconomic, were maximized by the choice.

KNOWLEDGE, RISK, AND SUBOPTIMAL BEHAVIOR

The atom smasher example illustrates that an objectively determined "best location" using one set of criteria might not be as good when looked at in another way. If you live in San Francisco and your mother-in-law lives in the Bronx, you may think that her location is optimal, but she may prefer to be near Telegraph Hill. Different viewpoints can produce quite different optimum location patterns.

Costs and benefits determine net benefits. But what are the costs? When is *risk* a cost? How should the risk of a flood be evaluated and introduced into a river flood plain industrial location problem? How we deal with costs, benefits, and their probabilities—which is what risk means—depends on whether we want to explain and predict locational behavior, or whether we want to predict long-run equilibrium location patterns.

Making Locational Decisions under Conditions of Uncertainty

People make location decisions on the basis of their information, their expectations, and their predictions. In part such knowledge is correct and based on accurate interpretation of real-world events. Facts about future events are unknown, and other facts are only partly known or are misunderstood. Decision-makers try to accommodate personal whims and business requirements at the same time. When a location decision must be made and implemented, they

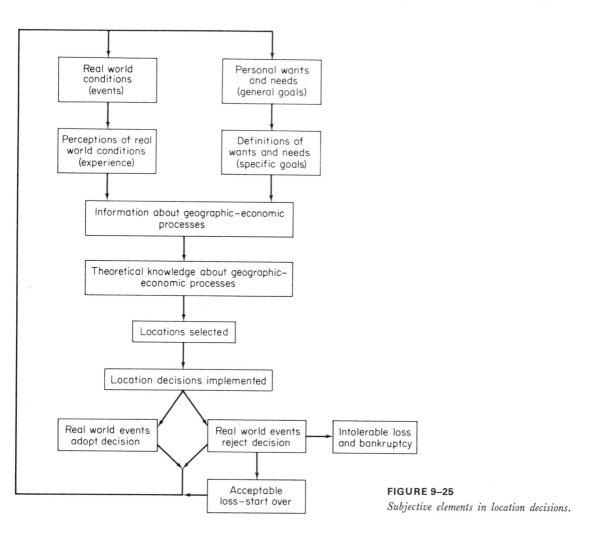

FIGURE 9–25

Subjective elements in location decisions.

adapt to the real world as they and only they see it (Fig. 9–25).

Certain real-world conditions operate independently of any one decision-maker or entrepreneur. Nature and the geographic-economic environment may be hostile, benign, or generous in the affairs of firms or families. A firm located at a particular place may be adopted by its environment or it may be eradicated. Sometimes better information can avert disaster; but then some businessmen are too short-sighted, bull-headed, or conservative in approaching risk to broaden their range of information sources and then act accordingly. For example, Gilbert White and his associates studied flood plain management and produced case studies of flood plains with recent histories of flooding. Surprisingly, flood plain occupants reveal a striking lack of conscious adjustment to flood danger. Entrepreneurs and householders may ignore the knowable dangers and locate on the flood plain anyway, but nature may step in and nullify their location decisions. How do we account for the behavior of individual flood plain occupants, as well as the long-run patterns of flood plain avoidance and occupation?

Types of Error in Location Decisions

Operationally, human decision-making is a process of chance discovery, partial knowledge, and selection among satisfactory alternative strategies, rather than a process of perfect knowledge leading to optimal choices. Constraints on location choices include:

1. Evaluation error: some things are known, others are not, and some of what is taken as known is erroneous.
2. Prediction error: some future events we can take for granted, but most future events vary in their predictability from zero to 1.0.
3. Risk and risk limits: some people cannot tolerate uncertainty, and thus certain alternatives must be foregone.
4. Performance threshold: the physical and cultural environment processes location decisions; some pass and some fail. The survival threshold is the performance level which must register for a decision to be viable and adopted by the environment.

Now assume that you are a decision-maker. Consider the series of location strategies in Table 9–1 and assume that your lowest acceptable outcome or performance threshold is fifteen. Such a value may be, say, a profit level for a year. Cover up columns V and VI with your hand. Which location strategy would you pick?

Each strategy includes choosing a location, setting a scale of operations, and deciding on a set of operating procedures (Table 9–1, col. I).

Each strategy carries an expected payoff level (col. II). Two sources of uncertainty surround each strategy: intrinsic adaption error and extrinsic adoption risk. Adaption error (col. III) is internal to the firm, coming from faulty evaluation and prediction when decision-makers either do not have all the facts or else interpret them incorrectly. Adaption error is the chance that the firm will do the wrong thing after it picks a particular strategy. The second major source of uncertainty is adoption risk (col. IV), the chance that an ever-changing environment will allow a strategy to succeed.

Let us assume that one of our strategies is to build a drive-in restaurant at a certain site. We set a scale of operations, perhaps acquire a franchise, hire help, advertise, decide on a menu, set prices, hours, etc. We also have to figure out what the public wants and how to provide it. As we make these decisions, all mistakes in judgment are lumped together under adaption error.

Outside our control are real-world conditions which may or may not be stable. A highway rerouting may cause our business to boom or bust. A competitor may open up another drive-in across the street. These are things beyond our control but they do affect our level of success. We lump all such extrinsic events under adoption risk, the chance we will succeed.

Some strategies appear at the outset to have good prospects for success. Others we expect

TABLE 9–1

Selecting Strategies in an Uncertain World

(I) Location Strategy	(II) Expected Payoff Level	(III) Anticipated (99 Percent Certain) Range in Payoff Variation Based on Evaluation and Prediction Error			(IV) Chance That Venture Will Succeed as Planned Given the Location and Operating Procedure	(V) Observed Outcomes t_1	(VI) Observed Outcomes t_2
A	24	12	to 36	(± 50%)	.25	19	20
B	19	15	to 23	(± 20%)	.60	21	14
C	13	9.75	to 16.25	(± 25%)	.90	14	16
D	17	14.5	to 19.5	(± 15%)	.80	17	15
E	7	6	to 8	(± 15%)	.70	8	9
F	14	10.5	to 17.5	(± 25%)	.75	15	13
G	10	5	to 15	(± 50%)	.50	5	7

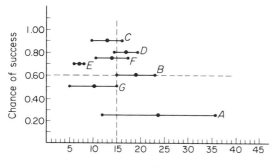

FIGURE 9–26

Evaluating alternative location strategies (data from Table 9–1, p. 326).

will have little chance of getting off the ground. If decision-makers are reluctant to get involved with anything deemed to carry a chance of success of less than 60 percent (high adoption risk), and yet they need at least an outcome of fifteen to survive, which of the alternatives from Table 9–1 will be chosen (Fig. 9–26)? Strategies A and G will probably be eliminated because they are too risky. C and E will be dropped because they will almost certainly return less than the payoff of fifteen needed for survival. Of the remaining three (B, D, and F) D and B look best. D is less risky but offers a lower expected payoff of 17; B is more risky but promises an expected payoff of 19. F's expected payoff is below 15 although actual payoffs may occasionally exceed 15.

It turns out that F would have been all right in period t_1, but to continue with F in period t_2 and beyond would be disastrous (Table 9–1). We expect therefore that decision-makers will tend to pick strategies B or D, and in the long run both will succeed even though if B is chosen there is an unexpected setback in period t_2 when payoffs drop below 15. On the other hand, a firm with solid long-term backing so that it can afford high risk ventures will clearly profit from strategy A. If A is picked, the system will certainly adopt the firm and the firm will prosper in the long term despite high risks over the short term.

Information Requirements in Location Analysis

In locational analysis we keep coming back to information flows and probabilities concerning the contingency of events (Fig. 9–25). Observed location choice behavior and observed location patterns diverge somewhat from the optimal even in the long run because of constraints on information flows, together with errors and difficulties in information acceptance and response. This is true whether the information is moving over back fences, among farmers in a rural area, or throughout the population by way of journals, newspapers, or television.

The information which moves along any channel may be slowed, interrupted, or modified in certain directions. Physical barriers such as lakes or mountains with few passes traditionally serve as obstacles in the spread of people and ideas. Political frontiers form semi-permeable boundaries today, although some of them are more permeable (Canada-United States) than others (China-U.S.S.R.). Information barriers can be socioeconomic, cultural, or personal. An unpopular information officer in a developing area will dispense less information than one who is liked and respected. But simple information transmission is not enough. People have to be tuned in. Acceptance of information varies with the medium and the message, but even in the United States, where so much information is spread in so many ways, personal contact often continues to be more effective in influencing human behavior than mass media. People learn from one another. If what they learn is incorrect, observed behavior and locational arrangements fall short of the feasible optimum.

In industrial decision-making, the information factory within a corporate headquarters is what keeps the place running. It operates in predictable ways and its approach to risk is especially revealing. White collar management people like to minimize risks to the firm and to themselves. They tend to have a "safety first" approach to their jobs, so they use conservative definitions of disaster levels, as do stockholders. Only when the firm is operating well above the disaster level can managers become more venturesome with profits than the investors would be. It is not the managers' money, and so long as the outcome remains above the safe minimum disaster level they pay no penalty except in capacity to expand the firm.

To summarize, if it is so difficult to optimize in location choice, why bother trying? For two reasons. Because in a subjective sense outcomes that satisfy provide a kind of psychological optimum even if economic maxima are not reached; and secondly, because the quest for optimum locations at all decision-making levels is a valuable means for discovering alternative satisfactory solutions. Even when costs alone (such as minimum aggregate transport costs) or net profits alone are used as criteria, location decision models are so complicated we are torn between reality and manageability, even though computers make things easier to handle.

COMPLEX LOCATION PROBLEMS

The high degree of self-sufficiency among frontier settlers and pre-Revolutionary farmers in America was basically a reflection of the limitations on human interaction over space. At the other extreme, Rome's size at its peak of vitality depended on a transport and communications system unrivaled in the world. Rome specialized as a control center, and exported military controls, political organization, law and order, and other services. Complementary areas supplied the city with the means of satisfying its needs, like grain from Egypt and the Ukraine. Specialization, interdependence, and interaction are the antithesis of generalization, isolation, and self-sufficiency. An isolated farmer generalizes because he cannot afford to specialize. If he needs a pair of shoes the lowest cost solution is to produce them at home; the cheapest inputs are his own labor and the materials like leather derived from his farming operation. The market is there also: himself. All other materials sources are too remote because of distance barriers.

Making a pair of shoes may take several days. The shoes therefore are rather expensive in terms of their modern labor equivalents. But considering his alternatives, the farmer chose the lowest priced alternative on a complicated aggregate cost surface. It does not pay for him to make shoes to sell somewhere else, nor does it pay him to buy shoes elsewhere or to make them elsewhere.

When subsistence farmers need other products besides shoes, they again confront a high aggregate transport cost surface around them and deep, steep-walled local pits where costs plunge at each farmstead. Modern transportation and communications systems bring down movement costs and make places more accessible to each other by creating cost-space convergence. Under primitive conditions movement is expensive and aggregate cost surfaces resemble rugged mountains, broken by sharp crevasses created by the few roads. Transportation improvement eliminates the valleys by bringing the mountains down. Cost surface configurations ultimately resemble the gentle swells and swales of a gently rolling prairie. Not only is there less local relief in the latter case, but the average elevation is lower too. When transport systems improve and networks fill in, all trips become cheaper and it is easier to get from each place to every other place.

Transport systems were built to promote interactions over space, and prevailing location theory models are largely transportation-oriented. To study the effect of cost distance in geographic space on the location of human activity let us examine the location problem in general terms rather than in terms of specific industry case studies. We shall consider only manufacturing activity and will assume that wages, interest, and other prices are equal everywhere. Only transportation charges will vary and these will vary with distance. Under such assumptions, with several inputs, one processing unit, and one market, three analytical approaches can be used to illustrate the search for the best location: isodapanes, mechanical models, and geometric solutions.

Isodapanes: Lines of Equal Aggregate Transport Cost

Let us assume that a product which is sold at only one market point is produced from two

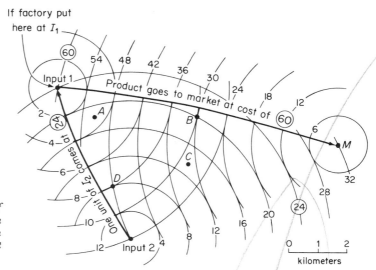

FIGURE 9-27

Two inputs and one market. For a set of sample points aggregate transport costs can be estimated and isodapanes can be drawn to identify the factory site with the lowest aggregate transport cost.

different inputs available at separated points away from the market. Where is the best factory location for input assembly, processing, and marketing? If one unit of each input is needed for one output unit, and transport charges on raw materials and on finished products vary with the distance they move over a transport surface, concentric isotims are defined for each input unit assembled and for each output unit marketed (Fig. 9–27). In the example given, isotims are concentric and equally spaced. Under a nonlinear (e.g., stepped or curvilinear) transport function they would not be equally spaced. If the movement surface were not isotropic, and movement were cheaper in some directions than others, the isotims would not be perfect circles.

Where should the factory be put to minimize the aggregate transport bill on each unit of marketed output? That bill comprises three parts: assembly at the factory of one unit of input I_1; assembly of one unit of input I_2; and the freight bill from factory to market M for one output unit. The units used to measure quantities of inputs and outputs may differ; one may be liters and the other kilograms. If we reduce everything to freight charges per kilometer per input unit or output unit then we can deal with disparate items in common terms.

If we put the factory at I_1 (Fig. 9–27), the

source of the first input, then we must bring in a unit of the second input from a distance of six kilometers at a charge of four dollars per kilometer. Because the factory is already at the source of input I_1 there is no freight paid for assembling the first input. After the two inputs are processed to yield one output unit, the product must be shipped to market ten kilometers away at a cost of six dollars per kilometer. Adding the three freight components we compute an aggregate transport bill per marketed output unit when the factory is located at location I_1:

$$0 + (6 \times 4) + (10 \times 6) = 84.$$

The value of the statistical surface which blankets the area of Fig. 9–27 equals 84 at point I_1. Other sample points on the aggregate transport cost surface are computed in the same way. With a sufficient sample of points, we can draw on the map a set of isodapanes, lines of equal aggregate transport costs. By examining the isodapanes, the lowest aggregate transport cost location can be found.

The isodapane method of locational analysis can easily be extended to include several inputs. If one or another input unit is exceptionally bulky and expensive to move, the lowest aggregate cost location will be near the relatively immobile item. If the finished product is hard to move, marketing costs are proportionately high and the optimum location occurs near the market. In this case we say that the particular processing activity is market-oriented.

If one of the inputs is very expensive to move, the lowest aggregate costs are incurred at a factor location near the source of that input because factory sites near the market would be prohibitively expensive. In nineteenth-century America, one raw material oriented activity was flour milling using water power. Of the three main items to be transported, wheat, water power, and flour, it was much easier to move a unit of flour than to move the unit of wheat that it came from. But it was almost infinitely easier to move the unit of wheat than it was to move the unit of water power needed to grind the wheat. Water power could be moved short distances with overhead cables, belts, and mill canals, but water power shipments usually were less than a kilometer until it became possible to move power great distances by changing it into electricity first.

Sometimes assembly costs go down because a more efficient processing method at the factory requires proportionately less of the input per unit of output than was previously needed. In the early years of iron and steel manufacturing, charcoal and coal were used very inefficiently as a heat source. The amount of coal needed to produce a ton of metal was so large that many furnaces and mills were built directly on the coal fields in the United States, Britain, and Germany. Then two things happened. New and more efficient metal production methods cut fuel needs (i.e., fewer tons of coal per ton of metal); and improvements in rail transportation meant reduced freight rates (lower charges per ton kilometer). As locational constraints were removed from the iron and steel industries, the differences between high points and low points on aggregate transport surfaces declined, isodapanes became more widely spaced, and iron and steel mills were built at new locations away from coal fields, for example in Gary, Indiana.

Mechanical Location Models

Some people have no difficulty grasping the idea of an aggregate cost surface with cost peaks at certain places and troughs elsewhere.

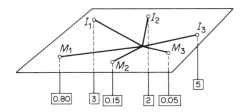

FIGURE 9–28

One unit of output is sold 80 percent in market M_1, 15 percent in M_2, and 5 percent in M_3, and uses up 3 units of input I_1, 2 of I_2, and 5 of I_3 when it is produced. If the weights in the mechanical model above are proportional to these values, the knot comes to rest at the optimum factory site, provided that all inputs and products pay the same freight rate per ton kilometer. If they pay different rates, each value above is multiplied by its respective freight rate per ton kilometer to yield six different net distance inputs, the values of which determine the weights on the strings.

They imagine that the isodapanes which define a statistical surface are the economic equivalent of the contour lines which describe the shape of the earth's surface on a topographic map. Other people find a mechanical analogy helpful in locational analysis. Let us assume that a product requires three inputs, I_1, I_2, and I_3, and that three, two, and five tons respectively are needed for one ton of product. The ton of product is sold in three markets, M_1, M_2, and M_3, in the following proportions: 80 percent, 15 percent, and 5 percent. Assume also that the assembly cost per ton kilometer is the same as the marketing cost per ton kilometer. Under these assumptions, where is the least cost location?

Imagine the following procedure. Attach to a table a map of the area. Drill holes in the map and table at each of the six locations: the three input sources and the three markets. Fit each hole with a frictionless pulley and pass a separate string through each hole. Knot one end of each string in a common knot above the table, and to the six loose ends hanging below the table attach weights proportional to the inputs needed or market share for a ton of marketed output. When the weights are suspended, the knot comes to rest at an equilibrium location which is the least cost location so long as the input amount and the market mixes remain unchanged (Fig. 9–28).

The solution in Fig. 9–28 is the correct one when we assume that the assembly costs per ton kilometer are the same as the marketing costs per ton kilometer. If, however, the transport rates per ton kilometer on inputs 1, 2, and 3 equal t_1, t_2, and t_3 respectively, and the transport rate on the output equals t_4 per ton kilometer, then the suspended weights in the mechanical model must be "weighted" by the transport rates. The new suspended weights are called "net distance inputs" and equal: $t_1 \cdot (3)$; $t_2 \cdot (2)$; $t_3 \cdot (5)$; $t_4 \cdot (.80)$; $t_4 \cdot (.15)$ and $t_4 \cdot (.05)$.

In Fig. 9–28 it is easy to see that if the manufacturing process were improved so as to require less of input number 2 per ton of output (say, instead of two tons, only one is needed), then the equilibrium position of the knot would move away from I_2. If market M_3 starts to grow very rapidly, and finally draws 50 percent of each ton of output, the knot would move toward M_3 and away from M_1 and perhaps M_2. Each of the six points exerts some

pull. The equilibrium knot position represents a resolution of the six conflicting forces. (It really works!)

Minimum Transport Cost Geometric Solutions

To illustrate a third form of analysis let us assume a location problem involving three inputs and one market. Let the weights attached to each point represent net distance inputs as follows: I_1: 5; I_2: 3; I_3: 6; and M: 7 (Fig. 9–29). Where is the best factory location?

If, as a first approximation to the optimum plant location, we select a random point A (Fig. 9–29) within the area defined by the four points, what happens? If we were holding at point A a knot attached to four forces pulling toward the three inputs and the one market (as we did in the mechanical model) and then we let the knot go, it would tend to be displaced toward the lowest cost location. We could do the same for point B. In fact, at any random point the resolution of the four pulls is in the *direction* of the optimum location. Thus, if the

FIGURE 9–29

Four forces pull on each potential factory site (A, B, etc.). If dashed arrow lengths are equal to net distance inputs, and arrow directions are parallel to a direct line from a site (say A) to each input and the market respectively, then the displacement direction from the sample site to the optimum site can be determined. If the direction of displacement is defined for two potential sites, then the optimal factory location is identified as that site falling at the intersection of the two direction arrows.

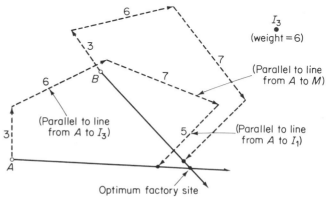

direction of displacement is given for any two random locations, lines from those points drawn in the displacement direction define the optimum location at the point where the lines intersect. We can pick a third random starting point and calculate the displacement direction as a check. All three lines will intersect at a common point. The common point of intersection is the lowest cost location.

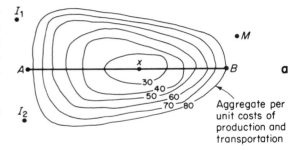

Location Costs, Location Benefits, and Relocation

The main value of simple location cost models is to dramatize how aggregate activity costs can vary from one place to another. Essentially the same argument can be used to estimate low cost locations for an office, a school, or a factory. It would be incorrect to assume that location decisions are based on money costs and non-money costs alone. We argued earlier that net benefits vary from place to place. It is legitimate to fasten all our attention on costs only when growth benefits are everywhere identical.

Geometric and mechanical models give simple, direct answers to location questions simply put. The isodapane model is not too neat, but it can handle complicated cases. Consider the structure of the isodapanes in Fig. 9–30a. If production costs are zero and we must pay only transport costs, then location X along traverse A-B is the optimum location considering only the aggregate transport costs per unit of marketed output. At X, aggregate transport costs per unit of marketed output are C_x. If a uniform price of P_0 prevails over the entire region, then maximum net returns are earned at X and net returns there equal P_0 minus C_x per unit. Such an idealized arrangement would not last long for several reasons:

1. P_0, the product price, can change as a result of competition from other factories producing the same product, and of changes in the prices of substitute products.
2. The minimum aggregate transport cost can change in magnitude and location as a result of adjustments in transportation charges as various modes compete and improve, and as new production techniques and tastes change the mix of inputs needed for each output unit.

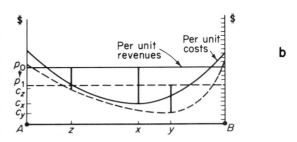

FIGURE 9–30

Shifts in prices, aggregate costs, and the location of optimum production locations.

3. Decision-makers have incomplete information about costs and benefits, and with better information they may adjust their operations to lower their costs.
4. Most decision-makers feel they have optimized if they have minimized risk and achieved some acceptable level of net money and nonmoney returns.

Now assume that price drops from P_0 to P_1 (Fig. 9–30b). Assume also that isodapanes shift to reflect a general reduction in transport costs as well as a relative decline in marketing costs compared to assembly costs. The optimal location is now Y instead of X. Net returns at the optimal location are reduced (i.e., P_0 minus C_x is greater than P_1 minus C_Y). Under these new circumstances what will happen to a producer at location Z who located his factory years ago at what was then an optimal site? Through time his operation has become progressively marginal. He will quit or relocate when P minus C drops below acceptable levels. This is what people mean when they speak of industrial location inertia or inertia in any locational arrangement: net benefits from present location continue to exceed short-run net benefits from relocation. Many of the personal and psychological costs and benefits involved are non-monetary, but they still are very important. For these reasons, the power of a small community to hold activities, families, firms, and so forth is often greater than its ability to attract new ones. Moreover, a community's power to encourage expansion of local enterprises is far greater than its ability to attract others.

Because some costs and benefits in the location problem are difficult to measure, and because others are often stated imprecisely or measured incorrectly, locational choices are seldom optimal in the strictest sense. Nevertheless, strict optimality may be little better than a reasonable approximation. When choosing a factory location, places near the optimum may be only slightly inefficient, and for all practical purposes such locations are certainly good enough. The fact that locations in the vicinity of the optimum are just about as good as the optimum itself is called the "principle of flat laxity." Locations away from the optimum are slightly inefficient; only places quite far away become substantially inefficient. If we could conveniently measure spatial inefficiency we could define a spatial inefficiency surface that would

TABLE 9–2

Regional Shifts in the Distribution of Employment in Electronic Equipment Manufacturing

Region	Employment in t_0	Percent in Each Region	Employment in t_1	Percent in Each Region
A	100	50	165	55
B	50	25	75	25
C	50	25	60	20
Total	200	100	300	100

rise bowl-like away from the optimum location—but with the bottom of the bowl almost flat (Fig. 9–31). This means that places near the optimal location are just about as efficient as the optimum itself.

Manufacturing and other trade and service industries migrate from one region to another, but such migration rarely involves moving factories, offices, or facilities. Instead, a distribution pattern changes because of differential growth rates among regions (Table 9–2). If an entire industry in region A has a growth rate faster than the industry, it will end up with a larger fraction of total industry employment at time t_1 than it had in time t_0, and we will speak of the industry shifting toward region A. Region C may also gain employment absolutely, but if its percentage growth rate lags behind the 50 percent industrial growth rate over all regions, then region C will end up with a smaller fraction of the industry in time t_1 than it had at time t_0, and we say that the industry is shifting out of region C. Through time, some industries like the automobile industry have concentrated in certain regions, while others, like iron and steel manufacturing, have dispersed into many regions. In the same industry one stage in the production process may cluster, like raw material processing, whereas in a later stage, like final assembly, dispersal may be common.

A large, open economic system like that in the United States encourages concentration, and decision-makers frequently conclude that concentration is profitable. If each of the fifty states erected steep tariff obstacles to movements of materials, labor, and products, industrial concentration would be more difficult or

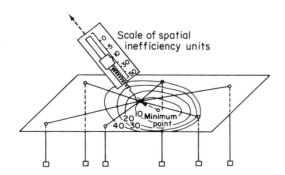

FIGURE 9–31

Aggregate transport costs per unit of marketed output will get increasingly higher at inefficient factory locations away from the minimum cost factory location. A map of the per unit cost increase is also a map of spatial inefficiency.

Scale of spatial inefficiency units

Minimum point

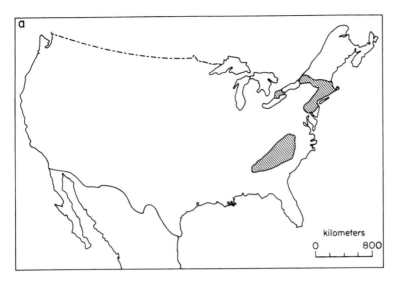

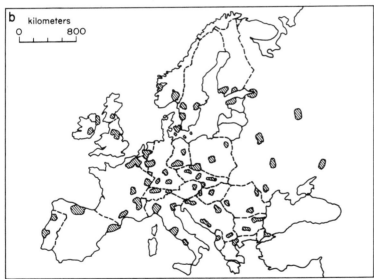

FIGURE 9-32

Boundaries in Europe create many local low points in aggregate cost surfaces for the textile industry. Free movement in the United States promotes industrial concentration at continental optima (adapted from Hamilton, Fig. 10–11).

perhaps impossible (Fig. 9–32). International boundaries interrupt cost surfaces by erecting what amount to step barriers on worker movement and freight flows. Within a country movement barriers still exist but they are either continuous or else each step increment in movement costs is relatively modest compared to the costs involved in international border crossings.

Assigning Activities to Locations; The Resulting Flows between Them

Contemporary human geography applies principles of spatial organization to a wide range of problems. Sometimes solving a problem requires special techniques. We cannot treat all of them here but we want to mention a few which represent real challenges, especially to business and government.

One important problem is called *the transportation problem*. Suppose we are given (1) a set of supply centers with known surpluses; (2) a set of market points with known demands; and (3) connecting transport routes with known capacities and transport costs. Then suppose we ask what is the least cost method of moving surpluses to markets? This is the transportation problem. It is not a complicated problem because all we consider are the surpluses, the demands, the flows, and the transport costs (Fig.9–33). We ask what is the best flow pattern, remembering that the farther we ship something, the higher is its delivered cost, and also that a surplus sent to one place cannot be shipped elsewhere. Once we have assigned activities to locations, the stage is set for organizing the flows among the locations. We will consider how to solve the tranportation problem in Chapter 13.

The *spatial price* equilibrium problem examines interregional shipments and receipts and the effect on flows of interregional price differences. Let us assume two regions; A has a surplus of commodity X, and B is a deficit region. The transport cost between the two regions is 50¢ per unit of X shipped. If region A has 100 units of X and region B has 25, and the price of X in the two regions is \$1.00 and \$2.50 respectively, how much X, will be shipped between region A and region B? The answer is that region A will ship to B so that the price in A rises and the price in B falls. Flow and price adjustment continue until region A's price per unit of X is equal to B's price plus transport costs. At that point there is no longer any incentive for region A to ship any more units of X to region B.

The speed with which prices adjust depends on the relation between price and quantity in the two regions. If the price and quantity are related as in Fig. 9–34, and if region A ships 25 units to region B, the price in A will go up to \$1.50, whereas the new supplies in region B will drive the price down from \$2.50 to \$2.00. After the 25 unit shipment, the 75 units in region A will sell for \$1.50 each. If the 75th unit were sent to region B it would become the 51st unit in B and would bring a price of \$1.98 from which a transport charge of 50¢ must be deducted. Clearly the \$1.50 in A is more than the \$1.98 — 50¢ in region B, so the prices in A and B stabilize at \$1.50 and \$2.00, separated by the per unit transport costs. In this way interregional price differentials (such as slightly higher prices west of the Rockies for eastern manufactured goods) emerge and persist after production points and customer locations have been established.

In a third spatial analysis problem we may have to determine both *location and flows*. For example, if customer locations are known and transport rates are given, where should a set of warehouses be located to minimize distributing costs from warehouses to customers? In this problem we have to assume that the amount of each customer's requirement is known, and that the capacity of each warehouse as well as its location is to be found. The number of warehouses is given.

One way to approach this problem is iteratively. First, select starting locations and starting capacities for the warehouses. Then allocate supplies to customers according to the transportation problem—that is, assuming that the

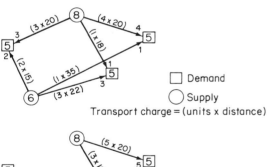

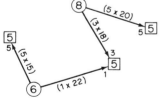

FIGURE 9–33

The transportation problem: given a set of supply points with available surpluses and a set of demand points with known requirements, ship the surpluses to the market while incurring the minimum total transport bill. Of the two assignment patterns shown above, which is the most efficient? Is there any other assignment which is more efficient than either of those shown?

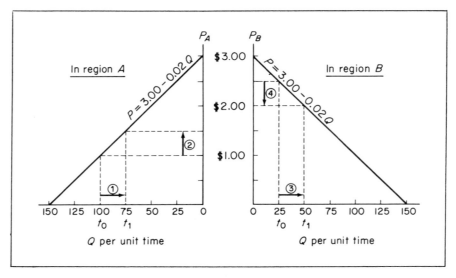

FIGURE 9–34

Price and quantity relations in a spatial price equilibrium. Initially, the price in A is $1 and the price in B is $2.50. If region A ships 25 units to region B in time period t_1, the price in A will rise to $1.50 and the additional supplies in B will cause the price in that region to drop in period t_1. Shipments continue from A to B until the two prices are separated by the cost of shipping a unit from A to B; in this case the per unit shipping cost is $0.50 (adapted in part from Garrison).

surpluses are at one set of points, and the demands are at another set of points, and the connecting transport routes, what is the least cost system of flows? Next, relocate the warehouses assuming the flows and warehouse capacities are given. Finally, reassign capacities among the warehouses according to the rule that each customer should be served from the closest warehouse. The three steps can be repeated one or more times with each iteration reducing the aggregate transport costs to a level a little below those from the previous trial. Gradually the locational patterns and flow assignments converge on an optimal solution.

Assignment problems occupy much of the decision-making apparatus of business and government. The problem of assigning n plants to n locations so as to maximize the combined profits of the plants is typical of such problems. The assignment of n public health centers to n locations in a metropolitan area so as to maximize levels of public health or to minimize

variations in health levels across a city is another example. Assigning workers to positions on an assembly line is a third case. In every instance we are interested in maximizing the combined net return of all the installations.

In the spatial assignment problem the units to be assigned are considered indivisible. Half a clinic or half a factory is not viable and cannot be assigned. Furthermore, interdependence is stressed because the assignment to one spot of a factory, or a clinic, or a worker affects the performance of units assigned elsewhere. To illustrate this point William Garrison devised an example of a firm with five plants to be assigned to five different locations. Each plant would return a different profit depending on its location according to a schedule (Table 9–3). The thing to notice in this problem of interdependent assignment is that a firm's profitability depends not on the sum of five separate enterprises, but on the location of each of the five with respect to the location of every other. Locations are never

TABLE 9–3

**The Profit Maximization Problem When Five
Factories Must Be Assigned to Five Locations
and Each Plant Earns a Different Profit
at Each Location, Depending on
the Arrangement of the Other Four**

Plants	Locations				
	1	*2*	*3*	*4*	*5*
A	17	14	18	20	19
B	21	16	19	22	15
C	18	24	20	16	19
D	19	15	21	17	23
E	24	19	16	18	20

Each plant earns a different profit at each location.

	1	*2*	*3*	*4*	*5*
A	17				
B			19		
C		24			
D				17	
E					20

One possible assignment pattern. Combined profits from the
five plants equal 97.

	1	*2*	*3*	*4*	*5*
A			18		
B		16			
C	18				
D				17	
E					20

Another possible assignment pattern. Combined profits
from this arrangement equal 89.

really independent. In the real world no firm stands alone. All have competitors and allies. In location decisions, other firms in the same industry and different industries have to be considered because all compete directly or indirectly for inputs, labor, land, transportation, and the consumer's dollar.

The problem then is to select from the many possible spatial assignment patterns the one pattern which maximizes the combined profits of all the firm's plants. There are five location options for assigning the first plant. Once one of the plants is assigned there are four choices for the second plant; then there are three choices for the third; and so forth. In all there are $5 \times 4 \times$

$3 \times 2 \times 1 = 120$ (or 5!) permutations for assigning five different factories to five different locations. Decision-makers try to identify the best allocation pattern from among the 120 patterns. If precise profitability information is available as in Table 9–3 the solution can be found by inspection after computing all the possibilities.

SUMMARY

The geographer's central question is: why are spatial distributions structured in the way that they are? This chapter approached the question of the location of human activities two ways: first, how do people locate things; and second, what kind of location patterns result? Recognizing that the private and the public interests may not coincide, and that social controls over location decisions are often needed, we then asked: how are location decisions made? In the ideal case they are made with perfect knowledge of costs and benefits at alternative locations, and they are made so as to maximize net benefits from operations at the optimal location. In the real case, perfect knowledge is replaced by error and an adverse reaction to risk, and the search for optimal locations is reduced to a search for a satisfactory location.

Sometimes people make mistakes in adapting to the environment, and their choice of a location turns out to be a poor one. Other times they fall short of a satisfactory solution because the world changes around them and fails to adopt what would otherwise have been sound location decisions. But whether the decision-maker fails to adapt or fails to be adopted, if his payoffs prove to be unsatisfactory, he loses as the environment sorts out decisions and specifies winners and losers. Thus, the reason spatial distributions are structured the way they are is partly a result of the changing organization of the world itself which permits some events to succeed while it quashes others.

A variety of mathematical tools have been devised for specifying the best location in different kinds of problems, including isodapane analysis, mechanical models, and various iterative procedures when many different alternatives must be compared one at a time. What is especially important is that location patterns ultimately control flow processes, and the flows in turn cause adjustments in the location of human activity. Spatial organization should therefore be understood in both a dynamic and a static sense.

Suggestions for Further Reading

ALEXANDER, J. W. "Measurement, Theory, and Planning," in *Economic Geography*, Part Nine. Englewood Cliffs, N. J.: Prentice-Hall, Inc., 1963, pp. 588–647.

BURRILL, M. F. "The Language of Geography," *Annals of the Association of American Geographers*, LVIII (1968), 1–11.

CAPLOVITZ, D. *The Poor Pay More: Consumer Practices of Low Income Families*. New York: The Free Press, 1967.

CHOYNOWSKI, M. "Maps Based on Probabilities," *Journal of the American Statistical Association*, LIV (1959), 385–88. Reprinted in B. J. L. Berry and D. F. Marble, eds., *Spatial Analysis: A Reader in Statistical Geography*, Chap. 5. Englewood Cliffs, N. J.: Prentice-Hall, Inc., 1968, pp. 180–83.

DOXIADIS, CONSTANTINOS A. *Ekistics: An Introduction to the Science of Human Settlements*. New York: Oxford University Press, Inc., 1968.

FRIEDMANN, J., and W. ALONSO, eds. "National Policy for Regional Development." in *Regional Development and Planning: A Reader*, Part IV. Cambridge, Mass.: The M.I.T. Press, 1964, pp. 489–700.

GALBRAITH, J. K. *The New Industrial State*. Boston: Houghton Mifflin Company, 1967.

GARNER, B. J. "Models of Urban Geography and Settlement Location," in R. J. Chorley and P. Haggett, eds., *Models in Geography*, Chap. 9. London: Methuen & Co., Ltd., 1967, pp. 303–60.

GOTTMANN, J. *Megalopolis: The Urbanized Northeastern Seaboard of the United States*. New York: The Twentieth Century Fund, 1961.

HAGGETT, P. "Surfaces," in *Locational Analysis in Human Geography*, Chap. 6. London: Edward Arnold Publishers, Ltd., 1966, pp. 153–82.

HAIG, R. M. "Toward an Understanding of the Metropolis," *Quarterly Journal of Economics* (February 1926), 179–206; and (May 1926), 402–34. Part 2,

"The Assignment of Activities to Areas in Urban Regions," is reprinted in R. H. T. Smith, Edward J. Taaffe, and Leslie J. King, eds., *Readings in Economic Geography: The Location of Economic Activity*. Chicago: Rand McNally & Co., 1968, pp. 44–57.

HALL, EDWARD T. *The Hidden Dimension*. Garden City, N.Y.: Doubleday & Company, Inc., 1966.

HOOVER, E. M. *The Location of Economic Activity*. New York: McGraw-Hill Book Company, 1963.

HOTELLING, H. "Stability in Competition," *The Economic Journal*, XXIX (1929), 41–57.

HOYT, HOMER. *The Structure and Growth of Residential Neighborhoods in American Cities*. Washington, D.C.: Federa lHousing Administration, 1939.

ISARD, W. *Location and Space Economy*. Cambridge, Mass.: The M.I.T. Press, 1956.

MCCARTY, H. H., and J. B. LINDBERG. *A Preface to Economic Geography*. Englewood Cliffs, N.J.: Prentice-Hall, Inc., 1966.

MCDANIEL, R., and M. E. ELIOT HURST. *A Systems Analytic Approach to Economic Geography*. Commission on College Geography, Publication Number 8. Washington, D.C.: Association of American Geographers, 1968.

MORE, R. J. "Hydrological Models and Geography," in R. J. Chorley and Peter Haggett, eds., *Models in Geography*, Chap. 5. London: Methuen & Co. Ltd., 1967, pp. 145–85.

SMITH, D. M. "A Theoretical Framework for Geographical Studies of Industrial Location," *Economic Geography*, XLII (1966), 95–113.

SMITH, R. H. T., EDWARD J. TAAFFE, and LESLIE J. KING, eds. *Readings in Economic Geography: The Location of Economic Activity*. Chicago: Rand McNally & Co., 1968.

SMITHIES, A. "Optimum Location in Spatial Competition," *The Journal of Political Economy*, XLIX (1941), 423–39.

SOMMER, R. *Personal Space: The Behavioral Basis of Design*. Englewood Cliffs, N.J.: Prentice-Hall, Inc., 1969.

ULLMAN, E. L. "Regional Development and the Geography of Concentration," *Papers and Proceedings of the Regional Science Association*, IV (1958), 179–98. Reprinted in J. Friedmann and W. Alonso, eds., *Regional Development and Planning: A Reader*, Chap. 8. Cambridge, Mass.: The M. I. T. Press, 1964, pp. 153–72.

WILSON, J. O. *Quality of Life in the United States: An Excursion into the New Frontier of Socio-Economic Indicators*. Kansas City, Mo.: Midwest Research Institute, 1968.

Works Cited or Mentioned

ADAMS, J. S. "Residential Structure of Midwestern Cities," *Annals of the Association of American Geographers*, LX (1970), 37–62.

CHINITZ, B. *Freight and the Metropolis*. Cambridge, Mass.: Harvard University Press, 1960.

GARRISON, W. L. "Spatial Structure of the Economy: I, II, and III," *Annals of the Association of American Geographers*, XLIX (1959), 232–39, 471–82; L (1960), 357–73. Reprinted in R. H. T. Smith, Edward J. Taaffe, and Leslie J. King, eds., *Readings in Economic Geography: The Location of Economic Activity*. Chicago: Rand McNally & Co., 1968, pp. 230–64.

HAMILTON, F. E. I. "Models of Industrial Location," in R. J. Chorley and P. Haggett, eds., *Models in Geography*. London: Methuen & Co., Ltd., 1967, pp. 361–424.

LÖSCH, A. *The Economics of Location*. New York: John Wiley & Sons, Inc., 1967.

VERNON, R. *Metropolis, 1985*. Garden City, N.Y.: Doubleday & Company, Inc., 1963.

VON THÜNEN, J. H. *Von Thünen's Isolated State*, trans. Carla M. Wartenberg, edited with an Introduction by Peter Hall. London: Pergamon Press Ltd., 1966.

WEBER, ALFRED. *Alfred Weber's Theory of the Location of Industries*, trans. C. J. Friedrich. Chicago: The University of Chicago Press, 1928.

WHITE, GILBERT. *Choice of Adjustments to Floods*. Chicago: Department of Geography, University of Chicago Research Paper No. 193, 1964.

CHAPTER 10

LOCATION AND
THE USE OF LAND

. . . and everything in its place.

Chapter 9 dealt with the question: given an activity or a flow pattern, what is its best location? Let us now look at location patterns and human spatial behavior in another way and ask: given a location, what is the best activity to carry on there? We can select a location with respect to an activity in which we are interested, or we can choose an activity for a certain location. The reason for the two approaches is that both help us make sense out of our experience of events in the world.

AGRICULTURE IN AN URBANIZING WORLD

Let us look first at agriculture, which is mankind's principal occupation. Agriculture comprises cultivating the soil, harvesting crops, and raising livestock. In agricultural areas, marked differences in crops, animals, organization of agricultural inputs, marketing procedures, prices, and land use intensity occur from place to place. In non-agricultural areas,

enormous tracts of land lie unoccupied by farmers because agriculture is too difficult or too unrewarding compared to alternative land uses. Because most of the land in inhabited parts of the world supports agriculture, and since most of mankind is supported directly by agriculture, it is appropriate that this prominent occupation and important use of land be described and explained. The general principles of agricultural land use can then be applied to the study of land used for other purposes.

A Classification of Agricultural Systems

The first step in an ordered inquiry into agricultural location patterns is a classification of agricultural activities. The traditional agricultural classification was presented by D. Whittlesey several decades ago and was based on four major dimensions of variation (Fig. 10–1). The four dimensions were: intensity, commerciality, crop specialty, and mobility. Each agricultural system scored somewhere between

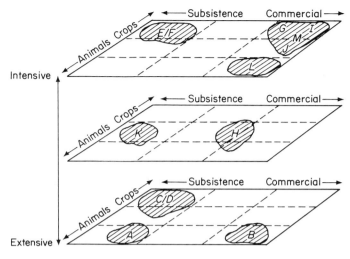

Key: A: Nomadic herding
      B: Livestock ranching
      C: Shifting cultivation
      D: Rudimental sedentary cultivation
      E: Intensive subsistence tillage: rice
      F: Intensive subsistence tillage: non-rice
      G: Plantation agriculture

H: Mediterranean agriculture
I: Pure crop farming
   (e.g., cotton)
J: Mixed farming
K: Subsistence crop and
   livestock farming
L: Dairy farming
M: Horticulture

FIGURE 10–1

Agricultural activity classification according to Whittlesey (1936).

high and low on each of the four dimensions shown in Table 10–1.

Because there are only two main kinds of migratory activities (nomadic herding and shifting cultivation), and these are relatively unimportant in terms of the numbers of people involved, they can be evaluated as subsets of their sedentary counterparts. A three-way

classification space is easy to visualize (Fig. 10–2). An activity such as cotton farming in Texas or California is located rather high on the intensity dimension, at the 100 percent mark on the crop specialty dimension, and at the 100 percent mark on the commerciality dimension. At another extreme lie the subsistence animal herders such as those who until recently roamed

TABLE 10–1

The Range of Variation in Attributes of Agricultural Systems

	Intensity	*Commerciality*	*Crop Specialty*	*Mobility*
High:	Many inputs of labor, capital, fertilizer, etc., per unit land area.	All output is sold on the market for cash.	All output is in the form of crops; no animals are raised.	Agricultural households or other activity units are migratory or nomadic; always on the move.
	.	.	.	.
	.	.	.	.
	.	.	.	.
Low:	Few inputs per unit land area.	All produce is consumed in kind by producer and his family.	All produce is in animal form; no crops raised.	Activity unit is sedentary; production permanently fixed at one spot.

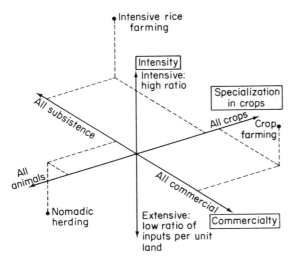

FIGURE 10–2

A three-way taxonomic space for Whittlesey's agricultural systems. Three of the thirteen types are shown located in the three-dimensional space.

dry lands in North Africa and Southwest Asia. Inputs per unit area are miniscule, emphasis is almost exclusively on animals, and much of what is produced is consumed on a day-to-day basis, with perhaps animal hides entering the market for cash. The rice tillage systems of South, Southeast, and East Asia differ again, but along the same three dimensions. Rice tillage is basically a crop enterprise with an admixture of draft and scavenger animals. Most of the output is used before it reaches the market; and in terms of labor, fertilizer, and capital inputs in the form of irrigation systems, it is an intensive enterprise in these regions.

In a modern taxonomy the four dimensions of variation implied by Whittlesey collapse to two: the first is a modernization dimension; the second refers to the crop-animal mix (Fig. 10–3). Along the modernization dimension several things vary together. At the lower end lie *traditional*, pre-industrial agricultural enterprises, mostly subsistence, and yielding low incomes

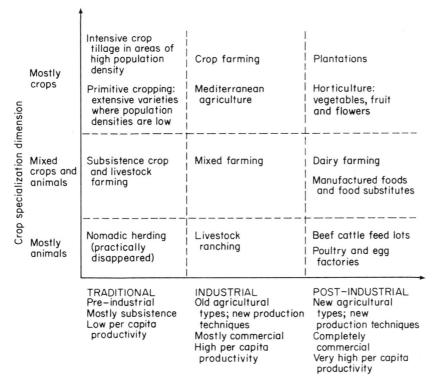

FIGURE 10–3

A two-way classification of contemporary agriculture and food production systems.

because of the very low per capita productivity. In a middle position along the modernization dimension we find the *modernized* versions of traditional agricultural types. The technology of the Industrial Revolution has been applied to very old agricultural forms, raising per capita productivity and therefore per capita incomes to relatively high levels. All activity is commercialized. At the frontiers of agricultural modernization we find not only new ways of dealing with age-old production problems, but also new types of agricultural production as well. These are the *post-industrial* forms, all commercialized, all yielding exceedingly high per capita returns, and all occupying the highest positions on the modernization dimension. Modern poultry and egg factories, beef feedlots, and artificial foods processed from seed and vegetable oils lead the list of modern food production forms whose link with traditional agriculture is slim indeed.

The Influence of Situation, Site, and Human Behavior

If an improved classification of agricultural systems is to be really useful, it should shed light on important practical or theoretical questions. Today, for example, we are less concerned than we once were about whether a farmer raises rice or wheat on his land. This distinction really is not as important as whether his efforts will enable him and his family to survive until tomorrow. When we consider the agricultural efforts of an entire nation in the underdeveloped world, our attention focuses on its chances for survival and prosperity. Accordingly, our new agricultural classifications emphasize per capita productivity and how productivity is changing through time.

Why is per capita productivity so low in some agricultural systems and so high in others? Two types of limits prevent a society at a place and time from increasing its maximum output per man. The first limit is the technological know-how that a society is able and willing to use; the other limit is the resource endowment per person (Fig. 10–4). The populated parts of the world can be mapped according to their positions in technological terms and in resource terms (Fig. 10–5). The resource endowment per person is a ratio indicating population pressure on the base of available resources. If the prevailing technological level is high in a place, technology can be exported to underdeveloped regions which are technology deficient. Besides describing and classifying the world's regions in terms of resource ratios and technological levels, the geographer's job is to explain why resource endowments and technological levels vary as they do with such profound regional

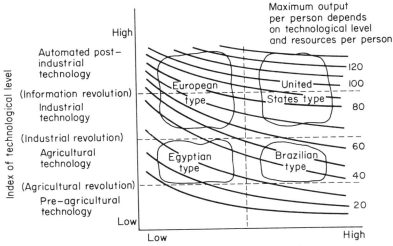

FIGURE 10–4

Technological levels, available resources per person, and ceilings on output per person in different regional economic systems.

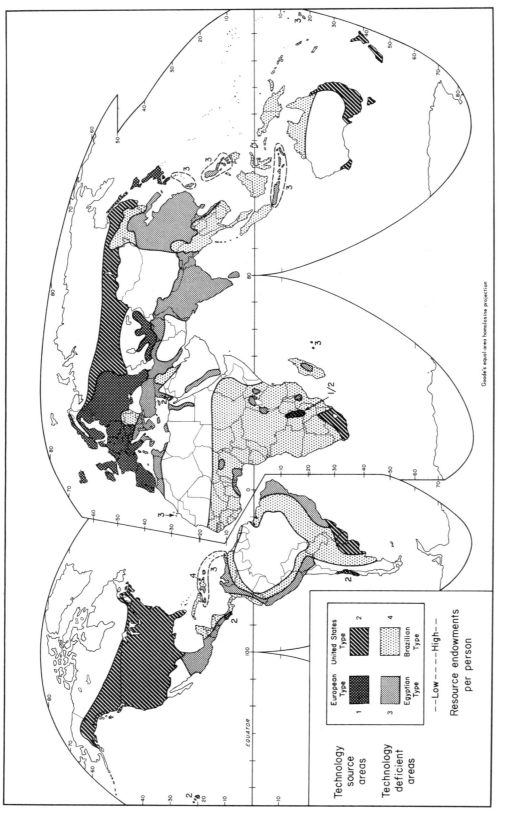

FIGURE 10-5

Generalized technology—resource regions of the populated world. The resource endowment per person indicates population pressure on available resources. Technological level is measured by whether technology is exported or imported (adapted from W. Zelinsky, Fig. 7, pp. 108–9 [based on Ackerman, Fig. 5; courtesy The University of Chicago, Department of Geography, Research Paper No. 80. Base map from Goode Base Map Series, copyright by The University of Chicago]).

disparities in levels of human prosperity and misery.

Technology levels at a place depend on the rate at which new ideas diffuse over the world from idea centers, and the promptness with which nations accept European industrial technology. Resource enowments at a place depend partly on nature's generosity, but even more on the technological level of the society that lives there. This is because resources are technically defined. Something is a resource only if it has value within a cultural-technological context. In the short run, fuels, ores, and human skills and energy which a country cannot tap are not, practically speaking, available resources.

In short, one set of places has had easy access to idea centers. They received new science and technology earlier than other places. Another set of places was dealt a good hand when nature distributed environmental resource potentials. Some nations like the United States, Canada, and the U.S.S.R. are well supplied with technology and they also have bountiful resource endowments. Some nations like Sweden, Switzerland, and Japan are exceptionally skilled technologically but have relatively few natural resources. Because they have developed their human resources to such high levels, these advanced nations are able to produce or import the resources they require. Per capita production levels are low in the underdeveloped world mainly because prevailing technological levels are so primitive. In countries like India and China the problem of low productivity is compounded by rather low resource endowments per person, but even in countries such as Brazil and The Congo, where natural resources abound, the technological know-how is so limited that productivity per person is severely retarded.

Thus, the differences across the world in (1) access to new technology, and (2) resource endowments per capita are well illustrated by the generalized technology-resource regions of the populated world (Fig. 10–5). These differences from place to place in technology and resources produce the variations in agricultural land use patterns. Like variations in most geographical patterns, agricultural land use depends on: (1) the relative location or situation of a

place in terms of its access to other places; (2) local site differences which comprise the inherent attributes of a place; and (3) patterns of human behavior. We have discussed access to new technologies and the sporadic occurrence of resources. Let us turn now to human behavior.

People use their environments according to their needs, their wants, and the know-how they absorb while growing up in the society that rears them. Habit—or custom or culture or whatever we choose to call learned behavior—is a fortuitous result of where a person is born and reared, and where a group happens to be located in the world. Some places are so remote from idea centers that new ideas have seldom if ever penetrated until recently. These are the cultural and technological ends of the earth. But besides overcoming distance, new ideas have to overcome cultural habits. New ideas can take hold in an area only if they are suitable for that area. Suitability depends on what people think, how receptive they are to new ideas, and whether people think the idea's implementation would be appropriate in the existing man-environment system. Ideas are adopted most rapidly after a series of examples demonstrates conclusively that they work. But hardly anyone wants to be the first to try the new idea. Their view is: "Why should we try something that is new and risky? We might be ruined. Even if the old way is not very productive, at least we know it will work every time." Thus, even after dramatic demonstrations by agricultural extension workers, the old ways hang on despite irrefutable evidence of their inferiority. Unless people's perceptions and anticipations are favorable, new ideas will languish.

As an example let us consider the physical (site), the locational (situation), and the behavioral (cultural) constraints controlling the agricultural events in a hypothetical region. It is dry in the northeast and the climate grades gradually into wet in the southwest. On one edge of the region is a metropolis which acts as a market for part of the region (Fig. 10–6). Region *A* supports mainly subsistence cultivation; virtually nothing is marketed because of the region's great distance from the metropolis and its ignorance concerning marketing possibilities. Region *B* engages in commercial

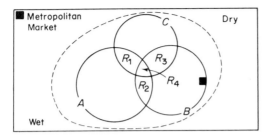

FIGURE 10–6

Agriculture in a hypothetical region. The activity at a place depends on site (intrinsic attributes of the place); situation (relative spatial position with respect to other places); and variations in perception and evaluation of alternatives by residents.

cultivation, especially for the metropolitan market located at the edge of the region. Region C, constrained by persistent drought and by a location remote from the metropolis, can neither cultivate crops nor profitably send to market any agricultural output. All its residents can do is practice a type of subsistence animal herding, perhaps goats. The predominant activity in each region is controlled by situation and site factors.

Several boundary zones exist where the three regions overlap. In the transition zones, mixtures of agricultural activities occur for a variety of situation, site, and behavioral reasons. For one thing, farmers differ in the amount of risk they will accept. In regions R_1 and R_3 (Fig. 10–6), for example, there is a risk of drought which some (farmers) will accept but which others (animal herders) will not tolerate. Some farmers carry on subsistence operations in region R_2 and others are commercial operators. This is partly because farmers differ in their awareness of marketing possibilities. Some farmers do not know of the possibilities and others, although they know, fail to act according to their knowledge. Market information diffuses away from the metropolis at different rates according to road density, and it diffuses away from the roads at much slower rates at the micro-regional level. As agricultural outputs move from farm to market, they encounter many of the same obstacles to movement that people

and ideas encounter. Crops are almost as difficult to market as ideas, and therefore some of the farmers in region R_4 will remain subsistence operators as others become commercial farmers. Finally, cultural variations among people make ideas move rapidly through some populations and slowly, if at all, through others. Thus, the reasons for mixtures of agricultural activities within the transition zones are the same as the reasons for geographical variations at the macro level. The three reasons are: (1) differences in relative locations in a communications space; (2) site differences; and (3) variations in human behavior habits. Let us turn, then, to an examination of how relative location can influence the use of the land. In the next chapter we will examine how the diffusion of new ideas brings about change in human behavior. In Chapters 12 and 13 we will discuss how man confronts his environment and puts it to work for him.

LOCATION AND THE USE OF RURAL LAND

Imagine an isolated homogeneous agricultural region consisting of a market town surrounded by a farming region. All farm output is sold in the single market town and all the farmers' non-agricultural wants and needs are supplied by the town. (These assumptions are based on those of the famous model devised in 1803 by a German economist-landowner, Johann Heinrich von Thünen.) What will be the production pattern and related land uses around the town?

Let us assume that the farmers are aware of only three production possibilities, and that local site resources will permit all three. They can go into the dairy business, the grain business, or they can raise livestock. Which one does each farmer pick for his particular farm? Or does each pick a mixture? On every farm in the region, a hectare of land devoted to each activity carries with it a set of revenues and a set of costs. Let us say that a hectare of land devoted to fresh milk production yields an output per year that brings revenues of $100 at the town market. The hectare's worth of output generates some production costs, say $30 per year, as well. The

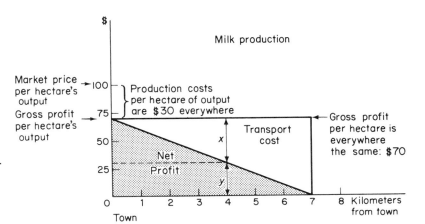

FIGURE 10–7

Net profits from milk production decline on farms increasingly distant from the market town. Net profit = 70 − 10x, where x = distance in kilometers from town. Thus, a dairy farm four kilometers from town earns a net profit of $30 per hectare, or 70 − (10 × 4). Net profits per hectare are zero at farms seven kilometers from town, and losses are incurred beyond seven kilometers.

gross profit per hectare is therefore $100 minus $30, or $70 per hectare per year. Thus, a dairy farm *at the edge* of town would earn a net profit of $70 per hectare.

Now consider the effect on net profit if dairying is carried on *away* from town. Since the farming region is homogeneous with respect to farm site quality and farming skill, a hectare of land anywhere in the region, if devoted to dairying, will yield output per year with a *gross* value of $100. Production costs per year remain at $30 so gross profits once more are $70 per hectare per year. But if the farm is located at a distance from the town, the milk output must be transported to market before it can be sold. The farther the farm is from the market, the larger is the portion of gross profit that is eaten up in movement costs. If it costs $10 per kilometer to ship to market the milk produced on one hectare (a hectare's worth of milk), then a dairy farm located one kilometer from town would earn a *net* profit per hectare equal to: revenue ($100) less production costs ($30) equals gross profits ($70); and gross profits ($70) minus transport costs ($10 × one kilometer) equals net profits ($60 per hectare for a dairy farm one kilometer from town). The relationship between revenues and costs for dairy farms at different distances from the town is concisely presented in graphic form in Fig. 10–7.

The farther a farm is located from its market, the larger is the portion of gross profits that must be paid out in transport costs per hectare of out-

put and the smaller is the net profit per hectare per year. At farms beyond seven kilometers from town, it would be silly for farmers to produce milk. Gross profits (market price minus production costs) per hectare equal $70, but it costs more than $70 to get the product to market. Net profits are reduced below zero. A farmer beyond seven kilometers who insisted on producing and marketing milk would not even break even. He would lose money.

Rural Land Use Patterns in the Isolated State: The Impact of Relative Location

Fortunately for the farmers in our region, there are land use options in addition to milk production. The land is also suitable for grain and meat production. Each of these activities has a corresponding set of revenues and costs, as shown in Table 10–2. When a farmer decides what to produce at his farm, he bases his decision on *net* profit per hectare. At every farm location there are three activity options and three net profit levels. If a farmer wants to maximize, he chooses the activity that returns the highest net profit per year.

We have already seen (Fig. 10–7) how a farmer's net profit levels for milk production differ depending on the distance between farm and market. Consider now the three options together (Fig. 10–8). One kilometer from town a farmer could carry on any of three activities and make a nice net profit on each of them. Which

TABLE 10–2

Annual Revenues and Costs per Hectare for Farmland Devoted to Three Activities

(1)	(2)	(3)	(4)	(5)	(6)
Crop or Land Use	Market Price per Hectare of Output per Year	Production Costs per Hectare of Output per Year	Gross Profit per Hectare of Output Per Year (Col. 2 − Col. 3)	Transport Rate per Kilometer per Hectare of Output	Limit of Profitable Production: Kilometers from Town (Col. 4/Col. 5)
Dairy	$100	$30	$70	$10	7 kilometers
Grain	65	20	45	3	15 kilometers
Meat	45	15	30	1	30 kilometers

one should he pick? If he wants a maximum net return per hectare per year he picks dairy farming. The graph in Fig. 10–8 summarizes net profit per activity for varying distances from the town. Net profit can be read directly from the graph. One kilometer out, the three activities compete for the land but the dairying wins because its net profit level is highest.

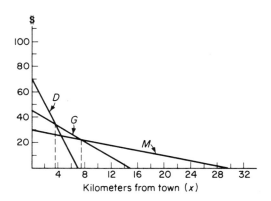

FIGURE 10–8

The von Thünen model assuming three alternative land uses. Each of the net profit curves plotted above can be defined as the gross profit minus a transport bill. The transport bill is the transport rate times the distance x *between a farm and the town:*

$$\text{Net profit for dairy } NP_D = 70 - 10x$$
$$\text{grain } NP_G = 45 - 3x$$
$$\text{meat } NP_M = 30 - x$$

The intersection of two net profit curves means that at that distance from town the two activities yield the same net profit. For example, $NP_D = NP_G$, *where* $70 - 10x = 45 - 3x$, *i.e., where* x = 3.6 *kilometers.* $NP_G = NP_M$, *where* $45 - 3x = 30 - x$, *i.e., where* x = 7.5 *kilometers.*

Milk is expensive to ship. It spoils if it gets old or is allowed to get warm. Modern refrigerated trucks have changed this in part, but traditionally milk could be produced only near towns and city markets because transport costs were so high. If transport costs are high, then net profits from dairy farming drop off sharply with increasing distances from market towns. In our example (Fig. 10–8), high transport costs cut into the net dairy profits so fast that at a distance of about 3.6 kilometers they have dropped to the point where they equal the per hectare net return from grain farming.

Now a hectare's output of grain has a lower market price than a hectare's output of milk, and although production costs are lower on grain, gross profit levels on dairy farming ($70 per hectare) are substantially higher than gross profits on grain farming ($45 per hectare). But transport rates on a hectare's worth of grain output are $3 per kilometer versus $10 per kilometer on milk. Consequently, the net profits on grain diminish relatively slowly with increasing distance between farm and market. At 3.6 kilometers, net profits on dairy farming equal net profits on grain farming. Between 3.6 kilometers and 7.5 kilometers grain farming yields higher net profits than either of the other activities. Beyond 7.5 kilometers meat production is most profitable. Meat is produced out to 30 kilometers. From 7.5 to 15 kilometers both grain and meat return positive net profits, but meat production produces higher net profits than grain. Farmers compare the profitable alternatives at their location, then they select the most profitable alternative.

In our simplified model, the profitability

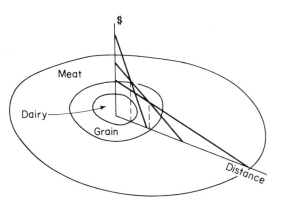

FIGURE 10–9

Land use rings around the market town if all assumptions are fulfilled.

of each activity at each location depends only on gross profit levels, distance and transport rates, and in this way location influences land use. Once we assume that a farm is located at a place, its activities depend on its location or distance from the market and the cost of overcoming distance. The rational, ideal, profit maximizing pattern of land uses presented in Fig. 10–8 yields a spatial pattern of concentric land use rings (Fig. 10–9). Beyond the thirty-kilometer limit, land is unused. None of the three possible enterprises can be carried on beyond thirty kilometers if farmers are to earn positive net profits.

Effects on Land Use Patterns of Changes in Revenues and Costs

A static concentric or annular land use pattern would result if rational farmers tried to maximize net profits per hectare in a homogeneous farming area serving an isolated town. Static conditions produce static concentric rings but conditions may change in three ways:

1. Revenues per hectare may rise or fall with changes in the town or changes out in the country or both. Towns can grow or shrink; tastes can change, and the demand for agricultural output will shift accordingly, raising or cutting per hectare revenues.
2. Production costs can vary through time also.

New production techniques, whether tools, fertilizers, or procedures, can cut costs. Bad weather or crop disease or increases in the costs of labor, machinery, and other inputs can raise production costs per hectare.
3. Transport costs vary through time, and the variation is usually downward. When von Thünen first presented his theory of agricultural land use from which this presentation derives, all commodities moved by wagon or on the farmer's back. Von Thünen's model was first presented in 1803. The inchoate theory was refined in his classic treatise: *The Isolated State* published in 1826. In the interval between 1803 and 1826 the steamboat and railroad were introduced, and since then land transportation has steadily improved, cutting transport costs lower and lower and causing radical adjustments in agricultural land use.

Each variable in the model, including revenues, production costs, and transport costs, is subjected to general inflation or deflation. Each variable can also change independently of the others. Both general changes and special changes initiate modifications in the concentric land use patterns (Fig. 10–10). Assume a revenue and cost structure that generates the original land use pattern at time t_0 (Fig. 10–10a). If all market prices double, revenues will double for every product and gross profits will rise also. Transport rates stay the same. This situation is shown in Table 10–3.

The *profit* zone for an activity is usually larger than the *production* zone for that activity. An output is chosen for production only where it yields a *higher* net profit than all the others. When all prices advance 100 percent, all production zones spread out. The limit of production is extended from 30 kilometers to 75 kilometers (Fig. 10–10b). Thus, the effect of generally higher prices is to bring previously submarginal land into profitable production. If all market prices were to drop, the margin of production would draw back toward the town and put the marginal farmers out of business.

If people get tired of eating grain products, because they prefer more meat and dairy products, because they think grain goods are too fattening, or for other reasons, the demand for grain may shift downward and revenues per hectare of grain may decline while dairy and meat revenues stay high. In the example shown

General changes

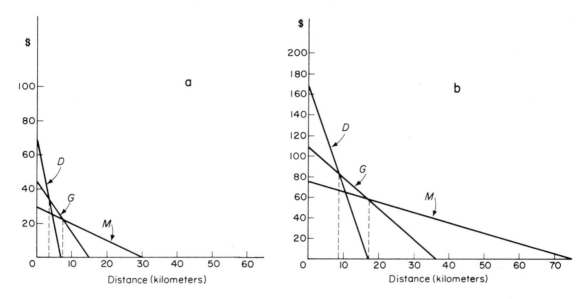

FIGURE 10–10

General and special land use changes within the von Thünen framework. (a) *The original situation.* (b) *All revenues per hectare double; gross profit levels and net profit levels adjust accordingly; land productivity remains unchanged and all three land use zones expand.* (c) *Gross profit on grain drops from* GP_G *to* GP'_G *due to a price decline or a production cost rise. Grain is no longer produced.* (d) *Transportation rate on meat drops 50 percent from 1.0 per kilometer to 0.5 per kilometer; the meat production zone expands outward and inward.*

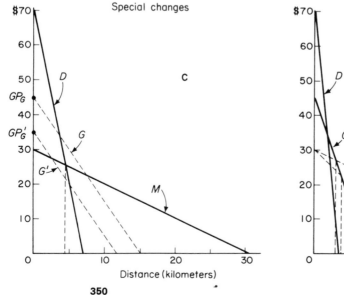

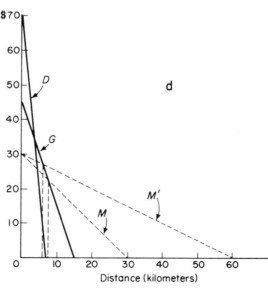

TABLE 10–3

Annual Revenues and Costs per Hectare after Market Prices Double

(1)	(2)	(3)	(4)	(5)	(6)
				Transport Rate per Kilometer per Hectare's Worth	
	Revenue Change	Production Cost		of Output	Limit of Profitable
Crop	$t_0 \rightarrow t_1$	(Unchanged)	Gross Profit		Production t_1
Dairy	$100 \rightarrow 200$	$30	$170	$10	17
Grain	$65 \rightarrow 130$	20	110	3	$36\frac{2}{3}$
Meat	$45 \rightarrow 90$	15	75	1	75

(Fig. 10–10c) net profits on grain farming diminish at all potential production points when gross profits drop from GP_G to GP'_G. After grain's zone of profit shrinks, at no point is grain more profitable than either dairy or meat raising. In such a case the grain production zone would be eliminated and the other two production zones would move in to replace it.

Transportation rate changes can also cause profit and production zones to expand and shrink. Consider a case where the transport rate on meat drops 50 percent due to an innovation. Perhaps a railroad is built or refrigerated freight cars are introduced (Fig. 10–10d). The lower freight rate means that even though gross profits at the market remain unchanged, net profits decline more slowly than they did before the rate reduction. Meat was once profitable out to 30 kilometers. After freight rates drop by 50 percent, the profitable zone for meat extends from zero to 60 kilometers, and the production zone extends from 6 to 60 kilometers instead of the former 7.5 kilometers to 30 kilometers. The meat zone widens inward toward town because the transport rate on meat went down while the grain transport rate stayed the same. With its newly acquired competitive advantage, the meat production zone can nibble 1.5 kilometers into the grain zone before, at 6 kilometers from town, grain and meat each produce equal net profits per hectare.

Pennsylvania and the Whiskey Rebellion

In early post-Independence times the commercial farmers of Pennsylvania lived in a variation of von Thünen's isolated state. Urban markets along the seaboard were supplied with food by interior farmers who shipped their produce overland. Transportation costs were high for everything and took a form similar to that indicated in Fig. 10–11.

Close to the market farmers engaged in dairy production (D), wood production for heating and construction (W), and horticulture (H). On the frontier, farmers produced grain (G) and especially wheat, which was harvested once a year and was relatively compact and easy to ship. Beyond the wheat frontier, meat was raised by grazing cattle and hogs in the woods. Live animals could transport themselves to market in the autumn and thereby overcome the difficult obstacles to interior travel. A few frontier farmers were so far from markets that they could enter the cash market only by raising grain and converting the grain to whiskey. The barrels of whiskey were high in value per unit weight, and a hectare's worth of grain in the form of whiskey could easily be piled on a mule's back and moved to market at relatively low per kilometer costs.

Frontier distilling went smoothly until the government decided to levy a special tax on whiskey. In September 1794 the frontier farmers rose up in rebellion against the tax and Federal troops were used to suppress them. The famous Whiskey Rebellion of the western Pennsylvania farmers is called to the attention of American history students as an early example of the use of Federal authority in the young Republic. But let us examine the economic geography of the Whiskey Rebellion. Frontier farmers produced whiskey because this was their *most* profitable cash producing activity, and for farmers beyond the meat profit limit, whiskey was the *only*

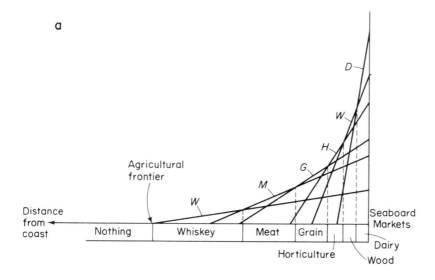

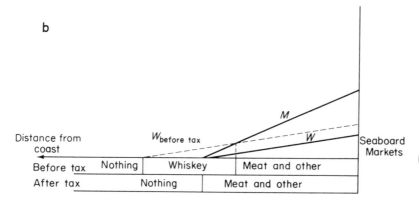

FIGURE 10–11

The circumstances leading to the Whiskey Rebellion.

profitable enterprise. Thus, when the government placed a tax on whiskey, the effect on net profits was the same as if whiskey prices had dropped or production costs had risen or some combination of the two.

Once the tax was levied, the net profit zone for whiskey production shrunk (Fig. 10–11*b*) and at no place was whiskey production more profitable than meat production. No doubt the rugged individualists beyond the frontier were looking for a good fight. Nevertheless, their anger was justified by the impact of such a discriminatory tax. In any case the situation was hardly permanent. As urban seaboard prices

rose and as better transport and settlement advanced westward, the whiskey frontier moved also and western Pennsylvania farmers turned to other pursuits.

Production Zones in the Fiji Islands

In the Fiji Islands the port town of Suva dominates the archipelago (Fig. 10–12). People in the outer islands come to Suva for educational, medical, and social services, and to purchase commodities. The main mode of travel is by wooden cutters and schooners of from ten to fifty tons. All shipping lines converge on Suva,

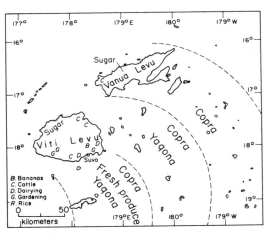

FIGURE 10–12

Cash production zones in the Fiji Islands (adapted from Couper, Fig. 2).

and it is difficult to travel from one island to another without going to Suva first, except when destinations happen to lie on the same line.

Because port facilities are so poor and boats are so small and slow, shipping is very expensive, and extremely high freight rates prevail between the islands and Suva. The time and cost barrier means that perishable cargoes are seldom shipped, and when they are it is only over short distances. Thus, the surplus foodstuffs in the islands cannot be safely marketed in Suva. The most distant islands, in fact, can safely ship only copra to Suva. Closer to Suva, yaquona (a shrubby pepper from which a beverage is made) and copra can both be marketed profitably. In the vicinity of Suva, fresh produce joins the list of profitable activities. The resulting distribution of significant cash production zones resembles the concentric ring pattern postulated for von Thünen's isolated state.

On the large island of Viti Levu, which is ringed by a main coastal road, production again adjusts to accessibility conditions and production is concentrated along the coastal areas rather than in remote interior areas which are not served by good roads. On the other main island, Vanua Levu, a main road along the north shore permits the profitable production

of cattle and sugar and easy overland transportation to the southernmost tip of the island, where these products are transferred to ships.

Location and Mineral Resource Exploitation: Von Thünen in Three Dimensions

Mining, another resource-based enterprise, has usually been regarded as the simplest case of the location problem. Coal, after all, cannot be mined except at coal deposits. However, a location question still arises on two levels: which mineral deposits will be tapped first? And within a mineral region, say a coal field, where will individual pits be located? Wilson considered this problem and its implications and then applied his model to the main New South Wales coal fields in southeastern Australia.

Imagine a hypothetical coal field containing a thick continuous coal seam (Fig. 10–13). From a surface outcrop the seam dips gently away to ever-increasing depths. The coal can be mined in either of two ways. Tunnels can be sent in laterally from the outcrop. As the seam gets deeper and deeper, tunnel mining gives way to vertical shaft or pit workings, and still further on, short shafts are followed by deep ones. Assume that all coal must move from the seam to the market in three steps: (1) from seam to tunnel mouth or pit head; (2) from tunnel mouth or pit head to an assembly and shipping point; and (3) from shipping point to market. It is not unreasonable to expect mine owners to locate operations on the coal field so as to minimize aggregate charges for all three steps. If transport costs increase with distance along each step of the way, the first mine will be located in the section of the coal outcrop that is nearest the market.

Such a coal mine location problem is something like a three-dimensional von Thünen problem with one crop. Coal deposits nearest the market are tapped first ($1a$, $1b$, $1c$ in that order). When market prices rise or production costs or transport costs drop, the range of profitable production points expand. Old tunnels go deeper ($1a$, $1b$, $1c$), new tunnels are dug ($2a$, $2b$, $2c$), and eventually shafts are sent downwards with each expansion phase (3, 4, then 5). Let us consider these expansion phases in turn.

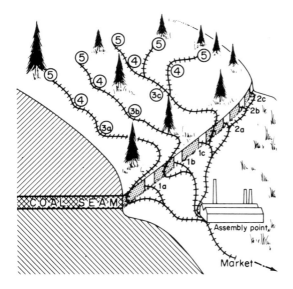

FIGURE 10–13

In coal field development, the deposits nearest the assembly point are tapped first. When they have been exhausted, deposits farther away and deeper in the ground are exploited (adapted from M.G.A. Wilson, "Changing Patterns of Pit Locations on the New South Wales Coalfields," Annals of the Association of American Geographers, LVIII [1968], Fig. 2, p. 79, by permission of the publisher).

As time passes, demand for coal rises, the early worked deposits are exhausted, and underground tunnel haulages become long and expensive. A second phase of mining activity begins with the opening up of a new batch of tunnel mines still on the outcrop but at a greater distance from the shipping point and the market (*2a, 2b, 2c*). Eventually, rising costs of surface haulage from tunnel mouths to the shipping point get so high that total costs are reduced by mining coal at greater depth but close to the shipping point (*3a, 3b, 3c*). The start of shaft mining through vertical or inclined shafts initiates a third coal field development phase when overland hauls from ever more distant tunnel mouths (phase 2) become excessive, and when haulages through tunnels (from phases 1 and 2) become excessively long and congested. In later stages of coal field development, new shafts are sunk to tap the shallow portions of the seam first until movements from

mine mouth to shipping point become so long it pays to sink still deeper shafts closer to the shipping point, trading increasing depth for shorter overland hauls. In the real world, there is seldom one coal seam of homogeneous quality, one assembly and shipping point, or one market. Thus, observed mining patterns appear highly irregular. Yet under the rational cost consciousness assumed in Wilson's ideal scheme, if we are given a location, the expected sequence of development should tend to occur according to the location of the coal and access costs.

In an exactly analogous way we could analyze the exploitation of the soft wood forests in the United States as lumbermen moved across New England, then the Lake states, and finally the Pacific Northwest. Forest stands closest to markets were exploited first, and with the passage of time the more remote forests were harvested. Similarly, in ocean fishing and in whaling proximate resources and shallow sea resources are exhausted before distant and deep sea efforts begin. In oil exploration, Colonel Drake's Pennsylvania discovery was followed by other wider and deeper searches. Recently, petroleum deposits deep under the sea and in remote northern Alaska have come into production. In all such cases the effective economic distance to market regulates the timing and extent of exploitation at the location of a resource.

LAND USE PATTERNS INSIDE THE CITY

A city is a locational arrangement of specialized people and activities designed to maximize transactions. At the local level a city is the best way for social and economic units to interact for maximum benefit to all. At the regional level, systems of cities come into existence to negotiate transactions between distant places, and to provide surrounding non-urban areas with goods and services they require.

Inside a city, location affects human activity and human activity modifies locational arrangements. Let us ask the topic question: how did the various space-using activities come to be located where they are in the city? Prevailing urban land use theory argues that a tract of urban land ends up supporting the activity that will pay the highest rental at that tract. If no user comes along, the highest rental is zero and the tract remains vacant. If several potential users are interested in the tract, the one who is willing to pay the highest rental will get the tract and put it to what is called the tract's *highest and best use.*

The rental is the price of a location. If a tract is owner-occupied, the owner in a real sense must "pay" himself each year a rental at least equal to the maximum rental he could charge another user. What the owner could collect if he rented to someone else represents an "opportunity cost" to the owner. It is this opportunity cost that the owner should use in computing what his own use of the tract is costing himself. If a tract is tenant-occupied, there is always a chance a potential tenant will come along and offer the landlord a higher rental. The tenant who is able and willing to pay the most should end up with the tract. If the highest and best use already occupies a tract, no rational, fully informed user will outbid it.

Why are users willing to pay a cash rental for tracts in the city? One reason is site features. There may be some intrinsic attribute of the tract which the user finds valuable. For a residential user it may be a view or a stand of lovely old oak trees; for a commercial user it may be a particular terrain configuration. An industrial user might find the land exceptionally well drained or maybe it contains on-site water, rock, clay, sand, or important material needed in a manufacturing process. For an office building, the site might be underlain by stable bedrock characteristics offering good foundation support. An advantage to one user might be an obstacle to other users.

Part of a tract's value depends on site attributes, but its relative location is usually far more important. By relative location we mean a tract's locational situation with respect to other activity areas inside and outside the city. If a tract of land can offer a potential user a substantial reduction in his total transport and communications bill, the potential user will gladly pay a higher rent to save money on interaction costs. Every activity has different transport requirements. A bread factory, for example, uses truckload after truckload of flour and ships out bread to all parts of the city. A steel mill brings in things like limestone, iron ore, and coke by the bargeload, perhaps scrap metal by the carload, and workers by bus and automobile. Then it ships out steel products by rail, barge, and truck. Steel mills therefore have a set of transport needs different from those of the bakery.

If Distances created no movement frictions, then one place would be as good as another when activities were located in and near cities. In a completely frictionless world, rentals could not be charged for intra-urban access or transport savings. Instead, rentals would reflect only the differences in site attributes.

Ceiling Rents and the Bidding Process

Within a city a ceiling rent exists at each site for each potential use. A ceiling rent for a use at a site is the maximum rental which that use would pay for that site. For any broad or narrowly defined use class a ceiling rent surface exists, defined by the ceiling rents at each tract for the use class under consideration. At a specific tract, say a hilly area at the edge of the city, the six highest ceiling rents might be:

agricultural	$50
mining	8
manufacturing	25
recreational	30
residential	85
retailing	15

In an open market involving informed landowners and potential users with good foresight, when the tract is sold or leased it will be put to residential use. A kilometer away on a flat tract next to a highway, a different set of ceiling rents prevails and the tract is purchased or leased for retailing:

agricultural	$ 65
mining	15
manufacturing	86
residential	60
retailing	175
general business and personal services	20

When a user discovers a tract which offers his activity some exceptionally attractive site and situational benefits, he tries to buy or rent the tract. Sizable benefits promise high profits, and with those profits comes the power to bid up the price of the tract, perhaps higher than any competing uses of a different kind, and as high as competing uses of the same kind. Obviously users must have correct information or errors and inefficiencies in land uses will result. If a tract owner rents to the first bidder who comes along instead of waiting for the highest bidder, the tract's advantages may never be exploited. If a tract owner thinks the ceiling rent is at a level which is higher than any bidder is willing to pay, then the tract will remain vacant, unless, of course, some misinformed user comes along and overestimates the value of the tract for his particular purposes.

The ceiling rents at each tract change through time. Ceiling rents for business and commerce traditionally have been highest in the middle of downtown areas, but their peak was reached a generation ago, and today cleared land near the downtown center is sometimes almost impossible to rent or sell. Ceiling rents were high at one time because accessibility of the downtown center to urban customers was excellent. Since rentals are the price of superior accessibility, the places most easily reached by the greatest number of customers could command the highest rentals.

The principle of "rent as the price of accessibility" is revealed in every American city. Because different activities have different interaction requirements and because different tracts have varying access advantages, activities tend to sort themselves out according to the bidding power of competitors in the land market. Some urban activities require access to all or part of the city. They will pay a premium for a good location on the *intra-urban* transport network. The downtown department store was the classic example of such an activity when streetcars were the main mode of intra-urban travel. A county general hospital and a police headquarters are other activities which need central locations and, thus, easy access to all parts of the city. The better the access, the higher the rent which will be paid by the highest and best uses.

Some urban activities need access to the surrounding region and to other cities, and so they will eagerly pay a premium for tracts along *inter-urban* transport lines. Factory districts have to assemble raw materials from rural and other urban areas, and outputs must be marketed in nearby rural areas and in other cities. Today the interregional transport system comprises water, rail, road, and air transport. Land use patterns in most American cities, however, were organized when water and rail routes alone were the most important intercity and interregional transportation systems.

Manufacturing and goods handling activities which need access to interregional transportation also need access to labor markets within the city. Thus, in bidding for sites, manufacturing activities have tried especially hard to get as close as possible to three types of places: (1) the city in which they operated; (2) a surrounding supply and market region; and (3) other cities. Within cities, therefore, ceiling rents for manufacturing traditionally were high along interregional transport lines, and low away from them, and they were very high where interregional transport lines came close to intra-urban transport lines (Fig. 10–14).

In areal and economic terms the three most important urban land using activities besides transportation are: (1) manufacturing and goods processing activities such as warehousing, wholesaling, and transshipment of goods; (2) local business, including retail trades, business services, and personal services; and (3) residence, which is by far the largest areal class of urban land use. In the absence of stringent zoning controls much of any city's land use pattern is the outcome of competition based on ceiling rent differences between industry and commerce. Residential land has usually been a residual use, something left over after the potent purchasing power of business and industry has preempted what it wants. Residential ceiling rents are usually substantially less than those offered by even the lowest bidding nonresidential users. As a consequence, it has been traditional in American cities for residential activity to settle on the periphery and for nonresidential uses to cluster at the core.

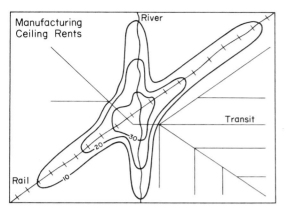

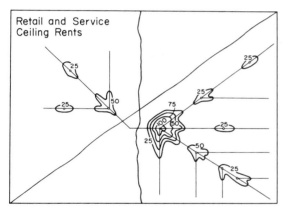

FIGURE 10–14

Intra-urban transport systems and ceiling rent surfaces for different activities.

Residential Land Use and Urban Transportation Eras

The assignment of activities to tracts in the city resembles the assignment of rural land uses to farming areas at different distances from the market. The downtown core of a typical city is a regional center at two spatial scales. It is the hub or nodal point for routes to rural hinterlands and to other cities. And it is also the place where intra-city transport lines converge. At one time a core location provided a firm with maximum accessibility (minimum aggregate inter-

action costs) of both types. The cash value of such superior access varied among the three use classes: industry, commerce, and residence. In the pre-streetcar era up to 1890, when everyone walked to work and to shop, ceiling rents declined from core to periphery for each activity class (Fig. 10–15). Commercial activities needed access to the whole city and could outbid all other uses. They therefore dominated the city center. Industry needed access to workers as well as to the regional transport hub. Both kinds of access were provided near the city center. Industry therefore occupied tracts adjacent to downtown. Residential land uses would have preferred locations near stores and jobs and would have paid a premium for tracts near the downtown center, but other uses pushed residents to the outer ring from which they walked to work.

A succession of innovations in intra-urban passenger transportation followed 1890. First came the electric streetcar, which dominated urban transportation until the 1920's. Then came the recreational automobile and the motorbus, which transformed commuting and

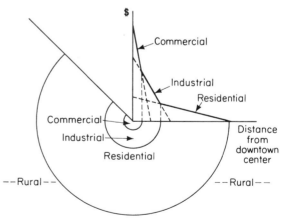

FIGURE 10–15

Urban activities compete with one another for locations which will reduce their transport costs. Transport cost savings enable users to bid up the price of tracts near the city center. The user making the highest bid obtains the tract for his use.

shopping habits up through World War II. After 1945, the freeway-automobile era removed practically all constraints on intra-urban passenger movements, and residential land use was possible almost anywhere in the metropolitan area (Fig. 10–16). Each revolution in passenger transportation decreased the desirability and the residential ceiling rents of inner areas and raised ceiling rents in a new band of rural land, thereby bidding it away from agriculture. With the introduction of the streetcar, people could get downtown to work and shop, and they were glad to get away from the congestion and high rents of the first residential ring which was established when everyone was forced to walk and therefore was required to live as close as possible to the downtown core.

Today, while the rich live in the suburbs on cheap land, the poor live in residential zone R_1 (Fig. 10–16) on land of somewhat higher value. Whether a family is rich or poor its budget is divided between (1) living costs; (2) housing costs; and (3) travel costs. After living costs are deducted from income, an allocation

is made between housing and travel. Inexpensive housing, usually smaller and older units, is concentrated near the downtown where the rentals per unit of land are high, and where net outlay per square meter of living space is high as well. Yet this is still the cheapest place to live, mainly because living space can be purchased in small units. Suburban families have enough money in their budgets after living costs have been deducted to pay for large amounts of suburban housing (which comes only in large units) in the form of big houses and yards or garden apartments.

Travel costs of both the poor and rich are reckoned in terms of time and money. For the inner-city poor, using public transportation is cheap in cents per kilometer but very expensive in time per kilometer. The suburban rich have more money than time. They pay a high money cost per kilometer, but they invest much less of their presumably more valuable time per kilometer of transportation than do the poor.

When the poor choose their housing and transportation arrangements, they have less money and more time than the prosperous

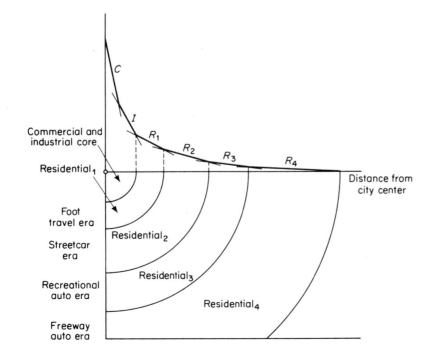

FIGURE 10–16

Passenger transportation within the city controls the population and housing density. As intra-urban movement becomes easier and easier, people are willing to pay less and less for central locations. Residential ceiling rents decline through time. The ceiling rents prevailing when a residential tract is first built up determine the density.

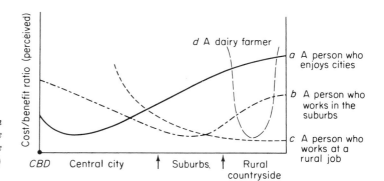

FIGURE 10–17

Variations from person to person in perceived cost/benefit ratios for alternative residential locations along a traverse from the central business district (CBD) into the rural countryside.

middle class, so they really are left with no choice as to where they can live. Even when their jobs move to the suburbs they are stuck in old inner-city neighborhoods. Even if the poor could find cheap housing in the suburbs they could not find public transportation to work. The dominant political force in suburban areas is a middle and higher income, automobile-oriented public which has vetoed both low cost housing *and* public transportation.

The choice of a place to live implies a local environment or site including the house and neighborhood, locational situation with respect to other places, and a set of moving costs. Like a leaf trying to reach the sunlight yet forced to remain attached to the tree branch, suburbs try to get as far as possible from the concrete, asphalt, smoke, noise, congestion, and racial and ethnic heterogeneity of the city while retaining maximum access to urban jobs, stores, schools, and amusements. A set of centrifugal forces throws them outward, and a set of centripetal forces draws them toward the urban system of income and consumption flows.

A residential location choice among alternatives is made on the basis of expected costs at alternative locations compared to expected benefits. If a person works downtown, the cost-to-benefit ratio might be lowest for him near the edge of downtown if he likes action, hates trees, and despises snow-covered driveways in winter (Fig. 10–17a). If a person's job is in the suburbs and his cost-to-benefits ratio rises for residential locations in the more urban or more rural settings, he probably will live close to his job (Fig. 10–17b). A person who works at a rural job

probably forgoes benefits and incurs high costs the closer he lives to the city center (Fig. 10–17c). A dairy farmer, for instance, is tied closely to his job and usually lives on his farm at his job (Fig. 10–17d).

After a family has selected a particular housing arrangement, in a specific neighborhood, at a certain location within the urban area, it is satisfied for a while. But sooner or later the family discovers that the benefits and costs of its housing arrangements start to change as the family enters a new stage of the life cycle. Tastes also vary. Some people like the vitality and variety of the inner city and others are intimidated by it. Purchasing power also affects a family's range of residential options. If a family's income rises, they may move to a nicer neighborhood. The poor also may become dissatisfied with their housing, but unlike the more generously endowed they are powerless to improve it.

Urban Transportation Improvements and the Elimination of Locational Advantages

The geography of American cities is changing because interaction needs continue to expand and because interaction possibilities have improved remarkably in recent decades. According to Edward Ullman, today's larger cities grow fastest because they permit a much greater number of fruitful internal and external interactions than do smaller places. Most of America's very small centers are stable or declining because they lack the interaction potential shared by the metropolises. Big centers can get bigger more easily because they have a larger mass to begin with. They can attract the innovators, the talent, relatives, friends, and other immigrants who are the prime resource of any city. A city's activities prosper in proportion to the number of similar as well as complementary activities that are located near one another. The cost savings that accrue when several activities operate at the same place are called *external economies*. Up to a point, the greater the volume and diversity of a city's activities, the more each of them can prosper. The specialties, facilities, and services in the metropolis are interdependent and mutually beneficial.

As a city grows in size and expands the interaction needs of its enterprises, it strangles if things are not properly located and if movements among interacting elements are too difficult. Indeed, some of the largest cities in the country are choking now, but the seizures so far have not proved fatal. Instead, transportation inside the city has actually improved as a result of the short haul advantages of automobiles which are self-loading and unloading, and motor trucks which can move to any address in the city. Moreover, the car and the truck allow the indefinite areal expansion of the metropolis. The only reason traffic congestion is so bad at rush hour is that we insist on allowing an unnecessary concentration of jobs downtown, and a concentration of the low-income housing in only one zone of the metropolis. Suburban areas can now compete with core areas in almost every respect. Nowadays, an enterprise need only be *somewhere* in the metropolis to be able to function at an efficient operating scale and to reap internal economies. Due to the ease of intraurban transportation and communication, any enterprise anywhere in the metropolis can take full advantage of the external economies that come with a metropolitan location.

The metropolitan growth industries are now the tertiary trade, repair, business service, and professional service sectors, plus the burgeoning quaternary information and control enterprises. New consumption patterns have accompanied higher and higher incomes, and when consumption preferences change, production patterns must adjust. Consumption shifts away from food, clothing, and shelter induce a parallel swing away from the industries based on raw materials and natural resources. The nineteenth-century economy built goods-producing factories; the enterprises that characterize the late twentieth century are quiet offices, libraries, computer centers, schools, laboratories, and communications centers. They produce what people consume. They can operate efficiently along any street or highway that a car can reach, and they are located in suburban areas to an increasing degree each year.

Population led the way to the suburbs while other activities stayed behind for a while to use up fixed capital. Activities such as retailing, personal services, local government, and education sprang up in the suburbs to serve neighborhood markets. When the net returns from staying in or near the downtown dropped below the payoffs from relocating in the suburbs, manufacturing, wholesaling, and information operations started moving outward. Suburban quarters offered easy free parking for employees, which was an attractive alternative to parking downtown at a dollar or more per day. One-story suburban buildings were traded for downtown elevators and congestion. Continuous production processes in one-floor factories were traded for cramped multi-story operations. Occasionally, even the real estate taxes were lower, either because public service levels did not need to be so high in the suburbs or because their costs were passed on to homeowners.

Professor Ullman argues that what happened inside cities because of the urban transportation revolution can be compared to what happened to world land use patterns in the nineteenth and twentieth centuries. When better transportation systems allowed formerly isolated farmlands to produce for world markets and to compete with less fertile areas nearer the markets, activity levels dropped sharply in the marginal areas near the market. In the parallel metropolitan case, better urban transportation systems have allowed formerly isolated suburban farmland to support residential and nonresidential activities and to compete with less useful areas near the city center. As a result, activity levels have dropped sharply in marginal areas near the downtown.

In Europe and depressed parts of the United States, rural economic decline is widespread in farming and mining areas rendered marginal by transport improvement. Cheaper food and fuel can be purchased from other parts of the world. The Common Market, established by the Treaty of Rome in which Italy, France, West Germany, Belgium, the Netherlands, and Luxembourg agreed to allow ultimately the free flow of labor, materials, and products among all regions, is an effort to foster beneficial adjustments in regional economic growth. Yet whereas the six governments subscribed to the principle of free-market adjustment, they also promote farm subsidies as a way of slowing down and easing the adjustment process.

In European and American cities central-city blight and deterioration are widespread in marginal industrial, commercial, and residential areas. Superior site amenities and transportation advantages can be obtained elsewhere in the metropolis, especially in the suburbs. The freedom of producers and consumers to relocate where they choose within the metropolis speeds up the rate of abandonment of downtown facilities and deteriorated inner-city residential zones. Yet government-subsidized urban redevelopment schemes and the location of public housing projects at the urban core are devices created to retard the metropolitan adjustment process and the desirable relocation of people and activities.

As frictions of movement are eliminated in the city, land values and land uses begin to reflect intrinsic site qualities more than locational or situational attributes. Tracts near the central business district command little locational value because their relative accessibility has declined. Furthermore, tracts near the center often are distinctly undesirable because they offer bleak surroundings, the tracts often are too small, and existing structures usually are too expensive to demolish. Moreover, in many cities, after extensive urban redevelopment activity, cleared land finds few enthusiastic buyers. Sometimes all that a cleared tract will bring on the land market is a price slightly below that of a new suburban tract with nicer surroundings.

In the old days, suburban houses clustered around the railroad station along the commuter rail line because access to and from a job in the city was all-important. A man who lived too far from the station could not get to his job on time. Today, residential development emphasizes aesthetic or site amenities, not access. The automobile has made most suburban sites almost equal in terms of access. When sites are equal in access terms, people are free to look for other things like site qualities. Waterfront properties in and around cities command a premium because many people like to live by the water. Wooded hill land has a similar value in morainic portions of metropolitan areas around the Great Lakes or in mountainous areas close to metropolitan centers near the east and west coasts of Anglo-America.

Once public housing and redevelopment authorities pay the full economic cost of clearing an inner-city tract for public housing, the investment per hectare is usually astronomical. Because this preliminary investment is so great, a high rise building is necessary to spread the high per hectare costs of the cleared land over as many housing units as possible. It is a vicious circle. The land is cleared at a tremendous cost; then, to recover the outlay of money, high density public housing is constructed. It is hard to see who gains; after all, the cheapest form of housing development is two-story housing on new land in the suburbs.

The high density city with a multi-story downtown core was needed at one time when everyone walked or used the streetcar. Now that few people do either, the *raison d'être* of the high density city has vanished. The downtown was a least net effort solution to an access problem. It was the center of greatest accessibility in a period when people had to move on foot or streetcar. Thus, looking ahead we can expect the downtown or central business district to become one of several intra-urban nuclei, perhaps the largest in a metropolitan area but not necessarily so. New business centers will develop, but many new activities will avoid congested centers entirely. Nothing but habit or tradition argues that a location at a center point is superior to an uncongested location elsewhere—perhaps along a circumferential belt line portion of the Interstate Highway System.

URBAN LOCATIONS
IN A RURAL LANDSCAPE

The most basic principles of economics concern the relations between demand and supply. Demand is a schedule describing the different amounts of a good or service that customers stand ready to buy over a period of time, given different market prices. Supply is a schedule denoting the various quantities of a good or service which an industry's firms are willing to put up for sale over a period of time under different prevailing market prices. Each schedule can be presented as a curve on a graph or in tabular form (Fig. 10–18). At the point where the two curves intersect—price=60, quantity=200 in our example—an equilibrium price and quantity situation obtains. The equilibrium

is considered to be quite stable. If a higher price, say 70, were to prevail for a while, customers would cut back their purchasing below 200 just as suppliers are putting out 250 to sell instead of 200. The subsequent glut on the market, indicated on the graph by the gap between the amount demanded and the amount supplied at a price of 70, would cause prices to tumble until the market was cleared of surplus. Eventually, price will settle down at 60 with a quantity of 200 being sold. Conversely, if the price should fall below 60, eager buyers enter the market anxious to buy more than 200 units, and finding less, they start bidding up the price until the amount demanded equals the amount supplied.

Neither the demand schedule nor the supply schedule stays in the same position permanently. Consumers' tastes are too volatile and producers are always finding new ways to do things, or are leaving one line of production activity to enter another. If such changes occur, shifts in demand and supply may follow. When we talk about a shift in demand we mean that consumer behavior as described by the demand schedule has changed. Let us say that originally demand was described by D and supply by S in Fig. 10–19a. Then demand increased—that is, it moved to the right, so that at a price of P_0, customers are willing to buy quantity Q_1, a substantial increase over Q_0. But since suppliers are only interested in selling Q_0 when the price is P_0, there is a shortage on the market equal to Q_1 minus Q_0 so the price is bid upward until it settles at P'. At the higher price sellers increase to quantity Q' the amount they are willing to sell and a new equilibrium is established. Supply can likewise shift (Fig. 10–19b). If supply advances, the equilibrium price is lowered and equilibrium supply increases. Through time, demand and supply are in constant flux (Fig. 10–19c).

Central Places and the Distribution
of Goods and Services

Whereas geographers recognize the usefulness of the equilibrium demand and supply model, they are troubled by the model's failure to consider the fact that the market for any good or service is an *area*, not a dimensionless point.

A demand schedule		A supply schedule	
Price	Quantity demanded per unit time	Price	Quantity supplied per unit time
100	0	100	400
80	100	80	300
60	200	60	200
40	300	40	100
20	400	20	0
0	500		

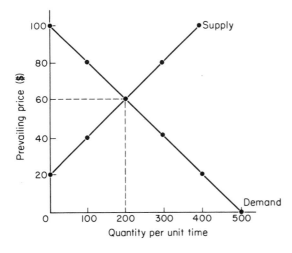

FIGURE 10–18

A demand schedule, a supply schedule, and the equilibrium price and quantity.

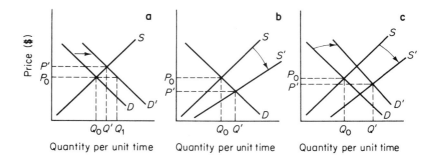

FIGURE 10–19

Supply and demand shifts.

Thus, in geographical analysis, a demand-supply model of the form given in Fig. 10–18 is understood as an average condition in a national or regional economy. The equilibrium price is understood as an *FOB* store or mill price. Customers who do not live next to the seller must pay more than the equilibrium price, the difference being attributable to transport costs between the factory or store and the customer.

Let us assume that demand and supply interact at the national level to determine a standard national price P_N, such as $2 for a haircut or $9 for a fifth of sour mash Tennessee whiskey. The prevailing prices mean that if you live next door to the barber shop or upstairs over the liquor store the cost to you for the service or good is the shop or store price. If you must travel any distance to either one, however, their delivered prices rise with your distance from the central place where they are sold (Fig. 10–20). You must pay the shop price plus transport costs.

Let us assume that Fig. 10–20a represents the national market structure for hot pizzas and Fig. 10–20b, defining an individual college student's demand for this product, describes the quantity purchased per person per unit time for a full range of delivered prices. Assume that delivery charges increase with distance whether the customer picks up his pizza or has it delivered. Under these assumptions, a student living above the pizza shop pays the established carry-out price P_N and buys q_1 pizzas per unit

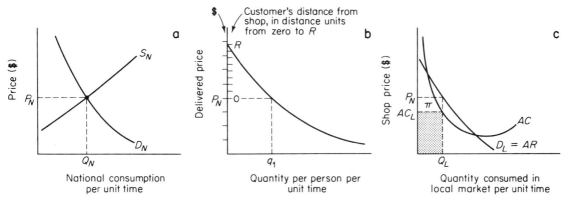

FIGURE 10–20

The effects of distance on delivered prices and on quantity consumed. (a) Demand and supply relations in the national market, with equilibrium price and national quantity established. (b) Individual demand schedule for a person inside a local market. He pays a delivered price per unit based on the established shop price plus a delivery charge. The higher the delivered price, the lower the consumption per unit time. (c) Costs and revenues for a shop selling within a local market area.

time. People living a distance away from the shop have to pay a delivered price higher than P_N and their consumption of pizza drops accordingly. At distance R the transport costs added to the shop's price raise delivered price so much that the quantity demanded per person drops to zero. Distance R is thus the radius or *range* of the pizza shop's market area.

The *threshold* is the minimum amount of sales needed per time period to bring a firm into existence and keep it in business. At an established pizza price of P_N and a maximum market penetration of R kilometers, will a pizza shop come into existence? That depends on how many people live within the circular zone R kilometers in diameter around the proposed shop site. Figure 10–20b defines the number of pizzas per person per unit time for different distances. What is needed is a population density figure for each point between zero and R kilometers. Within one kilometer of the shop site there are 3.14 square kilometers. If the population density is 4,000 per square kilometer, 12,560 people live within one kilometer. If R is 15 kilometers, then the area enclosed by a circle of R radius is 706.5 square kilometers. Each square kilometer has (1) a delivered price, (2) a quantity per person per unit time, and (3) a population. The total amount of pizza demanded per unit time, at a shop's price of P_N, within R kilometers can be computed. Assume that it equals Q_L (Fig. 10–20c), the quantity demanded by the local market when the shop price is P_N.

Will the pizza shop be established? Figure 10–20c gives average revenue information for different local outputs. If average revenue (price) is P_N and the local market wants to buy quantity Q_L, what average costs are incurred? Average costs must be equal to or less than price for the shop to open.

Price times quantity represents total revenue, which is also called the market. The threshold is the minimum market needed to bring a good or service into existence given the cost structure which prevails. In the pizza case, with a prevailing price P_N, an output of Q_L will be demanded by the local population. Average costs will equal AC_L and so the shop looks like a good bet. What about two shops across the street from one another? If they are identical they will

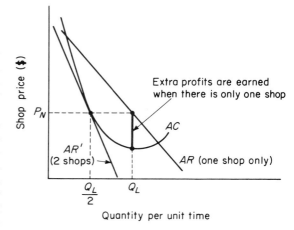

FIGURE 10–21

Threshold conditions present for either one shop or two shops. In the two-shop case, each shop faces a demand curve AR', *charges* P_n, *and sells* Q_L *per unit time.*

divide the market revenue and each will operate at higher average costs per unit of output. All extra profits will be eliminated, but both shops would survive under the conditions shown in Fig. 10–21.

Threshold, Range, and the Order of a Good or Service

The *threshold* for a firm selling a good or service is the minimum market (price times quantity) needed to bring it into existence and to keep it going. Threshold is often discussed in terms of a number of people, but counting people is only a substitute for measuring total effective demand at the store or shop. The *range* for a firm selling a good or a service is the average maximum distance people will go to purchase it. People want everyday needs like bread, milk, gasoline, or beer available at nearby stores and shops. No one would drive across the city for a loaf of bread. At ten o'clock in the evening when you are studying and run out of cigarettes, you might "walk a mile" for some cigarettes, but you would not walk much beyond that. The range for common everyday needs and wants is small.

In order for the sale of a good or a service to

be supported, threshold conditions must be met within the appropriate range. Threshold and range vary for each good and service. In nineteenth-century agricultural America, for example, the threshold for a country church was small, perhaps no more than a dozen-and-a-half families or so. Yet the range was limited as well. If the threshold could not be met within an area a few kilometers in radius, the church and pastor could not be maintained. If families were settled four to the section (one square mile or 2.6 square kilometers), and only half of them attended church, and a regular attendance of eighteen families was needed for threshold conditions, it would take an area only five kilometers square to support the church (Fig. 10–22). If the range (maximum distance people would go) for a weekly journey to worship was five kilometers the church could be supported easily. More than eighteen interested families live within five kilometers of the church.

In newly settled areas where travel was difficult, where population was more sparse, and perhaps the fraction of the population inclined

toward regular church attendance was smaller, churches were fewer and separated by considerable distance. It was often impossible for threshold conditions to be met within the range. On the frontier, the range was smaller than in more permanently settled Eastern areas which had better roads. People will go farther for any service or good if travel is easier.

If population and purchasing power are evenly spaced over the landscape, sales and service centers generally will spring up evenly spaced. For any areal market the most efficient system of marketing regions is a hexagonal lattice. Hexagons are the regular polygons closest to a circle in shape, but which still completely cover an area without any overlaps or unserved areas (Fig. 10–23). If threshold conditions are met within the range, a seller can locate and thrive at the center of his trade area.

What size are the hexagonal trade areas for each good and service? Spacing of sellers determines the size of trade areas. A seller's trade area must be large enough to keep him in business, yet not so large that peripheral customers are priced out of the market by excessive transportation charges. If thresholds can be met within customer buying ranges, sellers will appear as closely packed as they can to serve the area. If thresholds cannot be met within range, the good or service will not be available.

Broadway theater is an example of a service requiring an extremely high market threshold. A large population is needed because such a small fraction of the people at any one time wants to attend a play. Goods or services with high thresholds are termed *high order* central functions. Usually high order functions have wide ranges as well but Broadway theater is a partial exception. To see a play you must be there when the curtain rises and then return home the same night or total price of the play will become prohibitive. The play schedule and transport constraints mean that people who live more than about eighty kilometers from New York cannot attend the theater too often, so range is short. Yet thresholds are high. Of all cities in the United States, only in New York do so many people live within eighty kilometers of one point: the theater district in midtown Manhattan. This large potential clientele is supple-

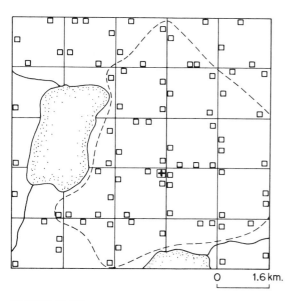

FIGURE 10–22

The approximate range for a country church in nineteenth-century rural America.

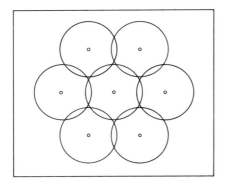

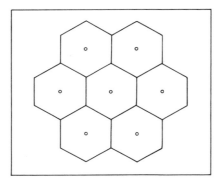

FIGURE 10–23

A central place cannot distribute goods and services beyond its tributary area. The radius of its trade area depends on the range of the functions performed in the central place (top). When evenly spaced centers compete with one another, the most efficient arrangement of trade area boundaries is a hexagonal lattice (bottom).

mented each day by enormous numbers of transients—tourists and business visitors—many of whom visit the theater sometime during their visits. So the theater district really serves two markets—a local one and a national one.

Several dozen theaters operate in New York City, but in Chicago and Los Angeles there are practically none by comparison. In those cities, the threshold population for the support of one theater is larger than in New York because a smaller fraction of the population likes to attend plays. Moreover, the average person in the Chicago area and especially in the Los

Angeles area lives much farther from the center of the city than does his counterpart in New York City. In the New York area, the average resident lives rather close to 42nd Street and Broadway.

High-Order Goods and Services and Itinerant Central Functions

The highest order goods and services may require the whole country to support them. The New York Stock Exchange and the Library of Congress are two such one-of-a-kind services. A specialized rare book dealer and a surgeon who performs a rare and costly operation may need the entire national market to keep them busy in their specialties. If the threshold for a central function comprises the whole nation, and the range is large enough to include everyone, all potential customers will be served.

It is just as easy for a person in San Diego to purchase stocks through the New York Stock Exchange as it is for someone in Brooklyn. Purchase costs do not increase with distance from the market place. Seldom is this the case, however. If a person in Lame Deer, Montana, wants to attend the Metropolitan Opera he is out of luck unless he happens to be visiting New York City. In a typical week, such a person is simply beyond the opera range. Yet if something is wanted or needed badly enough, and customers are too far from the central supply place, sometimes the seller can move instead of the buyer. History and geography are full of precedents. Itinerant peddlers on the frontier are but one example. Door-to-door salesmen are the modern version. If the customer will not come to the shop, the shop sometimes will move to the customer.

In part of the underdeveloped world periodic central places which operate only one day per week are quite common. Merchants are mobile rather than sedentary. By coordinating their movements from place to place a group of merchants can sell more than if they arrive alone at a town to sell their wares. The range for one good or one service might usually be very short, but when shopping lists contain several items, the range for the set of goods and services is substantially greater. In England such traditional arrangements persist into modern times, with the bus system responding by providing special bus services for rural areas on market day. In Belleville, Pennsylvania, the Old Order Amish farmers of Mifflin County gather every

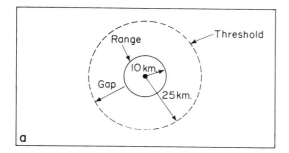

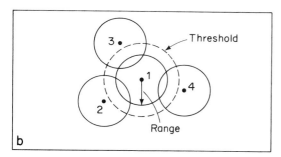

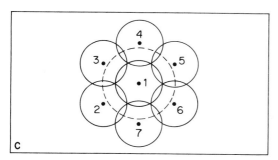

FIGURE 10–24

When the threshold cannot be met for a central function within the range of the function, the function will not normally exist at a central place (a). Sometimes the seller visits market fairs to expand his trade area and to eliminate the gap between threshold and range. The smaller the gap, the fewer days a merchant needs to be on the road to stay in business (b vs. c).

Wednesday to buy and sell at public auction their weekly needs and farm produce.

Let us examine the periodic market day, or market fair as it is sometimes called, according to the principles of threshold and range. Throughout much of the underdeveloped world, if a seller settles at one point (Fig. 10–24a), he can draw customers from perhaps a ten-kilometer radius because of poor transportation. Yet he may need all the trade within twenty-five kilometers to reach the threshold sales level necessary to keep him in business. If the situation were reversed and people were willing to travel up to twenty-five kilometers, yet the merchant only needed the customers from within ten kilometers to reach the threshold he needs to stay in business, there would be no problem. Under primitive transportation conditions, a merchant has two choices. He can go out of business, or he can move from place to place with the number of stops per week or month depending on the gap between threshold and what the range allows. Sometimes four stops per week are satisfactory (Fig. 10–24b). Patterns of six peripheral stops and one large fair at the major central place on the seventh day are not uncommon (Fig. 10–24c). In many parts of West Africa, itinerant merchants frequently visit two market fairs per day—one in the morning and the other in the afternoon.

The circuit courts of appeal in the United States were organized in a similar way. All parts of the country legally had access to federal appeals courts, yet in early days no small region generated enough appeals to keep a Supreme Court justice busy full-time in any one place. Instead, cases were collected and heard as each justice traveled around his circuit. The Metropolitan Opera and most symphony orchestras go on tours today for the same reason. Big top circuses once did the same thing and professional wrestling exhibitions continue to do so. Most smaller cities and towns can fill the house for one or two performances, but the local market is insufficient to support the activity on a continuing basis. Conversely, if many touring shows, including orchestras, theater groups, circuses, and ball clubs, had to perform only at home in a large metropolis, often they would go bankrupt.

The Marketing Principle and Lower-Order Functions

Assume that two high-order central places each offer a full line of goods and services and lie along a traverse *A-B*, separated by a considerable distance (Fig. 10–25). The number of contacts per unit time between a center and each point in its market area declines geometrically with distance from the center. Consumption drops because transportation costs to the central place raise delivered costs of goods and services to people who live away from the center. When delivered costs rise, consumption declines. At point *C* midway between centers *A* and *B* the influence of both centers combined is weakest. If central functions are added in sufficient numbers at *C*, they can make inroads on the spheres of influence or tributary market areas of both *A* and *B*. If *C* can meet threshold conditions for a whole set of goods and service functions the new center can prosper. Center *C* will draw trade from its adjacent hinterland. If there is not sufficient business to support all three centers, the new center or one of the old ones must fail.

An enterprise can sell any good or service if average revenue exceeds average costs as described earlier. If several stores and shops are founded and prosper at *C*, a new central place can be established. This is how smaller cities, towns, and villages managed to establish themselves between large cities when the central place systems in agricultural areas were established. Central places like center *C* were able to prosper by performing intermediate order functions; then the same process was repeated for still lower-order goods and services which typically have smaller thresholds and shorter ranges. The process might repeat for additional centers until the market area lying between centers is so small that thresholds cannot be met for any good or service. The fact that lower-order goods and services could establish themselves on the margins of higher-order trade areas is referred to as the *marketing principle* of spatial competition and central place location.

In central place discussions the four lowest orders of central places are called hamlets, villages, towns, and cities. A typical set of goods and services is available in each order of center (Table 10–4). Driving through a rural agricultural area between two cities, we normally pass through a series of central places that resembles the sequence in Fig. 10–26. Between two cities a small number of towns is interspersed among a larger number of villages. The tiniest central places, the hamlets, are encountered most frequently. Such a sequence is the result of the marketing principle and spatial competition between central places. Between two villages, for example, there is not enough purchasing power to support another village, but there may be plenty of demand to support a hamlet selling lower-order goods and services to a small neighborhood area.

Traffic levels and traffic controls in and near each center, such as speed limits, traffic lights, and stop signs, are proportional to the center's order in the central place hierarchy. In the United States most cities and towns were founded and located as central places in the eighteenth and nineteenth centuries. As a result, the spacing of central places today

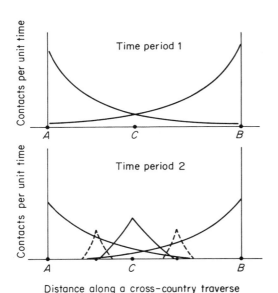

FIGURE 10–25

The emergence of central places at locations intermediate between existing centers, according to the marketing principle (adapted from Brush and Gauthier, Fig. 4).

TABLE 10–4

Typical Central Functions Offered by Each Order of Central Place

Order of central place	Central Functions			
	7th Order	*6th Order*	*5th Order*	*4th Order*
4th ORDER: CITY	all 7th	all 6th	all 5th	sheet metal works sporting goods store hospital regional high school
5th ORDER: TOWN	all 7th	all 6th	jewelry store department store dentist optometrist	(none)
6th ORDER: VILLAGE	all 7th	elementary school physician hardware store automobile dealer	(none)	(none)
7th ORDER: HAMLET (lowest order)	tavern gas station church grocery store	(none)	(none)	(none)

depends on horse and wagon transportation and the composition of consumer shopping lists in those days. Daily or weekly needs were handled at the local hamlet and village level respectively. Perhaps once a month a trip to town was necessary. Visits to the city might have been restricted to once or twice a year because high-order goods and services were seldom needed by rural folk.

Since the nineteenth century few new communities have been founded in the United States but many have disappeared. Customers no longer are constrained by horses and wagons, and besides they have more money to spend. Today because of the time-space convergence due to automobiles and good rural roads, a farmer can visit the sixth-order village center more easily than his grandfather could get to the nearest hamlet. He also has more money to spend and longer shopping lists. Many of yesterday's sixth-order goods and services have become today's everyday needs. The result has been the steady decline of the hamlet as a central place.

Inside American metropolitan areas the same

FIGURE 10–26

Four orders of central places along a highway traverse and the variations in traffic densities thereby created (adapted from Brush and Gauthier, Fig. 4).

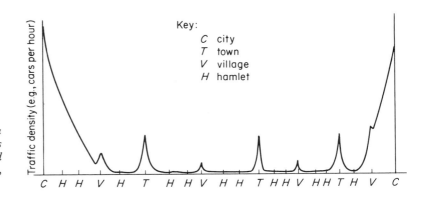

Key:
C city
T town
V village
H hamlet

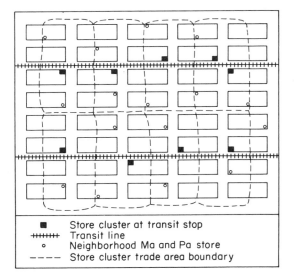

Store cluster at transit stop
Transit line
Neighborhood Ma and Pa store
Store cluster trade area boundary

FIGURE 10–27

Low-order central places inside the city during the streetcar era. Typical city blocks of about 2 hectares contained 25–60 families, depending on lot sizes and housing densities. The arrangements here are common in Midwestern cities.

thing has happened to the smallest central places, and for much the same reason. As mobility and prosperity improved after World War II, the ma and pa grocery store, supported largely by bread and milk sales to a hundred or so neighborhood families, passed out of existence. Ma and pa stores came into existence on the boundary of higher-order trade areas served by stores clustered at transit stops (Fig. 10–27). As incomes rose after the war, shopping lists became longer, people consumed less bread, and electric refrigerators eliminated the need to buy milk every day. Moreover, the introduction of the automobile made it easy to visit supermarkets which were built in abundance after 1950. These changes all spelled the demise of the small neighborhood grocery. Agricultural areas and metropolitan areas are both areal markets. The principles of central place location and central place viability apply continuously from the core of the city to the remote margins of settlement.

The Geographical Model of Central Places

The geographical distribution of central places displays remarkable regularities. In the ideal case, postulated by the German geographer Walter Christaller, if there is (1) an isotropic surface with equal movement ease in any direction, (2) a uniform distribution of population and purchasing power, and (3) uniform terrain and resource endowment, then settlements will spring up at evenly spaced points to serve tributary market areas with goods and services. Viewed areally, settlements will be regularly spaced. The spatial patterning of their locations will be a triangular lattice, and each center will serve a hexagonal trade area. If we consider the central places in terms of their relationships with one another, we observe that their organization follows a hierarchical pattern. Hamlets, the lowest-order central place, provide low-threshold, low-range goods and services to nearby hexagonal markets. A smaller number of villages come into existence to supply goods and services demanded so infrequently that one hamlet-level trade area alone could not support establishments providing them. Higher-order central places supply not only goods and services of their respective orders, but those supplied at smaller places as well (Table 10–4). Higher-order towns and cities offer a wider variety of goods and services and have bigger populations, more establishments, larger trade areas, and greater trade area populations than hamlets and villages. The hierarchical organization of central places is expressed in two ways: higher-order centers are evenly spaced but more widely spaced than lower-order centers; and an orderly nesting of market places ensures that every low-order place is served by every level of high-order place (Fig. 10–28).

The central place model developed by Christaller represents a hierarchical theory wherein each central place is a member of a functional order of centers. Each order performs a specific group of central functions and the centers in that order have populations which fall within a certain size range. Part of a central place's population depends for its employment on the goods and service business provided by sur-

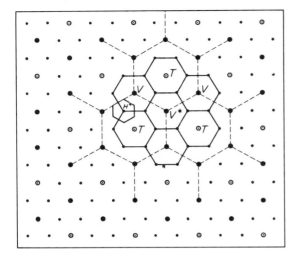

FIGURE 10–28

The location of hamlets, villages, and towns within the central place model. Centers of each order serve trade areas of the same order. Centers of each order define the vertices of the hexagonal trade areas for centers of the next higher order. Town locations, for example, define the vertices of the cities' hexagonal trade areas.

rounding markets. The remainder of a center's goods and service business and employment depends on the markets that are created by the center's own population. Each central place offers a variety of goods and services for two reasons. The costs of doing business decline when firms cluster; and people add shopping lists together to save time and effort when they shop. The hierarchy notion is reinforced by the lumpiness of stages in production and distribution systems. In the sequence: grocery store– wholesale grocer–jobber–manufacturer, there are discrete steps, not a smooth gradation in function from one level to the next. The sequence: grade school–junior high school–senior high school–junior college–four-year college–university, is an example of a stepped hierarchy from the service activities. Society's functional organization is hierarchical and the central place system is the spatial manifestation of such social order.

As we move up the central place hierarchy

each order's centers are more complete in their goods and service offerings. The large, complex, high-order centers possess all the functions of lower-order centers plus a group of functions that distinguishes them from lower-order, less complex central places. Discrete population levels for each order in the hierarchy arise because of the discrete groups of functions which cluster at centers of each order.

Many attempts have been made to verify the deductive central place model and results are encouraging. Small centers tend to be located midway between large places. Large centers tend to be evenly spaced within market areas. The mix of goods and services available at a central place resembles the mix available in other centers of the same order and size, and differs from the mix available at centers occupying adjacent orders in the central place hierarchy. John Brush discovered that hamlets serving the rural farm population in southwestern Wisconsin occurred at a mean distance of 8.8 kilometers from one another and from centers of higher order. Villages, which require wider market areas to meet their greater threshold requirements, were found at an average distance of 15.8 kilometers from one another and from towns, which also perform village-level functions. Towns, as expected, were still more widely spaced, and occurred on the average every 33.9 kilometers.

The marketing principle dictates that central places performing functions of the same order will be evenly spaced. Towns perform town functions in addition to village and hamlet functions. A wide market area is needed to support a town function but a village can survive and prosper between towns, and a hamlet can survive between villages. In the ideal case, the central places of various orders and their market areas form the nested hierarchy illustrated in Fig. 10–28. In that hierarchy a village V^* is able to thrive between three towns by carving out its own village-level trade area where it can successfully compete with the three towns for village-level trade. Additional villages are supported at similar locations between other sets of towns. At the hamlet level the process repeats itself once more. Each town and village serves a small local area with hamlet-level goods and services. The threshold for hamlet-level functions is so small that places such as *H** (Fig. 10–28) can meet threshold conditions and thrive between three adjacent higher-order centers (two villages and a town).

In Wisconsin, Brush also found what earlier researchers had suspected for a long time. In their shopping behavior people often assemble their shopping lists and visit just the town, rather than make hamlet and village type purchases at the nearest hamlet or village before visiting the town for higher-order purchases. The large center attracts people from long distances and takes part of the consumer's dollar that might otherwise have gone to support hamlets and villages near the town. Accordingly, near towns, hamlets and villages occur less frequently than expected, and in locations remote from large centers the hamlets are closer together. The same tendency was observed in the spacing of villages. They were found farther apart when they were in active competition with towns, and closer together in places where the gravitational pull of the largest centers was weakest.

A central place's order depends on the number of different goods and services it offers. Once central places are classified by order, their trade areas can be delimited by measuring highway traffic flows (Fig. 10–29). Converging traffic around a center reflects hinterland capture. Unfocused traffic patterns in interstitial areas correspond to market areas shared with several centers. Local traffic areas around towns and villages are well defined, and thereby outline market areas.

Central Place Systems in the United States

In the United States, the larger a central place's population, the greater the number of different functions offered there (Fig. 10–30*a*). Furthermore, the number of different functions in a center corresponds closely with the number of goods and service establishments performing them (Fig. 10–30*b*). Inside urban areas, shopping districts are equivalent to rural central places which serve lower-density market areas. Central

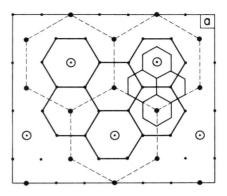

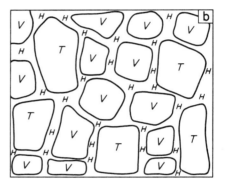

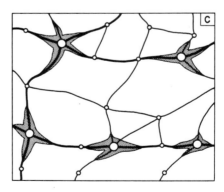

FIGURE 10–29

(a) *Theoretical arrangement of trade centers and trade areas for hamlets, villages, and towns.* (b) *Probable trade areas for village-level functions performed by villages and towns; note hamlets supported at competitive locations midway between larger centers.* (c) *Major roads serving villages and towns. Shading shows highway zones of heavy traffic near towns. Highway traffic density patterns define trade area boundaries.*

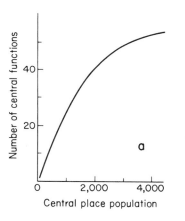
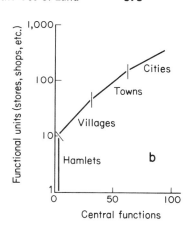

FIGURE 10–30

(a) *The number of functions offered at a central place increases with the central place's size (based on data from southern Illinois). (b) The number of functional units increases when the number of central functions rises (based on data from southwestern Iowa, adapted from Berry, Barnum, and Tennant, Fig. 7).*

place theory can be used to interpret the size distribution and spatial patterning of both kinds of centers. When intra-urban shopping districts are examined in terms of functions performed, number of establishments, and trade areas served, a hierarchy of business centers is revealed. Neighborhood centers sell low-order convenience goods to local customers. The larger community shopping center offers infrequently demanded high-order goods which need a market area several times larger than that which supports the neighborhood center.

The downtown central business district is the highest-order center in an urban area. In a smaller town the downtown may be the only center in the urban area, but in larger urban areas with two or more orders of centers, the number of orders and the spacing of centers depend directly on the city size, population densities, and purchasing power throughout the urban area. If population and purchasing power were equal at all points of the city, and if the entire city had been built at the same time, centers would be evenly spaced. But because cities were built over a period of years, during which the ease of intra-urban transportation steadily improved, population densities decline from a high point near the downtown to a low at the rural-suburban fringes. In response to the variation in population density, commercial districts in urban areas are closely spaced in the inner city and progressively more widespread farther out (Fig. 10–31).

Low-order central places within urban areas are regularly spaced with respect to population when it has been weighted according to purchasing power. On a regular map which portrays a distribution in absolute space, the spacing of centers appears uneven. We can, however, take a map drawn in absolute space where a kilometer on the ground represents a centimeter or some other unit on the map, and mentally transform it to a map drawn on another basis—for example, population times purchasing power. A map in absolute space tries to give equal areal representation on the map to every square kilometer on the earth. In an areal cartogram based on population weighted by purchasing power, each weighted population unit is equally represented. For example, a square centimeter on the cartogram might represent the purchasing power of a thousand people rather than a square kilometer.

Consider Fig. 10–31. If the urban area depicted on the map were transformed to an areal cartogram wherein each population unit received proportional areal representation, the low-density fringe of the urban area would shrink and suburban shopping centers would show up closer together. Closer to the downtown, high-density tracts would be transformed to receive proportionately large representation on the cartogram. The closely packed central shopping districts would spread out to a spacing arrangement similar to the transformed arrangement of suburban centers.

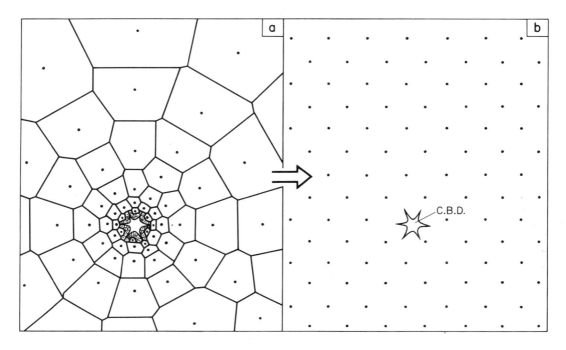

FIGURE 10–31

Spacing of trade centers in the city. (a) *Ideal spacing of centers in the city when land area in the city is given proportional representation on the map.* (b) *Ideal spacing of centers in the city when urban land area is represented on the areal cartogram in proportion to the size of the population in each neighborhood and its buying power.*

John Borchert studied the size and spacing of the twenty-four most important central places in the United States. Their sizes and functional complexity set them aside as the first-, second-, and third-order centers of the national central place hierarchy. The first-order center is New York City, the national metropolis. There are six second-order centers in addition to New York which is a first-order center as well as a second-order center: Boston, Philadelphia, Detroit, Chicago, San Francisco, and Los Angeles. Each second-order center acts as a regional metropolis for a vast portion of the United States market. The seven largest centers are joined by seventeen metropolitan centers to complete the set of high-order central places (Fig. 10–32). The twenty-four third-order centers serve twenty-four third-order trade areas. The trade areas of Washington, D.C.,

and Baltimore overlap extensively because they are located so close together.

The distribution of second- and third-order centers in the United States appears very uneven if we view the United States in absolute space. It is true that some of the East Coast cities got an early start and a competitive edge on newer cities inland. It is also true that American cities did not grow up on a homogeneous plane. But these temporal and physical irregularities are not very important. Let us see why. The seven first- and second-order centers appear to be scattered very unevenly across the country. In such a vast country as the United States, three of the centers are clustered on the East Coast: New York, Boston, and Philadelphia; two are on the West Coast: San Francisco and Los Angeles; and the other two lie close together on the Great Lakes: Chicago and Detroit. Why are there such

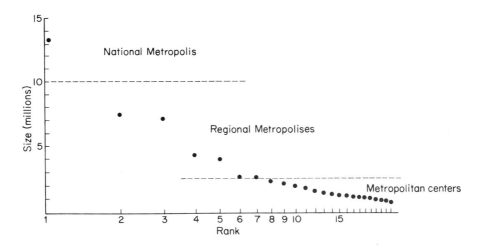

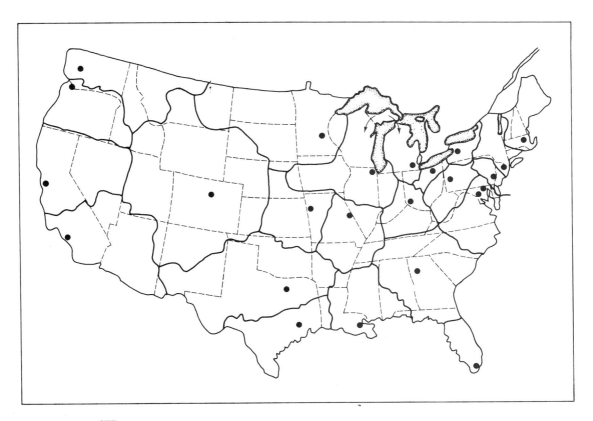

FIGURE 10–32

First-, second-, and third-order trade centers and trade areas
in the United States (suggested by J. R. Borchert).

massive gaps between clusters of second-order centers? The answer is, there aren't. The gaps exist only if we insist on viewing the United States in absolute space when we are talking about a socioeconomic process based on people and their purchasing power. Square kilometers should not be equated with people or their purchasing power when we consider the size and location of central places.

Two major regions of the United States have until recently been unimportant as market areas for second-order goods and services. First, the dry plains and Rocky Mountain region has supported prosperous individuals, but there are so few of them they were inconsequential in terms of second-order thresholds. Consequently, population multiplied by purchasing power yielded a small number throughout United States history. The second region unserved by second-order centers was the South. Over 30

percent of the national population lived in the South in 1960, and when the present second-order centers were founded and rose to prominence the South's population share was even higher. Yet the South has been a poor region in terms of per capita productivity and personal income per person. Thus, population multiplied by purchasing power yielded a number too small to generate second-order centers. With two larger portions of the national market area eliminated or diluted, the distribution of second-order centers appears much less uneven (Fig. 10–33). San Francisco and Los Angeles split the Western market. Chicago and Detroit divide the Midwest. Boston dominated New England. New York and Philadelphia took care of the rest. By deemphasizing the South, the Great Plains, and the mountainous west in Fig. 10–33, we can demonstrate that second-order centers are approximately equally spaced in terms of people

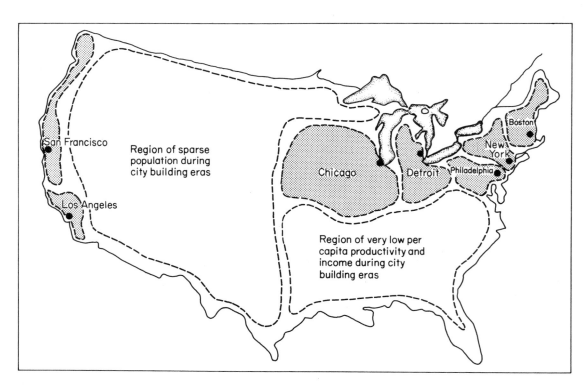

FIGURE 10–33

Second-order trade centers and trade areas in the United States.

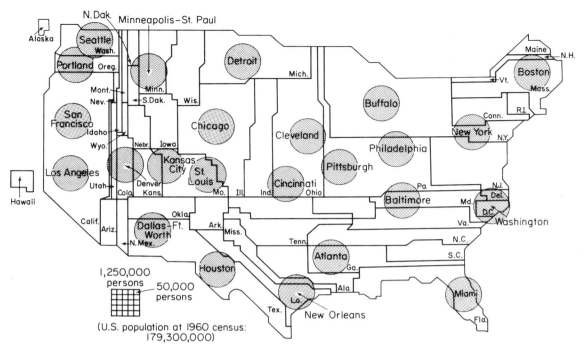

FIGURE 10–34

Third-order trade centers in the United States, evenly spaced when they are mapped in population space.

and purchasing power. We can go one step further, too, and show how third-order centers are spaced when they are plotted on an areal cartogram in which each state is drawn inside a United States outline map, but drawn so that its area is proportional to its population (Fig. 10–34). On the cartogram, there is no question about the even spacing of centers. The regularities anticipated by central place theory are clearly confirmed.

Central place theory describes only equilibrium spatial patterns and tendencies; it is not a dynamic theory explaining the actual processes by which centers are created at certain locations and then undergo change at those locations. In the United States since Independence, the system of central places has grown from a few small cities serving thirteen former colonies to a complicated network comprising thousands of centers from the national metropolis to the smallest hamlet.

From the time of its founding each central place has served rural populations and lower-order central places. Yet, from the beginning, rural settlement patterns were constantly in flux and urban centers grew at wildly different rates. The frontier moved west and no sooner had agricultural, forestry, and mining populations reached their widest extent than rural depopulation began. The agricultural service centers and transport nodes grew rapidly for a time, and then manufacturing centers rose to prominence. Recently, urban growth rates have been highest on the West coast, South coast, and Atlantic coast cities which many Americans feel are the nicest places in which to live.

As an example to illustrate the spatial-temporal adjustment process in central place terms let us consider the location of major league baseball teams at two different dates. From 1900 to 1953 the American and National League teams played in those metropolitan centers that were at the top of the urban hierarchy at the turn of the century. Ten cities had all sixteen teams. New York had three teams, while Boston, Philadelphia, Chicago, and St. Louis each had two (Fig. 10–35). Major league baseball is a high-threshold central function, so the number of teams in a city is a good index of the city's order in the central place hierarchy. The distribution of teams in 1900 represented the relative importance of centers and their accessibility to one another (by train); but by 1950 the central place hierarchy and its transport linkages in the United States had undergone some dramatic adjustments. The domi-

nance of Northeastern metropolises was no longer as absolute as it had been a half-century earlier. Instead, cities in the West and South had grown to become major rivals of the old urban hierarchy. After 1953, therefore, several teams were moved to reflect the new urban order. When the Braves moved from Boston to Milwaukee the dam broke and several problems were publicly recognized. For example, it was quickly apparent that whereas the threshold for a major league team had not changed greatly in half a century, the nation's population had doubled between 1900 and 1950. Moreover, people had more leisure, more mobility, and

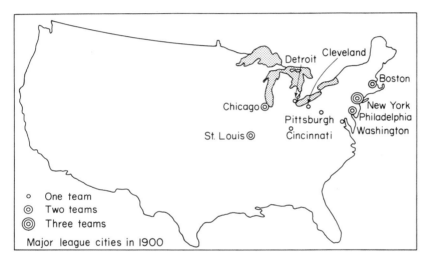

FIGURE 10–35

Professional baseball cities: 1900 and 1970.

more money for recreation in 1950 than in 1900. Both leagues therefore expanded from eight teams to twelve between 1954 and 1970 and increased the number of games played. The distribution of teams in 1970 reflected not only the twentieth-century changes in American urban structure and jet-liner transportation, but closer links to Canadian cities as well, as Montreal received a major league baseball franchise (Fig. 10-35).

Selecting ball park locations within metropolitan areas presented a location problem of another scale. Old Yankee Stadium in Manhattan was located so the baseball fans could come to games on public transportation. Across the country, during the streetcar era, baseball parks were located at a major streetcar intersection outside the downtown. Such a location represented a point of low aggregate travel. Nowadays people go to baseball, football, and basketball games by automobile. Most of the new ball parks built in recent years reflect this fact. Suburban locations are often picked to minimize aggregate travel effort of spectators and to provide plenty of parking.

PATTERNS OF CHANGE
ON THE MODERN URBAN SCENE

The modern world is an urbanizing world. The contemporary geographer who studies the earth as man's home turns more and more to the metropolitan arena. But the principles of modern geography are valid in both urban and rural applications. In the developed world right now and in the developing world very soon the geography of the present and future must stress the urban rather than the rural environment.

Modern urban societies organize their affairs at two different spatial scales. Brian Berry put it succinctly when he suggested that we could profitably examine cities as systems within systems of cities. What are the two spatial scales of urban study? One is the city as a point or a node within a regional area; the second considers the city as carrying on most of the business of modern society. Let us summarize our discussion of locational principles by reviewing how they apply to changing urban systems in

the United States at both the broad and the detailed scales. Consider first the locational distribution of cities in the United States. Cities were founded at different times and grew at different rates through time. Urban development occurs in time but on an areal stage. Some geographical forces which have affected urban location and urban growth are rather uniform over large areas, like precipitation, or sunshine, or soil fertility. Uniform areal forces fostered an even distribution of cities. Other geographical forces which affected urban location and growth occur sporadically over the land. If cities were founded and prospered on the basis of some sporadically located resource base, such cities will be located in highly erratic fashion. Along the New England coast are towns and cities founded as fishing ports. Across the northern Great Lakes cut-over region of the United States are many towns begun during logging days. Michigan has towns at copper ore deposits; Minnesota has iron mining towns; Illinois has lead mining towns; and dozens of cities and towns in Oklahoma and Texas exist where they do only because of the oil and gas deposits located sporadically underneath them.

Some geographical forces that influenced urban location and growth are competitive. From the marketing principle in central place theory we learned that in a homogeneous rural market area, centers of a given order will be evenly spaced in an optimum distribution network. Frequently, instead of urban activities scattering because of competition, they are drawn together. Many manufacturing enterprises are agglomerated activities and they benefit from clustering just as retail firms selling identical goods and services from central places may benefit from separation.

There is much more order in the locational distribution of cities than appears at first sight, but this fact is not discovered until the order is looked for. When we search for an order or a pattern in the locational distribution of cities we must be equipped with a set of constructs in order to recognize the patterns which the constructs permit us to see. When we examine a system of cities, the most useful constructs concern the three roles—(1) collecting, (2) processing, and (3) distribution—which cities

Three urban roles

I. Every city collects from a resource base.

II. Every city is a processer located at a transport-communications node.

III. Every city is a central place, distributing goods and services to smaller places and rural markets.

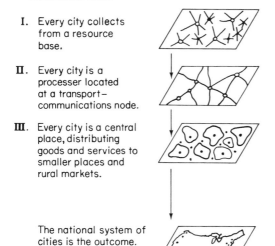

The national system of cities is the outcome.

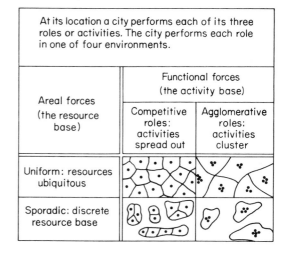

At its location a city performs each of its three roles or activities. The city performs each role in one of four environments.		
Areal forces (the resource base)	Functional forces (the activity base)	
	Competitive roles: activities spread out	Agglomerative roles: activities cluster
Uniform: resources ubiquitous		
Sporadic: discrete resource base		

FIGURE 10–36

The national system of cities performs three roles. Each city performs its roles within various areal and functional environments.

perform and the locational requirements of these three roles. Each role is performed according to the areal (uniform, sporadic) and functional (competitive, agglomerative) forces that prevail (Fig. 10–36).

Regularities in the physical world (climate, soil, terrain, etc.) and in the human world (technological know-how, language, religion, political organization, etc.) promote a uniform distribution of collector cities. Sporadic irregularities in the physical and human world encourage an irregular pattern of town and city locations. Places founded primarily as collectors have a spatial distribution reflecting the physical and human resource base of the areas where they are located. Processing cities need access to inputs and to markets. They must be well placed with respect to transportation and communications networks regardless of whether they process goods (a factory), people (a university), or information (a government office). Processing activities tend to cluster so they can use facilities in common, cut costs, and thereby reap external economies. Because the inputs, transportation nodes, and communications nodes needed by processors occur sporadically, cities and towns specializing in processing display a highly irregular locational pattern.

A third urban archetype is the city distributing goods, services, and intelligence to other centers and to rural clients and markets. If markets are uniformly spread over the land, distributing centers will be spaced uniformly as well. If markets are located sporadically, distributors will bypass some areas in favor of others. Occasionally a number of different central functions disseminating goods, services, information, or control will cluster because clustering is mutually advantageous. Government agencies, for example, have to supply all areas in their jurisdiction with government of some sort. Government offices often form a cluster when all of them serve the same jurisdiction. Banks and insurance companies also find it advantageous to cluster, even though their service areas may not exactly coincide.

By focusing on collecting, processing, and distribution we can describe the proportion of a city's spatial and temporal behavior that is traceable to each role. By examining for each activity its areal environment (uniform or sporadic resources?) and functional environment (competitive or agglomerative roles?) we can explain why cities are located in one pattern rather than another.

Inside every urban area principles of spatial organization regulate locational structure and spatial behavior. Producers build cities and consumers use cities. The city is a stage which is constantly reshaped by the actors who use it. In the central business district the national or regional system of cities intersects the city as a local spatial system. This downtown center is a regional and a local transportation and communications hub at the same time. Activities which found it to their advantage to locate in the same city also found it advantageous to occupy adjacent tracts within the city.

The downtown of the typical American city comprises several activity areas (Fig. 10–37). During the nineteenth century when most of our downtown cores were being erected, adjacent to each activity area lived a population group which tended to dominate the nearby activity area. Activity areas included retailing, wholesaling, financial activities, transportation, fresh food, warehousing, administration, and so forth. In one city the Germans and Poles might have worked predominantly in manufacturing; the Irish might have dominated transportation and administration; the Jews might be found in wholesaling and retailing; the Italians in food handling, and so on. Immigrants preferred residence in neighborhoods with members of their own nationality groups. If members of an ethnic group dominated a highly localized activity, the group's choice of a residential neighborhood adjacent to their jobs was a least net effort solution, especially before 1890 when almost everyone walked to and from work.

From a starting point in 1890, the city spread outward in every possible direction. The push

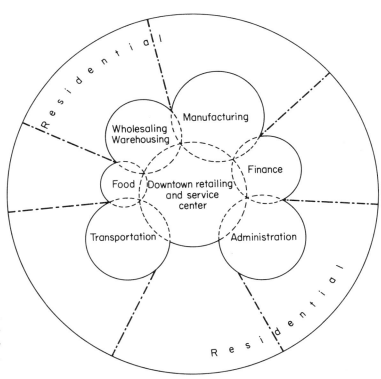

FIGURE 10–37

The American city up to 1890, including activity areas of the central business district and adjacent high density residential sectors (suggested by D. Ward).

outward by residents was prompted by (1) continued foreign and rural immigration; (2) natural population change based on births exceeding deaths; (3) more disposable income to spend on housing; and (4) improved mobility based first on the streetcar and then on the automobile, making it unnecessary to live next door to one's job. As housing demand rose and fell with fluctuations in the business cycle, land developers and home builders erected band after band of new housing, putting each increment on the edge of the previously built-up city. In the walking era and streetcar period high density housing prevailed, but as people's preferences for low density housing were met in the automobile era, residence and jobs could be linked no matter where a person lived.

In terms of each consumer's behavior, the city traditionally has usually been a residential sector (Fig. 10–38). Jobs, stores, government, doctors, and so forth were located downtown, and relatives and neighbors were nearby. For a household living between the downtown and the suburbs, housing toward the downtown was older, cheaper, and more cramped, but if they moved outward, this would mean a newer house, bigger yard, and more room. As people moved up in the world then they typically

moved outward in short jumps. Conversely, a death, divorce, or abrupt financial reverse forces drastic reversals and long inward moves. Lateral moves are uncommon because housing in the same residential ring is similar to what is already occupied, and people usually move to acquire a different kind of housing unit and often a different neighborhood. Lateral moves are also uncommon if they take a household into a different ethnic environment. Consequently, ethnic groups move outward together as they move up in the world. They seldom invade laterally the territory of another ethnic group. A group may, however, because of rapid population increase or income gains push outward faster than peripheral groups are willing to vacate.

On the housing construction side, residential builders put up housing wherever they think it will sell, and the market for new housing varies significantly from one sector to another. Population change and improvements in purchasing power vary from sector to sector, so some housing sectors expand outward faster than do others. The result is that the city expands eccentrically due to differences in housing demand and supply from one sector to another. As we saw earlier, each boom in the business cycle added a new band of housing around an urban area. In each recession or depression housing construction practically halted. Extended booms created wide rings of new housing and short booms produced narrow rings. Whereas the width of each residential growth ring depended on the intensity and duration of the building boom, the shape of the housing increment was inflenced by the transport systems of the times. The streetcar era and the modern freeway-automobile era promoted star-shaped urban expansion. The pre-1890 walking period and the era of the bus and the recreational automobile before World War II fostered compact, circular urban growth.

If an ethnic group prospers but continues to grow in size, it will provide vigorous support to an expanding housing market and stimulate a rapid sectoral push outward (Fig. 10–39, case A). If population change is rapid in a sector but economic success is limited, densities can only

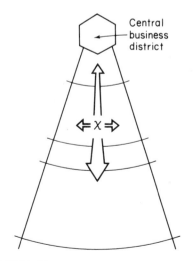

FIGURE 10–38

One's home town is really a sector, not the whole city.

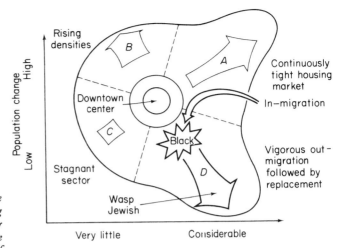

FIGURE 10–39

Producers and consumers of housing determine the directions of urban areal expansion. The housing submarkets are stagnant or active in a sector depending on a combination of population change in the sector and the relative economic success of the sector's residents.

rise because the group can ill afford new houses (Fig. 10–39, case B). If the group within a sector fails to expand its population or its income, a stagnant sector is created (Fig. 10–39, case C). If the group achieves superior prosperity yet does not expand in population, it creates demand for plenty of new housing but leaves a vacuum behind it as it moves outward (Fig. 10–39, case D). The overcrowding in sector B will not be alleviated by housing surpluses in sector D because urban residents do not usually consider living in residential areas on the opposite side of the city. Instead, the vacancies near the downtown center in sector D are filled by new immigrant groups.

In many American cities, sector D is best exemplified by housing sectors with prosperous Jewish populations on the suburban margin and recently arrived black populations in the sector's decrepit inner neighborhoods. If we move outward in an average sector, we encounter a minimum of ethnic variation. Instead, variation exists primarily in age, quality, and density of housing, and in stage in the life cycle. Moving around a housing ring, residences are similar in style and condition, but pronounced ethnic and income differences emerge. In a general sense, then, the rings and sectors define a city's neighborhoods. In each neigh-

borhood the physical landscape is very homogeneous; the cultural landscape is equally homogeneous in terms of family size, income, education, and ethnic composition. It is the city's neighborhoods, formed in this way, which provide the market area for the hierarchy of lower-order central places inside the metropolis. Such commercial nuclei are exactly analogous to their rural counterparts. They are located at transport nodes throughout the city and serve adjoining market areas.

To summarize, we can note once again that geography's central question is: why are spatial distributions structured the way they are? In Chapter 9 we approached this question by asking how people locate events and what kinds of location patterns result. The basic concern in that chapter was: given an activity, what is its best location? This chapter began with the question: given a location, what is the best activity to pursue there? In a rural setting this is a question about the competing uses for land—non-agricultural, agricultural, or urban. Inside the city the question is: which competing land use can pay the most for each tract of land?

Whatever the spatial organization of land uses in one period, it is seldom permanent. Location is the major determinant of land use and location is constantly redefined by transportation and communication innovation. As the relative locations of places are redefined by the time-space convergence wrought by ever-greater ease of interaction, land uses are regularly adjusted and geography's central question must be asked once more.

Suggestions for Further Reading

ADAMS, J. S. "Directional Bias in Intra-Urban Migration," *Economic Geography*, XLV (October 1969), 302–23.

———. "Residential Structure of Midwestern Cities," *Annals of the Association of American Geographers*, LX (March 1970), 37–62.

ALONSO, W. *Location and Land Use: Toward a General Theory of Land Rent*. Cambridge, Mass.: Harvard University Press, 1964.

ANDERSON, T. "Social and Economic Factors Affecting the Location of Residential Neighborhoods," *Papers and Proceedings of the Regional Science Association*, IX (1962), 161–70.

BERRY, BRIAN J. L., and FRANK E. HORTON. *Geographic Perspectives on Urban Systems, with Integrated Readings*. Englewood Cliffs, N.J.: Prentice-Hall, Inc., 1970.

BORCHERT, J. R. "American Metropolitan Evolution," *Geographical Review*, LVII (1967), 301–32.

BUNGE, W. "Experimental and Theoretical Central Place," in *Theoretical Geography*, Chap. 6. Lund: Gleerup, 1966, pp. 134–76.

CHISHOLM, MICHAEL. *Rural Settlement and Land Use: An Essay in Location*. New York: John Wiley & Sons, Inc., 1967.

HARVEY, D. "Theoretical Concepts and the Analysis of Agricultural Land Use Patterns in Geography," *Annals of the Association of American Geographers*, LVI (1966), 361–74.

HENSHALL, J. D. "Models of Agricultural Activity," in R. J. Chorley and P. Haggett, eds., *Models in Geography*, Chap. 11. London: Methuen & Co., Ltd., 1967, pp. 425–58.

LÖSCH, A. *The Economics of Location*. New York: John Wiley & Sons, Inc., 1967.

MEIER, R. L. *A Communication Theory of Urban Growth*. Cambridge, Mass.: The M.I.T. Press, 1962.

———. "The Metropolis as a Transaction-Maximizing System," *Daedalus*, XCVII (Fall 1968), 1292–1313.

MORRILL, R. L. "The Development of Spatial Distributions of Towns in Sweden: An Historical-Predictive Approach," *Annals of the Association of American Geographers*, LIII (1963), 1–14.

———. "The Negro Ghetto: Problems and Alternatives," *Geographical Review*, LV (1965), 339–61. Reprinted in Brian J. L. Berry and Frank E. Horton, *Geographic Perspectives on Urban Systems, with Integrated Readings*. Englewood Cliffs, N.J.: Prentice-Hall, Inc., 1970, pp. 419–34.

WOLPERT, JULIAN. "The Decision Process in Spatial Context," *Annals of the Association of American Geographers*, LIV (1964), 537–58.

Works Cited or Mentioned

ACKERMAN, EDWARD. "Population and Natural Resources," in P. M. Houser and O. D. Duncan, eds., *The Study of Population*. Chicago: The University of Chicago Press, 1959, pp. 621–48.

BERRY, B. J. L. *Geography of Market Centers and Retail Distribution*. Englewood Cliffs, N.J.: Prentice-Hall, Inc., 1967.

———. *Theories of Urban Location: An Introductory Essay*. Commission on College Geography, Resource Paper No. 1. Washington, D. C.: Association of American Geographers, 1968.

———, H. G. BARNUM, and R. J. TENANT. "Retail Location and Consumer Behavior," *Papers and Proceedings of the Regional Science Association*, IX (1962), 65–106.

BORCHERT, J. R. "The Twin Cities Urbanized Area: Past, Present, Future," *Geographical Review*, LI (1961), 47–70.

BOSERUP, E. *The Conditions of Agricultural Growth*. London: George Allen and Unwin, Ltd., 1965.

BRUSH, J. E., and H. L. GAUTHIER, JR. *Service Centers and Consumer Trips*. Department of Geography, Research Paper No. 113. Chicago: The University of Chicago, 1968.

CHRISTALLER, W. *Central Places in Southern Germany*, trans. C. W. Baskin. Englewood Cliffs, N.J.: Prentice-Hall, Inc., 1966.

COUPER, A. D. "Rationalizing Sea Transport Services in an Archipelago: An Application of Simple Space Theory," *Tijdschrift voor Economische en Sociale Geografie*, LVIII (1967), 203–8.

HALL, P., ed. *Von Thünen's Isolated State*. Oxford: Pergammon Press, 1966.

HARRIS, C. D., and E. L. ULLMAN. "The Nature of

Cities," *Annals of the American Academy of Political and Social Science*, CCXLII (1945), 7–17 (also available as Bobbs-Merrill Reprint No. G–85).

WARD, D. "The Emergence of Central Immigrant Ghettos in American Cities, 1840–1920," *Annals of the Association of American Geographers*, LVIII (1968), 343–59.

WHITTLESEY, D. "Major Agricultural Regions of the Earth," *Annals of the Association of American Geographers*, XXVI (1936), 199–240.

WILSON, M. G. A. "Changing Patterns of Pit Location on the New South Wales Coalfields," *Annals of the Association of American Geographers*, LVIII (1968), 78–90.

ZELINSKY, W. *A Prologue to Population Geography.* Englewood Cliffs, N.J.: Prentice-Hall, Inc., 1966.

PART IV

Spatial Diffusion Processes

CHAPTER 11

SPATIAL DIFFUSION:
MESHING SPACE AND TIME*

He that will not apply new remedies must expect new evils;
for time is the greatest innovator.

—*Francis Bacon*

THE DYNAMICS OF SPATIAL PATTERNS

That man and his works exist in space and time is so obvious that it hardly seems worth mentioning. But upon closer examination this simple fact is seen to contain such conceptual richness and intellectual challenge that it underlies one of Geography's most exciting contemporary fields. When we consider man and his works in space and time, we can no longer think about static structures and relationships. Rather, we focus explicitly upon the dynamics of spatial patterns, so that diffusion processes acting over space and through time become the core of our concern.

The mechanisms of spatial diffusion are little understood, and there is much exciting work to be done along the entire continuum from

general theory to the solution of practical problems. The unfolding of man's patterns over geographic space and through time is a fascinating thing to watch and study, and once you have thought about the processes at work in these fundamental dimensions of human existence, you can never wholly return to "pre-diffusion" thinking. This is crucial, for whether you plan to become a social scientist, or simply an informed and responsible citizen, the world of 2000 is desperately going to need men and women with a clear understanding of man's use of space over time.

As we begin to think about various types of spatial diffusion, two things are immediately obvious. First, anything that moves must be carried in some way. Secondly, the rate at which some things move over geographic space will be influenced by other things that get in the way. Thus we must consider initially the carriers and the barriers that can influence particular movements.

---

*The bulk of this chapter is adapted from Peter Gould, *Spatial Diffusion*, Commission on College Geography Resource Paper No. 4, 1969, by permission of the Association of American Geographers.

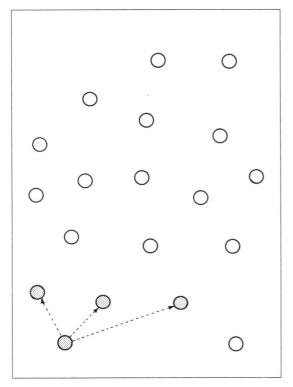

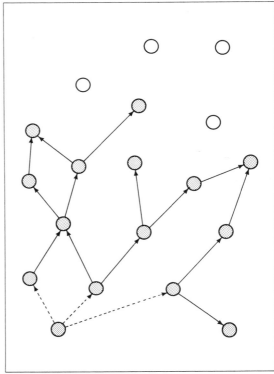

FIGURE 11–1

Expansion diffusion: the initial stage (adapted from Brown and Moore, Fig. 3).

FIGURE 11–2

Expansion diffusion: a later stage (adapted from Brown and Moore, Fig. 3).

The Carriers . . .

Let us think about a very common type of diffusion problem in which an idea spreads through a group of people. We can think of a rumor running like wildfire through a student population, or perhaps a new farming technique moving through an agricultural area. Initially, only a few people will have the idea or rumor (Fig. 11–1), but soon the idea is communicated to friends and neighbors. The new knowers tell their acquaintances in turn (Fig. 11–2), and gradually the idea spreads through the population. We shall call such a process an *expansion diffusion*, for as an idea is communicated by a person who knows about it to one who does not, the total number of knowers increases through time.

Not all diffusion processes are of the expansion type. In many cases, an initial group of people or carriers will themselves move (Figs. 11–3 and 11–4), so that they are diffused through time and over space to a new set of locations. We shall call this type of process *relocation diffusion*. The commonest example, of course, is that of migration, groups of people moving their residences from one place to another. The whole geography of early settlement in the United States can be regarded as the diffusion of new immigrants across the face of America. The process continues today. If we could take time lapse movies from a geographic space laboratory located permanently over the United States, and run through half a century in five minutes, we would see a number of relocation diffusion processes going on simul-

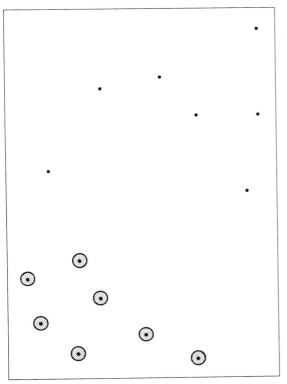

FIGURE 11-3

Relocation diffusion: the initial stage (adapted from Brown and Moore, Fig. 3).

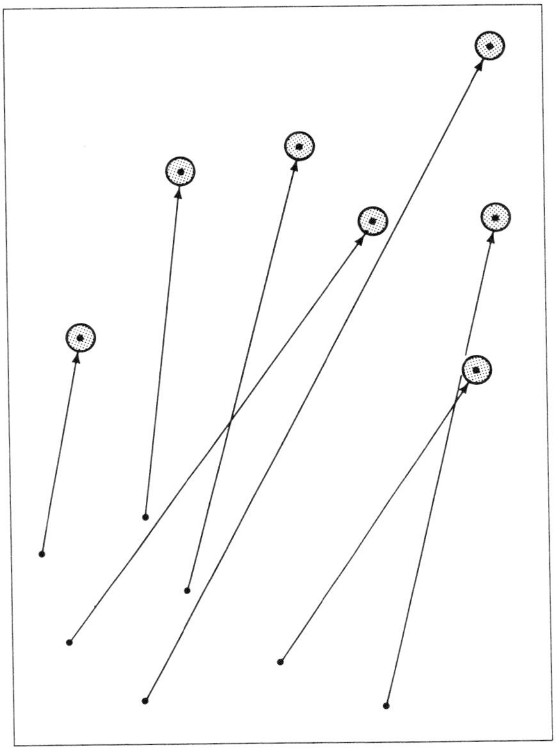

FIGURE 11-4

Relocation diffusion: a later stage (adapted from Brown and Moore, Fig. 3).

taneously at different speeds and scales. A slow but gathering tide of people from the South moves to the cities of the North, while rural Midwesterners leave their farms and move to California and the big cities. Superimposed upon these slow processes of relocation diffusion acting over decades are the faster seasonal flows of vacationers, diffusing to sun and snow in winter and shore and lake in summer. Still faster movements on the film would be the rapid pulses of daily commuters to and from the bedroom suburbs that surround our cities. Taken altogether we would see a constantly shifting sequence of patterns as *homo americus* moves ever faster and farther in the temporal and spatial flux.

Processes of expansion diffusion can also be considered in other terms. Suppose a disease

is diffusing through a population by direct contact; that is, one person must actually touch, or be extremely close to another before the disease can be transmitted. Examples would be the venereal diseases that sometimes reach epidemic proportions in many of the large cities. We call such diseases *contagious*, and the term has been somewhat loosely borrowed to characterize rather similar diffusion processes. *Contagious diffusion* always expands, and it is strongly influenced by the frictional effect of distance. Many ideas and diseases are passed to people very close to those who already have them.

Simple geographic distance is not always the strongest influence in a diffusion process, for some ideas and innovations seem to leap over many intervening people and places. Such

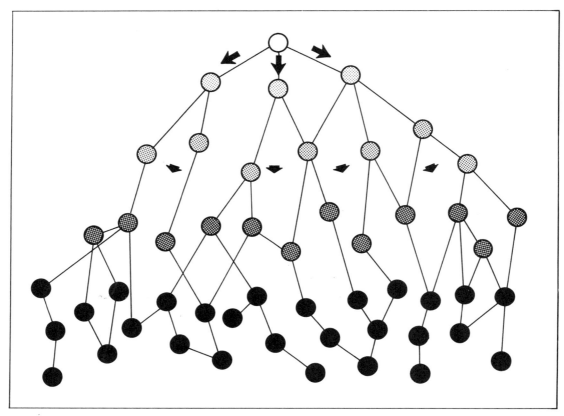

FIGURE 11–5

The process of hierarchical diffusion.

leap-frogging usually characterizes processes of *hierarchical diffusion*, in which large places or important people tend to get the news first, transmitting it later to others lower down the hierarchy (Fig. 11–5). Hierarchical diffusion occurs because many things diffuse in relative space—in which big cities, linked by very strong information flows, are actually "closer" than they are in simple geographic space. Many clothing fashions, for example, originate in the major fashion centers of New York, Paris, and Rome, and then diffuse to other towns that form major regional nodes in Europe and the United States. From these the new skirt lengths and other "vital statistics" diffuse to the provincial centers and the small towns in the rural areas, often encountering raised eyebrows in

the process! Most people have observed the contrasts between fashions in the metropolitan areas and those in the rural towns that are due to this common diffusion lag. Similarly, at the personal level of diffusion, many farming innovations are adopted first by the larger, wealthier, and locally more important farmers in a region, and only then do the ideas trickle down the local social hierarchy to others in the area.

It is obvious that real diffusion processes do not always fit neatly into one category or another. As we have seen already, expansion diffusions are always of the contagious type. As an idea or an innovation spreads gradually outward from a core area by a process of contagion, the sum total of adopters also grows.

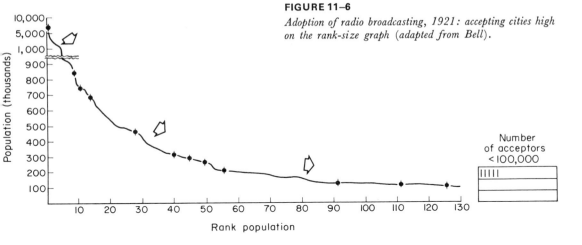

FIGURE 11–6

Adoption of radio broadcasting, 1921: accepting cities high on the rank-size graph (adapted from Bell).

Our cities often appear to diffuse outward by such a process, expanding along the urban fringe and main arterial routes. But expansion diffusion can also move through one of the many hierarchies that structure both geographic space and society. A cultural idea, for example, can cascade down the highbrow-lowbrow hierarchy, even as the lodges of many fraternal organizations trickle down to the small towns from the larger national and regional centers.

Most relocation diffusion also seems to be of a hierarchical nature, as people and institutions move up and down the layered structures

characterizing much of modern life. The diffusion of academic talent, for example, is largely of a hierarchical nature. The game of academic musical chairs is not played at one common level, but diffuses intellectual talent upward and downward to maintain and reinforce the very structure that channels and guides the movement.

Even from these relatively simple examples we can see that most diffusion processes are mixtures of more basic types. In many problems several modes of diffusion may operate simultaneously and in changing intensities. In the

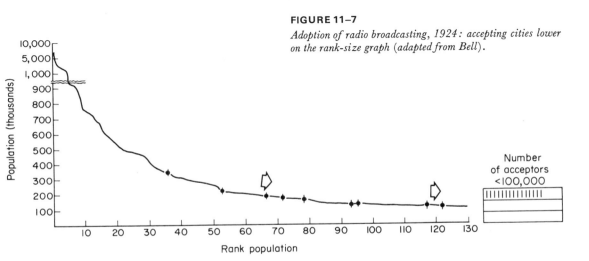

FIGURE 11–7

Adoption of radio broadcasting, 1924: accepting cities lower on the rank-size graph (adapted from Bell).

FIGURE 11-8
Cities adopting radio broadcasting in 1921 (adapted from Bell).

early years of the diffusion of radio broadcasting in the United States, for example, there were strong hierarchical and contagious components operating at the same time. If we plot the populations of American cities against their rank order (Figs. 11–6 and 11–7), and connect the values with a line, we can use the graph as a base for plotting the places that "adopted" radio broadcasting during a particular year. Notice that a break of scale occurs on the vertical axis, and that a tally of cities with populations less than 100,000 and below the 130th in rank has been kept in a score box. Otherwise the graph would be many feet long. Broadcasting started in Pittsburgh and by 1921 only five places with fewer than 100,000 people had stations, though many of the larger cities possessed this modern innovation. In contrast, most of the new acceptances by 1924 were in

in towns below 100,000, and all the cities over 360,000 had received radio broadcasting stations before this date.

While we can think of the innovation "sliding down" the rank-size graph as it trickles through the urban hierarchy, it is also clear that some contagious diffusion was occurring at the same time. In 1921 (Fig. 11–8) many of the large metropolitan places, such as New York, Boston, Detroit, St. Louis, San Francisco, Los Angeles, and Seattle, had radio stations, together with some of the smaller cities such as Dallas, Des Moines, Kansas City, and Knoxville. By 1924 (Fig. 11–9) many of the towns around the innovating cities have "caught the radio bug," the contagious effect being particularly noticeable around Kansas City and Des Moines where nearby towns lie between the earlier stations. Similarly, around San Francisco,

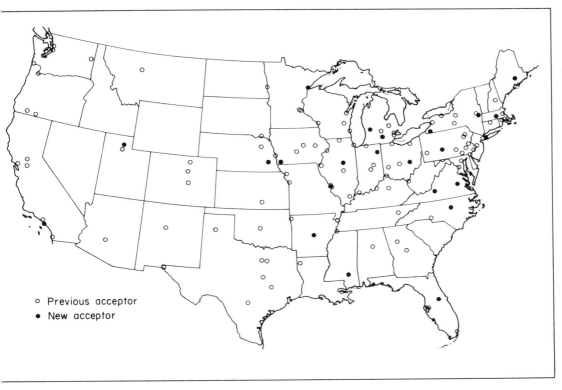

FIGURE 11–9
Cities adopting radio broadcasting in 1924 (adapted from Bell).

Seattle, Detroit, Cleveland, and Philadelphia the contagious effect has started, with radio stations in a number of small towns far down the urban hierarchy but close to one of the early innovating centers. Thus the hierarchical process appears more important in the earliest years, while the contagious process assumes greater control once the basic patterns are established. We can think of innovating "stones" being dropped through a hierarchical chute into the still millpond of America's space, each one producing a wavelike pulse of contagious innovation to cities in its area of influence.

Thinking of diffusion processes in terms of waves is a common verbal and conceptual analogy which orginated in an early work entitled *The Propagation of Innovation Waves* by Torsten Hägerstrand, the Swedish pioneer of modern diffusion studies. One of his early studies

in Sweden investigated the diffusion of automobile ownership which seemed to sweep like a wave across the southern province of Skåne. Noting that Skåne often gets innovations from the rest of Europe before other parts of the country, Hägerstrand recorded the growth of automobile ownership in an hexagonal network of observation cells (Fig. 11–10). In 1920 (Fig. 11–11), car ownership was scattered in a few towns, but two years later (Fig. 11–12), the "wave of innovation" had washed strongly across the area from Denmark and its capital city of Copenhagen. By 1924 (Fig. 11–13), the new horseless carriage had penetrated nearly every corner of the province.

The wave analogy, like all analogies, must be used with care and circumspection. It is a useful conceptual device for thinking about certain types of diffusion processes provided it is not

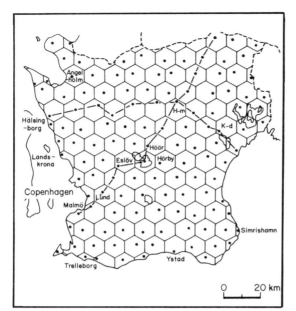

FIGURE 11–10

Skåne, southern Sweden: the network of observational cells (adapted from Hägerstrand, 1952, Fig. 1).

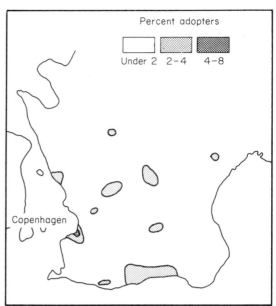

FIGURE 11–11

Adoption of automobiles in Skåne, 1920 (adapted from Hägerstrand, 1952, Fig. 1).

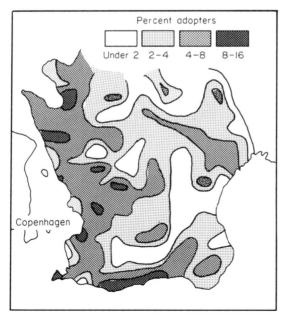

FIGURE 11–12

Adoption of automobiles in Skåne, 1922 (adapted from Hägerstrand, 1952, Fig. 1).

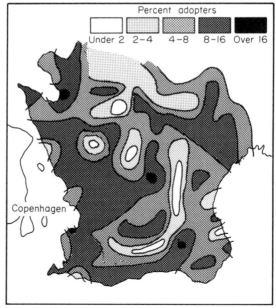

FIGURE 11–13

Adoption of automobiles in Skåne, 1924 (adapted from Hägerstrand, 1952, Fig. 1).

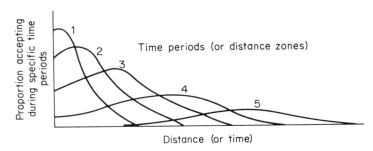

FIGURE 11–14

Waves of innovation losing strength with distance from the source area (adapted from Morrill, 1968, Fig. 10).

taken too literally in all cases. For some types of diffusion the wave is a useful model of the actual diffusion process, and it seems to stand up well when the process involves individual decisions to adopt an innovation.

Like all waves, innovation pulses across a landscape tend to lose their strength as they move away from the source of the disturbance. If we plot the proportion of people accepting a new idea against their distance from the source (Fig. 11–14), we can see how an innovation wave gradually loses its strength over time. During the first time period, many people close to the origin accept the innovation, while only a few further away make the decision to adopt. In the second time period or "generation," the wave has moved outward, but the crest is lower and the energy of the innovation pulse is spread over a wider area. During successive periods the crest continues to move away from the source area, but with ever-diminishing intensity, until the wave has completely spent its force. The same sequence also characterizes many pioneer settlement waves. Land is taken up rapidly close to the source of the migration in the early time periods, but later the crest of the pioneer wave moves outward. During the final periods there are only a few new settlements near the origin, and most of them are simply a filling-in of marginal areas left behind after the first enthusiastic pulse.

... and the Barriers

Diffusion processes are influenced by more than the basic characteristics we have considered so far. Usually they do not move over smooth and homogeneous surfaces, for we know that geographic space is seldom close to the ideal transportation surface where movement is equally easy in all directions. Many things get in the way, slow down, and alter the course of other things that are diffusing through an area. The unfolding patterns of innovation, migration, and urban growth that we see in sequences of maps are never even and symmetrical, like a circular wine stain expanding on a tablecloth. Rather they are channeled more quickly in some directions than others as barriers to diffusion slow down and warp the pure forms, which are more familiar perhaps to the physical than the social scientist.

Barriers in the way of a diffusion wave can have three basic effects. Upon hitting an *absorbing* barrier, a pulse of innovation is stopped cold. In the vicinity of such a barrier

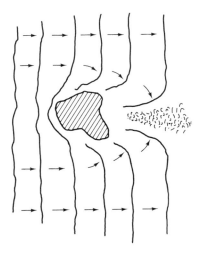

FIGURE 11–15

Ocean waves meeting an absolute island barrier.

all the energy is completely absorbed, so that the process of diffusion is halted. While completely absorbing barriers are rare, impenetrable swamps and unscalable mountains can stop the diffusion of migration waves, although settlement waves tend to flow around such barriers if they can. In the same way, ocean waves tend to be refracted around islands that stand as absolute barriers (Fig. 11–15). Considerable mixing and turbulence often occur on the leeward side of such barriers as two segments of a wave clash before forming again and going on their way. In the South Pacific such lines of turbulence are often phosphorescent at night, and are used by canoe captains to guide them home.

Sometimes an innovation wave will hit a barrier and then bounce off it. Such barriers are termed *reflecting*, and they can often channel the energy of a diffusion process and intensify it in a local area. For example, a wave train approaching two reflecting barriers (Fig. 11–16) will be slowed down initially as part of the energy is reflected inward. But like a weir in a river, the build-up of energy will eventually be released explosively so that the wave front is pushed out and distorted. We can speculate that seacoasts may form reflecting barriers for flows of communication generated by people who live on the land-water boundary (Fig. 11–17).

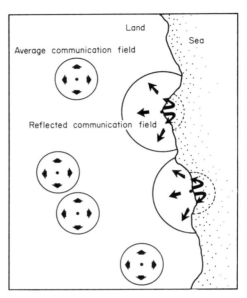

FIGURE 11–17
Personal communication fields reflected by seacoasts.

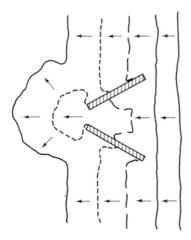

FIGURE 11–16
Waves meeting two reflecting barriers.

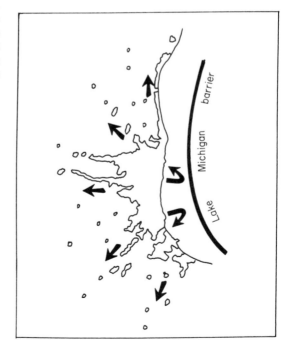

FIGURE 11–18
Chicago reflecting from the Lake Michigan barrier.

If we assume that there is a certain communication field around the average person that fulfills his personal need for normal interaction with others, then a person living on the coast has half of his field cut off—unless he wants to talk to the fishes. We really do not know, but it is possible that he will compensate for this severance of his field by expanding its radius on the landward side. In a sense, his need for an average amount of communication bounces off the reflecting seacoast barrier and extends his communication field further inland. Similarly Chicago's expansive energy has been reflected by the Lake Michigan barrier. Having been thwarted in the east, the limits now reach far into the rich agricultural land to the west (Fig. 11–18).

Sometimes a barrier may play both absorbing and reflecting roles, depending upon the mechanism of the diffusion process. Suppose, for example, a forest of maple trees is extending north into a new area bordered on the west by a lake (Fig. 11–19). Its seeds are windblown, and near the lake some of these will fall in the water and be lost—that is, they are absorbed by the barrier. The reproductive energy of the forest will be reduced locally, and the boundary might lag behind the rest of the advancing front. On the other hand, the lake will have a quite different effect upon an advancing forest of walnut trees, whose nuts are carried and buried in the ground by squirrels (Fig. 11–20). If the squirrels near the lake work just as hard as those further away, their efforts will tend to be reflected by the lake so that the western edge of the walnut forest will advance ahead of the rest. This illustration, though very simple and even fanciful, makes an important point. Barriers must be functionally defined: when we talk about them we must also consider the type of diffusion process with which they interact.

Pure absorbing and reflecting barriers are rare. In most cases, barriers are *permeable* rather than absolute, allowing part of the energy of a diffusion pulse to go through, but generally slowing down the process in the local area (Fig. 11–21). A long, narrow lake, for example, may not completely stop communication between people who live on opposite sides (Fig. 11–22), but the intensity of communication

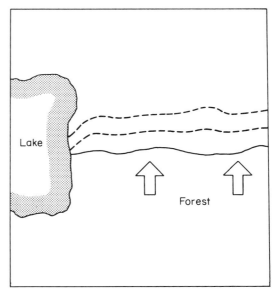

FIGURE 11–19

An advancing maple forest by an absorbing lake (adapted from Nystuen, Fig. 1a).

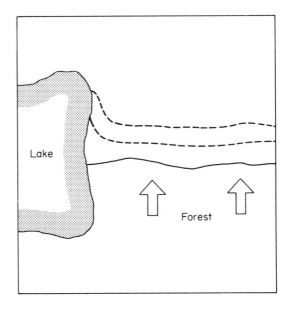

FIGURE 11–20

An advancing walnut forest by a reflecting lake (adapted from Nystuen, Fig. 1b).

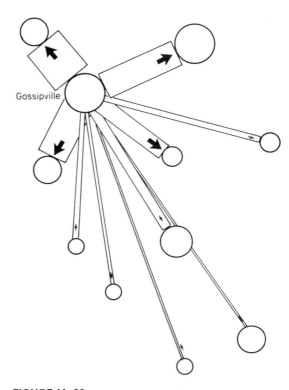

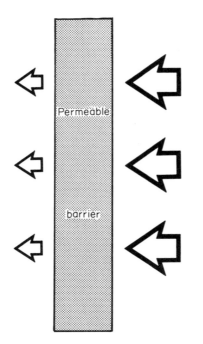

FIGURE 11–21

A permeable barrier reducing the flow through it (adapted from Nystuen, Fig. 1c).

FIGURE 11–23

Communication flows from Gossipville before hinterland severance.

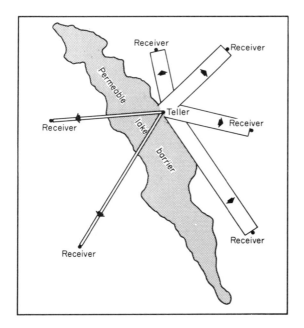

FIGURE 11–22

Communication flows reduced by a permeable lake barrier.

flows across the lake, say daily telephone gossip, will be much less than with people on the same side. Similarly, a political boundary can seldom be completely sealed, and although it may absorb much of the energy of potential interaction across it, a trickle usually gets through. Suppose we consider the telephone traffic of Gossipville with other towns in the region (Fig. 11–23), and plot on a graph the number of calls generated by various towns against the distance they are away (Fig. 11–24a). As distance increases, the intensity of the interaction declines. Now suppose a political boundary is placed through the area (Fig. 11–25). It is permeable, so rather than completely cutting telephone traffic it simply reduces it below what it was before (Fig. 11–24b). How can we measure the barrier effect? One way is to try to line up the two pieces of the

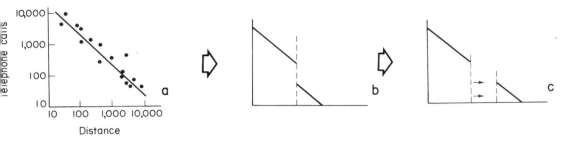

FIGURE 11–24

Measuring the effect of a political boundary upon flows of telephone calls (adapted from Nystuen, Figs. 2a–c).

graph by displacing the lower segment to the right. The amount we have to displace the segment now becomes our measure of the barrier effect in terms of distance units (Fig. 11–24c). In this way, we could say that a particular permeable barrier has roughly the same retarding effect on communications as a certain distance. Diffusion through the region will also be retarded. Even where a political boundary is very permeable, as between the United States and Canada, there is always some barrier effect. In Europe, political boundaries also often mark changes in language, so the barrier effects are much stronger, slowing communication and diffusion processes down much more.

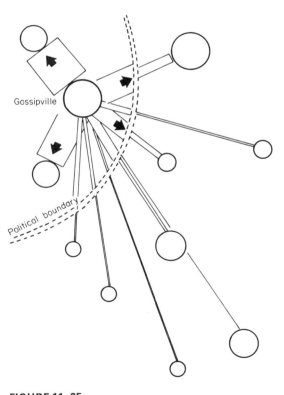

FIGURE 11–25

Communication flows from Gossipville after placement of political boundary.

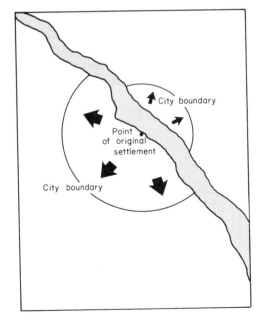

FIGURE 11–26

Effect of permeable river barrier upon the diffusion of a city.

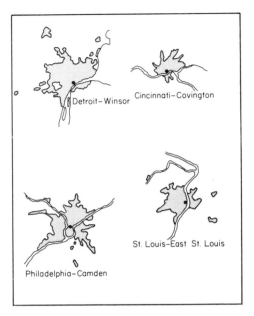

FIGURE 11–27

Barrier effects of rivers upon some North American cities.

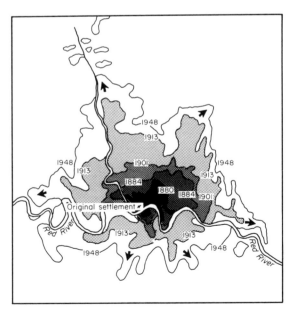

FIGURE 11–28

Diffusion of Winnepeg from site of original settlement on the Red River.

Rivers are often permeable barriers to the diffusion of urbanized areas, which move in a contagious fashion from the original core of settlement. Few cities start on both banks of a river, and the side that receives the first settlement always seems to have an advantage (Fig. 11–26). Many riverine cities in North America display the "truncated circle" effect as the process of urban diffusion is much more rapid on the side that got the initial head start (Fig. 11–27). Winnipeg is typical of such cities. It started as a small pioneer settlement on the banks of the Red River in the middle of the nineteenth century (Fig. 11–28). By 1880 the urban area had expanded considerably on the side of the founding settlement, but the opposite bank had received only a small dab of urbanization. The original advantage is maintained even today as the city has exploded on one side but has developed much more slowly on the other because the river forms a permeable barrier to the diffusion of urbanization.

While we have considered the general properties of barriers in fairly abstract ways, they actually come in many different varieties. The most obvious, perhaps, are the *physical* barriers —the mountains, deserts, swamps, lakes, and oceans. At one time they may have been totally absorbing, but today their permeability is increasing rapidly, and for some types of travel and communication they no longer exist. In earlier days, however, when the technology of transportation and communication was at a much lower level, physical barriers were often important, slowing down the rates at which ideas, innovations, and people could diffuse over the land. In the eastern United States, for example (Fig. 11–29), the Appalachians slowed down severely the rates of travel from New York City in 1800. While five days coaching could get one traveler far along the Hudson-Mohawk corridor, another person moving west would be slowed down by the succession of ridges and valleys in Central Pennsylvania. Within this folded mountain area, the long narrow valleys between the high barrier ridges still influence patterns of communication and human interaction. Marriage ties within the same valley, for example, seem to be much more probable than across the ridges of low per-

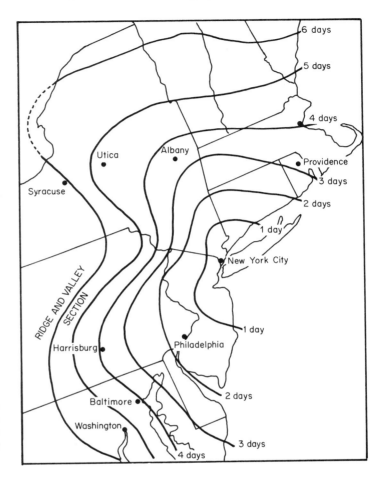

6 days

5 days

4 days

Utica

Albany

Providence

3 days

Syracuse

2 days

1 day

New York City

RIDGE AND VALLEY SECTION

1 day

Harrisburg

Philadelphia

2 days

Baltimore

Washington

3 days

4 days

FIGURE 11–29

Travel times by stage coach from New York in 1800 (from R. Brown's adaptation of map in Paullin; courtesy of American Geographical Society and Carnegie Institution of Washington, D.C.).

meability to other valleys (Fig. 11–30), and today distinctive groups of family names lie along the same alignments. Similarly, adoptions of modern farming equipment are shaped by the distorted information fields. Innovations like hay driers diffuse much more quickly along the narrow valleys than "across the grain" of the land.

Barriers to diffusion processes may be far more subtle than the physical ones that are so easily seen upon the landscape. Often they are *cultural* in nature and take many different forms. Where ideas spread by one person telling another, differences in language can greatly retard their diffusion. The intensity of telephone traffic generated by Montreal, for example,

depends not only upon the size and distance of the receiving centers, but also upon whether the towns are in Quebec or Ontario. The linguistic barrier at the boundary of the French and English parts of Canada is very marked. Europe is also wracked by linguistic barriers, so that new ideas, whether cultural or technological, diffuse far less quickly there than in the United States. A number of prominent Europeans have commented upon the distressing political consequences of such barriers, noting that only the United States seems capable of taking advantage of the Common Market, for American thinking simply overrides such barrier problems.

Linguistic barriers are some of the most stable and long lasting in their effects, although a

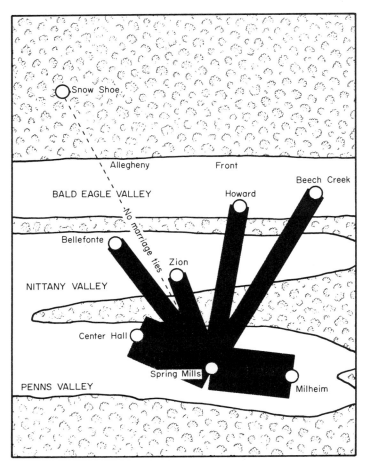

FIGURE 11–30

Intensity of marriage ties in ridge and valley section of Pennsylvania (courtesy Peirce Lewis).

language may itself diffuse slowly through an area. English has diffused rapidly during the past century to many parts of the world, and is now the major language for most commercial and technical discussions. On the other hand, language diffusion may take many generations. Along the boundary of Sweden and Finland, for example, the Finnish language seems to be gradually creeping in. Finland subsidizes its northern pioneer agriculturalists, while Sweden does not. A preponderance of young and presumably lonely males on the Swedish side, and a more even balance between the sexes on the Finnish side, mean that many young Swedish men near the border marry Finnish girls. The first language of the children, however, is that of the mother, and Finnish

appears to be slowly diffusing southward into Sweden.

Religious and *political* barriers may also thwart or slow down the diffusion of innovations. A number of international organizations of a fraternal nature are forbidden in some countries on political grounds, and Rotary International has yet to diffuse to some of the countries of Europe with totalitarian regimes. Other innovations, like physical and chemical methods of birth control (hopefully absolute barriers to a diffusion process themselves), have their rates of acceptance slowed down by religious barriers. Some of the great religious controversies of contemporary times involve the permeability and ultimate removal of such cultural barriers. In many cases the effects of barriers change under pressure from the process of diffusion itself.

Where a course of diffusion is dependent upon human beings making individual decisions to adopt or reject an idea, the major barriers may lie in the minds of men. In many cases, therefore, we are dealing with *psychological* barriers which can retard the course of an innovation. Considered individually, these barriers may appear to be either completely absorbing or completely permeable, but when we consider the adoption decisions of many people the question of psychological barriers appears more complicated. In any area through which a new idea diffuses there will always be some early *innovators* who adopt first (Fig. 11–31). Once they set the example they are quickly followed by a group called the *early majority*, and their example brings in the *late majority* in turn. Finally come the *laggards*, at the tail end when nearly everyone else has adopted the new practice.

The normal curve describing the distribution of innovators and laggards can also be expressed as a logistic curve. Plotting the proportion of people adopting an innovation along the vertical axis and the time of adoption along the horizontal axis (Fig. 11–32), we start on the left and gradually accumulate the proportion of adopters as we move to the right. When T is small, at the start of the diffusion, we shall only accumulate a very small proportion of early in-

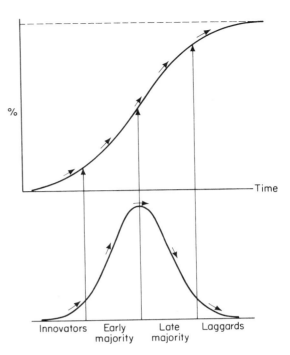

FIGURE 11–32

Accumulating the distribution of innovation acceptors.

novators. But suddenly the innovation takes hold and the proportion of adopters rises quickly as the early majority comes in. At this point, a further move to the right of the normal curve brings in the late majority, but the diffusion is obviously slowing down as the cumulative curve on the graph begins to turn the other way. Eventually the laggards come in. Having broken out of the source area, the wave bursts across the landscape and finally crashes upon the most remote part of the region.

We can summarize all these barrier effects by considering a very small spatial system of five nodes (Fig. 11–33), in which things may diffuse along the lines of communication connecting each node directly to all the others. We can draw up a five-by-five table or matrix containing the proportion of times, or probabilities, that one node will communicate with others in the system. The nodes may be people, towns, or even census districts or countries—the same principles

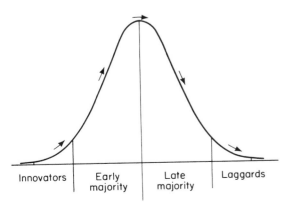

FIGURE 11–31

Distribution of innovation acceptors.

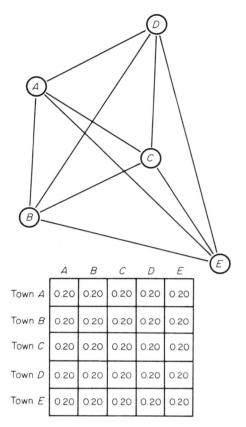

	A	B	C	D	E
Town A	0.20	0.20	0.20	0.20	0.20
Town B	0.20	0.20	0.20	0.20	0.20
Town C	0.20	0.20	0.20	0.20	0.20
Town D	0.20	0.20	0.20	0.20	0.20
Town E	0.20	0.20	0.20	0.20	0.20

FIGURE 11–33

Communication matrix for equal nodes and no distance effects.

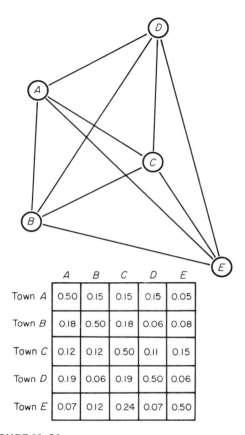

	A	B	C	D	E
Town A	0.50	0.15	0.15	0.15	0.05
Town B	0.18	0.50	0.18	0.06	0.08
Town C	0.12	0.12	0.50	0.11	0.15
Town D	0.19	0.06	0.19	0.50	0.06
Town E	0.07	0.12	0.24	0.07	0.50

FIGURE 11–34

Communication matrix for equal nodes with frictional effects of distance.

hold. If the nodes are of equal size, and distance has no retarding effect upon communication, then we might expect each node to communicate with all the others in roughly equal proportions. The diffusion of a message at any particular time period would be equally likely. Notice that the rows always sum to 1.0, for we assume a closed system here, a set of people or villages cut off from the rest of society.

If distance has an effect upon communication (Fig. 11–34), we would expect new ideas and messages to move with greater probability between nodes close together than between those farther apart. We can describe such frictional effects upon message flows by simply changing the values of the probabilities of interaction between a pair (a row and a column)

in the matrix. Town A, for example, is roughly equidistant from B, C, and D and has roughly the same chance of passing a message to them at any one time ($p = 0.15$). Town E is farther away, and its chance of getting a message from A is much smaller ($p = 0.05$). If the nodes are of different sizes, then a big center like D will generate and receive more messages than its smaller neighbors (Fig. 11–35). In the interaction matrix, the probabilities of generating and receiving change accordingly, and D's column has very large values compared to the others.

Finally, we can introduce our three sorts of barriers (Fig. 11–36). Town B is behind an absorbing barrier, and as it is severed from the system its row and column in the matrix both contain zeros. Town A cannot communicate

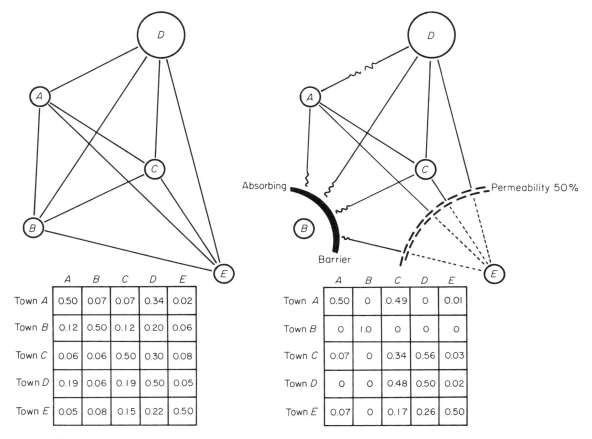

FIGURE 11–35

Communication matrix for unequal nodes and effects of distance present.

	A	B	C	D	E
Town A	0.50	0.07	0.07	0.34	0.02
Town B	0.12	0.50	0.12	0.20	0.06
Town C	0.06	0.06	0.50	0.30	0.08
Town D	0.19	0.06	0.19	0.50	0.05
Town E	0.05	0.08	0.15	0.22	0.50

FIGURE 11–36

Communication matrix for system with permeable and absorbing barriers.

	A	B	C	D	E
Town A	0.50	0	0.49	0	0.01
Town B	0	1.0	0	0	0
Town C	0.07	0	0.34	0.56	0.03
Town D	0	0	0.48	0.50	0.02
Town E	0.07	0	0.17	0.26	0.50

directly with *D*, but it can send its messages to *C*, which sends them on to *D* (equivalent to English and German families communicating during the Second World War via the Red Cross in Switzerland). Thus the proportion of messages between *A* and *D* is added to those normally going between *A* and *C* and between *C* and *D*, raising the values in the matrix accordingly. Town *E* is on the other side of a permeable barrier which cuts the chance of a message down by one half. The barrier effect is reflected in the matrix by very small values in *E*'s row and column. Thus, the interaction or diffusion matrix summarizes most of our discussion. Contagious diffusion is likely where a population is fairly homogeneous and distance effects are strong. Hierarchical diffusion is more common

when transmitting and receiving nodes are of greatly different sizes, and communication flows are more intense between the giants than between the small fry. Distance effects change the probabilities of interaction, and so warp and channel the diffusion of ideas as communication is cut down between distant places. Finally, absorbing, reflecting, and permeable barriers, which may be psychological, cultural, political, or physical, can all be expressed in terms of the elements of the interaction matrix. Absorbing barriers sever pieces from a system of spatial interaction, and turn rows and columns to zero; permeable barriers reduce some probabilities, and thus twist and warp the relative communication space through which messages and ideas must diffuse.

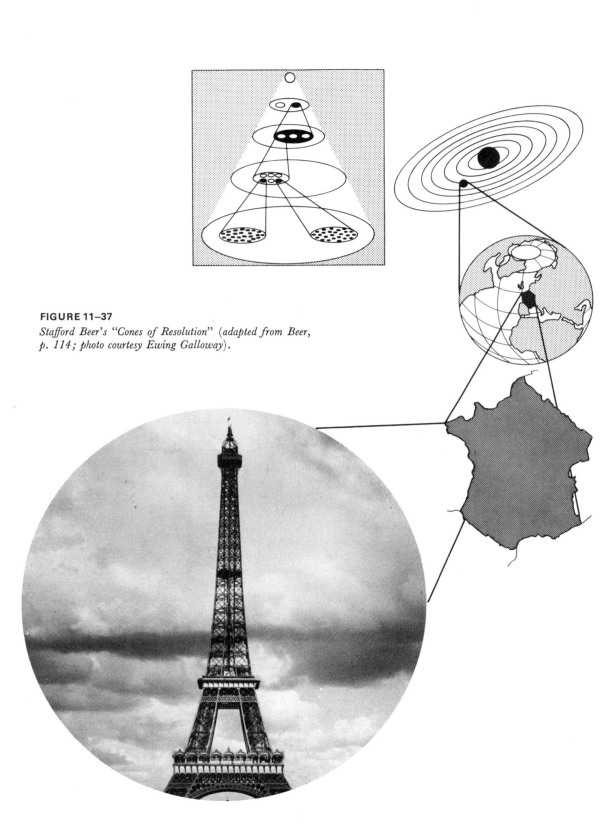

FIGURE 11–37

Stafford Beer's "Cones of Resolution" (adapted from Beer, p. 114; photo courtesy Ewing Galloway).

LEVELS, SCALES, AND CONES OF RESOLUTION

Processes of spatial diffusion occur at many geographic scales, and we should not be surprised if we require several models to help us clarify these difficult processes operating at a variety of levels. Shifts in scale are very common in all the sciences. As we become dissatisfied with very general statements giving gross overviews of a subject, we direct our attention to smaller and smaller pieces of the problem, shifting along a continuum from the macro to the micro viewpoint. The evolution of many sciences with names prefixed by *micro-* attests to the shifts in scale that have taken place throughout science during the past half century. Stafford Beer, a prominent operations research worker, has coined the provocative phrase "cones of resolution," implying that problems can be considered at many different levels and in varying degrees of detail (Fig. 11–37). In any subject, the pendulum usually swings back and forth between the extremes, and with people working at all scales of inquiry, the "cones of resolution" are eventually filled with a hierarchy of models. More general models, high up in the cones, are supported by a number of others at lower levels.

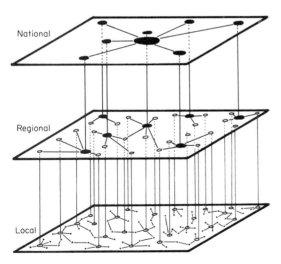

FIGURE 11–38

National, regional, and local planes of diffusion.

Geographic diffusion processes may be considered in the same light (Fig. 11–38). At the micro scale on the lowest plane, ideas and innovations spread through social communication networks linking individuals to one another. But considered at the regional level, a different network of communication may come into play in the middle plane, probably closely aligned to the pattern of linkages between central places. Finally, on the upper plane, at the national or even international level, macro flows of information, warped and shaped by great metropolitan fields, diplomatic relationships, political considerations, and so on, guide the course and intensity of diffusion processes. We shall consider each of these levels in turn, starting at the lowest plane, where our models have the greatest degree of spatial resolution, and ending at the macro level.

Diffusion at the Individual or Micro Level

Many diffusion processes operate at the micro level, where innovations and ideas diffuse from individual to individual. Small hand tractors for plowing and puddling rice fields spread from one Japanese rice farmer to another, often driving out more traditional methods of cultivation. As in Europe at an earlier time period, the horse in Japan disappears in a simultaneous process of reverse diffusion. As we saw in our discussion of geographic relationships, tractors diffuse from a distinct node of innovation in the flat wheatlands of North Dakota. At the same time, hybrid corn takes hold in Iowa to spread like a prairie fire through the Corn Belt. Membership in a progressive farmers club on Kilimanjaro, allowing the coffee farmers to auction their high quality production at better prices, spreads across the face of the great mountain, channeled by the feeder roads, and warped by the traditional hierarchical structure of chief and headman. From a family planning clinic in India, the I.U.D. diffuses from one woman to another as an innovation in birth control. In Colombia, new varieties of high-yielding beans diffuse from selected "seed villages." All these examples characterize diffusion processes which require individual,

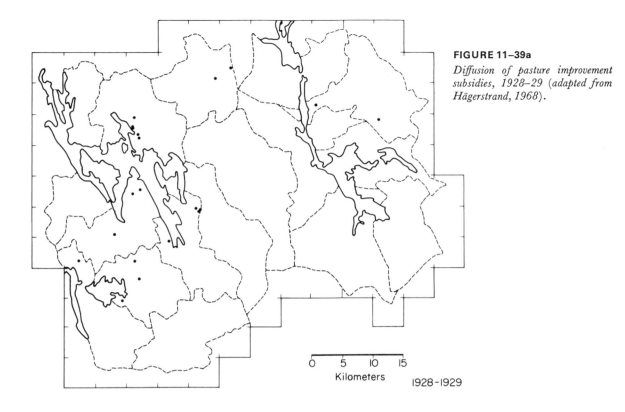

FIGURE 11–39a

Diffusion of pasture improvement subsidies, 1928–29 (adapted from Hägerstrand, 1968).

0   5   10   15
Kilometers   1928-1929

face-to-face communication before adoption takes place.

Most of the pioneer work on modeling diffusion processes comes from Sweden, where the spread of many innovations has been examined in considerable detail by Hägerstrand. For example, in the late 1920s the Swedish government tried to persuade farmers to abandon the old custom of allowing cattle to graze in the open woodlands during the summer months. The grazing cattle damaged the young growth, and a subsidy was offered to help the farmers fence and generally improve their pasture lands. In the early years (Figs. 11–39*a* and *b*) distinct clusters of farms adopting the pasture improvement subsidy formed in the western part of the region, with one or two solitary innovators in the east. By 1931 (Fig. 11–39*c*), the innovation had diffused rapidly in the west, moving out from the initial clusters of the earlier years. The eastern portion of the area lagged behind, and only caught up in the next two years (Figs. 11–39*d* and *e*), while the pattern of adoption continued to thicken and develop in the west.

When we examine such a sequence of maps, we are struck by the degree of spatial regularity in the unfolding pattern. It is almost as though a photographic plate were being developed. There seems to be a "latent diffusion image" present in the early maps, waiting to appear after a certain lapse of development time. It was precisely from the very careful examination of such map sequences that the first models of spatial diffusion were developed. We are going to consider one that has become a classic, for it gives us great insight into a very complex spatial process. Moreover, you can use it yourself as a base from which to make up examples of your own to get a feel for the way in which a spatially dynamic model works.

Let us consider the pieces of the problem, and simplify some of them to make this first step in model building as easy as possible. The process of simplification is not just important, it is essential: it is easy to make things difficult, but very hard to make them simple. If we can simplify a problem in the beginning by making some assumptions, we can always make it more

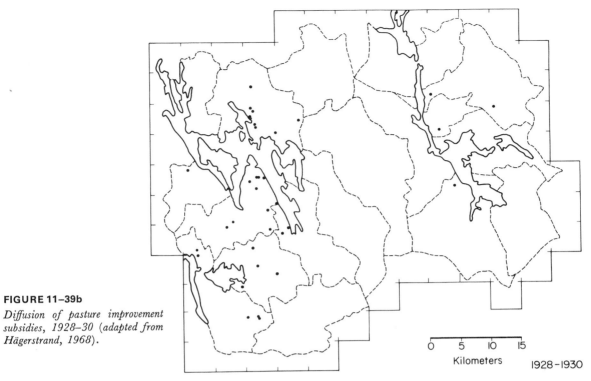

FIGURE 11–39b

Diffusion of pasture improvement subsidies, 1928–30 (adapted from Hägerstrand, 1968).

1928–1930

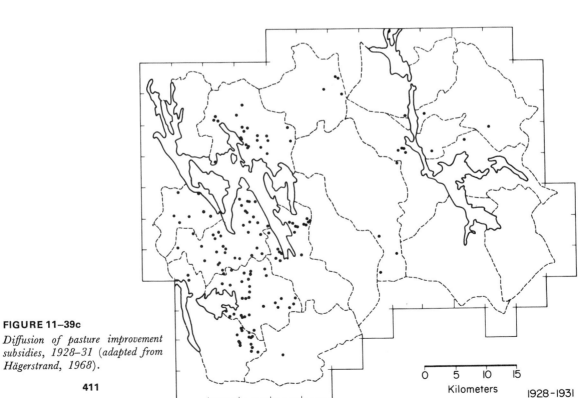

FIGURE 11–39c

Diffusion of pasture improvement subsidies, 1928–31 (adapted from Hägerstrand, 1968).

411

1928–1931

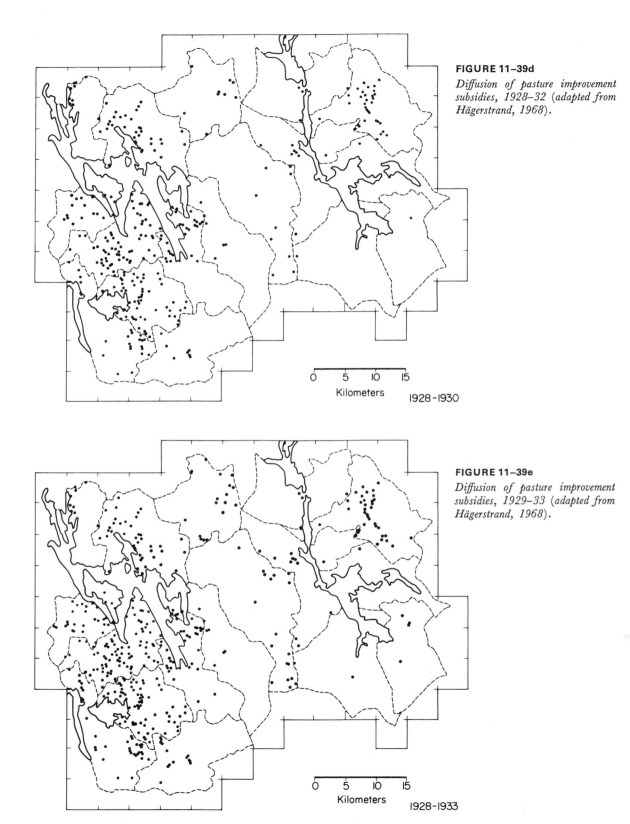

FIGURE 11–39d

Diffusion of pasture improvement subsidies, 1928–32 (adapted from Hägerstrand, 1968).

0   5   10   15
Kilometers
1928-1930

FIGURE 11–39e

Diffusion of pasture improvement subsidies, 1929–33 (adapted from Hägerstrand, 1968).

0   5   10   15
Kilometers
1928-1933

complicated later if we are dissatisfied by our first attempt.

Our first assumption, then, is that the potential adopters are spread evenly over the land, and that a message or an innovation can move with equal ease in any direction. We can lay a grid over the map of the region with the same number of potential adopters in each grid cell. Thus our model starts with the assumption of a homogeneous transportation surface, or, if we like, under *isotropic* conditions. The next question is: how will messages move? We shall assume that news about an innovation moves only by one person coming into face-to-face contact with another and telling him about it. Communication, in other words, is by pairwise telling, and as soon as a person hears the news we assume that he will adopt the innovation.

But now we come to the core question, the answer to which forms the most important building block of the entire model: how will a person who knows about an innovation be paired with someone who does not? Our final assumption is that the probability of a teller communicating with a receiver depends only on the distance between them. The probability of a message passing between two people is high when they are close together, and small when they are far apart. Each person is at the center of a communication field that is strong close to him but gets weaker as distance increases (Fig. 11–40). The problem is, how can we measure the probability of a person passing a message over a certain distance? If we take a sample of telephone calls in a region, and plot the log of the number of calls against the log of the distance over which they move (Fig. 11–41), we will get some notion of the way in which the friction of distance cuts down interpersonal communication. The line shows us the general decline in the intensity of communication with distance, and defines for us the mean information field (MIF) for the people in the area. In very simple societies, the slope of the MIF is usually quite steep, indicating that most of the time the people only communicate over short distances, and that long-distance communication is unlikely. In other, very mobile societies, the slope of the MIF is shal-

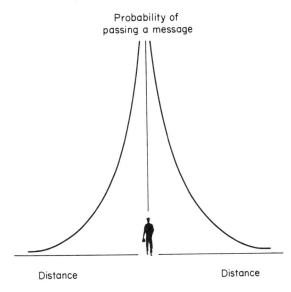

Probability of
passing a message

Distance   Distance

FIGURE 11–40

A personal communication field.

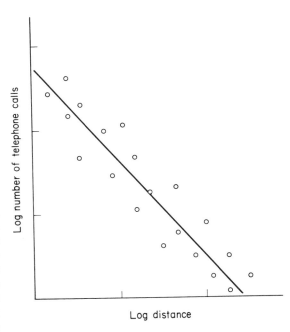

Log number of telephone calls

Log distance

FIGURE 11–41

A straight-line relationship: log of calls against log of distance.

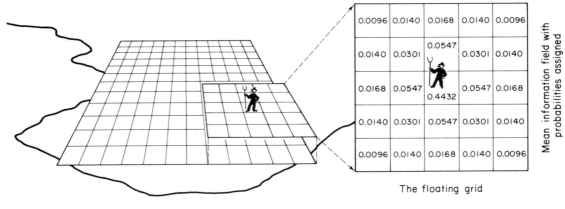

The floating grid

FIGURE 11–42

The mean information field as a floating grid of twenty-five cells with the probabilities of communication assigned.

lower, indicating that the probability of communicating over long distances is greater.

The problem now is to put all the pieces of the model together. First we must translate our estimate of the MIF into operational form. We shall assume that the mean information field takes the form of a small, square grid of twenty-five cells, the sizes of which are the same as those of the grid lying over the map (Fig. 11–42). To each of the twenty-five cells in the MIF we assign a probability, assuming that the teller is located in the middle cell. Thus the center cell receives a value of 0.4432, so that every time the teller passes a message there are 4432 chances out of 10,000 that the message will go to someone else in the same middle cell. On the other hand, the corner cells are far away, and the chances of a message passing to people in these remote locations is much smaller. A probability value of only 0.0096 is given, or 96 chances in 10,000. Notice that the sum of all the probabilities is 1.0, or complete certainty, implying that whenever a message is passed it must fall within the mean information field.

To make the probability assignments in each of the cells operational we accumulate them by starting in the upper left hand corner and moving row by row to the lower right hand cell. To each cell we assign some four-digit

numbers corresponding to the probability of the cell receiving a message from the center (Fig. 11–43). Thus, in the upper left cell, we have the interval 0000–0095, which contains 96 four-digit numbers, because $p = 0.0096$. The next has the numbers 0096–0235 assigned, giving it 140 numbers corresponding to the number of chances in 10,000 of receiving a message. This cumulative process continues until we reach the last cell in the lower right hand corner with the numbers 9903–9999.

We now have all the pieces of the model, and all we need is a driving force—an engine—to power it. Our source of energy is going to be a stream of four-digit, randomly chosen numbers. We can imagine the MIF floating over each of the map cells in turn, stopping at a cell whenever it comes across one of the first few innovators. We assign these from actual data at the start of the analysis. Whenever it stops over a teller a four-digit number is drawn at random, and this number locates the cell of the next adopter. For example, if we had the numbers 0000 to 9999 in a hat, and drew out 7561, then we would know that the innovator had passed the message to the adjacent cell lying just to the east of the middle one (Fig. 11–43). A person in this cell immediately adopts the innovation and the MIF "having writ moves on," continuing to scan each of the cells of the map in turn, stopping over each innovator, and generating a new adopter of the innovation. One complete pass over the map constitutes one time period or generation of the diffusion process. Then the MIF starts again, stopping twice as often during the second generation, four times as often during the third generation, eight times during the fourth, and so on. Of course, random numbers are not really pulled out of hats: random number tables are available in published form, and today these models are programmed for large computers which have special ways of generating random number series.

As one generation of telling follows another, a pattern of diffusion develops from the initial assignment of the first few innovators to the population cells. This is why we call this type of diffusion model a *simulation* model, for it simulates a process acting through time and over space. Notice, also, that every time we run the model we get a slightly different result; the diffusion pattern from one run will not match another run exactly. This is because our model is probabilistic in nature, for we use probabilities in the mean information field, and generate every new innovator by the chance (or probabilistic) process of drawing random numbers. This way of powering a simulation model is known as the *Monte Carlo* method, so our complete diffusion model is known as a Monte Carlo simulation model under isotropic conditions. The model is very simple, and can easily be

run by hand with a map of grid cells, a table of random numbers, and a small plastic overlay with 25 cells for a MIF. As it stands, however, it is not terribly realistic. Innovations do not diffuse at a geometric rate in the 2–4–8–16–32 fashion, but begin to slow down as saturation levels are reached. One way to make the model more realistic would be to identify and label the individual adopters in every one of the grid cells on the map. Let us suppose that there are ten potential adopters in each cell and we call them Mr. Zero, Mr. One, and so on up to Mr. Nine. Every time a teller got the MIF, and generated a new message to another cell, we could draw a second, single-digit random number between 0–9. If the person, say Mr. Seven, has not been "hit" before we would make him adopt then and there. But if he had been drawn before and already had the message, then the second hit would be wasted upon him. The MIF would move on without generating a new adopter that time. In the early stages the pattern of diffusion would develop quickly, but at later time periods we would be getting close to saturation and many of the pulses of information would be wasted on the people who already had the news. Thus the diffusion process would slow down, producing the much more realistic S-shaped diffusion curve of adoption. Of course, adding this additional requirement to the model would involve a lot of bookkeeping, but since simulation models are now programmed for computers this is no longer a problem.

Another way of making the model more realistic is by allowing the population to vary from cell to cell on the map—a seemingly simple addition to the model which produces some very tedious arithmetic operations. Suppose, for example, that the MIF lands on a teller surrounded by an unevenly distributed population. Realistically, the probability in each cell of the MIF should change, being weighted by the number of people in each of the underlying map cells who are potential adopters. We can weight the probabilities in the MIF by multiplying them by the corresponding populations on the map (Fig. 11–44). Having multiplied through, the sum of the "weighted probabilities" is now greater than 1.0, so we must divide each value by this total to *normalize* our MIF. When this is done,

0–95	96–235	236–403	404–543	544–639
640–779	780–1080	1081–1627	1628–1928	1929–2068
2069–2236	2237–2783	2784–7214	7215–7761	7762–7929
7930–8069	8070–8370	8371–8917	8918–9218	9219–9358
9359–9454	9455–9594	9595–9762	9763–9902	9903–9999

FIGURE 11–43

Accumulated intervals for mean information field.

0.0096	0.0140	0.0168	0.0140	0.0096
0.0140	0.0301	0.0547	0.0301	0.0140
0.0168	0.0547	0.4432	0.0547	0.0168
0.0140	0.0301	0.0547	0.0301	0.0140
0.0096	0.0140	0.0168	0.0140	0.0096

Mean information field

X

6	9	4	12	1
1	3	2	2	16
8	23	4	7	2
5	21	9	12	6
2	6	11	26	5

Underlying population grid

=

0.0576	0.1260	0.0672	0.1680	0.0096
0.0140	0.0903	0.1094	0.0602	0.2240
0.1344	1.2581	1.7728	0.3829	0.0336
0.0700	0.6321	0.4923	0.3612	0.0840
0.0192	0.0840	0.1848	0.3640	0.0480

Σ 6.8477

the probabilities add up once again to 1.0, and we can assign the corresponding intervals of four-digit numbers once again. If you think that this has to be done for every teller before every telling, you can see that this is an extremely time-consuming process. Again, computers come to our aid to perform such dull, but necessary arithmetic.

Barriers may also be introduced. Between some grid cells on the map we can assign absolute or absorbing barriers, so that when a teller tries to generate a message to a cell lying on the leeward side of a barrier the message is blocked. Or we can assign barriers to our map with varying degrees of permeability—again on a probabilistic basis. For example, if a barrier is 50 percent permeable, and a teller generates a message across it, we can flip a penny and allow the telling only if it lands heads. Alternatively, we can generate another random number, and allow the telling only if the number is an even one. Assigning a number of absorbing and permeable barriers slows down the course of a diffusion simulation, matching the process in the real world under similar conditions.

Finally, we can add psychological barriers to our model. If we assume that the chances of adoption change according to the number of tellings, then we can convert a probability distribution of tellings into a number of two-digit intervals (Fig. 11–45). With a computer keeping track of both the person in the cell and the number of times he has been hit previously, we can generate a two-digit random number

before deciding whether the new adoption has taken place. Suppose, for example, that a teller contacts Mr. Three in a particular cell, and he has been hit once before. The first time the number 89 was drawn, but because it did not lie in the interval 0–4 Mr. Three did not adopt. The second time, however, the two-digit number 24 comes up. This lies in the interval 5–24 so Mr. Three adopts the innovation and becomes a teller himself at the time of the next generation.

How well do such models simulate actual diffusion processes? Let us examine the spread of pasture improvement subsidies once again (Figs. 11–39*b*, *c*, *d*, and *e*), and simplify the maps by recording the adopters in each cell

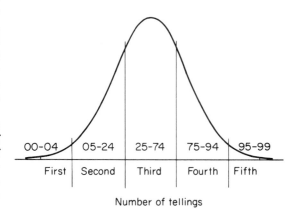

FIGURE 11–45

Accumulated intervals for the number of tellings.

Normalized probabilities

$$= \begin{array}{|c|c|c|c|c|}
\hline
0.0084 & 0.0184 & 0.0098 & 0.0245 & 0.0014 \\
\hline
0.0020 & 0.0131 & 0.0159 & 0.0087 & 0.0327 \\
\hline
0.0196 & 0.1837 & 0.2588 & 0.0559 & 0.0049 \\
\hline
0.0102 & 0.0923 & 0.0718 & 0.0527 & 0.0122 \\
\hline
0.0028 & 0.0122 & 0.0269 & 0.0531 & 0.0070 \\
\hline
\end{array}$$

FIGURE 11–44

Weighting the probabilities with the underlying population.

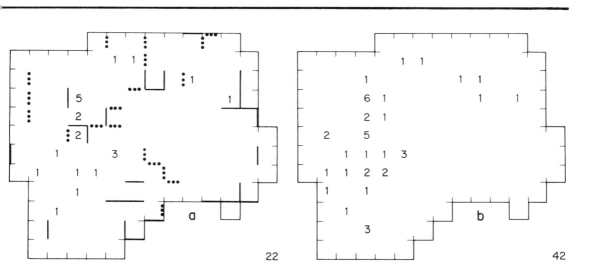

FIGURE 11–46

Actual diffusion of pasture subsidies (adapted from Hägerstrand, 1965, Fig. 3, p. 376).

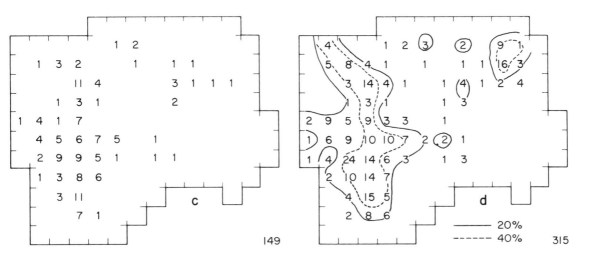

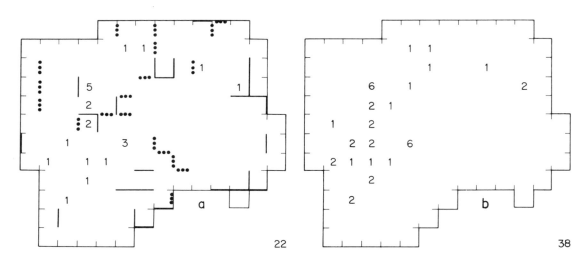

FIGURE 11–47

Simulation of pasture subsidy diffusion (adapted from Hägerstrand, 1965, Fig. 6, p. 380).

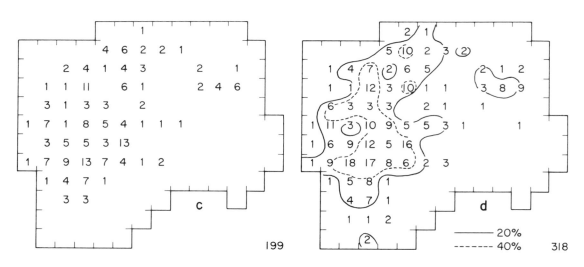

(Figs. 11–46a, b, c, and d) adding some absolute barriers (solid lines) and others of 50 percent permeability (dotted lines). The barriers represent the effects of the long lakes that lie in the region and thwart communication. When a computer simulated the diffusion of pasture subsidies under these conditions, the patterns developed by the Monte Carlo method closely resembled the actual conditions (Figs. 11–47a, b, c, and d). A comparison of the final output

(Fig. 11–47d) with the last map in the real sequence indicates how closely the simulated and the actual patterns coincide. The isolines enclose areas where 20 percent and 40 percent of the potential adopters accepted the pasture innovation, and both patterns display the major clusters resulting from a rapid diffusion in the west. The eastern outliers also correspond reasonably closely. The testing of spatial simulation models poses difficult, and still unsolved,

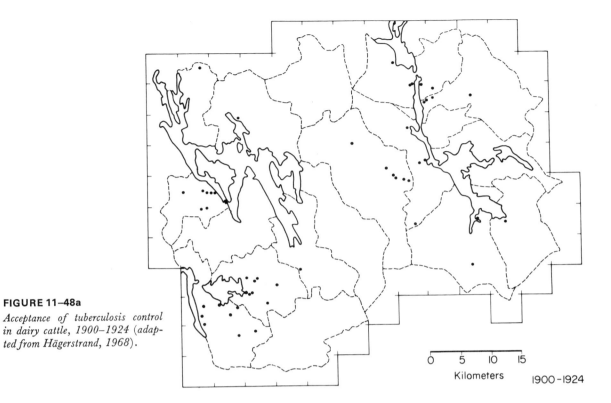

FIGURE 11–48a

Acceptance of tuberculosis control in dairy cattle, 1900–1924 (adapted from Hägerstrand, 1968).

1900–1924

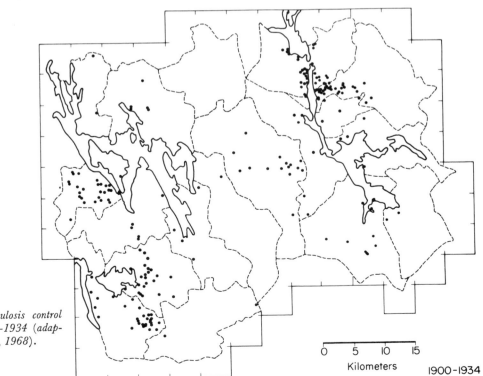

FIGURE 11–48b

Acceptance of tuberculosis control in dairy cattle, 1900–1934 (adapted from Hägerstrand, 1968).

419

1900–1934

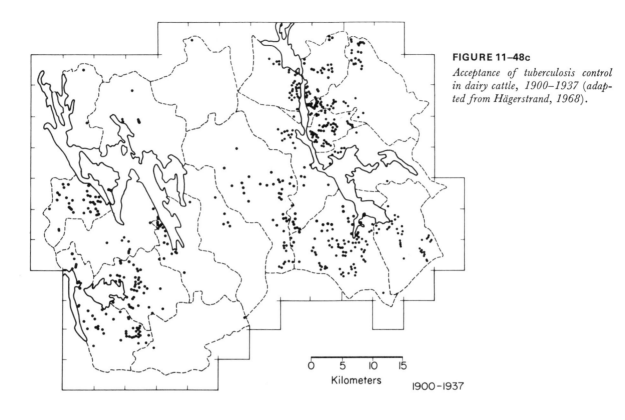

FIGURE 11–48c

Acceptance of tuberculosis control in dairy cattle, 1900–1937 (adapted from Hägerstrand, 1968).

1900–1937

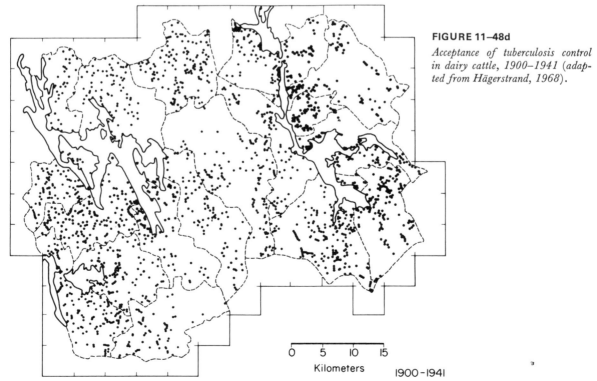

FIGURE 11–48d

Acceptance of tuberculosis control in dairy cattle, 1900–1941 (adapted from Hägerstrand, 1968).

1900–1941

problems in an area of applied mathematics known as inferential statistics. However, it seems likely that our present visual comparisons and judgments will stand up well when more sophisticated methods come along.

No matter what the starting configuration, many patterns of diffusion at the micro level consistently display an uncanny regularity and order as they gradually unfold upon the landscape. In the same area as the pasture subsidy, Hägerstrand also examined the spread of other innovations such as control of tuberculosis in dairy cattle, soil mapping, automobile ownership, postal checking accounts, and telephones. The acceptance of tuberculosis control (Fig. 11–48*a*) involved considerable sacrifice on the part of the farmers, and the diffusion was slower than for the pasture subsidy. Distinct spurts in acceptances occurred as price supports were passed for the purchase of TB-free milk. By the end of 1924, nearly sixty farms had tuberculin-tested dairy herds, and these formed distinct clusters in the southwest and northeast, with a few scattered innovators. A large area empty of acceptors slices diagonally across the area from north to south. By 1934, after ten years of diffusion, the innovation had made considerable progress (Fig. 11–48*b*). Over two hundred farms in the area had accepted it, the original clusters had thickened and expanded to form the typical nebula-like pattern so characteristic of innovation movements, and a "bridge" of acceptors had formed across the middle area as the farms were battered from two sides. The pattern for 1937 (Fig. 11–48*c*) confirms the regularity of the process as the innovation pushed outward from the original core areas, particularly in the southeast where farmers accepted with a rush after the price of TB-free milk went up. By 1941 (Fig. 11–48*d*), the process was almost complete as the "laggard areas," far from the initial clusters, finally accepted the innovation.

Just in case you think that only Swedish spatial behavior is orderly and predictable, let us turn to the United States as a final example of diffusion at the individual or micro level. The northern high plains of Colorado immediately to the east of Denver have traditionally been an area of dryland farming and cattle grazing. During the late 1940s and early 1950s many of the farmers tried to enlarge their cattle feeding facilities, but a series of droughts hammered the area and getting sufficient feed for the large herds became a severe problem. Some of the farmers began to turn to irrigation, installing wells and pumps to tap the groundwater, even though this often involved heavy outlays of capital up to $40,000. By 1948, forty-one wells had been sunk (Fig. 11–49*a*). People throughout the area knew that pump irrigation was possible, but decisions to adopt the new form of farming appear to have come only after face-to-face discussions among the farmers. Once again, we are concerned with the problem of establishing the mean information field for the people of an area. In Colorado barbecues are big social events and people come to them from kilometers around. By establishing patterns of barbecue attendance, and supplementing the information with telephone call data, it was possible to estimate the effect of distance upon social communication in this area of scattered farms.

Using procedures similar to Hägerstrand's, and starting with the 1948 configuration, the geographer Bowden simulated the diffusion of 410 wells (Fig. 11–49*b*), the same as the number of actual wells that had been adopted by 1962 (Fig. 11–49*c*). So close was the average of ten simulated runs to the pattern in the real world, that the model was used as a predictive device. The actual 1962 configuration was used as a new starting point to simulate the expected pattern in 1990 (Fig. 11–49*d*). An additional "rule of the game" was also provided, specifying that no more than sixteen wells could be drilled in any one township (*S-S-S* area on the map). By 1990, a total of 1644 wells are expected, drawing 20 million acre feet of water every year. Only time will tell whether the actual course of diffusion will proceed in this way, but one aspect of the model has proven itself already: the rule of no more than sixteen wells per township has been adopted as a result of the geographical study. Sometimes roles are reversed, and the real world makes itself conform to the geographer's model!

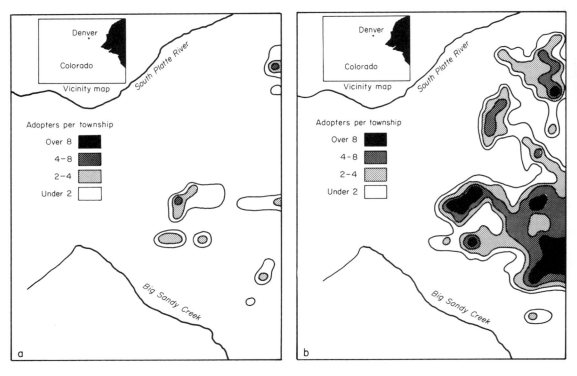

FIGURE 11–49

(a) *Irrigation wells per township: actual pattern, 1948;* (b) *simulated pattern of 410 wells;* (c) *actual pattern of 410 wells by 1962;* and (d) *simulated prediction for 1990 (adapted from Bowden, Figs. 10, 11, 22, and 27).*

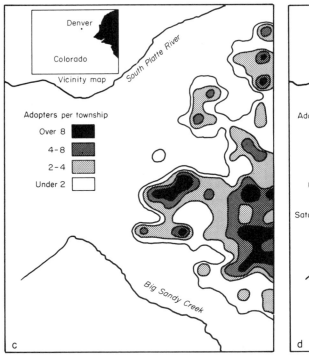

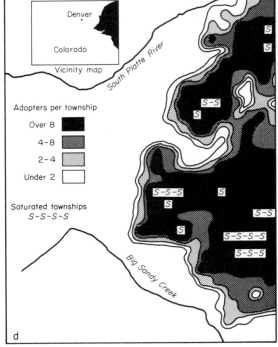

Changing the Scale:
Urban Aspects of Diffusion

As we change scale and move up the cone of resolution, agglomerations of many individuals become the next focus of our diffusion processes. The modern world is becoming increasingly urbanized, as the vast compactions of men, together with their concrete, steel, and asphalt works, gradually ooze from the earlier cores of intense settlement. "Ooze," of course, is a relative term, for its use to describe the flow of cities into the countryside depends upon our perception of time. Sometimes the diffusion of a city seems almost explosive. In 1840 London was a city not more than eight kilometers across with about two-thirds of it on the north bank of the Thames (Fig. 11–50a). Even in 1860 (Fig. 11–50b) it still appears very compact, although these twenty years only proved to be the bucolic lull before the urban storm. By 1880 (Fig. 11–50c), with the Industrial Revolution sparking the expansion and commuter railways reaching out like tentacles into the countryside, London started to explode. By 1900 (Fig. 11–50d) it was developing dormitory towns ahead of its advancing waves, but these were later enveloped by the great urban mass. The year 1914 (Fig. 11–50e) already saw many of the dormitory towns that had once been villages linked into continuous urbanized strips. As more and more fields and woodlands were devoured by the metropolitan giant, people tried to develop institutional barriers against the seemingly inexorable diffusion process. Parklands were set aside and "Green Belts" laid out where building was strictly controlled. Today (Fig. 11–50f), the built-up area is sixty-five kilometers across, and commuter line tentacles influence development as far as eighty kilometers from the center. Legally the towns surrounding London are separate and distinct; geographically they are but pieces in the ongoing process of spatial diffusion. In America the same process is even less controlled, and today many of the original nuclei on the Eastern Seaboard have diffused outward, coalesced, and given birth to a new phenomenon in man's history—Megalopolis.

Within cities similar processes of diffusion occur. In America many of the distinct ethnic patterns of residence are the result of gradual movements out from a core area—often against strong psychological barriers of racism. In a part of Seattle, for example (Fig. 11–51a), a distinct Negro ghetto developed from a small core area consisting, in 1940, of a score of blocks at the junction of two main thoroughfares. With natural increase and in-migration the burgeoning population expanded outward, and by 1960 (Fig. 11–51b) the ghetto had diffused in a very regular and compact fashion to embrace about 140 residential blocks. "Rules of the Game" may be a hideous misnomer for labeling the factors underlying a spatial process with such distressing human consequences, but when a number of well-specified forces are linked together in a simulation model they do seem to account for the spread of the ghetto. Using a mean information field to generate the probability of house-searching contacts, estimates of in-migration and natural increases, and rules for specifying the number of randomly generated contacts required for "block-busting," Richard Morrill simulated the diffusion of the ghetto over a quarter of a century (Figs. 11–51c and d). There appears to be a very close correspondence between the simulated and actual pattern, indicating that most of the general aspects of the process have been incorporated into the model. A typical two-year "generation" in one of the simulations over a ten-year census period shows how each of the old locations generates new contacts and entries into blocks that have been "busted" (Fig. 11–51e). To the west, resistance from whites was very high and many contacts had to be generated before entry was allowed. A final simulation for the period 1960–64 (Fig. 11–51f) indicates how the barrier of prejudice biases the spread of the ghetto to the east and north—areas where resistance of white real estate operators and residents was less than it was in the west.

Block-busting is an aspect of spatial diffusion that implies the notion of a *threshold*. Whereas psychological barriers at the micro level were never absolute, for we allowed the possibility of acceptance even on the first contact, in the ghetto simulation the rules specified that *at least* a certain number of contacts had to be made before block-busting occurred. A definite threshold

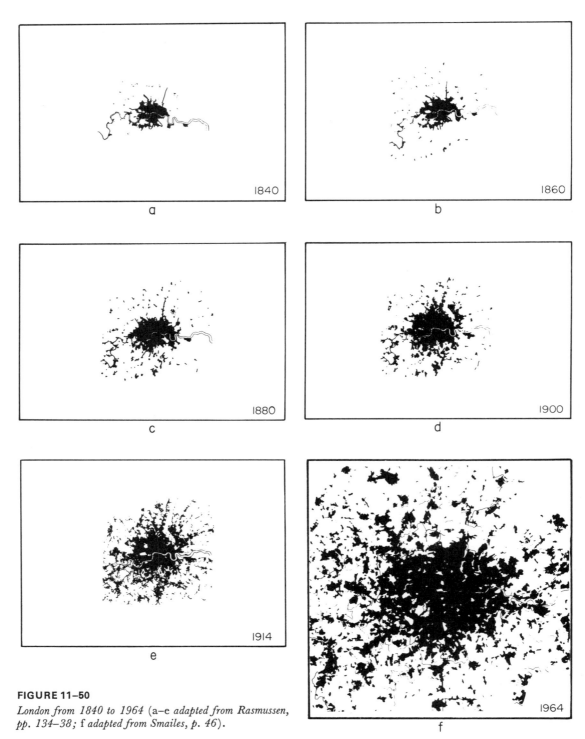

FIGURE 11–50

London from 1840 to 1964 (a–e adapted from Rasmussen, pp. 134–38; f adapted from Smailes, p. 46).

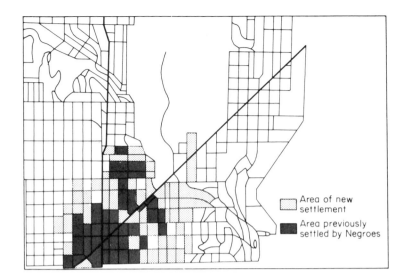

FIGURE 11–51a

Seattle: ghetto growth, 1940–50 (adapted from Morrill, 1965, Fig. 9).

FIGURE 11–51b

Seattle: ghetto growth, 1950–60 (adapted from Morrill, 1965, Fig. 10).

FIGURE 11–51c

Seattle: simulated expansion of ghetto, 1940–50 (adapted from Morrill, 1965, Fig. 11).

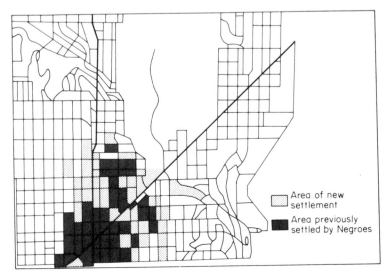

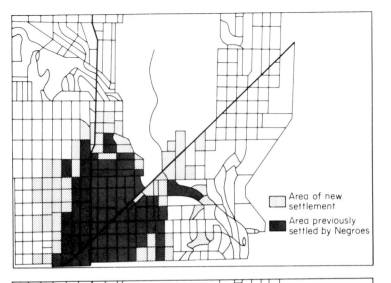

Area of new settlement

Area previously settled by Negroes

FIGURE 11–51d

Seattle: simulated expansion of ghetto, 1950–60 (adapted from Morrill, 1965, Fig 12).

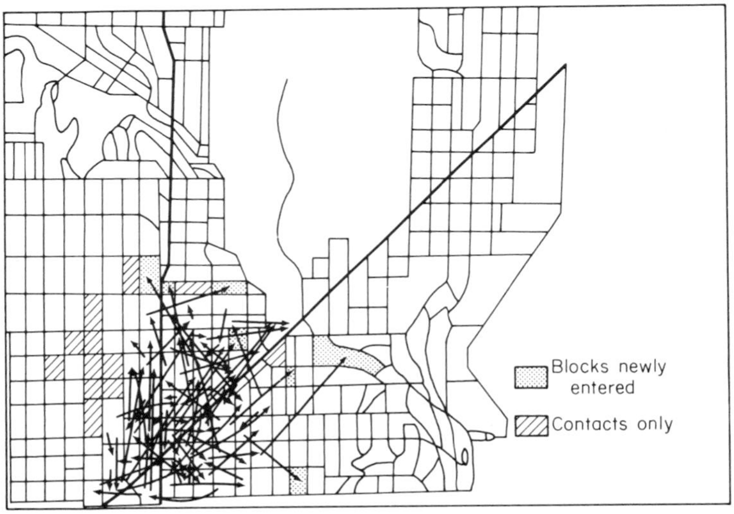

Blocks newly entered

Contacts only

FIGURE 11–51e

Seattle: a two-year stage in a ten-year simulation (adapted from Morrill, 1965, Fig. 13).

FIGURE 11–51f

Seattle: simulated expansion of ghetto, 1960–64 (adapted from Morrill, 1965, Fig. 14).

Heavy entry in 1960

Light entry in 1960

New entry, 1960–1964

Contact, but no entry

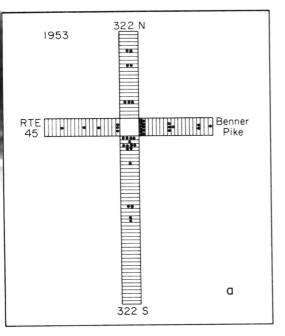

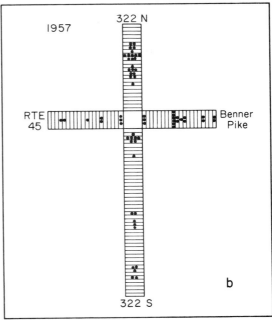

FIGURE 11–52

Actual diffusion of billboards from State College, 1953–65 (adapted from Colenutt).

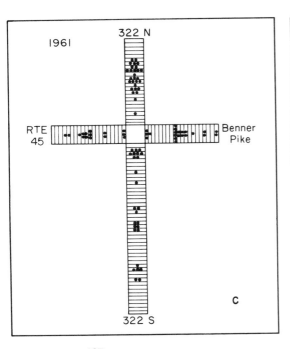

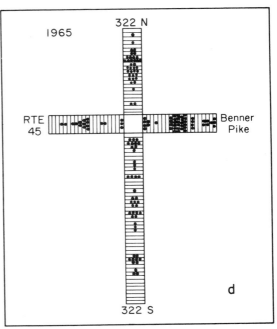

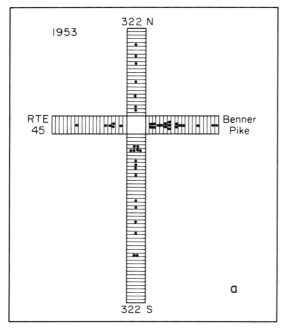

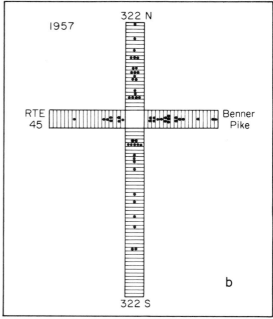

FIGURE 11–53

Simulated diffusion of billboards, 1953–65 (adapted from Colenutt).

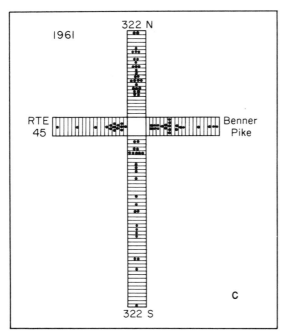

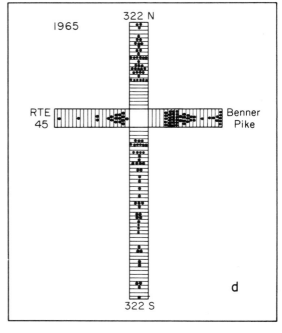

value had to be reached before the barrier crumbled. Somewhat similar problems were encountered in a study of linear diffusion of billboards along the four main arteries leading from a university town in Pennsylvania that grew rapidly after the Second World War (Figs. 11–52*a–d*). In 1953, billboards tended to cluster close to the town where traffic density was highest. Four years later the town fringe expanded north and south, and billboards were relocated near the head of the advancing urban wave. In 1961, the former clusters were much thicker, and by 1965 the beautiful approaches to the town resembled the typical "billboard alleys" that lead into most of America's urban centers. Probabilistic rules required to simulate such a diffusion process were complex, and included (1) probabilities assigned according to distance from the center, (2) removal rates conditioned by the growth of the town, (3) saturation limits, (4) "barrier farmers" who preferred a beautiful countryside, and (5) a series of threshold values for each parcel of land along the highway. The threshold values changed as the diffusion process was simulated by the computer to allow for the effects of spacing and "visual competition." In the first simulated time period (Figs. 11–53*a–d*), the model generated billboards over a greater length of road than in the real world, but as the process continued the conformity between the two sequences became much stronger. By 1965, the patterns on Route 322 North and the Benner Pike are closely approximated, although the other two routes conform less well. The model overpredicts the density of billboards close to the town on Route 45, because a small airport has a barrier effect upon the actual course of diffusion. On 322 South a series of small dormitory villages distorts the pure field effect of distance from the main town, and secondary nodes of billboard diffusion appear.

Another Change of Scale:
Diffusion at the Regional Level

As we move further up the cone of resolution, we tend to blur and smooth individual effects by lumping many human decisions together. We saw before how the diffusion of pioneer settle-ment over an area can be extremely regular, even though it is the result of a myriad of quite unique decisions by individuals and their families. Thus, we may be able to find considerable regularity and order in the aggregate patterns of spatial behavior.

Diffusion processes involving many decisions over large areas are very complex, and the patterns we see on the map are often confusing. We say such cartographic patterns are "noisy" —a very useful bit of jargon borrowed from electrical engineering where sharp incoming signals may be blurred and distorted by extraneous background "noise." To separate signals from noise we need filters. As we saw previously in the case of settlement in Pennsylvania, the regional trend surface can often perform this function, filtering out the noisy local effects that obscure crisp regional trends or signals. Around the turn of the century, for example, Czarist Russia experienced many agrarian riots which seem to have started in the southeastern Ukraine and Baltic provinces. These were precisely the areas with small plot sizes, high tenure rates, and an extreme polarization between the peasantry and nobility, so that feelings of deprivation tended to be very high. When average riot times in the forty-eight provinces are plotted in a three-dimensional graph against location, a simple trend surface uncovers the broad patterns of the diffusion of rioting (Fig. 11–54). The two core areas are disclosed, and the broad spacing of the waves indicates that peasant riots spread rapidly between these areas of intense social discontent. From the southern source in the Ukraine, the cry for agrarian reform spread in almost even ripples, like those produced by a stone dropped in a pond, until localized areas of resistance were met in the more remote northwest. A definite barrier effect seems to have operated near the Baltic provinces, for the wave crests are much closer together as the movement leaves this second source area.

Another example of filtering a complex diffusion process to disclose major regional trends comes from northern Tanzania. The area just south of Lake Victoria experienced the rapid spread of cotton marketing cooperatives over a fifteen-year period. The movement started quite

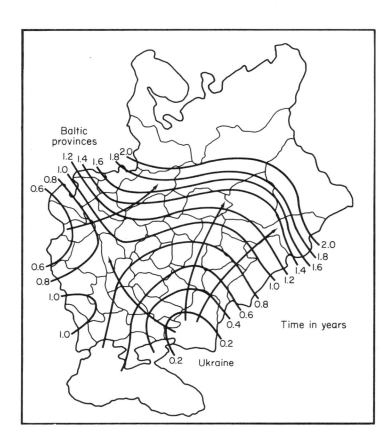

FIGURE 11–54

Diffusion of agrarian riots in Russia, 1905–7 (from Kevin Cox and George Demko, "Agrarian Structure and Peasant Discontent in the Russian Revolution of 1905," The East Lakes Geographer, Henry L. Hunker, ed., III [October 1967], Fig. 2, p. 11, with permission of the authors and editor).

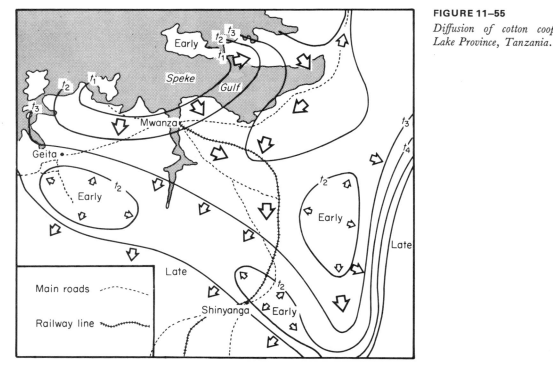

FIGURE 11–55

Diffusion of cotton cooperatives, Lake Province, Tanzania.

spontaneously in the peninsula and island district of Ukerewe (Fig. 11–55), but the local dhow and fishing traffic carried the idea across the Speke Gulf to the major lake port of Mwanza where another early cluster appeared. A slightly later group started just to the east of the major railway line. As soon as the innovation caught hold, the British Trusteeship administration tried to control the spread for both political and administrative reasons. From its earliest years the movement was used as a political channel, because the leadership of such a powerful institution quickly came into the hands of better educated, urbanized Tanzanians. TANU, the party that led Tanzania to independence, used the network of cooperative communications with great skill, and the TANU flag was often seen outside many cooperative societies in the pre-independence years. Thus we have an example of a political diffusion process following closely on the heels of one triggered by economic considerations.

To control the cooperative movement the Trusteeship administration required a system of licensing. Once a license had been issued to a new cooperative, attempts were made to train the treasurers and other officers in the rules of bookkeeping and simple management, for the sums of money involved were often very large. Unfortunately, the monetary temptations also injected further noise into the diffusion process. While overall standards of honesty were extraordinarily high, many cases of theft occurred. Individual cooperatives split up so that sets of new officers could be formed who would then have all the prestige accruing to important local men as well as the chance of dipping into the till. Because many cooperatives were started later in areas already passed over by the initial wave of innovation, the overall pattern becomes obscured. At the same time, migration was taking place in the area as new cotton lands were taken up in the eastern part of the region.

We must also remember that a cotton cooperative is a very small, single-function central place serving the farmers around it. As such, it is a space-competitive institution, for cooperatives will tend to locate far enough away from an established society to gain the minimum,

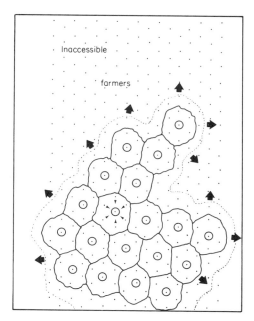

FIGURE 11–56

Planned cooperative diffusion packing space tightly.

or threshold membership required. A completely planned and controlled diffusion of cooperatives will pack space tightly from the beginning (Fig. 11–56), with no "waste space" between them to be filled later. But most cooperative diffusions are not controlled (Fig. 11–57), and while in the short run more farmers may be influenced (dotted line), the overall process may turn out to be spatially inefficient. Farmers in areas caught between existing cooperatives will have to start small and inefficient societies later, or the people will have to carry their cotton a long way for marketing. Thus a system allowed to generate new locations unimpeded may have to develop inefficient cooperatives later to "fill the gaps."

All these considerations—central place, demographic, political, financial, and administrative—are noise obscuring the smooth regional signals. By fitting trend surfaces, however (Fig. 11–55), we can filter out the noisy elements and see the main pattern. From the early centers in Ukerewe District and around Mwanza, the cooperative idea spread partic-

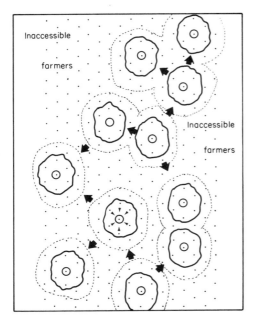

FIGURE 11–57

Uncontrolled diffusion of cooperatives with inefficient spatial gaps.

ularly rapidly down the railway, hopping to the large town of Shinyanga, which became a fourth transmitting node. Similarly, in the west, the idea leapt to an administrative center Geita, and a fifth innovation center developed. Thus at this regional level the diffusion process, already obscured by numerous institutional factors, also involves both hierarchical and contagious effects.

When institutions like cooperatives diffuse through an area, we must not forget the tremendous changes they can bring in their wake. Cooperatives in Tanzania stopped unscrupulous buying practices by cotton agents and provided a system of information nodes from which new seeds and cultivation practices could be transmitted to the farmers. Incomes rose sharply in the area, although most of the money went for immediate consumption rather than for long-term capital investment. Similar cases can be found all over the underdeveloped world. During the 1940s in eastern Nigeria,

for example, small palm oil presses diffused like a forest fire through the area, and by 1953 over three thousand were in operation. Later Pioneer Oil Mills spread just as quickly, and handled large quantities of palm oil kernels extremely efficiently. Pressed yields were four times larger than those obtained from the old, inefficient way of extracting oil by boiling, and the quality also was greatly improved. Incomes went up accordingly. But the innovation had an impact far beyond the economic. Young men became rich, and tensions between generations became severe as a new class of entrepreneurs challenged the traditional bases of obedience. The women's traditional right to the hard kernels was lost when these were cracked and pressed by the mechanized mills, and many women rioted when their economic status changed. Thus waves of innovation may leave behind them eddies of social disruption and conflict that continue to swirl for a long time after the excitement of the initial impact has past.

Diffusion at the National and International Levels

The final shift in scale brings us to the national and international levels, where innovations diffuse over long reaches of space and time. In past millennia, many innovations spread very slowly over a "sticky" geographic space. Despite the great technological changes of the past century, we sometimes forget the impact of new forms of transportation and communication upon diffusion rates. As technological innovations in transportation spread, they also speed up the diffusion of other innovations by shrinking the space in which new ideas must move. We can see this when we compare two innovations separated by millennia in time but diffusing over the same area.

Ten thousand years ago agriculture—the simple act of planting and harvesting—was an innovation that moved slowly from a hearth area near the present-day borders of Turkey, Iraq, and Iran (Fig. 11–58). Archaeologists, using radio carbon dating, have determined the earliest evidences of agriculture at about eighty sites in Europe and the Middle East,

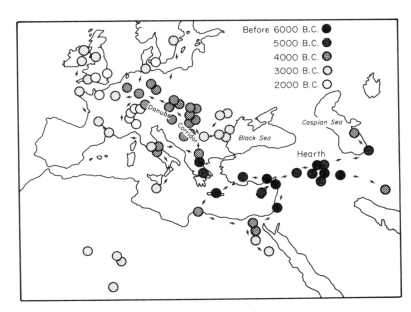

FIGURE 11–58

Diffusion of agriculture, based on radio carbon dating of archaeological sites (adapted from J. G. D. Clark).

and if these dates are plotted a very clear course of diffusion is disclosed. The earliest evidence dates agriculture at the hearth around 8650 B.C., and by 5000 B.C. the innovation had spread to Turkey and Israel, and leapt across short stretches of water to Crete and Greece. The movement highlights a definite east-west channel, and during the next thousand years the alignment extended south to Egypt and Libya, east to the Caspian Sea, and west through the Danube and Rhine corridors. Italy also received the innovation, almost certainly across the Adriatic Sea from Greece. By 3000 B.C. the innovation had spread far and wide. The Nile provided a channel to the south, many coastal sites of northwestern Europe had agriculture by this time, and the ideas spread east to the Black Sea. Another thousand years passed and the gaps were filled in. After six thousand years, the main diffusion wave crashed upon the British Isles—the outermost fringes of civilization.

Four millennia later Britain replied with a technological idea of her own. In 1825 the Stockton-Darlington railway line carried its first brave passengers, and by 1840 all of England's major cities were linked by steel rails (Fig. 11–59). The innovation pulse leapt across

the channel to three innovation centers on the continent—Amsterdam, Prague, and Lyon. By 1846 the new rail lines laced western Europe, and from Russia's Window to the West, St. Petersburg (now Leningrad), a line was started to Moscow. The innovation wave continued in the 1850s and '60s through Italy, Spain, southern Sweden, and eastern Europe. A bulge southeast from the early node at Prague reminds us that the Danube corridor still had a channeling effect upon human innovations six thousand years after the agricultural wave surged through in the other direction. In the 1870s, Russia started its explosive rail-building program, and the pulse continued through the Balkans, Denmark, and Sweden. By the First World War, the wave had even reached the Albanian backwater and moved through the Middle East.

We can show the diffusion of such place-linking and space-shrinking innovations in the form of an isochronic map implying that the process is controlled mainly by contagion. The hierarchical influence, while discernible in the appearance of small nodes ahead of the main wave, is minimal. In the twentieth century, however, the diffusion of many innovations at the international level seems to be structured much more tightly by hierarchical processes

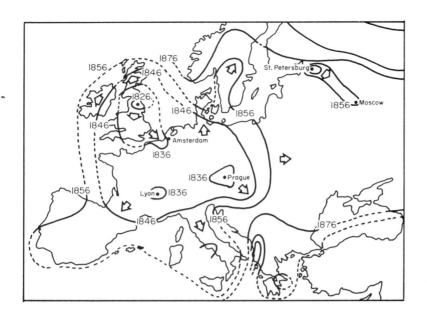

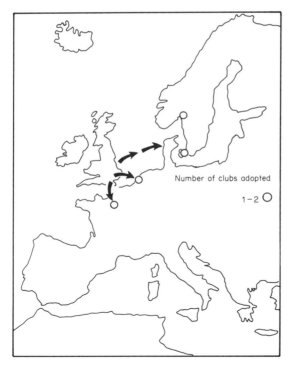

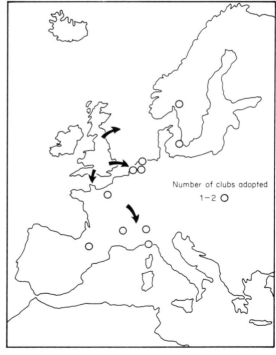

FIGURE 11–60

Diffusion of Rotary Clubs, 1922 (adapted from Hägerstrand, 1966).

FIGURE 11–61

Diffusion of Rotary Clubs, 1923 (adapted from Hägerstrand, 1966).

relying upon the huge, space-binding flows of information between major urban areas. Many diffusion processes now take place in relative, rather than geographic space. The diffusion of Rotary International through Europe, for example, demonstrates the "space-hopping" influence of the central place structure, particularly in the early years before the innovation seeped by contagious processes from the major regional transmitting centers to the surrounding areas.

Rotary International started in Chicago in 1905, and then jumped the Atlantic to Dublin and London in 1911. After the First World War, clubs in continental Europe started in the major cities close to Britain, and by 1922 the national primate cities of Paris, Amsterdam, Copenhagen, and Oslo had accepted the innovation (Fig. 11–60). The next year a second node

appeared in southern France and northern Italy in the large commercial centers of Genoa, Milan, Lyon, and Toulouse (Fig. 11–61). The density also increased in the Low Countries, around the initial entry point into the continent. Two years later (Fig. 11–62), six clubs were in operation in Holland and Belgium, and the innovation diffused rapidly in Switzerland and northern Italy to form a distinct core area from which the innovation hopped down the Italian boot to Rome and Naples and east to Trieste and Vienna. Scandinavia felt the pulse from the original Copenhagen and Oslo nodes, and in a large jump a club appeared in Lisbon— undoubtedly following the strong commercial ties to England's oldest ally, Portugal. By 1930 contagion was quite apparent, for the crisp hierarchical pattern was blurred as the neighborhood effect took over (Fig. 11–63). Dense

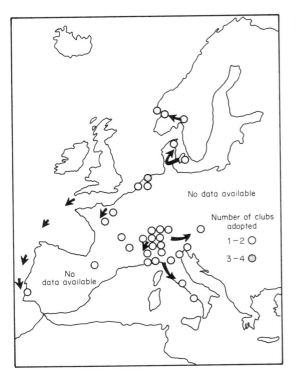

FIGURE 11–62

Diffusion of Rotary Clubs, 1925 (adapted from Häger-strand, 1966).

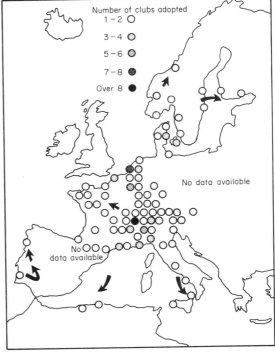

FIGURE 11–63

Diffusion of Rotary Clubs, 1930 (adapted from Häger-strand, 1966).

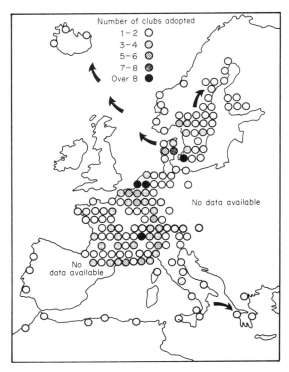

FIGURE 11-64

Diffusion of Rotary Clubs, 1940 (adapted from Häger-strand, 1966).

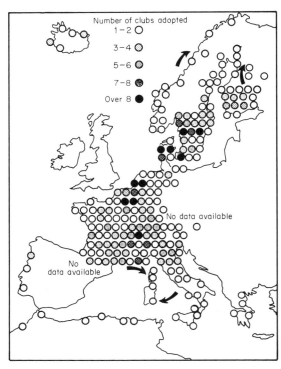

FIGURE 11-65

Diffusion of Rotary Clubs, 1950 (adapted from Häger-strand, 1966).

patterns appeared in the industrial areas of Holland, Belgium, and northern Italy, and nine clubs were founded around that most international of cities, Geneva. Finland and Germany started to adopt the idea, but the diffusion in Germany was slow as Nazism emerged and the country looked inward rather than out to international fraternal organizations. French commercial ties brought the innovation to the shores of North Africa; and Oporto, the heart of the old British port trade, adopted from Lisbon. In the next decade contagious effects dominated to increase the density in the original core area (Fig. 11-64). But on the periphery hierarchical effects still appeared. Athens adopted and retransmitted the innovation to two other commercial centers in Greece; Reykjavik did the same in Iceland, after receiving the pulse from

Denmark. The wave was slowed by the political barrier in Germany, Austria, and Czechoslovakia, and finally stopped by the Second World War. After the war, the innovation diffused as though a dam had been broken (Fig. 11-65). Dozens of clubs appeared to form a major crescent starting in Finland and ending in northern Italy, but many of the edges around the core were sharp. The Iron Curtain formed an absolute barrier to the east. The relatively poor area of the Mezzogiorno of southern Italy only had a few scattered clubs, while in France the more rural, northwestern portion of the country had a similar sparse pattern.

Looking at the diffusion of radio broadcasting in the United States and Rotary Clubs in Europe, it is clear that major urban linkages structure geographic space today, controlling the initial patterns of adoption before contagious effects take over and blur the clear hierarchical forces. Further evidence for the importance of hierarchical effects comes from the United States, as America's embryonic urban system crystallized from the spatial flux of the earlier and less structured nineteenth century.

Prior to the development of immunization methods, cholera was a dreaded intestinal disease that swept across America in three great epidemic waves. An Asian pandemic reached Europe in 1826, and after penetrating the whole continent crossed the Atlantic to North America in 1832. At that time, the United States was a frontier nation with little structure to its urban system. Water transportation was king upon the routes along the Hudson-Erie canal and the vast Ohio-Missis-

sippi system. The epidemic started in Montreal, Quebec, and New York City, and two cholera prongs diffused along the Great Lakes and the Hudson-Mohawk valleys to meet in Ohio in July (Fig. 11–66). From there the lake traffic carried the disease to Chicago (a town of less than 4,000), while another prong moved rapidly down the Ohio and Mississippi Rivers to New Orleans. A third seeped down the east coast from New York, reaching Charleston, South Carolina, in late October. Size of city had virtually nothing to do with the movement, and a plot of city size against the time the cholera first appeared shows no relationship whatsoever to the urban structure. On the other hand, the timing of the epidemic's appearance is a clear function of distance from the source areas (Fig. 11–67a), and we can distinguish between three sets of observations depending

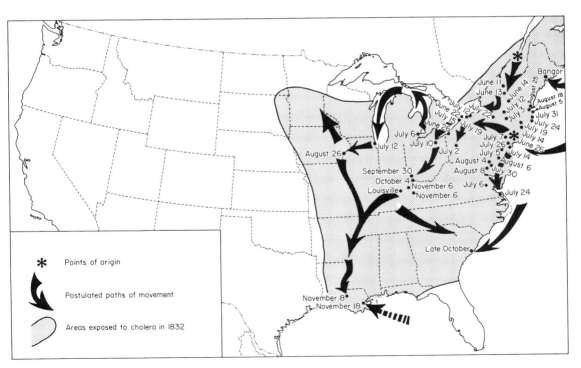

FIGURE 11–66

Diffusion of cholera, 1832 (adapted from Pyle, Fig. 1).

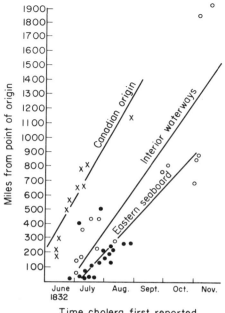

a

Time cholera first reported

upon the source and routeways. All indicate a clear relationship between the time of arrival and distance.

A second cholera epidemic diffused through the United States in 1849, but by this time the space was somewhat more structured by the emerging urban foci, as well as by the railways that were already lacing pieces of the country together. Entry again came from Europe, and New York City and New Orleans received the dread disease within nine days of one another (Fig. 11–68). Thus two widely separated points became the transmission centers. From New

FIGURE 11–67

Relationships of cholera epidemic to distance and city size: (a) cholera reports and distance from source in 1832; (b) cholera reports and city size, 1849; (c) cholera reports and city size, 1866 (adapted from Pyle, Figs. 1a, 2, 2a).

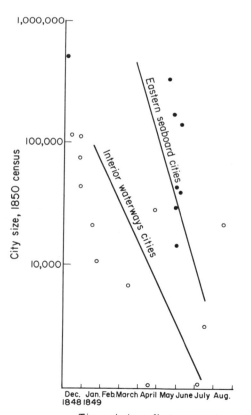

b

Time cholera first reported

c

Time cholera first reported

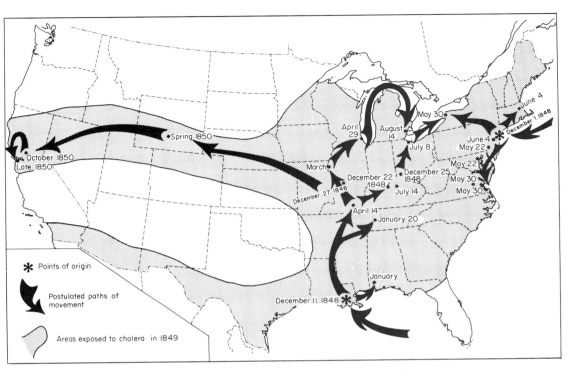

FIGURE 11–68
Diffusion of cholera, 1849 (adapted from Pyle, Fig. 2).

Orleans the disease swept up the Mississippi with the river traffic, forking at the Ohio-Mississippi junction only to meet again at the shore towns of Lake Erie. Two years later cholera appeared on the Pacific, having been carried to Sacramento with the pioneers along the California Trail. In New York, the disease lingered during the cold winter months in the slum areas, and then burst out in the spring to jump down the hierarchy along the Hudson-Mohawk corridor and the eastern seaboard. Unlike the earlier epidemic, city size was now a clear factor in the spread of cholera (Fig. 11–67*b*). Distinguishing between the seaboard and interior cities, a movement from the large cities down the hierarchy to the smaller ones is quite evident.

The last epidemic of 1866 took place in a much more tightly structured geographic space (Fig. 11–69), as nearly all the rapidly growing urban places were linked by railways. Movement

was now much easier, particularly between the major towns and cities, and when cholera struck New York City again it trickled rapidly down the hierarchy to Detroit, Chicago, and Cincinnati. From Chicago it arrived at St. Louis, met the wave down the Ohio, and surged on to New Orleans. The effect of the urban hierarchy was never more apparent (Fig. 11–67*c*), as there was a clear relationship this time to city size. Moving with the great flows of goods, people, and information from one intense concentration of human activity to another, the cities became the new transmitting nodes developing "cholera fields" that finally blanketed America.

More than diseases flowed across early America. As the major alignments of urban places and transportation routes strengthened during the course of settlement, many basic functions and institutions diffused across the land. An important institution for a developing

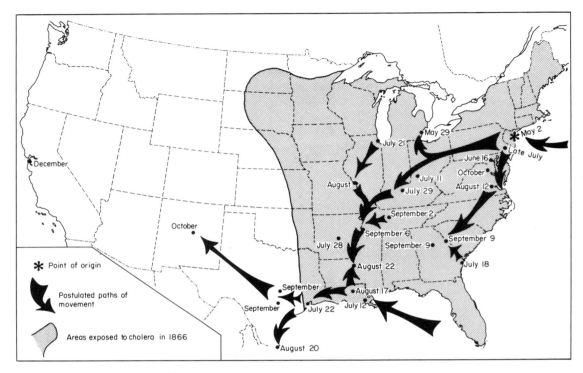

FIGURE 11–69
Diffusion of cholera, 1866 (adapted from Pyle, Fig. 3).

country is banking, for the location as well as the availability of credit is an essential consideration in a development process based upon private entrepreneurial drive. Banks diffused along two distinct alignments (Fig. 11–70), starting just after the War of Independence at Boston, New York, Philadelphia, and Baltimore. By 1810 many towns on the northeastern seaboard had adopted banking, and the linear pattern split along the coastal and Fall Line towns of the South, and along the Ohio River system into Kentucky. By 1830 (Fig. 11–71) banking facilities had expanded considerably, moving with the tide of settlement that formed another major diffusion process from the eastern seaboard. An embryonic hierarchy could be seen on the eastern seaboard, and major alignments appeared northward along the Connecticut and Hudson-Mohawk river valleys. The Ohio Valley alignment also strengthened, forming a baseline for the diffusion of banks north and

south, with settlement into Ohio and Kentucky from Cincinnati and Louisville. On the Mississippi, river ports like St. Louis, Vicksburg, Baton Rouge, and New Orleans adopted the new economic institution, and more banks spread in Georgia in the wake of plantation agriculture to provide credit facilities and commercial ties to the financial hearth in the northeast.

Ten years later (Fig. 11–72), the rise of small industrial centers in the North required considerable banking expansion, especially along the New York-Albany-Buffalo axis. The Fall Line cities also formed a sharp line inland, roughly parallel to the East Coast, while farther west banking was adopted in towns settled during the previous decade from the Ohio-Mississippi axis. Finally, at the start of the Civil War in 1861 (Fig. 11–73), most of the new bank adoptions lay in the Midwest, especially in the areas previously swept by the frontier

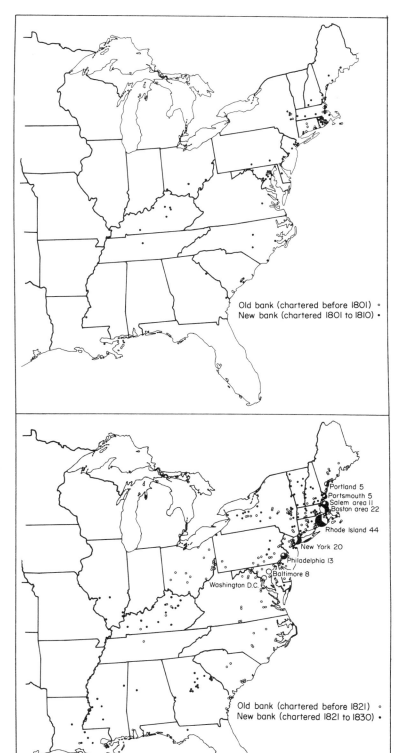

FIGURE 11–70

Diffusion of banking in early America, 1810 (adapted from Girling).

FIGURE 11–71

Diffusion of banking in early America, 1830 (adapted from Girling).

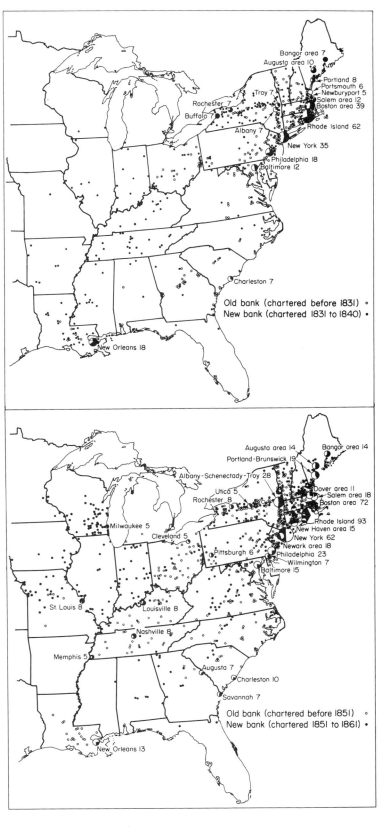

FIGURE 11–72

Diffusion of banking in early America, 1840 (adapted from Girling).

FIGURE 11–73

Diffusion of banking in early America, 1861 (adapted from Girling).

ten years before. A clear advancing wave could also be seen in the Southeast, as settlement advanced finally into the rugged barrier country of the Appalachians.

The overall patterns of banking adoption are not always sharp. Filling in gaps behind the frontier of settlement is common, and many legal problems at the state level complicated the timing of bank adoption. Furthermore, Europe, through the Bank of England, had considerable influence during this period of American history. A depression in Europe often resulted in the calling of funds from major banks on the East Coast. These, in turn, called for funds from small banks on the frontier, many of which were quite unable to meet sudden obligations. Bank failure rates were generally high in the early years, and often soared during times of depression. Thus the diffusion waves often overextended themselves during good times, only to ebb and fail when the times were hard.

Not all the innovations of early America were financial in nature. One of the most striking aspects of the early years of the Republic was the Classic Revival, a notion that the new democracy of Washington and Jefferson embodied all the ideals and visions of ancient Greece and Rome. Its most visible expression is in the architectural styles of the nineteenth century, but another revealing measure is in the frequency of classical place-names given to the towns that formed as swirling eddies behind the great diffusing wave of settlement that swept across America.

Before 1790 (Fig. 11–74a), only a few towns had names that tied them to the ideals of the classical world two millennia before. But

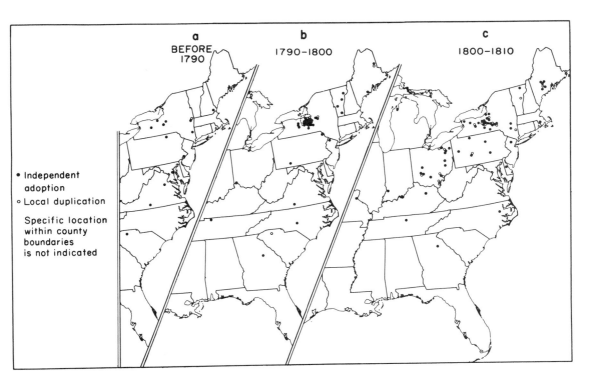

FIGURE 11–74

Adoption of classical place names: (a) *before 1790;* (b) *1790–1800;* (c) *1800–1810 (courtesy of Wilbur Zelinsky, Fig. 3).*

immediately after the Revolution, as educated men looked to earlier democratic ideals, an intense node appeared in western New York State (Fig. 11–74*b*). This distinct cultural hearth was also the area of most intense architectural development of the classical style, as well as of many other cultural innovations of early America. It is from this hearth area that we can trace the diffusion of classical names indicating the way in which these ancient ideals were carried by the early settlers. By 1810 the western node of New York State was still prominent (Fig. 11–74*c*), and indications of the first ripple outward appeared as many of the raw, frontier towns in Ohio adopted such names as Euclid and Caesar. Three decades later (Fig. 11–75), the wave had moved westward, with an intense node just behind the frontier in southern Michigan. At the same time, the classical ideal had diffused to the outermost fringes of settlement in the West. In the original hearth, the adoption-settlement phase was almost over. In the decade prior to the Civil War (Fig. 11–76), the classical wave swept even further west, reaching the easternmost parts of Kansas and Nebraska and pushing north into Minnesota. Two distinct nodes appeared at the end of the California and Oregon Trails as early pioneers took the idea to Sacramento and the Willamette Valley. These were the first outliers of a frontier that became increasingly fragmented and hollow after 1860 (Fig. 11–77). By 1890 the main wave had pushed into the Dakotas, and in the Southeast a sudden surge of classical naming occurred. But then the frontier closed, and opportunities for naming towns became increasingly rare. In the twentieth century the innovation wave carrying the classic democratic ideals dissipated and faded away.

In early America innovation waves quite

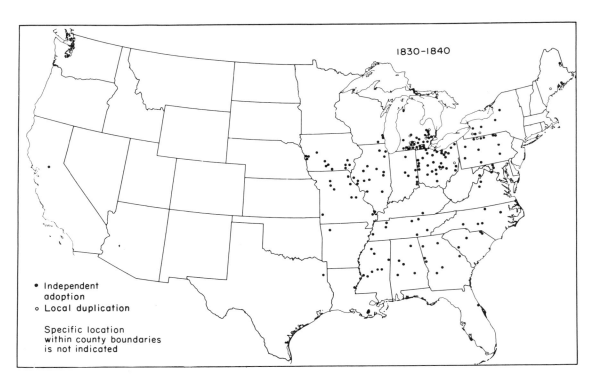

FIGURE 11–75

Adoption of classical place names, 1830–40 (courtesy of Wilbur Zelinsky, Fig. 5).

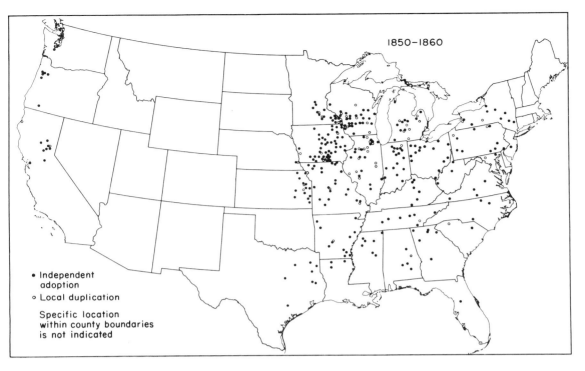

FIGURE 11–76

Adoption of classical place names, 1850–60 (courtesy of Wilbur Zelinsky, Fig. 7).

FIGURE 11–77

Adoption of classical place names, 1880–90 (courtesy of Wilbur Zelinsky, Fig. 10).

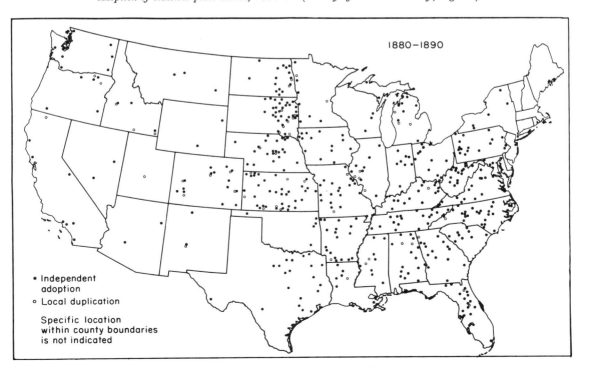

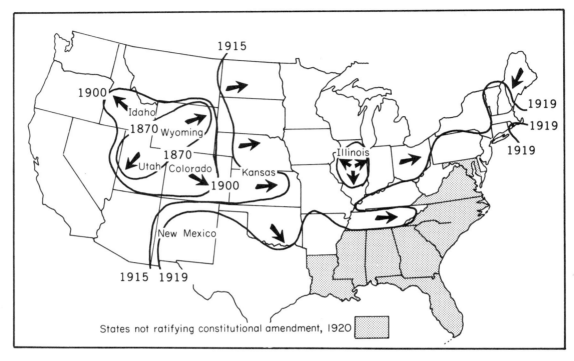

FIGURE 11-78

Diffusion of presidential suffrage for women, 1870-1920.

naturally had their origins in the eastern portion of the country. After the frontier of settlement passed, and even as it was moving through some areas, new innovations appeared in the West and moved eastward in a series of counter-diffusions. Many of the innovations involved individual rights, as frontier men and women chafed under archaic and undemocratic laws carried earlier from the East. The right of women to vote for the President of the United States, for example, was first granted in Wyoming in 1869 (Fig. 11–78). The innovation was quickly adopted by the legislators in neighboring Utah, but a quarter of a century passed before the men of contiguous Colorado and Idaho yielded. By the First World War, most of the western states had granted the right of presidential suffrage, and diffusion eastward was evident along a main corridor of national communication through Kansas, with a leap to Illinois in 1913. The First World War saw a great deal of activity by suffragettes, and if the world that emerged was "safe for democracy" many legislators thought it might also be safe for women to vote. By 1919 the suffrage wave had pushed eastward, distorted by a distinct barrier effect that started in conservative New England and ended in New Mexico. It is within this intransigent belt that the nine states that failed to ratify the constitutional amendment of 1920 lie. The amendment finally broke the barrier effect, and women across the land received the innovation passed into law by the more enlightened people of Wyoming half a century earlier.

A number of America's reform movements have sprung from the western node of innovation. Divorce laws, for example, have always been more enlightened in the West than in the East. While it is difficult to trace the diffusion of very complicated and subtly different divorce statutes, we can use divorce rates as a surrogate measure—that is, a measure standing in for the thing we actually want to observe (Fig. 11–79). If we take the first years in which the divorce rate exceeds 0.75 per thousand, Wyoming again appears as an innovation node, together with Washington, Oregon, and Nevada. We must remember that these areas were raw lumbering country in the 1870s, and "mail-order brides" were imported by the carload. By 1880 divorce rates indicate that the laws had eased over most of the West, although a distinct barrier effect was clear in Roman Catholic New Mexico. In the East a node of social enlightenment appeared in Maine and New Hampshire, but it failed to get very far as it ran into opposition from legislators in the other eastern states. By 1890 rates were up in Kansas, Oklahoma, Arkansas, and Texas, and an "innovation" node appeared once again ahead of the wave in Illinois.

The two courses of diffusion from the West show strikingly similar patterns. The same nodes appear at the start of the process, and the eastern thrusts seem to match each other even to the appearance of a secondary node in Illinois. The barriers are also virtually identical. We might speculate that many reform movements follow similar paths of diffusion, and that they will continue to do so in the future. Abortion reform, for example, started in Colorado and may also spread along channels that seem to be very stable over time.

Many of the innovations we have observed diffusing through the United States moved at a time when America was still an underdeveloped country. Since most of the world is still grossly underdeveloped, and because new innovations are a vital aspect of any development process, our final example is of spatial diffusion in the developing West African country of Sierra Leone.

Sierra Leone experienced the modernizing

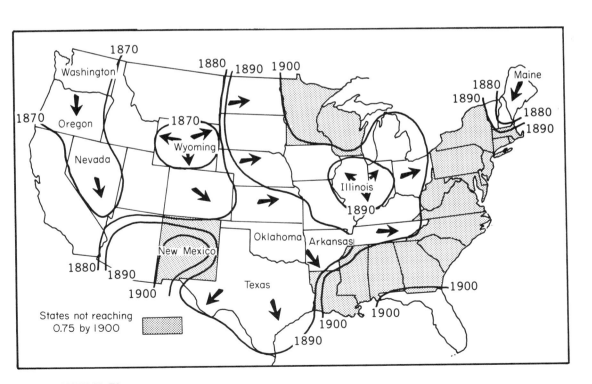

FIGURE 11–79

Diffusion of divorce reform using the surrogate measure of the date at which rates first exceeded 0.75 per thousand (courtesy A. Manheim).

influences of British colonial rule during the first six decades of the twentieth century. An important innovation was the political one of local government and administration, which allowed a large number of people to participate in local affairs and small-scale development schemes. A new system of local government was not imposed upon the 140 chiefdoms of Sierra Leone by the colonial administration, but was allowed to diffuse from two core areas. The final decision to adopt local government ordinances lay with the people.

If we return to the idea of a three-dimensional graph with space and time coordinates, and plot each small chiefdom as a point whose location is determined by latitude, longitude, and time of adoption, we can fit an increasingly complex series of diffusion surfaces (Figs. 11–80*a–d*). The first time-space surface is a plane, which indicates that the simplest and

most general trend is from the coast inland (*a*). As we allow the surface to warp, first two (*b*) then three (*c*) and finally four times (*d*), it fits the scatter of chiefdoms ever more closely. Thus the final surface (*d*) filters out all the local anomalies and noise and indicates the smoothed and general trend of the diffusion process. From the eastern and western core areas of model chiefdoms the innovation of local administration spread rapidly through the area where the demonstration effect was strongest, and where the news could be carried quickly by people traveling over the main railway line that formed a major axis of development in the country. From the basic east-west alignment the idea diffused south, until by 1945 nearly every chiefdom had adopted the innovation. In the strongly Moslem and more conservative north, resistance was much greater except along a branch of the railway that was used as a main line of penetra-

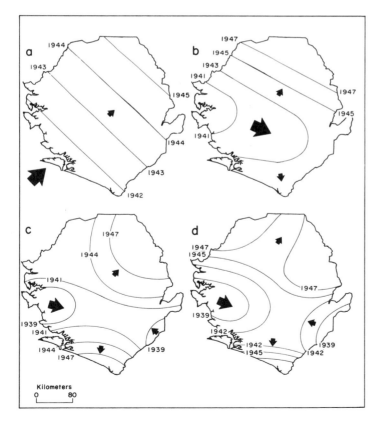

FIGURE 11–80

Diffusion of native administration in Sierra Leone (adapted from Riddell, Fig. 10).

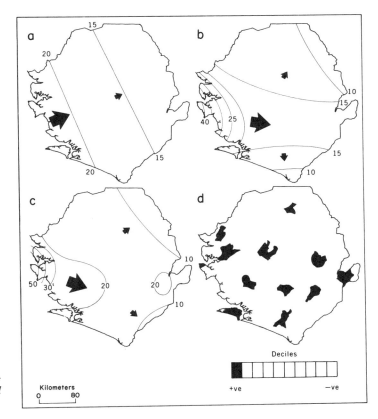

FIGURE 11–81

Diffusion and development of the modernization surface in Sierra Leone (adapted from Riddell, Fig. 26).

tion. Thus the relationship between diffusing innovations and the underlying structure of the transportation network is strong and critical. Accessibility must itself diffuse, and only then can innovations flow over the accessibility surface formed by the roads and railways that structure the space through which new ideas must move.

When we use the term *modernization*, we invoke a composite image of many social, economic, and political developments moving together, changing the minds of men as well as the human landscape. In Sierra Leone summary indices of this most complex process have been devised to measure the spatial variation in modernization across the country (Figs. 11–81*a*–*d*). Using trend surfaces to filter major national patterns from the local effects, the simple plane of modernization slopes gently from the coast toward the interior (*a*). Allowing the plane to bend (*b*) discloses the wedge effect of the railway slicing the country into two parts along the east-west axis, while more complicated warpings (*c*) show the build-up of modernization in the capital city on the coast (within the 50 isoline). Notice particularly the way the developing surface of modernization parallels the diffusion surface of local government administration, indicating that this overall, composite surface tightly structures the national space that controls in turn the way new ideas and innovations move. Interestingly, if the chiefdoms lying far above this last modernization surface are plotted as residuals (*d*), the rudimentary urban hierarchy of Sierra Leone is highlighted. Thus we end with a mixed process that can be decomposed into hierarchical and regional components, in the same way that diffusion processes themselves can be characterized in hierarchical and contagious terms.

The Frontiers of Diffusion Research

The links between spatial and other aspects of human behavior are only now being forged as geographers and other members of the human sciences link hands and work together on clusters of problems that cannot be solved by men and women within the narrow focus of traditional academic fields. The models far down the cones of resolution must be linked across the fields, even as within Geography itself much effort is needed to develop and tie together the few useful abstractions we possess. Our ability to model diffusion processes at the regional and national levels is limited, in part because of severe data problems that were compounded, until very recently, by the lack of large and rapid computing facilities. This barrier has now been broken, but the barrier of human thought and intellect remains.

At the applied and empirical levels there is also much to do, although the links between application and theory must never be weakened. The ties must be strong because theory building and empirical testing are complementary processes. Also, there is nothing so applicable as good theory. But when field studies are undertaken without a strong background of theory, and when theory develops divorced from the world, profundity is likely to fly out the window. Many future applications of the theory of spatial diffusion will have strong humanitarian overtones, for if we can gain a thorough understanding of these spatio-temporal processes our ability to advise is greatly magnified. What are the crucial points to tap in a network of social communication to maximize the diffusion rate? Remember that the root of crucial is *crux*—a cross. Crossing points in a network represent places where decisions may be made. To increase the rate at which things diffuse through a network, we need to know much more about these "spatial decision" points. What are the poles of spatial accessibility into which we can plug new ideas, medicines, and innovations in a poor but developing country? How stable are the main channels over time? Some of these are questions you may like to answer yourself.

Suggestions for Further Reading

BOWDEN, LEONARD W. *Diffusion of the Decision to Irrigate.* Chicago: Department of Geography, The University of Chicago, 1965. Department of Geography Research Paper No. 97.

HÄGERSTRAND, TORSTEN. "Aspects of the Spatial Structure of Social Communication and the Diffusion of Information," *Papers of the Regional Science Association*, XVI (1966), 27–42.

———. *Diffusion of Innovations*, trans. Allan Pred. Chicago: The University of Chicago Press, 1968.

———. "A Monte-Carlo Approach to Diffusion," *European Journal of Sociology*, VI (1965), 43–67. Reprinted in Brian J. L. Berry and Duane F. Marble, eds., *Spatial Analysis: A Reader in Statistical Geography*, Part V, Chap. 7. Englewood Cliffs, N.J.: Prentice-Hall, Inc., 1968.

MORRILL, RICHARD. "The Negro Ghetto: Problems and Alternatives," *Geographical Review*, LV (1965), 339–61. Reprinted in Brian J. L. Berry and Frank E. Horton. *Geographic Perspectives on Urban Systems, with Integrated Readings.* Englewood Cliffs, N.J.: Prentice-Hall, Inc., 1970, pp. 419–34.

NYSTUEN, JOHN D. "Boundary Shapes and Boundary Problems," *Peace Research Society Papers*, VIII (1967), 107–28.

PYLE, G. F. "Diffusion of Cholera in the United States," *Geographical Analysis*, I: 1 (January 1969), 59–75.

Works Cited or Mentioned

BEER, STAFFORD. *Management Science.* Garden City, N. Y.: Doubleday & Company, Inc., 1968.

BELL, WILLIAM. "The Diffusion of Radio and Television Broadcasting Stations in the United States," unpublished master's thesis, Department of Geography, The Pennsylvania State University, 1965.

BOWDEN, LEONARD W. *Diffusion of the Decision to Irrigate.* Chicago: Department of Geography, The University of Chicago, 1965. Department of Geography Research Paper No. 97.

BROWN, LAWRENCE A. *Diffusion Dynamics.* Lund: Gleerup, 1968. Lund Studies in Geography, Series B, No. 29.

———. *Diffusion Processes and Location.* Philadelphia: Regional Science Research Institute, 1968. Regional Science Research Institute Bibliography Series No. 4.

————, and E. G. MOORE. "Diffusion Research: A Perspective," in C. Board, R. J. Chorley, Peter Haggett, and D. R. Stoddart, eds., *Progress in Geography* Chap. 4. London: Edward Arnold Publishers, Ltd., 1969, pp. 120–57.

CLARK, J. G. D. *World Prehistory: A New Outline.* Cambridge: Cambridge University Press, 1969.

COLENUTT, ROBERT. "Linear Diffusion in an Urban Setting," unpublished master's thesis, Department of Geography, The Pennsylvania State University, 1966.

COX, KEVIN, and GEORGE DEMKO. "Agrarian Structure and Peasant Discontent in the Russian Revolution of 1905." *East Lakes Geographer*, III (October 1967), 3–20.

GIRLING, PETER D. "The Diffusion of Banks in the United States from 1781 to 1861," unpublished master's thesis, Department of Geography, The Pennsylvania State University, 1968.

GODLUND, SVEN. *Ein Innovationsverlauf in Europa, Dargestellt in Einer Vorläufigen Untersuchung Über die Ausbreitung Der Eisenbahnninnovation.* Lund: Gleerup, 1952. Lund Studies in Geography, Series B, No. 6.

HÄGERSTRAND, TORSTEN. "Aspects of the Spatial Structure of Social Communication and the Diffusion of Information," *Papers of the Regional Science Association*, XVI (1966), 27–42.

————. *Diffusion of Innovations*, trans. Allan Pred. Chicago: The University of Chicago Press, 1968.

————. "A Monte Carlo Approach to Diffusion," *European Journal of Sociology* VI (1965), 43–67.

————. *The Propagation of Innovation Waves.* Lund: Gleerup, 1952. Lund Studies in Geography, Series B, No. 4.

HUDSON, J. C. "Diffusion in a Central Place System," *Geographical Analysis*, I: 1 (January 1969), 45–58.

MACKAY, J. ROSS. "The Interactance Hypothesis and Boundaries in Canada: A Preliminary Study," *The Canadian Geographer*, XI (1958), 1–8. Reprinted in Brian J. L. Berry and Duane F. Marble, eds., *Spatial Analysis: A Reader in Statistical Geography.*

Englewood Cliffs, N. J.: Prentice-Hall, Inc., 1968, pp. 122–29.

MORRILL, RICHARD. "The Negro Ghetto: Problems and Alternatives," *Geographical Review*, LV (1965), 339–61.

————. "Waves of Spatial Diffusion," *Journal of Regional Science*, VIII (1968), 1–18.

NYSTUEN, JOHN D. "Boundary Shapes and Boundary Problems," *Peace Research Society Papers*, VIII (1967), 107–28.

OKIGBO, PIUS. "Social Consequences of Economic Development in West Africa," *Annals of the American Academy of Political and Social Science*, CCCV (May 1956), 125–33.

PAULLIN, C. O. *Atlas of the Historical Geography of the United States.* New York: Carnegie Institution and American Geographical Society, 1932.

PYLE, G. F. "Diffusion of Cholera in the United States," *Geographical Analysis*, I: 1 (January 1969), 59–75.

RASMUSSEN, S. E. *London: The Unique City.* Hammondsworth, England: Penguin Books, Inc., 1960.

RIDDELL, JOHN BARRY. "Structure, Diffusion and Response: The Spatial Dynamics of Modernization in Sierra Leone," unpublished doctoral thesis, Department of Geography, The Pennsylvania State University, 1969.

SMAILES, A. E. "The Site, Growth, and Changing Face of London," in R. Clayton, ed., *The Geography of Greater London.* London: George Philip and Son, 1964.

YAPA, L. S. "A Spaciometric Model for Diffusion of Innovations: Simulation Experiments," unpublished doctoral thesis, Department of Geography, Syracuse University, 1969.

YUILL, R. S. "A Simulation Study of Barrier Effects on Spatial Diffusion Problems," Michigan Inter-University Community of Mathematical Geographers, Discussion Paper No. 5, April 1965.

ZELINSKY, W. "Classical Town Names in the United States," *Geographical Review*, LVII (1967), 463–95.

PART V

Spatial Organization and the Decision Process

CHAPTER 12

INDIVIDUAL SPATIAL DECISIONS IN A NORMATIVE FRAMEWORK

Theory may be compared with reality for various ends. . . . Comparison now has to be drawn not to test theory, but to test reality. Now it must be determined whether reality is rational.

—*August Lösch*

DECISIONS ON THE LANDSCAPE

Whenever we look out of the window at the human landscape, or examine the work of man in a compressed and modeled space in the form of a map on our desk, we observe the geographical expression of a myriad of human decisions. Even the common road map is nothing more than the cartographic expression of the decisions of men. The roads, railways, and airlines that lace the face of the earth are the products of a long and often-repeated process of decision-making etched into the landscape, as millions of individual human beings have decided to make journeys for a variety of quite individual reasons. Where many men make the same decision to journey between points, we see an expression of those decisions in the form of dense commuter lines to our large cities and the thick flows of road, rail, and air traffic. Where the decisions of only a few overlap, the cartographic expression is the foot trail of the hiker

or the beaten paths of a simple agricultural space economy.

Land use patterns are also the geographic expression of many decisions by individuals. Whether the agricultural landscape is the seemingly chaotic quilt of plots cleared from the bush in Africa, the highly manicured and geometric countryside of northern Europe, or the vast expanses of grain found on the collective farms of Kazakstan, patterns of sprouting and ripening crops reflect the decisions of men made months before. Even human settlements, whether great metropolises or tiny villages, represent the constant reinforcement of residential decisions made by individual men and women.

Most of the time we have been considering spatial patterns, processes, and structures at fairly large or macro levels. While we went to the individual or micro level when we discussed diffusion problems, most of our models provide intellectual frameworks for larger scales of inquiry. The gravity model, for example, aggregated many individual human beings into a mass which was considered to have a point location. Central place theory illuminated the patterns formed by distinct aggregations of people. Changing the scales of observation is a very common practice, and it is time to do so once again, to see if we can disaggregate many of the "human lumps."

The Normative Viewpoint

In focusing upon individual spatial decisions, we shall start by adopting a normative viewpoint. For the moment our models will not attempt to describe how men actually behave and decide, but how they should behave and make decisions if they wish to achieve certain well-defined objectives. We have already worked with simple normative constructions, for the Weberian weight triangle that optimized a location under a set of simplifying assumptions was an example of such a model. Similarly, the traveling salesman solutions specified normative routes under carefully specified conditions. As we work within normative frameworks it is always important to remember that a number of simplifying assumptions have been made. We shall usually assume we are dealing with *homo economicus*, or Economic Man, who has perfect information and uses it with complete rationality to maximize economic output or minimize inputs. As we saw in Chapter 2, these assumptions about information, rationality, and optimizing behavior may be unrealistic, but if we can determine the decisions that should be made under such simplified conditions we can derive optimal patterns against which actual ones may be gauged. Thus normative models can be used as rulers against which reality may be measured —providing we can specify reality.

Sometimes we can relax the very severe assumption of perfect information. But in doing so, and in ascribing to an individual the inability to obtain perfect knowledge, we must adopt other assumptions under which individuals may actually decide upon alternative courses of action. One only has to think about a pattern of land use in a marginal environment to raise the question of the assumptions under which

individual farmers decide to grow certain crops. They may view Nature as a benign and bountiful Earth Mother, or they may regard her as a scheming, vindictive bitch who is trying to minimize their returns from farming. Men always have imperfect information because they cannot foresee the future, and in marginal environments, where their uncertainty is great, they often try to minimize the risks they take. Note, however, that minimization of something is just the reverse of maximization. Instead of trying to maximize output under conditions of great risk, farmers may wish to minimize the risk and so put a floor of income below which they will not fall even though Mother Nature does her worst.

Attitudes to Risk and Uncertainty

An individual's decision in uncertain situations will also depend upon his attitude toward risk. One individual may be extremely conservative and cautious, while another, feeling that life itself is a gamble, may take pleasure in

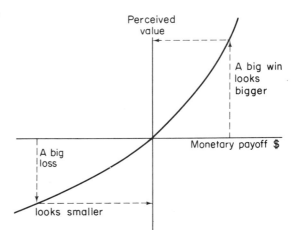

FIGURE 12–1

The utility function of the reckless decision-maker (from J. G. Kemeny and G. L. Thompson, "Attitudes and Game Outcomes," in Contributions to the Theory of Games, *Vol. 3, Fig. 1, p. 284, eds. M. Dresher, A. W. Tucker, and P. Wolfe, No. 39 of Annals of Mathematical Studies, copyright © 1957 by Princeton University Press. Reprinted by permission of Princeton University Press).*

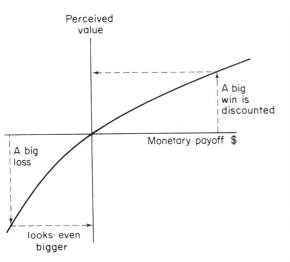

FIGURE 12–2

The utility function of the cautious decision-maker (from J. G. Kemeny and G. L. Thompson, "Attitudes and Game Outcomes," in Contributions to the Theory of Games, *Vol. 3, Fig. 1, p. 284, eds. M. Dresher, A. W. Tucker, and P. Wolfe, No. 39 of Annals of Mathematical Studies, copyright © 1957 by Princeton University Press. Reprinted by permission of Princeton University Press).*

making risky decisions in an environment of uncertainty. Moreover, the actual value or utility that people place upon different amounts of money will vary according to their attitudes in various decision-making situations. When we graph the actual monetary payoff against the perceived value or utility it has for an individual, the utility functions will vary according to the attitudes of the decision-makers. To the reckless person (Fig. 12–1), a large monetary payoff looks even larger, while big losses tend to be discounted. On the other hand, the cautious player (Fig. 12–2) tries to avoid large losses and tends to disparage and discount large wins as he faces further problems of making decisions for the future. The poor person (Fig. 12–3) will tend to exaggerate large wins or losses, while the rich person will do the reverse by underestimating large positive or negative payoffs (Fig. 12–4). Since real men are complicated mixtures of these very simple descriptive types, a common utility function is one that changes its slope depending upon its position in the payoff-utility space (Fig. 12–5). Many people are reckless when payoffs and losses are small but become very cautious when large positive or negative

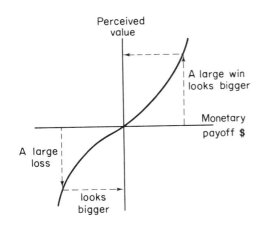

FIGURE 12–3

The utility function of a poor person (from J. G. Kemeny and G. L. Thompson, "Attitudes and Game Outcomes," in Contributions to the Theory of Games, *Vol. 3, Fig. 1, p. 284, eds. M. Dresher, A. W. Tucker, and P. Wolfe, No. 39 of Annals of Mathematical Studies, copyright © 1957 by Princeton University Press. Reprinted by permission of Princeton University Press).*

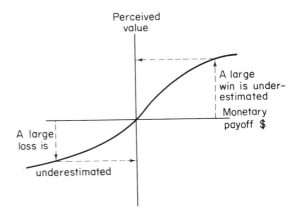

FIGURE 12–4

The utility function of a rich person (from J. G. Kemeny and G. L. Thompson "Attitudes and Game Outcomes," in Contributions to the Theory of Games, *Vol. 3, Fig. 1, p. 284, eds. M. Dresher, A. W. Tucker, and P. Wolfe, No. 39 of Annals of Mathematical Studies, copyright © 1957 by Princeton University Press. Reprinted by permission of Princeton University Press).*

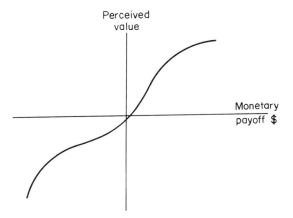

Perceived
value

Monetary
payoff $

FIGURE 12–5

A common utility function derived from a mixture of segments from the reckless and cautious functions (from J. G. Kemeny and G. L. Thompson, "Attitudes and Game Outcomes," in Contributions to the Theory of Games, Vol. 3, Fig. 1, p. 284, eds. M. Dresher, A. W. Tucker, and P. Wolfe, No. 39 of Annals of Mathematical Studies, copyright © 1957 by Princeton University Press. Reprinted by permission of Princeton University Press).

sums of money are involved. Geographic patterns that express many men's decisions will reflect complicated mixtures of utility functions. Whether we consider urban or rural areas, any pattern of land use is going to mirror such attitudes. But in seeking general statements with some explanatory power, we may have to ignore the unique, individual psyches of our spatial decision-makers.

Normative Decisions to Maximize Income

As an example of normative decision making with spatial implications consider an African farmer in the Kikuyu Highlands of Kenya who has just obtained a twelve-acre farm under the Resettlement Scheme. To simplify things let us suppose he has a choice of growing either a cash crop, like Arabica coffee, or subsistence crops for the local market. Because we are in a normative framework, we assume that he wishes to maximize his total cash income. Of course, even if he is a pure economic man, acting with perfect rationality upon complete information,

he is not going to become wealthy over night. The amount he can produce is going to be limited by the land available to him and the number of hours of labor that he and his two wives can devote to agricultural production. We shall also assume that an international coffee agreement places a restriction upon the amount of Arabica coffee he is allowed to produce. Faced with these constraints upon his production, and knowing that coffee is worth one hundred cents per pound, while subsistence crops are only worth eighty-six, the farmer has a difficult decision to make as he tries to judge the proportions of the two crops he should grow during the coming year. Let us take up the elements in the problem one by one.

He has a twelve-acre farm, but under the terms of the Resettlement Scheme he must always rest one quarter of his land each year, so nine acres are actually available to him. To raise a pound of subsistence crops requires .0012 of an acre, so if he put the nine acres available for production completely into subsistence crops for the local market he could produce 9/.0012 or 7,500 pounds. That is the upper limit of subsistence production, for anything more would require him to use more land than the nine acres available. Cash crops use up .0015 of an acre per pound, so if the farm were devoted entirely to coffee production he would get 9/.0015 or 6,000 pounds per year.

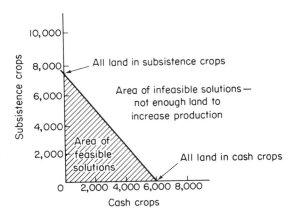

FIGURE 12–6

The land constraint on the production of the African farmer.

We can plot these quantities on a graph whose axes represent outputs of cash and subsistence crops (Fig. 12–6). He may, of course, choose various mixtures of cash and subsistence crops between these two limits, so that his output of the two may vary along the line joining the two extreme positions. Any point below and to the left of this line represents a perfectly feasible solution, although less than he could get from choosing a point somewhere actually on the line. On the other hand, any point above and to the right of the line represents an impossible choice for him because he does not have enough land available. It is for this reason that we call land a *constraint* upon both his production and his desired objective to maximize output. Such a constraint may also be expressed as an equation, or simple algebraic statement of the form:

$$.0015(P_c) + .0012(P_s) \leq 9 \text{ acres}$$

where: P_c = pounds of coffee produced for cash, and

P_s = pounds of subsistence crops.

Another constraint is the amount of labor that the farmer and his wives can devote to farming. Let us suppose that 4,500 hours are available each year, with subsistence crops requiring 0.9 hours per pound, while cash crops require 0.5 hours per pound. If all the hours were devoted to the production of subsistence crops the output would be 4,500/0.9 or 5,000 pounds,

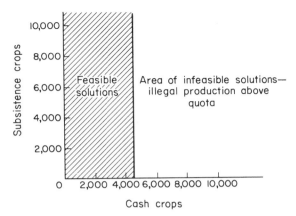

FIGURE 12–8

The coffee quota constraint on the cash crop production of the African farmer.

while the same number of hours devoted to cash crops would produce 4,500/0.5 or 9,000 pounds. These two limits may also be plotted on a graph (Fig. 12–7), and the line joining them represents all the varying choices between the two extremes. Once again, any point below and to the left of the line represents a feasible use of labor, although in this region of the graph, the further they moved from the line, the more spare time for other duties or leisure activities the family would have. Any point above the line represents an infeasible solution, unless they take time away from other affairs and so loosen the labor constraint. Again, we can express this statement in equational form:

$$0.5(P_c) + 0.9(P_s) \leq 4500 \text{ hours}$$

There is also an international agreement in effect limiting the production of coffee. Assume that the government assigns 2.2 acres for each adult in a family, so that the farmer may not devote more than 6.6 acres for cash crops. A pound of cash crops still requires .0015 of an acre, so if he decides to use his quota up to the limit he may produce 6.6/.0015 or 4,400 pounds of coffee. This constraint may also be shown graphically (Fig. 12–8), and is represented by a vertical line constraining only the production of cash crops. A point to the left of this line is within his quota limit, while points to the right

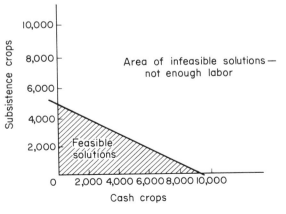

FIGURE 12–7

The labor constraint on the production of the African farmer.

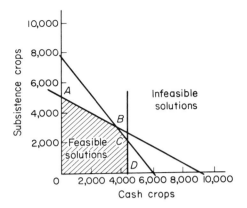

FIGURE 12–9

The land, labor, and coffee quota constraints combining to delimit the area of feasible solutions.

of it are infeasible because they are illegal. The equation expressing this constraint is simply:

$$.0015(P_c) \leq 6.6 \text{ alloted acres}$$

Since all of the constraints are working together and must be considered by the farmer as he tries to choose the best proportion of crops for the coming year, we should combine all the graphs into one (Fig. 12–9). When all the constraints are taken together they delimit an area of feasible solutions for the farmer. Along the boundary of this area the labor constraint is operating between A and B, the land constraint between B and C, while the quota constraint limits production along C and D. Thus the farmer's decision to maximize his output has been reduced to the question of choosing the best point within the feasible area.

Since he is trying to maximize his income, rather than simply pounds of production, he must take into account the relative prices of subsistence and cash crops. We can express his objective to make as much money as he can in the form of an equation. The pounds of cash crops (P_c) and subsistence crops (P_s) are multiplied by their respective prices to give the farmer's annual income Z. Thus we can write:

$$100(P_c) + 86(P_s) = Z$$

The farmer's decision about the proportion of cash and subsistence crops must be such that Z is maximized, subject only to the fact that land, labor, and the coffee quota are constraining his farm production. It is for this reason that we call the equation an *objective function*— literally, a function describing his objective. This may also be plotted on our graph (Fig. 12–10). To plot the objective function, we take some value it can assume, say 1,000 shillings. If the farmer grows only cash crops at 1.0 shillings per pound, he will need 1,000 pounds to get an income of 1,000 shillings. On the other hand, if he grows only subsistence crops at 0.86 of a shilling per pound, he will need 1,163 pounds to get 1,000 shillings. Thus the objective function will have a slope of 1,163/1,000 or 1.16, which is exactly the same as the ratio of the two prices 1.00/0.86.

Since the farmer wishes to place in the objective function the optimal amounts of subsistence and cash crops, the problem becomes one of moving the objective function on our graph as far from the origin as we can toward the boundary of the area containing all the feasible solutions. We can see that the farthest point to

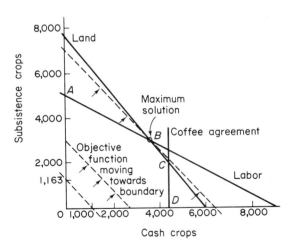

FIGURE 12–10

The land, labor, and coffee quota constraints combined with the objective function that is being driven to the boundary of the convex area of all possible and feasible solutions.

which the farmer may drive his objective function is the point B, which gives him the solution to his problem. If he grows 3,600 pounds of cash crops, and 3,000 pounds of subsistence crops he will maximize the possible income from his farm.

We can think of the farmer's problem as a search around the boundary of the feasible area on the graph for the point that will optimize his objective function. We might even imagine him starting at the origin and walking around the boundary carrying the objective function as a ruler in his hand. At each point on the boundary he inserts the values for cash and subsistence crops into his objective function and calculates the income he would obtain. His income is zero at the origin of the graph, because no subsistence or cash crops are produced. It rises as he moves toward A, where he inserts 5,000 pounds of subsistence crops, but zero cash crops. By walking on to B the value of Z in his objective function rises still further. However, when he gets to C he finds that his income has declined slightly, while at D it is lowered even more.

The area containing all the feasible solutions is called a *convex set*, and the process of groping around the boundary using an objective function as a ruler is called *linear programming*. All the information we have expressed in words, graphs, and equations may also be expressed in tabular form (Table 12–1). In the summary form we can see how units of cash crops and subsistence crops use up varying amounts of land, labor, and the quota, which are not unlimited resources, but available only in the amounts shown. Very efficient ways of searching

for the optimum have been developed working with such tables, which are known as *Simplex Tableaus*. Most problems involving large numbers of possible choices and constraints are solved by computers today, but the basic principles of searching around the boundary remain as simple as in the small problem we have examined here.

Spatial Aspects of Many Optimizing Decisions

Such a normative or optimizing view of an individual farmer's decisions has obvious implications for agricultural planning in a world that is getting hungrier every day. Though the problem has been simplified, many of the small farms in the Kenya resettlement schemes have been analyzed within such a framework to help farmers achieve greater output. In the United States many farmers come to their state universities and agricultural colleges for such analysis and advice. Thus the patterns of the landscape that exist today reflect human decisions made under strongly normative assumptions.

A normative viewpoint also allows us to go beyond the bounds of an individual's farm, and ask some important questions about larger patterns of land use within a region. In a region containing many farms of varying sizes and characteristics, there will be different optimal solutions for each farm. If we calculate these for a large number of farms, we can plot the maximum income available to each farmer and so construct a map showing the normative income surface (Fig. 12–11*a*). By careful fieldwork, interviewing, and estimation, we can also construct the actual income surface for the area (Fig. 12–11*b*). If the actual surface is then subtracted from the maximum one calculated under the severe normative assumptions of optimization, a third map can be made showing the degree to which farmers failed to approach the optimum available to them (Fig. 12–11*c*). Such a map has great utility for spatial planning, particularly in underdeveloped areas where limited numbers of agricultural officers must be assigned. Those areas with large gaps between

TABLE 12–1

Simplex Tableau for the African Farmer

Things That Can Be Produced

Cash Crops	Subsistence Crops	Constraints on the Production
0.0015	0.0012	Land—9 acres
0.5	0.9	Labor—4500 hours
0.0015	0	Coffee quota—6.6 acres
1.0	0.86	Price in shillings per pound

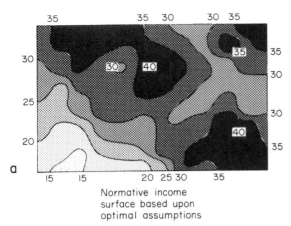

Normative income
surface based upon
optimal assumptions

a

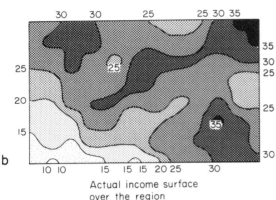

Actual income surface
over the region

b

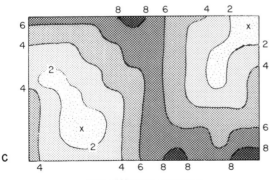

The difference between
the normative and
actual income surfaces

c

FIGURE 12–11

Deriving the difference between actual and optimal income surfaces to pinpoint areas for intensive agricultural extension work. Agricultural extension stations are at the locations marked (X).

FIGURE 12–12

Middle Sweden: variation in the potential productivity of farm labor under normative assumptions (adapted from Wolpert, Fig. 3; reproduced by permission from Annals of the Association of American Geographers, *LIV [1964]).*

the actual incomes and the maximum possible would be pinpointed for intensive agricultural extension work.

Such an analysis has already been made for a large farming region in Sweden, a highly developed country with a relatively efficient farming system. Under several severe normative assumptions, a large number of farms were analyzed to determine the potential productivity of farm labor (Fig. 12–12). The potential productivity surface reflects the monetary return for an hour of labor if the farmers decided to use all the farm resources optimally and if the environment is held to average values of rainfall and temperature. Using data collected by a Swedish government survey, actual labor productivity values were then plotted and a second surface constructed (Fig. 12–13). The whole surface is lower than the potential surface, for no farmers actually organize their farm activities to reach the values derived under the normative conditions. With information about the actual and potential productivity, a third surface was constructed over the region indicating the degree to which farmers in the area approximated the optimal production schedules (Fig. 12–14). In the northwest, for example, farms were organized so that an actual hour of labor expended returned over 80 percent of the potential. On the other hand, farms in a small area to the east were so poorly organized that an hour of labor returned less than half the income that could have been achieved if optimal decisions had been made. Thus the potential

FIGURE 12–13

Middle Sweden: variations in actual productivity of farm labor (adapted from Wolpert, Fig. 4; reproduced by permission from Annals of the Association of American Geographers, *LIV [1964]).*

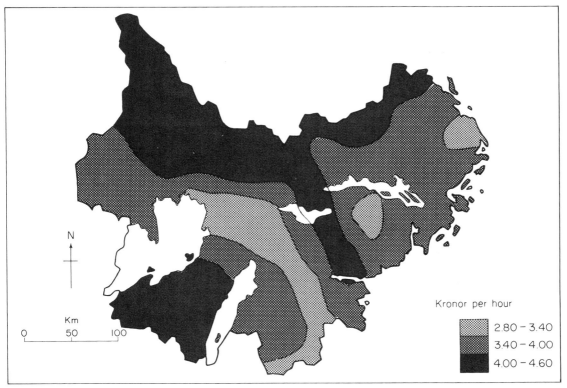

Kronor per hour

2.80 – 3.40
3.40 – 4.00
4.00 – 4.60

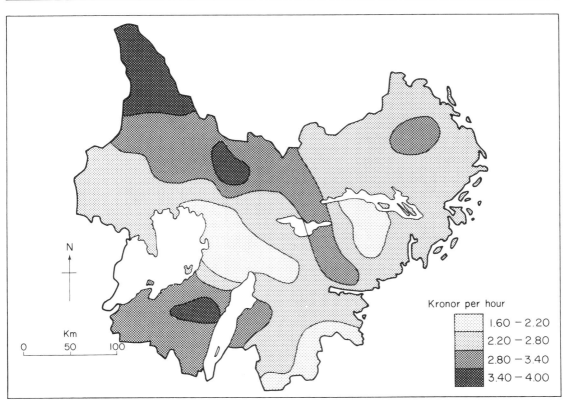

Kronor per hour

1.60 – 2.20
2.20 – 2.80
2.80 – 3.40
3.40 – 4.00

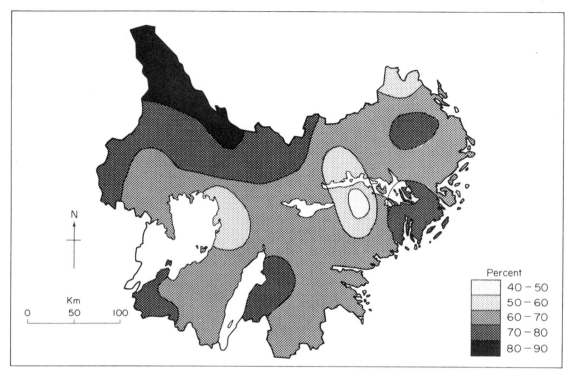

FIGURE 12–14

Middle Sweden: variations in the degree to which the productivity of farm labor approaches the maximum possible under normative assumptions of farm organization (adapted from Wolpert, Fig. 5; reproduced by permission from Annals of the Association of American Geographers, *LIV [1964]).*

surface is a standard against which actual decisions may be measured. The farmers of middle Sweden are not perfect economic men acting under conditions of perfect information. Organizing farm production is a difficult business, and many farmers are content to plod along at a lower rate of production. Such behavior has been termed *satisficing* behavior, implying that many people make decisions to reach a threshold level of satisfaction. Upon achieving it, they feel little incentive to drive for maximum economic returns. Flows of information also vary over space, and the final surface reflects a lag in the diffusion of new technological innovations. Knowledge of new techniques diffuses from the major centers of agricultural innovation in the Stockholm area to agricultural

offices throughout the region. Large farmers get the news first, and only later are their decisions copied by smaller farmers near them. Information trickles down through a hierarchy of official agricultural centers, and then spreads in a contagious fashion from farm to farm.

Whether we consider land use patterns in an underdeveloped or in a more advanced country, the gap between the actual and potential surfaces must be closed. This may mean raising the aspiration levels of the farmers or greatly increasing the flows of information throughout the region. Such tasks are often very difficult, but normative models may help to pinpoint areas where limited resources of farm extension personnel may be assigned most effectively.

Normative Decisions to Minimize Costs

In considering individual decisions in a normative framework, we are not confined to problems of maximization. Rather than maximizing output by searching for an optimal point, men may wish to minimize inputs instead. The problem of minimization is simply the reverse of maximization, and the same normative models may be used to find and describe the optimal decisions.

Let us consider one of the world's most serious problems, that of hunger, malnutrition, and dietary deficiency. In areas of the world where incomes are very low and knowledge about nutrition is not widespread, it may be helpful to devise human diets that meet basic nutritional requirements at the lowest cost. Notice that we have used the phrase "at the lowest cost," implying that our objective now is to minimize something while meeting, at the same time, some basic requirements or constraints.

Let us assume that we are constructing minimum cost human diets in Guatemala. We need some basic information about the sorts of foods available at a marketplace, their nutritional content, and their prices for some standard unit of weight, say 100 grams. Such information can be shown in the form of a table to which we may add the basic nutritional requirements that we are trying to meet at the least cost (Table 12–2).

TABLE 12–2

Simplex Tableau for Human Diet Problem

Foods That Can Enter the Diet			Nutritional Requirements to Be Met
Eggs	Oranges	Tortillas	
160	69	201	Calories—2700
11.3	0.8	5.5	Protein—65
.0005	7.5	0	Vitamin C—75
9¢	3¢	1¢	Cost per 100 grams

For example, eggs are very good suppliers of protein, but their vitamin C content is practically zero. They are also quite expensive items in a diet, particularly when compared to tortillas, which supply many calories and some protein at a very low cost. Oranges supply relatively large quantities of vitamin C, but their protein content is very small. Thus our problem is to choose certain quantities of the available foods to meet the calorie, protein, and vitamin requirements at the least cost. The familiar Simplex Tableau lists the foods that can enter a solution, together with the constraints that must be met. For this simplified problem we can obtain the solution with paper and pencil, but such methods will obviously fail us in any realistic problem with a dozen nutritional requirements to be met and scores of foods at different prices available to fulfill them.

In our example we can meet the vitamin C requirement by bringing ten units of oranges into the solution. In using oranges, however, we also gain some calories (690) and a little protein (8). To make up the remaining calories and proteins we may choose eggs or tortillas. Tortillas are preferable because eggs are so expensive and our aim is to minimize the overall cost. If tortillas are to meet the remaining protein requirement $(65 - 8 = 57)$ we must bring 57/5.5 or 10.36 units into solution. Notice that as we meet the protein requirement we also slightly more than meet the calorie requirement. This is because our constraints were of the "at least" variety, for in this example we only want to set a floor of minimum requirements below which we will not accept a solution. Thus the cost of the diet meeting at least these minimum requirements is:

10 units of oranges $\times$ 3¢ =	30.00¢
10.36 units of tortillas $\times$ 1¢ =	10.36¢
Total	40.36¢

Moving from the Normative to the Descriptive

If geographers went around the world devising minimum cost diets in this fashion, it is unlikely that many individuals would actually decide to adopt their solutions. Minimum cost diets tend to be unpalatable and dull, even though they do meet all the requirements of the human body at the lowest cost. Because the human family clings to a spectrum of cultures and values, people have different food preferences which change only very slowly. Rice-eaters often dislike the wheat and maize of other cultures, and methods of food preparation familiar to some may be so strange to others that new foods will go unused even under conditions of severe deprivation. American frontier cattlemen killed the buffalo, and promptly imported British strains of cattle to replace them. In East Africa today, the natural grazing animals are also being replaced by other meat sources.

When we consider the effect of such things as people's food preferences, we must begin to move from the purely normative viewpoint toward models that describe and take into account such non-economic but very human characteristics. Let us suppose, in the example above, that people had a great liking for eggs, and would sacrifice much to have a hundred grams in their diet each day. To move the problem described by the first tableau (Table 12–2) away from the prescriptive toward the descriptive, we must require, purely on cultural grounds, that a new food preference constraint be met. This would appear as a new row in the tableau (Table 12–3). The new tableau is structured so that the only column that can enter the solution to fulfill the cultural craving for eggs is the egg column itself. We assume that no amounts of oranges or tortillas are going to make up for this food preference! Under these new conditions a single unit of a hundred grams of eggs must enter the solution at a cost of nine cents. In meeting the cultural requirement for eggs, however, we also gain some calories (160) and proteins (11.3) and a very small amount of vitamin C (0.0005). We now require slightly fewer units of oranges than before (9.9995 instead of 10) to meet the vitamin C constraint,

and these will have a slightly lower cost of just under thirty cents. As the units of oranges enter the solution they bring with them some calories (69 each unit) and a small amount of vegetable protein (0.8 per unit), so we require fewer tortillas than before to make up the calorie requirement. As we bring tortillas into the solution, to meet the remaining calorie requirement, we happen to overfulfill the protein requirement by nearly five grams. Thus the total cost of the diet meeting both the nutritional and cultural constraints is:

$$
\begin{array}{llr}
1.0000 \text{ unit of eggs} & \times\ 9\cent = & 9.0000\cent \\
9.9995 \text{ units of oranges} & \times\ 3\cent = & 29.9995\cent \\
9.2040 \text{ units of tortillas} & \times\ 1\cent = & 9.2040\cent \\
\hline
\textit{Total} & & 48.2035\cent
\end{array}
$$

The cost of the diet has risen by nearly eight cents, for the people will certainly have to pay for their food preferences as the solution moves away from the minimum point determined under the strict normative conditions.

Minimum Cost Surfaces over Geographic Space

By calculating the costs of human diets meeting basic nutritional requirements at a large number of points in southern Guatemala, we can construct a surface varying over the country (Fig. 12–15). Such variation from place to place is due to a number of complex and interacting variables. Prices for foods vary according to supply and demand. Particular foods are grown in some places and not in others because the environmental diversity of southern Guatemala is great. Foods also vary considerably in their nutritional ability to meet minimum human requirements. While some of the big towns like Guatemala City are peaks on the cost surface, because of the intense urban demand for foodstuffs, generally the surface is low, implying that an adult male can obtain a well balanced diet for about 10¢ a day. However, food preferences in Guatemala are strong, and they vary greatly between the predominantly Indian population of the West and the Ladino population of the East. When different cultural constraints are added at each point, the overall cost surface is raised considerably (Fig. 12–16).

TABLE 12–3
Simplex Tableau Modified by Cultural Constraint

Foods That Can Enter the Diet			Nutritional Requirements to Be Met
Eggs	Oranges	Tortillas	
160	69	201	Calories—2700
11.3	0.8	5.5	Protein—65
.0005	7.5	0	Vitamin C—75
100	0	0	Craving for eggs—100
9¢	3¢	1¢	Cost per 100 grams

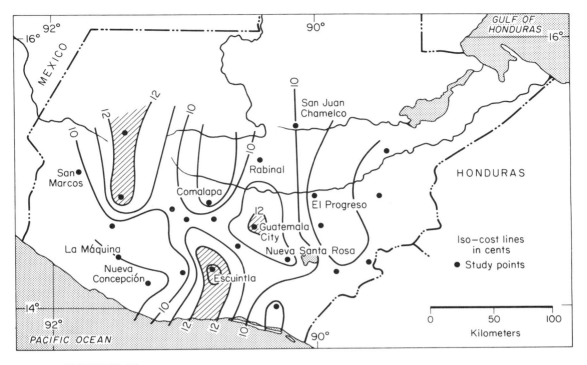

FIGURE 12–15

Southern Guatemala: the surface of minimum cost diets (adapted from Gould and Sparks, Fig. 3, p. 68).

FIGURE 12–16

Southern Guatemala: the cost surface after food preferences have been considered (adapted from Gould and Sparks, Fig. 7, p. 73).

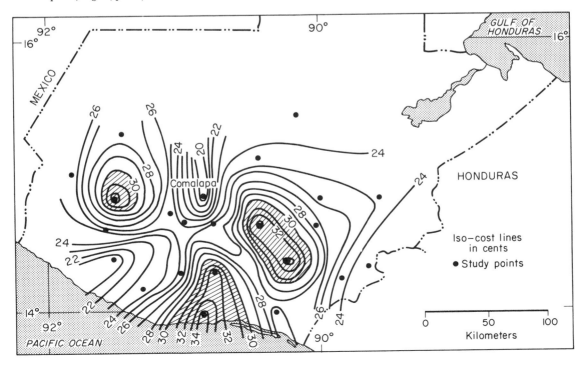

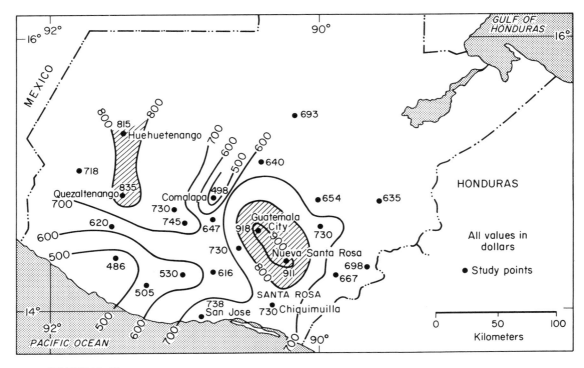

FIGURE 12–17

Southern Guatemala: surface of costs required to sustain a family for a year (adapted from Gould and Sparks, Fig. 15, p. 81).

The main peaks of urban demand still appear, but generally the overall cost of a diet meeting the cultural as well as the nutritional constraints has been tripled.

Using solutions derived from linear programming models, from household budgetary surveys, and from information about average family sizes in Guatemala, we can construct a final surface indicating required family incomes over the country. Values represent the annual income a family needs to house and clothe all its members while eating palatable meals that meet standard nutritional requirements (Fig. 12–17). Once again, towns like Guatemala City and Nueva Santa Rosa stand out as peaks, and it is precisely in this area that there are large discrepancies between the actual incomes and those required under the assumptions of the minimum cost

model. It is also in this region that the highest rates of childhood malnutrition and death from dietary deficiencies are reported. Many children suffer from *kwashiorkor*, a serious disease brought on by extreme protein deficiency. As we learn more and more about the effects of poor nutrition upon young children, particularly upon their intellectual development which can be permanently retarded by early malnutrition, it is clear that such scientific analyses have strong humanitarian implications for the spatial assignment of resources in an area, and can be of considerable use in securing the well-being of its future citizens. America, too, has its pockets of starving children whose development, both physically and mentally, will be permanently retarded unless enlightened dietary policies are discovered and applied.

Decisions to Minimize the Cost of Distance

The problem of minimizing the value of an objective function when it is subject to constraints can be placed directly in a spatial context. For example, if we wish to minimize the total cost of movement in a transportation system, an objective function expresses the fundamental "principle of least effort." Such a principle implies that under normative conditions of complete rationality and full information human beings will decide to ship goods and move themselves so as to minimize the overall costs of movement in the whole system.

Let us take a very simple example (Fig. 12–18), where the problem is to ship the surplus production of two towns, A and B, to meet the deficits or requirements of two others, C and D.

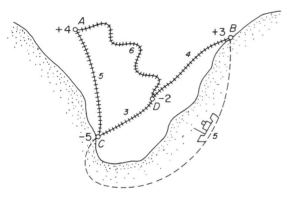

FIGURE 12–18

Shipping surpluses from towns A *and* B *to towns* C *and* D *by rail and coastal steamer.*

Surpluses may be shipped to D by railway at the costs shown in the diagram. Town C may also receive shipments by rail, but because it is on the coast it may also receive some goods from B by water transport. Our objective is to satisfy the requirements at C and D with shipments from A and B at the least cost. Our objective function must therefore express our desire to minimize the total cost of movement. Indicating with an arrow amounts moving between a surplus and a deficit area, for example

$(A \rightarrow C)$, and using the costs between them, our objective function becomes:

$$5(A \rightarrow C) + 6(A \rightarrow D) + 5(B \rightarrow C) + 4(B \rightarrow D) = Z \text{ minimize}$$

We cannot put any values we like in the objective function as we try to find the minimum value of Z because the process is subject to certain constraints. For example, all the surpluses must exactly equal the deficits. This is really a mathematical requirement to get a minimum solution. If the surpluses are somewhat greater than the deficits, we can always add storage bins to our problem as destinations for the excess amounts. This means that the quantities shipped by A, to C and D, must equal 4.0 exactly, the number of surplus units available for shipment at A. We can express this condition in the form of the following equation:

$$(A \rightarrow C) + (A \rightarrow D) \qquad = 4$$

In the same way, the amount shipped from B to C and D must exactly equal three, so that:

$$(B \rightarrow C) + (B \rightarrow D) = 3$$

Now looking at it from the deficit side, the amount that C receives from A and B must equal 5.0, while the amount that D receives from the two surplus areas must equal 2.0. We can express these requirements or constraints like this:

$$(A \rightarrow C) + \qquad (B \rightarrow C) \qquad = 5$$
$$(A \rightarrow D) + \qquad (B \rightarrow D) = 2$$

Now we can gather all of these constraint equations together (Table 12–4) and see that the

TABLE 12–4

Simplex Tableau for Transportation Problem

Flows between Cities That Can Enter Solution				
$A \rightarrow C$	$A \rightarrow D$	$B \rightarrow C$	$B \rightarrow D$	Requirements to Be Met
1.0	1.0	0.0	0.0	= 4 A's shipments
0.0	0.0	1.0	1.0	= 3 B's shipments
1.0	0.0	1.0	0.0	= 5 C's receipts
0.0	1.0	0.0	1.0	= 2 D's receipts
5	6	5	4	Cost of shipping 1 ton

transportation problem can be put into exactly the same form as the other problems of maximization and minimization we have met before.

We have, once again, a normative problem arranged in a Simplex Tableau. The 1's in the tableau can be interpreted in the following way. Consider the column $(A \rightarrow C)$, and assume that a flow of one ton moves from A to C at a cost of 5. A flow of one ton from A to C helps us to meet the requirement or constraint that A must ship 4 tons. At the same time it helps us meet another constraint, namely that C must receive 5 tons. So every time a single ton moves between these towns the 1's in the $(A \rightarrow C)$ column indicate that the flow meets a unit of the corresponding requirements. Obviously an $(A \rightarrow C)$ flow does not help us meet the requirements at B or D, so zeros are opposite these rows in the $(A \rightarrow C)$ column.

We have four possible ways of shipping goods, and the problem is to determine the degree to which these possibilities should be brought into solution to minimize our objective function while at the same time meeting all the constraining conditions. This is exactly analogous to the problem of minimizing the cost of a human diet by bringing various foods into solution to meet certain constraining requirements.

Suppose we take advantage of the relatively cheap water transportation and let B ship all three units to C (Fig. 12–19). City A would now

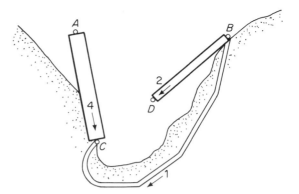

Flows from towns A *and* B *meeting deficits at* C *and* D *at a total cost of 33.*

have to fulfill the rest of the requirement at C, while shipping its remaining surplus over the tortuous railway to D. Thus our objective function would be:

$$(5 \times 2) + (6 \times 2) + (5 \times 3) + (4 \times 0) = 37$$

giving the overall cost of movement.

Is thirty-seven the minimum solution? As we started to make shipping decisions, we were drawn to the relatively low costs of water transportation, and immediately brought the largest shipment possible from B to C into solution. Unfortunately, this meant that D had to obtain its requirements from A over the relatively high cost railway route through difficult terrain. Suppose instead, we let B ship 2 units to D, and its remaining 1 unit by water to C (Fig. 12–20). Town A can then fulfill the remaining requirement at C, so our objective function becomes:

$$(5 \times 4) + (6 \times 0) + (5 \times 1) + (4 \times 2) = 33$$

This solution considerably reduces the overall cost, and is, in fact, the lowest that we can devise.

Placing transportation decisions in such a normative framework allows us to evaluate alternative courses of action to improve and upgrade the transportation system. For example, if ocean freight rates were lowered from 5.0 to 4.0, would this allow B to send all its surplus to C and lower the overall cost of the solution?

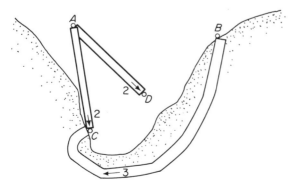

Flows from surplus towns A *and* B *meeting deficits at* C *and* D *at a total cost of 37.*

Inserting the values in our objective function once again:

$$(5 \times 2) + (6 \times 2) + (4 \times 3) + (4 \times 0) = 34$$

We see that the overall cost has been raised from the previous minimum we established. Of course, if we kept the previous pattern of flows, and lowered ocean freight rates to 4.0, our overall cost would be driven down still further.

Finally, let us consider the effect of a new constraint on this very simple transportation system. Assume that the railway line between B and D has a very narrow gauge and uses antiquated equipment that places an effective upper limit of 1.0 unit of commodity flow along the line. This additional constraint may be added to our simplex tableau (Table 12–5) in the form of the equation:

$$(B \to D) \leq 1.0$$

which states that movement between B and D must be less than or equal to one. Since city

TABLE 12–5

Simplex Tableau for Transportation Problem with Capacity Constraint Added

Flows between Cities That Can Enter Solution				
$A \to C$	$A \to D$	$B \to C$	$B \to D$	Requirements to Be Met
1.0	1.0	0	0	= 4 A's shipments
0	0	1.0	1.0	= 3 B's shipments
1.0	0	1.0	0	= 5 C's receipts
0	1.0	0	1.0	= 2 D's receipts
0	0	0	1.0	≤ 1 $B \to D$ constraint
5	6	5	4	Cost of shipping 1 ton

D can only receive one unit from B now, it must get the remaining requirement from A. Thus our objective function becomes:

$$(5 \times 3) + (6 \times 1) + (4 \times 2) + (5 \times 1) = 34$$

and the overall cost of movement has been raised from the previous 33.0. As we would expect, a country with capacity constraints that reduce the flows of goods and commodities over its transportation system has to pay a price for such inefficiencies. Transportation facilities

built at one time to meet smaller and simpler needs may severely hinder development in later years. As transportation requirements increase, they produce congestion, delays, and limitations for which severe and accumulating costs are paid. Many early railway lines built in the underdeveloped world during the turn of the century have become severe constraints on later progress. Numbers of narrow gauge railways with tortuous turns have been abandoned, or the routeways have been severely modified as larger gauges have been built.

Unfortunately, we cannot show even the simplest two-by-two transportation problem in graphical form. Where only two alternative choices are available (as in the case of the African farmer), the constraints and the objective function can be shown in a space of two dimensions. But our two-by-two transportation problem generates four possible alternative shipments, so we would need a graph in four dimensions to show the constraining equations. From the tableau (Table 12–4), you can see that the number of constraints is always at least the number of origins and destinations in the problem, while the number of alternative ways of shipping is the number of origins *times* the number of destinations. Thus making decisions about transportation movements involves the evaluation of very large numbers of alternative courses of action. Even in a relatively small problem with five origins and ten destinations, this means we would have fifteen constraint equations with fifty terms in each. It is little wonder that special methods have been devised to solve transportation problems, and that large computers are required to make rational decisions in this geographic problem area.

Evaluating Alternative Decisions

Suppose we had a country divided into five surplus regions ($A–E$), and nine deficit regions ($O–W$), together with exports overseas (X). Although we cannot show all the constraint equations with their fifty terms, we can summarize all the information shown in Fig. 12–21 in tabular form (Table 12–6). To find the minimum solution to this problem by hand would be almost impossible. So many alternative ways

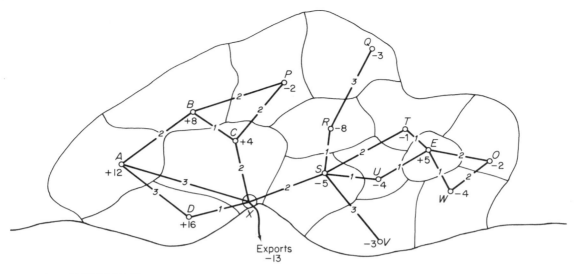

FIGURE 12–21

A country with five surplus and ten deficit regions with costs of transportation over each road link.

TABLE 12–6

Costs of Shipping Goods between the Five Surplus and Ten Deficit Regions

Regions with Surpluses	Regions with Deficits									
	O −2	P −2	Q −3	R −8	S −5	T −1	U −4	V −3	W −4	X −13
A + 12	9	4	9	6	5	7	6	8	8	13
B + 8	9	2	9	6	5	7	6	8	8	13
C + 4	8	2	8	5	4	6	5	7	7	12
D + 16	7	5	7	4	3	5	4	6	6	11
E + 5	2	8	6	3	2	1	1	5	1	14

exist of shipping the surpluses to the deficits that the human brain cannot take them into account. Even the most skilled juggler cannot keep too many balls in the air at once.

If we allow a computer to solve the problem for us, the overall cost is 307 with a distinctive pattern of commodity flow (Fig. 12–22). Let us put ourselves in a position of a Minister of Transportation in a poor, underdeveloped area who is trying to improve road transportation with very limited funds. Only three alternatives come within the range of the

finances available. A new gravel road may be built between *P* and *R ;* another gravel road may be constructed from *C* to *S ;* or it would be possible to use the funds to upgrade the roads between *A*, *D*, *X*, *S*, *T*, *E*, *W*, and *O*, from gravel to tarred standard. Transportation over the road between *P* and *R* would cost 1.0 unit, and would obviously make the deficit regions *Q* and *R* more accessible to surplus region *B*. Cost over a new road between *C* and *S* would be 3.0 units, and would increase the accessibility of the deficit regions to surplus regions *B* and *C*. Upgrading the gravel road to tarred standard from *A* all the way through to *O* would have the effect of cutting present transportation costs in half along the route. Under the normative assumption of rational decision-making to minimize overall costs of movement, which of the alternative plans would yield the most benefits? Building the road between *P* and *R* (Fig. 12–23) clearly gives *B* a greater advantage in shipping to the deficit regions *P*, *Q*, and *R*. But other deficit regions like *V*, formerly supplied by surplus region *B*, must now fulfill their requirements from other regions such as *A*. This means in turn that *A* cannot ship so much

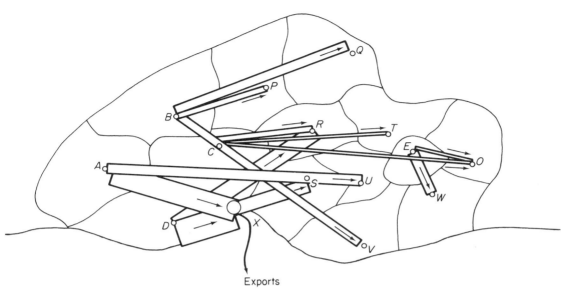

FIGURE 12–22

Minimum cost flows from the surplus regions to the deficit and export regions. Total cost 307.

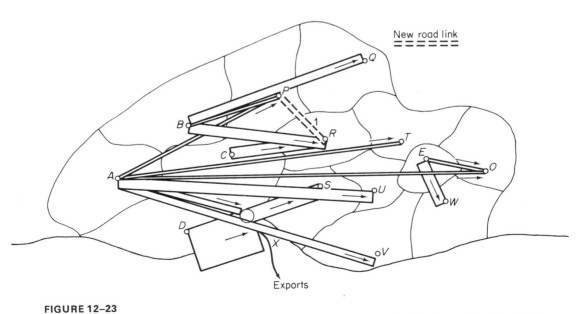

FIGURE 12–23

Commodity flows with the addition of the new link from P to R in the network. Total cost reduced from 307 to 280.

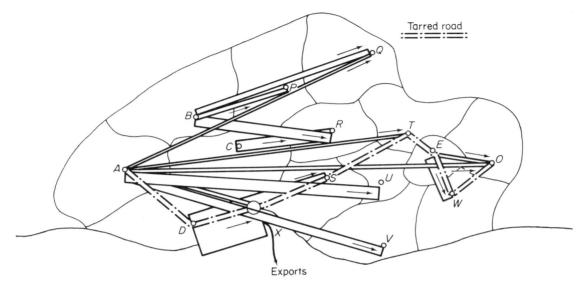

Tarred road

FIGURE 12-24

Commodity flows when the road from A to O is tarred, cutting costs by 50 percent over the upgraded links. Total cost reduced from 307 to 257.5.

to the port for export, so the export requirement must be made up by shipments from *D*. But, by meeting the export requirements, *D* is now no longer in a position to ship to deficit region *R*. This, however, is satisfactory because *R* is now receiving goods from *B* over the new link between *P* and *R*. These changes in the flow pattern, as a result of building this new road, illustrate the important point that a shift in one flow induces a ripple of shifts throughout the entire transportation system. The whole system is interdependent, and as we seek the overall minimum cost, seemingly small additions to the network may have radical effects upon the patterns of commodity flow throughout the country.

If transportation funds are used to build a road between *C* and *S*, the overall cost is reduced only to 297 units, compared to 280 with the *P* to *R* link. Clearly, this is a less desirable course of action. However, if the road between *A* and *O* is upgraded to tarred standard, cutting the costs in half over each tarred link, the overall cost declines to 257.5 units, far below the minimum cost with either of the new links in the

network (Fig. 12-24). Under this alternative plan, the commodity flow pattern closely resembles that derived under the *P* to *R* alternative, with the exception that *A* now supplies a small amount of *Q*'s requirement. The tarring program greatly increases the locational advantage of all the regions through which the tarred road goes, and even allows surplus region *A* to reach along the tarred portion and up the gravel road to supply *Q*. Thus a decision to upgrade a portion of a transportation network, and thereby radically lower the costs over the improved section, sends ripples of benefits throughout the whole system rather than just the areas receiving the improvements.

Paying a Price for Location

Changes in the quality and configuration of a transportation network benefit all parts of the system, but some places benefit more than others. By virtue of their positions relative to all the other surplus and deficit areas, some regions have distinct locational advantages. We can express such advantages in terms of locational rents which may be thought of as subsidies or penalties given to some places by virtue of the particular pattern of production and transportation. In the problem above, region E obviously has the best location compared to all the other surplus areas. It is surrounded by deficit regions to which it may ship its surplus over short, low-cost links. On the other hand, region A is relatively far from the deficit regions, and we can think of the people in A paying a spatial penalty simply because of their remote location with respect to all the other areas in the country.

Linear programming methods allow us to evaluate the relative locational advantages of the regions (Table 12–7). When we solve a

TABLE 12–7

The Penalties Paid by the Shipping Regions by Virtue of Their Relative Locations within the Pattern of Surpluses and Deficits

Shipping Region	Before Tarring	After Tarring	Change in Relative Advantage
A	−7	−4.5	+2.5
B	−7	−5.5	+1.5
C	−6	−4.5	+1.5
D	−5	−3.0	+2.0
E	0	0	0

transportation problem, we get a great deal of additional information about the system. Not only can we derive the pattern of shipments minimizing the overall cost, but we can also obtain the values of the spatial rewards and penalties of every shipping and receiving region. Regions A and B pay the biggest penalty relative to E, but the other western areas also pay locational penalties compared to E sitting smugly in the middle of the deficit regions to the east. After the road tarring program has been completed, however, the relative locational advantages shift as transportation improvements shrink and adjust the cost space, bringing some surplus and deficit regions closer together than they were before. Region E still maintains its great locational advantage, but the penalty gap between it and region B, which now pays the highest cost for inaccessibility, has been narrowed. While all

regions have benefited to some extent, the tarring program has given a considerable accessibility boost to region A, and only slightly less of a boost to D. We would expect this shift because the tarred road actually runs through these regions, linking them much more tightly to the deficit areas. Finally, regions B and C have also benefited, though somewhat less because they did not actually receive any tarred links under the program.

Surfaces of Locational Penalties

That some places should have locational advantages over others is obvious, and the discussion of the von Thünen model in Chapter 10 demonstrated that farmers near the market had a definite advantage over those further away because their profits were not eaten up by high transport costs. Suppose we have the values for many points in a large area. From these we can construct locational rent surfaces showing the relative advantage that places have compared to the worst location in the region.

In the Upper Midwest (Fig. 12–25) the enormous wheat production moves in response to demands from both the national and international markets. Farmers in the eastern part of the region have a locational advantage since the production from their farms can be easily tapped by the Great Lake ports. Farmers in the extreme west have even larger locational rents, for their farms are highly accessible to the port and markets of the west coast. Since the region is "drained" of wheat by railways focusing both east and west, it is not surprising that the central portion of the region is the least accessible, and farmers in the central Dakotas are penalized relative to those more fortunately located.

As we examine surfaces describing the spatial variation in relative locational advantage, we must remember once again that we are in a normative framework. Costs are assumed to be a reasonably simple (although not necessarily linear) function of distance, but alterations in a freight structure can change the relative locational advantage of some points quite drastically. If people decide to blanket an area with the same freight rate structure, they

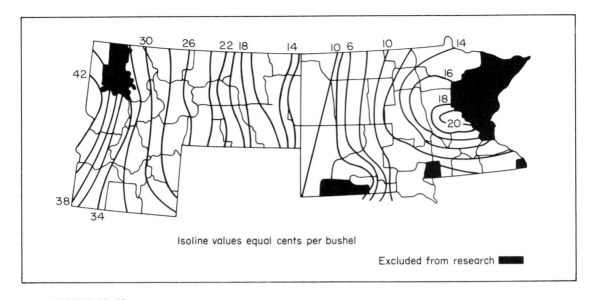

Isoline values equal cents per bushel

Excluded from research ▮▮▮▮

FIGURE 12–25

Location rent surface for wheat in the Upper Midwest showing the advantage that the eastern and western areas have by virtue of their relative location (adapted from Maxfield, Fig. 4; reproduced by permission from The Professional Geographer, *XXI [*July 1968*]).*

can completely wipe out the natural locational advantage of some points with respect to the market requiring their goods. In such cases, points nearer the markets in effect subsidize those further away. We have only to think of postal rates in the United States to see how one price for all letters means that short movements are subsidizing long ones. Similarly, many cities have a single bus fare which works to the advantage of long-distance riders who are subsidized by those traveling over the short distances. In terms of cost alone, and discounting the value placed on employees' travel time, an office building could be placed anywhere in a city blanketed by a single fare structure, for in these very simple terms no point has a locational advantage over another.

A number of geographical problems concern the making of decisions to minimize overall costs of movement when surpluses are assigned to deficits. Many such problems have practical implications at the local level. For example, schools in an area may be considered as points with pupil deficits, while the homes of pupils are considered as sources of surplus children. Given the locations and various capacities of the schools in a region, we may wish to draw school district boundaries by assigning pupils to schools to minimize the overall cost of movement.

In Grant County, Wisconsin, for example, thirteen school districts developed over time through trial-and-error adjustments based upon local political boundaries (Fig. 12–26). Using a normative transportation model to decide the allocation of students to schools, the optimal boundaries for each school district were drawn (Fig. 12–27). The new boundaries minimized the overall cost of transporting the children each day to school. A comparison of the actual districts to those minimizing the costs of movement (Fig. 12–28) indicates that about 18 percent of the home areas should be reassigned to different schools. Some schools, such as Lancaster's with a large capacity, should extend their pupil hinterlands; while other school areas, such as Hazel Green, should contract. Generally, the normative model requires a series of small

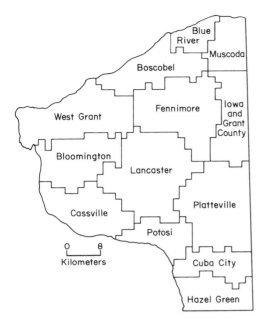

FIGURE 12–26

The actual pattern of school districts in Grant County, Wisconsin (adapted from Yeates, Fig. 7–11, p. 110).

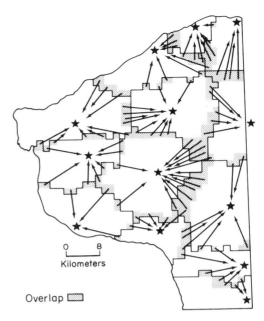

FIGURE 12–28

Changes in the school districts to meet the criterion of least-cost assignment (adapted from Yeates, Fig. 7–13, p. 112).

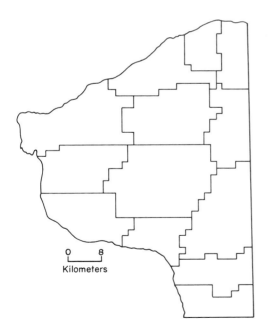

FIGURE 12–27

School district boundaries derived from a least-cost assignment of school children (adapted from Yeates, Fig. 7–12, p. 111).

trading-off adjustments between the districts to produce the least-cost solution. Though the adjustments are small, adopting them could save the county nearly $4000 per year on its school bus program—funds sorely needed for other educational requirements.

The drawing of school boundaries under such normative assumptions has implications beyond the financial. There is fairly strong evidence that the performance of young children in school may be related to the distance they feel they move from their homes to their school classrooms. A young child of six or seven feels comfortable in the small space around his home, but may become increasingly nervous as the distance from this familiar space increases. Working with eight hundred children in England, the psychologist Lee noted the way a young child's adjustment to school declined with the distance he traveled (Fig. 12–29). The walking curve was higher because a child always felt he could return home by himself. Children who came by bus were dependent upon others and felt greater stress because they felt so far removed in time from their homes. As psychic stress increases, performance at school deteriorates.

477

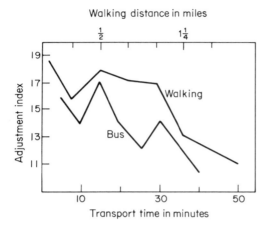

FIGURE 12–29

The decline in children's adjustment to school with the time/distance of the journey (adapted from Lee, p. 106).

Thus it would appear that there are good psychological as well as economic reasons for minimizing the school journeys of young children.

In a similar way, boundaries between many sorts of centers that administer particular services may be drawn upon the same normative grounds. Family planning clinics, for example, might be set up in a rural area of India, and the administrative boundaries drawn between them to minimize the travel of the women seeking birth control advice. In the United States, the boundaries between social security offices might be similarly delimited to minimize the effort of the people they serve. Perhaps, as population patterns change, the administrative boundaries could be altered after every census. In this way the important spatial aspects of administration become dynamic factors rather than archaic legacies hindering contemporary programs and increasing overall costs.

A number of the normative problems involve questions of agricultural production, the costs of human diets, and the assignment of surpluses to deficit regions. Increasingly these problems are expanding beyond the regional and national scope, involving the whole world and the human family. By the year 2000 geographic satellites with radiation and chemical sensors may well be scanning all agricultural regions on a continuous basis, feeding information into large computers which will constantly revise and refine estimations of food production on a worldwide basis. Surplus and deficit regions will be identified instantaneously, and flows of foodstuffs programmed to minimize the overall costs of movement. If this occurs, it will be because individual members of the human family have decided to work together within a normative framework for humanitarian reasons. Perhaps the very perspective we are getting now of "Spaceship Earth" as a tiny speck in the universe will help men to see that they are part of the human family. For this larger, aggregate family many of the biggest spatial problems remain unsolved.

THE THEORY OF GAMES AS A CONCEPTUAL FRAMEWORK

Making optimal decisions is always a difficult task because men cannot foretell exactly what the future will bring. Frequently the outcome of a decision is not immediately apparent, and in the interval between the decision and its final result many unforeseen things can happen. Sometimes fortunate events produce payoffs far exceeding our expectations. At other times, unfortunate things happen which reduce or even nullify the effect of a decision made at an earlier time. As men make decisions for the future under conditions of uncertainty, they may well adopt quite conservative attitudes toward the environment in which they live and work. Where risk is great and disaster possible, men may behave very cautiously just to be on the safe side.

In considering spatial decision-making under conditions of great risk and uncertainty, it is useful to adopt the normative framework of the *theory of games*. In such a context, the environment, or nature, will be considered as a vindictive, minimizing player in a game played by men against her. At first sight such an assumption may seem particularly severe and unrealistic, for we often think of nature as inanimate and quite indifferent to men's desires and wishes.

However, it is not difficult to think of numerous situations where men are very uncertain about the future and have little knowledge about their opponent. In such cases, they may well assume the worst and act to minimize their risks. The pioneer wheat farmers of the United States, Canada, and southern Australia made decisions of considerable risk in almost total ignorance of the environment. In the early years of pioneer farming there were none of the extensive rainfall records and soil surveys that there are today. In the Soviet Union many of the ambitious schemes to extend grain farming into the highly marginal environments of the Virgin Lands have met with disaster. Men made the wrong decisions in the face of an apparently vindictive player of which they had insufficient knowledge. Whether we consider a simple subsistence economy or the extensive patterns of commercial agriculture around the world, some men must make decisions in marginal areas where environmental variability is very high and the uncertainty of future events is very great.

Strategies in a Game

The theory of games is founded upon principles familiar to us since childhood. In almost any game two or more people face each other and choose certain strategies to win. A simple game like tick-tack-toe involves two players making alternative decisions to line up three similar symbols by choosing from an ever-smaller number of available strategies. A game of chess also involves two players making alternative decisions, but in this ancient game the number of alternative strategies is so large that it is difficult to know the best one to choose at any given point.

We can represent the essential, underlying idea of such games in the form of a payoff matrix (Fig. 12–30). Two players, A and B, are playing against one another. Each has two possible strategies to choose from, and the payoff to player A is shown in each cell of the matrix. We assume, to keep this first example simple, that player A always wins and that he tries to win as much as he can. Player B always loses but does his best to minimize his losses. For example, if both players choose their first

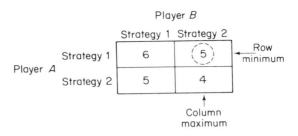

FIGURE 12–30

Payoffs from Player A to Player B in a two-by-two zero sum game with a saddle point.

strategies, player A obtains a payoff of six at the expense of B. If they both decide upon the second strategy, player A only wins four at B's expense. The wins of player A are always exactly balanced by the losses to player B, and because the wins and losses sum to zero such a game is called a two-person zero-sum game.

Choosing the Best Strategies

What is the best strategy that player A can adopt to maximize his payoff from the game? Clearly the first strategy is preferable, for at the very least he will win five with a possibility of increasing it to six. Looking at it from B's point of view, it is clear that the second strategy is the most preferable, since five is the most that can be lost, with a possibility that the loss may be reduced to four. Thus player A should always choose the first strategy, while player B should always choose the second. In this case player A will win five each time the game is played. Notice that the value of five in the upper right hand cell is the lowest value in the row and the highest value in the column in which it appears. Such a row minimum that is also a column maximum is termed a *saddle point*. If we think of a saddle (Fig. 12–31), there is one point that is simultaneously a lowest high and a highest low. Such a point in a payoff matrix, called the *minimax* point, indicates that each player has a "best" strategy. Any departure away from such a point by one player will mean the other can always take advantage of him. For example, if player A chooses the second strategy he will not get as much as before. Similarly, if player B

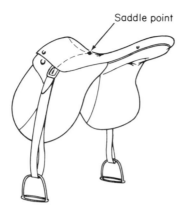

Saddle point

FIGURE 12–31

The concept of a saddle point: the lowest high and the highest low.

choose the first strategy his losses will be greater than the second strategy disclosed by the saddle, or minimax point.

Let us consider a second payoff matrix (Fig. 12–32), in which the values have altered somewhat. If the players look for a saddle point in this game, remembering that a saddle point is a value that is simultaneously the minimum of a row and the maximum of a column, they will look in vain. We can imagine how the course of a sequence of games might proceed. Player *A* chooses the first strategy while *B* chooses the second. *B*'s losses are very high at six, so the next time he chooses strategy one. Now his losses are only three but *A* is greatly dissatisfied with such paltry winnings. Accordingly, *A* shifts to the second strategy and raises his payoff to five, whereupon *B* shifts back to the second strategy to reduce his losses to four. Thus the game proceeds, each player chasing

| | | Player *B* | |
		Strategy 1	Strategy 2
Player *A*	Strategy 1	3	6
	Strategy 2	5	4

FIGURE 12–32

A Two-person zero-sum game with no saddle point.

the other around the payoff matrix, for the game is totally unstable without a saddle point. Thus, there is no single best strategy for either player, and some other criterion must be found by Player *A* to choose strategies that will maximize the payoff from the game.

Mixing Strategies

If a unique strategy for maximizing his payoffs is not available, player *A* must mix the two strategies available to him. But this poses two difficult questions: first, the proportion of the time he should choose to play each one; and, secondly, the way they should be mixed together. If we take the absolute difference between the payoff values obtainable from using each strategy (Fig. 12–33), and assign the proportion derived from each strategy to the alternate one, we will have the solution to the first question. In this example, player *A* should choose Strategy 1 a quarter of the time, while devoting the remaining three quarters to Strategy 2. As for the second question, the strategies must be mixed in a random pattern, perhaps by flipping two pennies and only choosing Strategy 1 when they both come down heads. By choosing the strategies at random, player *B* can never discern any order or pattern in the selections of player *A*. If player *A* did not mix his strategies randomly, player *B* might discover patterns and regularities in the sequence of choices, and with this information alter his own choice of strategies to minimize his loss still further. We can find the best mixture of strategies for *B* in the same way (Fig. 12–34). Taking the absolute difference between the payoffs in the column of the matrix, player *B* should mix his strategies fifty-fifty to minimize his losses to player *A*.

Given the proportion of time that player *A* should choose each strategy, we can calculate the minimum value of the game to him. This is the value to him over the long run assuming player *B* is always playing to minimize his losses. Considering the payoffs in Column 1, player *A* will get a payoff of 5 three-quarters of the time, and a payoff of 3 for the remaining quarter. Thus the value of the game will be:

$$(5 \times \tfrac{3}{4}) + (3 \times \tfrac{1}{4}) = 4\tfrac{1}{2}$$

FIGURE 12–33

The solution for Player A mixing his strategies.

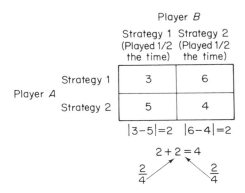

FIGURE 12–34

The solution for Player B mixing his strategies.

It makes no difference which column we work with, for if we use the payoff values in the second, then the result is:

$$(6 \times \tfrac{1}{4}) + (4 \times \tfrac{3}{4}) = 4\tfrac{1}{2}$$

Thus the value of the game to A is the same whichever column of payoff values is used.

Choosing Locations as Strategies in a Game

When men make decisions to locate economic activities, the payoffs they receive from their locational choices may often depend upon factors in the environment in which the decisions were made. Consider a Jamaican village where fishing is carried out from large canoes by lowering baited fishing pots attached to ropes and wooden identification floats to the bottom of the sea. The captains of the canoes may place their fishing pots in various locations (Fig. 12–35). If the pots are placed in the calm waters of the lagoon they will seldom be lost. However, the fish caught in the lagoon are of a much lower quality than those caught out beyond the reef in the open water. The local people recognize such quality differences in the local market, and the prices for lagoon fish are lower than those for the open water varieties. At first sight, it seems that the canoe fishermen should place all their pots beyond the reef in the open sea to catch the high quality varieties. The only trouble is that occasionally an offshore current runs in a quite unpredictable fashion, battering the pots on the sea bottom and dragging the floats below the water so they become sodden and sink. Prices for the open sea fish are higher, but the pot losses ·of the canoe fishermen must also be considered as locations are chosen. Market prices for the two sorts of fish also vary with the current. When no current is running and the canoes fishing in the deep water have a good day, their fine quality fish drive the prices of the lagoon varieties down still further. On the other hand, fishermen in the lagoon find that prices rise for their fish when the deep sea fishermen cannot compete in the market on those days

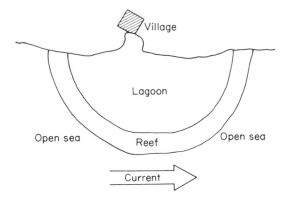

FIGURE 12–35

The Jamaican village and the alternative locational strategies for the fishing pots.

	Environment	
	Current	No current
All pots in	17.3	11.5
Fishermen In − out	5.2	17.0
All pots out	−4.4	20.6

FIGURE 12–36

The 3 × 2 payoff matrix for the Jamaican fishermen playing against the offshore current (adapted from Davenport, p. 8).

when the offshore current is running strongly.

We can consider all the canoe captains in the village as a single player in a game against the environment. Three locational strategies are available to them. They may place all their pots in the lagoon; they may put some in and some out; or they may wish to put all the fishing pots beyond the reef in the deep sea (Fig. 12–36). If we consider the environment as a vindictive player against the fishermen, two strategies are available which may be labeled simply Current and No Current. If the fishermen put all their pots in the lagoon and the current is running, they will average 17.3, for there is no competition from the high quality, deep sea fish in the marketplace. But if no current is running, then the deep sea fishermen drive the price of lagoon fish down, and the fishing canoes with captains who were conservative and cautious only obtain 11.5. Canoes that put some in and some out do fairly well when no current is running, but quite poorly when the current runs and they suffer pot losses. Finally, canoes that place all their pots out to sea do extremely well with 20.6, but when the current runs their pot losses result in out-of-pocket costs. We represent these losses in the payoff matrix by −4.4.

	Environment	
	Current	No current
In − out	5.2	17.0
Fishermen All out	−4.4	20.6

FIGURE 12–37

One pair of strategies from the 3 × 2 game showing a saddle point.

Since they do not know when the environment will choose the Current or No Current strategies, which locations should the fishermen use? Let us consider pairs of strategies, looking first at the In-Out and All-Out options as a two-by-two payoff matrix (Fig. 12–37). Notice that the value 5.2 is a row minimum at the same time that it is a column maximum. This saddle point in the In-Out row dominates the All-Out choice completely. We can therefore eliminate the latter from consideration and turn to the next two-by-two payoff matrix with the All-In and In-Out strategies for the fishermen (Fig. 12–38). Since there is no saddle point in this payoff matrix, it is clear that the fishermen must mix these strategies to maximize their payoff from the fishing game against nature. From the difference in the payoff values, 67 percent of the fishing canoes should place their pots in the lagoon, while 33 percent should choose the In-Out strategy. In this way the maximum payoff will be achieved by the village as a whole. The remarkable thing is that careful fieldwork by the anthropologist Davenport has disclosed that 69 percent of the fishing captains did decide to stay within the lagoon, while only 31 percent ventured out into deep water. Thus it would seem that the village, over a long period of trial and error, has moved toward the minimax solution in this game against nature.

Consolidating Farms in Areas of Flooding

Mixing locational strategies may make extremely good sense in very variable environments where extreme events can cause disaster. In Japan, for example, traditional patterns of land tenure and inheritance often resulted in a farmer owning a number of small plots scattered around his valley. After the Second World War, considerable consolidation of landholding was forced upon many of the farmers, the idea being that large plots would be more efficiently worked, with less time wasted in moving from one small plot to another. In some areas the decision to make each farmer work at a single location was an effective one. But in some valleys, where sudden floods were experienced, the consolidation of many small plots into a single

FIGURE 12–38

The second pair of strategies from the 3 × 2 game, indicating that mixed strategies are required.

		Environment		
		Current	No current	
Fishermen	All in (played 67%)	17.3	11.5	|17.3 − 11.5| = 5.8
	In – out (played 33%)	5.2	17.0	|5.2 − 17.0| = 11.8

Total 17.6

large farm meant that a farmer's efforts could be completely wiped out if a flood came. Prior to the consolidation program, when the same farmer had a number of small plots scattered over the bottom and sides of the valley, he was never completely ruined by such an environmental disaster. Looking at the question of farm consolidation from the farmer's point of view, we can represent the problem in the form of a simple payoff matrix (Fig. 12–39). If all the small, traditional plots are consolidated and no flood comes, the payoff to the farmer is clearly very large. However, if his consolidated farm lies in the valley bottom and a flood comes, his losses are so severe that he must pay out of his meager savings to feed and support his family. If he maintains the traditional pattern of fragmented landholdings, he will not do as well as under consolidation when there is no flood. But when the flood comes some of his small plots escape destruction, and he may have enough to keep him and his family going until better times come. Notice that the value of 5 in the payoff matrix is a row minimum and a column maximum, a saddle or minimax point that indicates the total dominance of the Split-Up strategy.

Hedging one's bets against the environment by choosing a number of locations is a very common strategy in environments that are marginal or disaster prone. In the Rift Valley of Kenya, for example, the Suk plant their crops in various altitudinal zones from the *Keogh*, or hot country on the valley floor, to the *Tourku*, or cold country along the valley rim (Fig. 12–40). By refusing to place all their locational eggs in one basket they not only smooth out the supply of a variety of foodstuffs, but ensure themselves of a basic minimum if the fields on the hot valley floor are ruined by drought.

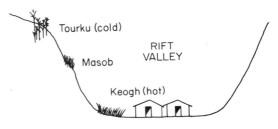

FIGURE 12–40

Alternative altitudinal strategies open to the Suk farmers of the Kenya Rift Valley.

The Spatial Assignment of Investment Decisions

Farmers like the Japanese and the Suk of Kenya make locational decisions about investing their time and energy in farming activities. Investment decisions always have a spatial dimension, and the payoffs accruing to such investment choices frequently depend on unpredictable strategies chosen by the environment. Since there are seldom sufficient investment funds to do everything at once, decisions must be made to split up the available resources and assign them to projects to achieve certain goals. Most investment decisions in underdeveloped countries involve not only choices between sectors of the economy, but also decisions of a spatial or locational nature involving

		Environment	
		Flood	No flood
Japanese farmer	Consolidation	−10	20
	Split-up	5	15

FIGURE 12–39

Japanese farmers choosing landholding patterns in valleys subject to flash flooding.

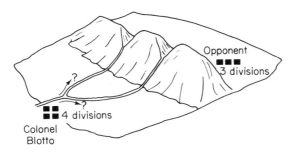

FIGURE 12–41

Colonel Blotto trying to win the passes from his opponent.

the actual assignment of investment to certain regions at the expense of others.

As we consider the general problem of dividing limited resources to achieve multiple objectives, we can think of the analogous problem in the military campaign of Colonel Blotto —a legendary figure in the theory of games. Colonel Blotto has four divisions with which he must try to capture two passes across the mountains (Fig. 12–41). His opponent has only three divisions with which to defend the passes. Each player in this "game" may decide to use his forces in different ways. The superiority of at least one division at a pass is required to win it (+1) while equal forces represent a tie and no gain (0). We may represent the strategies open to each player in the form of a payoff matrix

(Fig. 12–42). For example, if Colonel Blotto decides to split his forces three and one against his opponent's forces which are split two and one, we must recognize two possibilities. If Blotto's three divisions meet the one of his opponent, Blotto will win one pass (+1). However, he will lose the other, where he has only one division overwhelmed by the two of his opponent (−1). In this case the win (+1) and the loss (−1) cancel out (0). On the other hand, if his three divisions meet the two of his opponent he will win one pass (+1) without losing the other where the single divisions from both sides meet in a tie (0). Since either of these possibilities are equally likely, Blotto will win a pass half the time so we may assign a payoff of that value to him in the matrix. Should Blotto decide to split his forces two and two, he will always win a pass against his opponent's two and one strategy, but will always win and lose a pass for no gain against the opponent's three and zero strategy.

At the junction of Blotto's three and one, and his opponent's three and zero strategy, there is a value of one half. There is no *smaller* value than this in the row, and no *larger* value in the column. Thus we have a saddle point, and Colonel Blotto should always choose to divide his forces in this way to maximize the outcome from this game.

As another example of spatial investment decisions we can imagine a Minister of Transportation in an underdeveloped country trying to decide upon the allocation of limited road maintenance funds in two regions A and B (Fig. 12–43). Sometimes the severe monsoon

	Opponent	
	$3 + 0$	$2 + 1$
$4 + 0$	$\frac{1+0}{2} = \frac{1}{2}$	$\frac{0+0}{2} = 0$
$3 + 1$	$\frac{1+0}{2} = \frac{1}{2}$	$\frac{1+0}{2} = \frac{1}{2}$
$2 + 2$	$\frac{0+0}{2} = 0$	$\frac{1+1}{2} = 1$

Colonel Blotto (row label for $3+1$)

FIGURE 12–42

Payoff matrix for Colonel Blotto and his opponent (from J. G. Kemeny and G. L. Thompson, "Attitudes and Game Outcomes," in Contributions to the Theory of Games, *Vol. 3, p. 287, eds. M. Dresher, A. W. Tucker, and P. Wolfe, No. 39 of Annals of Mathematical Studies, copyright © 1957 by Princeton University Press. Reprinted by permission of Princeton University Press).*

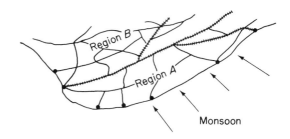

FIGURE 12–43

Two regions, A and B, whose road systems must be maintained against the annual monsoon rainfall.

		Monsoon	
		Heavy rain in A Low rain in B	Moderate rain in A Moderate rain in B
Minister of Transportation	Heavy investment in A Low investment in B	$A = 50$ $B = 50$ Total 100	$A = 75$ $B = 45$ Total 120
	Moderate investment in A Moderate investment in B	$A = 40$ $B = 70$ Total (110)	$A = 50$ $B = 65$ Total 115

FIGURE 12–44

Payoff matrix for the assignment of road maintenance funds to Regions A and B.

season produces catastrophic floods in *A* while leaving *B* relatively untouched. In other years the monsoon is less severe in *A* and spreads part of the rainfall over region *B*. Faced with the problem of assigning the limited road maintenance funds, our geographically trained Minister might structure the problem in the form of a payoff matrix (Fig. 12–44). As a result, he may, like Colonel Blotto, decide upon a division of resources that will maximize his payoff from the game and keep the transportation system going. The payoff matrix would imply that a division of investment funds to maintain roads at gravel standards is a better strategy than giving one region tarred roads to the almost total neglect of the other.

Decisions against Nature and Other Men

When men make individual locational decisions, their payoffs may be affected not only by the environment but also by the decisions that other men in the game have made. For example, since the Second World War winter recreational facilities in Pennsylvania have expanded rapidly. Large numbers of ski resorts dot the state (Fig. 12–45), but the problem faced by a single man in locating such a recreational facility may not be an easy one. Snow conditions vary throughout the state, and those ideal snowfalls that are heavy and stay on the ground are not necessarily close to the major

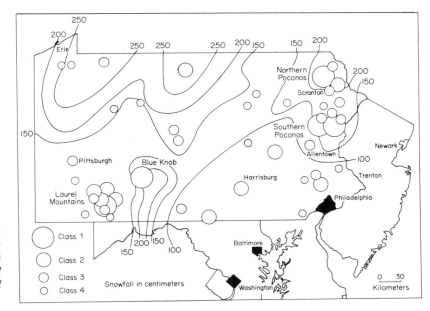

FIGURE 12–45

The locational game of ski resorts in Pennsylvania. An operator faces the problem of locating with respect to snowfall, major markets, and other resort operators (adapted from Langdale).

markets of Pittsburgh, Philadelphia, Baltimore, and Washington. A decision by a ski tow operator to locate in areas of heavy and reliable snowfall moves him away from these major markets. Conversely, a move toward the major sources of skiers may mean entering areas of lower and less reliable snowfall so that payoffs are greatly reduced. Moreover, where many operators are setting up ski tows, the payoff accruing to any one of them hinges not only upon the environment and the snowfall strategy chosen by nature, but also upon the degree of competition from other operators in the area. Thus we have a locational game with many players against each other as well as against the environment. Location becomes a crucial variable that can make or break a ski resort's operations. Fortunately, new technology can help modify the strategies open to the environment. Modern snowmaking equipment can greatly decrease the uncertainty faced by earlier ski resort operators, for nature's vindictive strategies are reduced in number as a result. Today resort operators with modern equipment are only concerned with the question of temperature variations and the cost of snowmaking equipment, for they can ignore the problem of insufficient natural snowfall.

Land Use Patterns as Strategy Decisions

The strategies open to men as they try to wrest a living from the environment may not always involve changes in location. In a farming area at any particular location, a number of choices may be open to the farmers as they select crops to grow each year. Consider, for example, farmers in the barren middle zone of Ghana (Fig. 12–46), a zone of high environmental variability where rainfall is particularly uncertain. Let us suppose that the farmers of a small village may grow yams, cassava, maize, millet, and hill rice for their staple foodstuffs. In our game theoretic framework, these may be considered as five strategies. Similarly, and to simplify the problem, let us suppose the environment has only two moisture strategies, wet years and dry years. The strategies may be arranged in the form of the now familiar payoff

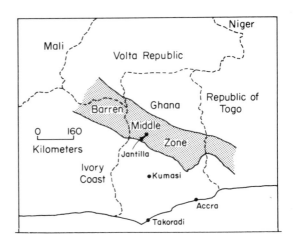

FIGURE 12–46

The barren middle zone of Ghana: an area of extreme environmental variability (adapted from Gould, 1963).

matrix (Fig. 12–47). We assume that some crops do better in wet years than others, for which too much moisture results in rotting and lowered yields. The payoff matrix describes a five-by-two game. However, where one player has only two strategies available to him the game can always be reduced to a two-by-two game. Rather than take every possible pair of rows in turn, we can obtain the critical pair of strategies for the farmer graphically (Fig. 12–48). If we draw two scales from 0 to 100, plot the values of each of the farmers' strategies on alternate axes, and then connect the points, the lowest point on the uppermost boundary will indicate which crops the farmers should grow to maximize their payoff from the game. The

			Environment moisture choices	
			Wet years	Dry years
		Yams	82	11
Farmers of Jantilla	Crop choice	Maize	61	49
		Cassava	12	38
		Millet	43	32
		Hill rice	30	71

FIGURE 12–47

A payoff matrix for farmers in the barren middle zone.

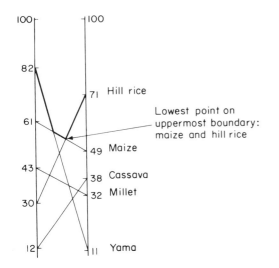

FIGURE 12-48

Locating the critical pair of strategies in the 5 × 2 zero-sum game (adapted from Gould, 1963).

maize 77.4 percent of the years and hill rice for the remaining time, mixing the years in a random fashion, or they may plant these proportions every year. As long as there are men around to plant the fields each year, it makes no difference in the long run which course is taken. A mixture of strategies in the right proportion each year, or a mixture of years in the right proportion will produce the same payoff over the course of a century or two. But men with children, who have experienced famine before, are not very likely to take such theoretical, long-term views. It is no use telling a man who has put all his fields into hill rice when a wet year turns up that his returns from the game of life will even out if he just hangs around long enough. Clearly, the farmers should mix their strategies between maize and hill rice each year, and in this way hedge their bets against nature at every planting season.

Strategic Decisions and Locations:
Coffee on Kilimanjaro

Sometimes the decision to select a particular strategy may depend upon the location of the farming activity. In East Africa, the snow-capped mountain of Kilimanjaro rises in a great cone from a dry plain at one thousand meters to a summit at nearly six thousand meters. Such a range of altitude produces extreme environmental contrast, from the dry scrub savanna at the base through dense tropical forests to alpine meadows above the treeline. Rainfall in the lower plains may average twenty-five centimeters per year while the middle slopes receive well over two hundred and fifty centimeters.

Since the 1920s the Chagga farmers have been growing Arabica coffee for commercial use, mixing the coffee with banana stands whose broad leaves provide the necessary shade. It has been found that by spraying a coffee tree with a copper solution, the yields of coffee may be raised. The copper solution makes the coffee

lowest point on the upper boundary is where the maize and hill rice strategies cross. Notice how the diagram closely resembles some of the graphical figures in the linear programming solutions we discussed before. Indeed, all zero-sum games may be recast as simple linear programming problems, and usually larger and more difficult games are solved in this way.

Focusing our decision upon the maize and hill rice strategies only (Fig. 12-49), we can solve the two-by-two game in the usual way. Thus the farmers should grow maize 77.4 percent of the time and hill rice the remaining 22.6 percent. If they do this the farmers can assure themselves the maximum return or payoff over the long run. However, the farmers also face a further problem, for they exist in time as well as in space. To maximize their returns in the game against the highly variable environment, they face two possibilities. They can put all their land into

FIGURE 12-49

The solution to the mixed strategy problem for the Ghanian farmer (adapted from Gould, 1963).

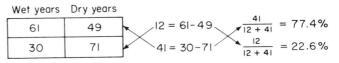

	Wet years	Dry years
Maize	61	49
Hill rice	30	71

$$12 = 61 - 49$$
$$41 = 30 - 71$$
$$\frac{41}{12 + 41} = 77.4\%$$
$$\frac{12}{12 + 41} = 22.6\%$$

Environment

		High moisture	Low moisture
Chagga farmer	Spray	+100	−40
	No spray	+60	+50

FIGURE 12–50

Payoff matrix for a Chagga farmer trying to decide about spraying his coffee bushes with copper solution.

tree retain its leaves, and with a larger area available for photosynthesis the coffee bush produces more and larger coffee berries. Unfortunately, if there is insufficient moisture, considerable damage may be done instead. Under dry conditions the coffee tree drops its leaves to cut down transpiration and so protects itself from drought, but when a tree has been sprayed with copper solution it cannot use this natural protective mechanism, and by retaining its leaves it transpires too much moisture and eventually kills itself. Thus the decision to spray or not to spray has a payoff that depends upon whether the environment provides sufficient moisture during a particular year. We can consider the problem facing a coffee farmer as a zero-sum game against the environment in payoff matrix form (Fig. 12–50). If a farmer chooses to spray his coffee bushes, and if there is plenty of moisture, he obtains a very high yield and a big payoff of 100. On the other hand, if the environment chooses a low moisture strategy, the coffee trees are killed and the farmer receives a negative payoff of −40. Adopting the No-Spray strategy means lower yields in a year of high moisture, but only a relatively small reduction in the drier years. If the farmer is totally ignorant about the environment, he should obviously adopt the No-Spray strategy, for the payoff of 50 is a saddle point in the payoff matrix.

Learning about Nature and Reducing Uncertainty

Fortunately the farmers and the personnel of a local coffee research station do not face

such a decision problem under conditions of complete uncertainty. For many decades rainfall records have been kept at a large number of stations on the mountain, and the way the probabilities of receiving sufficient moisture vary up and down the mountain may be accurately estimated (Fig. 12–51). With knowledge about the probability with which nature will choose a moisture strategy, a farmer can weight the payoff matrix to determine the choice of his own strategy in the game. Farm *A*, high on the mountain, will always have sufficient moisture, and therefore should always choose the Spray strategy. Over the course of ten years, a farmer at this location can calculate:

	High Moisture $p = 1.0$	Low Moisture $p = 0.0$
Spray	$1.0(100) = 100$	$0.0(-40) = 0$
No spray	$1.0(60) = 60$	$0.0(50) = 0$

Spray $100 + 0 = 100 \times 10$ years $= 1000$
No spray $60 + 0 = \ 60 \times 10$ years $= \ 600$

A farmer spraying would get 1000, while a farmer not adopting the spray strategy would only get 600.

Farm *B*, lower down the mountain where there is an 80 percent chance of obtaining sufficient moisture, can also weight the payoff matrix with probabilities:

	High Moisture $p = 0.80$	Low Moisture $p = 0.20$
Spray	$0.80(100) = 80$	$0.20(-40) = -8$
No spray	$0.80(60) = 48$	$0.20(50) = 10$

Spray $80 - 8 = 72 \times 10$ years $= 720$
No spray $48 + 10 = 58 \times 10$ years $= 580$

Clearly, at this location, it is still worth spraying the coffee trees to maximize the payoff from the game, even though in some years of insufficient moisture penalties will have to be paid.

Finally, at Farm *C* low on the mountain there is only a 60 percent chance of getting sufficient moisture. Weighting the payoff matrix a third time:

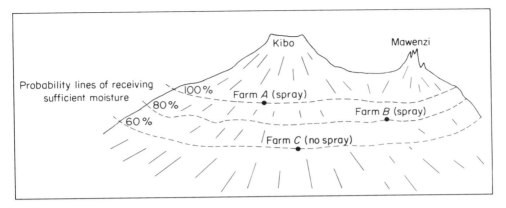

FIGURE 12–51

Chagga farmers on Kilimanjaro weighting payoff matrices with moisture probabilities to decide upon spraying strategies.

	High Moisture $p = 0.60$	Low Moisture $p = 0.40$
Spray	$0.60(100) = 60$	$0.40(-40) = -16$
No spray	$0.60(60) = 36$	$0.40(50) = 20$

Spray	$60 - 16 = 44 \times 10$ years $= 440$
No spray	$36 + 20 = 56 \times 10$ years $= 560$

If a farmer sprayed his coffee trees with copper solution he would only obtain a payoff of 440 over ten years, as opposed to 560 if he adopted the No-Spray strategy. Thus, somewhere between Farm *B* and Farm *C* there is a make-or-break boundary line above which the Spray strategy should be adopted.

In the face of total uncertainty about the environment the Spray strategy would never be adopted. As knowledge about the environment accumulates man can learn to make decisions taking his accumulated knowledge of nature into account. Finally, he may also try to alter the strategies available to nature. On the southern slopes of Kilimanjaro, the Chagga farmers have covered the face of the mountain with a lacework of ingeniously constructed irrigation furrows. The volcanic mass of the mountain behaves like a sponge, soaking up the torrential rainfall of the higher slopes and releasing it in a myriad of springs lower down.

These are cunningly channeled and guided by the farmers to insure a plentiful supply of water to their coffee and banana farms. In this way they reduce the uncertainty of the natural environment, and so can make decisions under more predictable conditions in their game against nature.

Suggestions for Further Reading

CLAYTON, ERIC. *Agrarian Development in Peasant Economics.* New York: Pergamon, 1964.

GOULD, PETER. "Man Against His Environment: A Game Theoretic Framework," *Annals of the Association of American Geographers*, LIII (1963), 290–97.

———. "Wheat on Kilimanjaro: The Perception of Choice within Game and Learning Model Frameworks," *General Systems* (1965), 157–66.

STOCKTON, R. STANSBURY. *Introduction to Linear Programming.* Boston: Allyn and Bacon, 1963.

WILLIAMS, J. D. *The Compleat Stratgist.* New York McGraw-Hill Book Company, 1954.

Works Cited or Mentioned

DAVENPORT, WILLIAM. "Jamaican Fishing: A Game Theory Analysis," *Yale University Publications in Anthropology* (1960), 3–11.

GOULD, PETER, and JACK P. SPARKS. "The Geographical Context of Human Diets in Southern Guatemala," *Geographical Review*, LIX: 1 (January 1969), 58–82.

KEMENY, J. G., and H. L. THOMPSON. "Attitudes and Game Outcomes," in *Contributions to the Theory of Games*, Vol. 3, eds. M. Dresher, A. W. Tucker, and P. Wolfe. No. 39 of Annals of Mathematical Studies. Princeton: Princeton University Press, 1957.

LANGDALE, JOHN V. "The Skiing Industry of Pennsylvania," unpublished master's thesis, Department of Geography, The Pennsylvania State University, 1968.

LEE, TERRENCE. "On the Relations Between the School Journey and Social and Emotional Adjustment in Rural Infant Children," *The British Journal of Educational Psychology*, XXVII: 2 (June 1957), 101–14.

MAXFIELD, DONALD. "An Interpretation of the Primal and the Dual Solution of Linear Programming," *Professional Geographer*, XXI: 4 (July 1969), 255–63.

WOLPERT, JULIAN. "The Decision Process in Spatial Context," *Annals of the Association of American Geographers*, LIV (1964), 537–58.

YEATES, MAURICE H. *An Introduction to Quantitative Analysis in Economic Geography*. New York: McGraw-Hill Book Company, 1968.

CHAPTER 13

INDIVIDUAL SPATIAL DECISIONS IN A DESCRIPTIVE FRAMEWORK

There is an inevitable divergence, attributable to the imperfections of the human mind, between the world as it is and the world as men perceive it.
—J. William Fulbright

SEARCHING IN A PROBABILISTIC WORLD

The normative framework is a useful one in which to consider the impact of individual spatial decisions. Many geographical problems can be considerably clarified by making some simplifying assumptions that take a difficult problem into an easier, normative area for solution. Abstracting problems into an optimizing framework often lets us use well developed bodies of mathematics and theory, and the normative constructs give us rulers against which real decisions may be measured.

When we move away from normative structures, toward more descriptive frameworks that try to account for actual decisions and patterns of spatial behavior, we also move along the continuum from deterministic to probabilistic models. The spatial behavior of individual men is never determined absolutely, although at a highly aggregate level very strong regularities may be discovered. These regularities may lead us to believe that spatial behavior is equally

predictable at finer scales, but at less aggregated levels we are usually forced to assign ranges of probabilities to many individual decisions. Thus the solutions to normative models at the deterministic end of the continuum are always smeared with probabilities at finer scales of observation because the severe assumptions of rational behavior and perfect information are never met. If a man wishes to maximize his output, then he will inevitably follow a wholly rational and determined path to reach his maximum or minimum objective. No room is left in the pure normative model for doubt, lack of information, or satisficing behavior.

As we consider spatial decisions in more descriptive frameworks we move toward probabilistic models because they attempt to incorporate many of the things that the normative models assumed away, particularly uncertainty and lack of information. The trouble is that as we move somewhat closer to reality we often must pay a severe price in the coin of complexity. Even under simplifying assumptions of their own, probability models are often more difficult to devise and manipulate than are their normative brothers. Mathematically they are usually more difficult, and we must often be content with very simple symbolic statements buttressed by descriptions in plain language. Some welcome such a shift, but in using language for the sake of public clarity we may lose the ability to manipulate symbols, as well as the new insights and solutions to our problems which often are discovered through such manipulations.

Searching and Optimality

The normative models we considered previously gave us optimal scale points against which actual decisions by men could be measured. Linear programming provided a farmer with the best decision to maximize his output, while the theory of games assumed that decisions would be made to minimize risk. The Weberian weight triangle optimized a location by minimizing costs of movement, and even central place structures were considered in light of geometrical figures that represented most efficient, equilibrium solutions. The line between such normative models and others that make less stringent assumptions is not an easy one to draw. Even if men with very strong motivations do strive to reach optimal condi-

tions, they never have perfect information about the environment in which their decisions are being made. Often, as optimal solutions are sought, men search for information about an environment whose conditions are changing even as they seek information about it.

The notion of searching for an objective is a very general one that underlies many spatial activities. All animals, from the amoeba to man, search space with varying degrees of skill to achieve certain objectives. Plankton search the space around them for food, flailing a tiny three-dimensional world and seizing upon any morsel that comes into contact with their feelers. Hunters and gatherers live most of their lives in a process of search for food and water. The Age of Exploration itself was characterized by extremely purposeful space searching activities, just as men have always searched for job opportunities and new houses. Today we are searching beyond the bounds of our planetary home, but a consideration of these search activities must be left to the first book on lunography.

Searching Space

Space may be searched in a very systematic or in a quite haphazard fashion. In many areas of applied mathematics, patterns of search are often of the former sort. You may describe, for example, the relationship between the

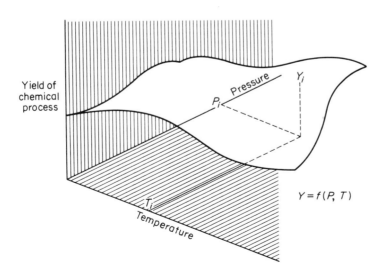

$$Y = f(P, T)$$

FIGURE 13–1

A response surface relating the yield of a chemical process to temperature and pressure.

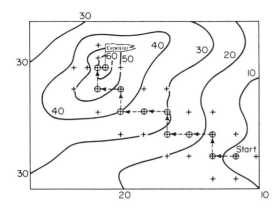

FIGURE 13–2

The method of steepest ascent; establishing values in the immediate area around a known position moving step by step up the steepest gradient to the top.

yield of a chemical process under different conditions of pressure and temperature in the form of a surface lying in a space of three dimensions (Fig. 13–1). Here the output or yield responds to varying pressure and temperature combinations, and we call the resulting wavelike figure a *response surface*. The exact configuration of such a surface is usually unknown because many thousands of experiments are needed to establish its form. To find the best combination of pressure and temperature giving us the greatest yield may involve very difficult problems of searching with little information under conditions of considerable uncertainty. Providing such surfaces only have a single peak, there are efficient mathematical search procedures that make use of common geographic

notions (Fig. 13–2). Any hill climber knows that a summit is reached most quickly by moving at right angles to the contour lines up the path of steepest ascent. In the same way, the searcher on a response surface can conduct a few experiments (+) immediately around a point of known value ($\oplus$), and then move in the direction of the steepest gradient. By moving short steps and sampling as he goes, he can gradually search his way to the peak—the point giving the highest yield of the process.

Many ingenious methods have been devised for searching mathematical spaces. When each minute of search must be paid for in time and money, very efficient methods of searching are clearly desirable. There may be similar occasions when geographic space must be searched as efficiently and effectively as possible. Enemy submarines and aircraft must be found quickly, and similar patterns of efficient search may be employed to find lifeboats lost at sea or hunters lost in the woods. Actual search patterns, whether those of animals or men, do not always cover the most ground in the shortest amount of time. While certain marine amphipods seem to use principles of steepest ascent with a natural facility beyond the grasp of many men (Figs. 13–3 and 13–4), other animals appear to use random patterns that are very wasteful because they cover the same ground again and again. Such random patterns also characterize the paths of bored students on a weekend as they search a town for some satisfying amusement.

Sometimes random and systematic search procedures are combined. An initial "random" hit may produce some interesting information, which then makes a closer and much more

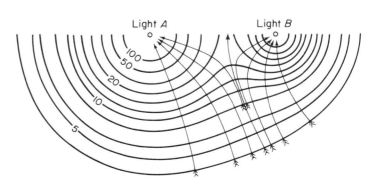

FIGURE 13–3

Search paths across lines of equal light intensity, by marine amphipods (adapted from Fraenkel and Gunn, Fig. 62, p. 139).

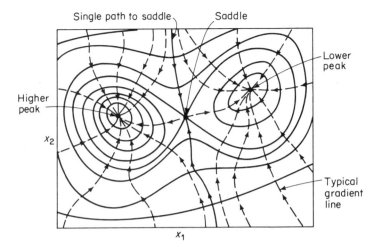

FIGURE 13–4

Search paths across a response surface by a mathematician (adapted from Wilde, Fig. 4–11, p. 120).

systematic search in the immediate area worth-while. Much of the exploration for oil and other minerals seems to be described by such combinations of random and systematic search. Indeed, the record of oil and gas drillings has been so poor that purely random methods of searching by sampling have been proposed instead of the present "guesstimating" procedures. Given the likelihood of oil fields of certain sizes being found, methods of random drilling may even be more effective than the so-called scientific ways used today.

Searching and Survival

In a probabilistic world in which searching and learning procedures are an integral part of our models, we must adapt and alter our own patterns of thinking. To become familiar with such probabilistic concepts, we shall use a very famous model designed by Herbert Simon. We have a simple space consisting of straight line paths joining an even lattice of sixty-four points (Fig. 13–5). A man moves along the paths in search of food—a single goal that lets him survive. As in the old fable of the grasshopper and the ant, we assume that our searcher is satisfied simply to get by from day to day without thought of tomorrow. His aspiration level is fixed. The piles of food are distributed at random on the even lattice of points, and providing the

man's search for food is successful he survives. Failure to find the caches of food means that in the long run he will starve. We can also assume the piles regenerate after an interval, but with no memory the man must always search for his supper.

Whether the man survives or not depends upon four things. The first two describe the properties of the environment in which the search takes place, while the others characterize the abilities of the searcher himself. In a very real sense, we can think of this problem as a simple Man-Environment system. We consider, first, those aspects of the environment which

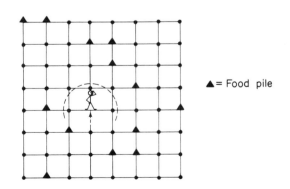

▲ = Food pile

FIGURE 13–5

A space of sixty-four lattice points with thirteen food piles to be searched by a man who can see one lattice point ahead.

will affect the probability of our searcher surviving.

The most obvious aspect of the environment affecting survival is the density of the food piles in the area. As an extreme case, every point on the lattice could have a food pile, and the problem of search would be reduced to a triviality. In the real world, we would say that such an environment is very bountiful, perhaps a lush tropical paradise where food can be easily gathered and picked off the trees. At the other extreme only a few piles of food would be present. We would characterize such an environment as harsh, possibly like the Australian desert where the round of life for the aborigine is one of constant search for food and water. In our particular example, let us suppose that thirteen of the sixty-four points have food, so that the proportion is 0.2031 or approximately 20 percent. We will give this quantity the symbol p.

The other thing in the environment that affects the chances of survival is the average number of paths diverging from each point on the lattice. In a very direct and spatial sense, the number of paths represents the number of alternative moves or decisions available to the searcher at any particular time. Notice that there are four cornerpoints with only two paths leading from each of them. Since our searcher arrives at a cornerpoint along one of two possible paths, he only has *one* other path leading from the point as an alternative (Fig. 13–6). At any other boundary point he arrives from one direction and has *two* alternative paths to take. He can either move along the outside boundary, or he may take the path leading in toward the center of the space. Thus four points on the lattice provide him with one alternative path to chose from, while twenty-four points provide him with two paths. The remaining thirty-six points are all interior ones, so that if he arrives at one of these he always is faced with *three* alternative paths diverging from his location. We can summarize the branching information (Table 13–1), noting that the space presents the searcher with a total of 160 alternatives. Thus on the average 160/64 or 2.5 alternatives are available at each move. We

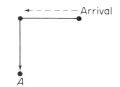

Arrival at a corner point
with only one alternative, A

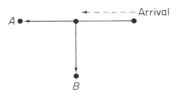

Arrival at a boundary point with two
alternatives available, A and B

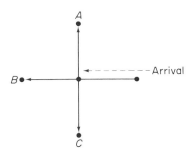

Arrival at an interior point with three
alternatives available, A, B, and C

FIGURE 13–6

The alternatives available to the searcher as he arrives at various points on the lattice.

TABLE 13–1

The Branching Decisions Available to the Searcher

Corner points	4 × 1 =	4
Boundary points	24 × 2 =	48
Interior points	36 × 3 =	108
Total		160
Average at each point	$\frac{160}{64}$ = 2.5	

will call this average number of paths diverging from a point d.

Two things affecting the probability of survival also characterize the searcher himself. One of these is the storage capacity of his body measured in terms of the number of moves he can make before he starves to death. The question of storage capacity faces all living things and seems to depend upon the ubiquity of the resources. A frog, whose natural habitat is wet and marshy ground, has a very small capacity to store water. On the other hand, a camel lives in an environment of very scarce water, and has developed a large storage capacity to deal with the problem of scarcity. Man also stores vital resources in varying amounts, depending upon how common they are. Food is sometimes very scarce, so we store food energy in the form of fat during good times. While water may sometimes be scarce, it is normally more common than food and our ability to store it declines accordingly. Air, the most ubiquitous vital resource, is hardly stored at all, and a few minutes without it results in death. We shall measure the storage capacity of the searcher in terms of the number of moves he can make before starving. This quantity will be given the symbol H.

The final element in the problem is the ability of the searcher to see ahead. If his perception is limited so that he can see only one lattice point away from his current location, then his ability to search is reduced and the probability of surviving may be lowered. On the other hand, his ability to survive may be increased by enlarging his knowledge of the local space by expanding the radius of his search horizon. We shall measure the searcher's ability to perceive the food dumps by the number of moves or lattice points he can see down the paths leading from his own location, symbolizing this number by v.

Let us summarize all the information we have about the problem (Table 13–2), and consider some actual numerical values for each of the variables. In our example p, the density of the food piles, is 0.2031, while the average number of paths diverging from the points d is 2.5. To keep the problem simple, we shall set v, the number of steps he can see ahead, to 1.0. Since each diverging path leads to a lattice

TABLE 13–2

Variables Characterizing the Man-Environment System

Variables	Symbol	Numerical Value
The Environment		
Richness of environment: proportion of branch points with food	p	$\frac{13}{64} = 0.2031$
Decision alternatives: average number of diverging paths	d	$\frac{160}{64} = 2.5$
The Searcher		
Storage capacity: number of moves without food before death	H	3.0
Ability to see ahead	v	1.0

point, it means that at any location there will be an average of d^1 or 2.5 points in view. Notice that if he can see on the average two lattice points ahead then d^2 (2.5 × 2.5) or 6.25 new points will come into view at each move. This is easier to see if we examine a simple tree or branching diagram (Fig. 13–7), and make the average number of branching points d a whole number, say two. If his radius of search is only

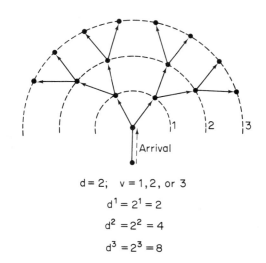

$$d = 2; \quad v = 1, 2, \text{ or } 3$$
$$d^1 = 2^1 = 2$$
$$d^2 = 2^2 = 4$$
$$d^3 = 2^3 = 8$$

FIGURE 13–7

The number of new points coming into view after a move depends upon the rate of branching d *and the radius of search* v *and equals* d^v.

one lattice point ahead, then after a move only d^1 or 2.0 new points will appear. If he can see two points ahead then d^2 or 4.0 new points will appear, while 8.0 or d^3 new points will come into view if he can see three steps ahead. In general terms, the searcher will always be able to bring d^v new points into view at every move. This means that if he makes m moves, then md^v new points will appear.

The searcher can make H moves without food before his survival in the Man-Environment System ends. To keep this particular example simple we have set H at 3.0, so that if he does not reach food within three consecutive moves he will starve to death. Notice that even if the searcher sees a food pile after making a move, he will need v more moves to reach it. This means that in our model we will require v to be smaller than H to keep it sensible.

We shall let the probability of survival be S, so that the probability of not surviving is $\$ = (1 - S)$. This, of course, is the same as the probability of having no food points in view in $(H - v)$ moves. In our particular example, $\$$ will be the probability of not seeing a food dump in $(3 - 1)$ or 2 moves.

Now suppose we let P_0 be the probability that none of the d^v new points that come into view after a move will have food. Then:

$$P_0 = (1 - p)^{d^v}$$

This expression simply says that the probability that none of the new d^v points coming into view after each move will have food is the proportion of empty food locations raised to the power of the number of points coming into view. Such a probability expression is easier to understand if we consider a very simple example of searching for food along a line (Fig. 13–8). Suppose we have ten points on the line, with three of them containing food, so that p is 0.30. In this one-dimensional space each point has only a single branch leading from it, so d is 1.0. We assume that the searcher can only see one point ahead so v is also 1.0. Thus the probability that a move will not bring a new food point into view is simply the proportion of empty points or $(1 - 0.30)$ or 0.7 This is exactly the same as:

$$P_0 = (1 - p)^{d^v}$$

or

$$P_0 = (1 - 0.3)^{1^1}$$
$$= 0.7$$

Notice in this very simple example of searching along one dimension that increasing the value of v does not bring any more *new* points into view after a particular move. Increasing the radius of perception means the searcher on the line can see more points at any one time, but a new step along the line still only brings a single new point into view. The chance that this particular new point will contain a food dump is 0.3, so the probability that it will not contain food is still $(1 - 0.3)$ or 0.7.

Returning to our original, two-dimensional example, the probability P_0 that no new points will contain food will be:

$$P_0 = (1 - 0.2031)^{2.5^1} = (0.7969)^{2.5} = 0.5671$$

Thus the probability of not seeing food after each new move is about fifty-seven out of one hundred, or a 57 percent chance. Remember that this is the probability after a single move, so if a searcher makes $(H - v)$ moves, then the probability of not seeing and reaching a food dump, or in other words the chance of not surviving, is:

$$\$ = (P_0)^{(H-v)}$$

In our example, this would be $(0.5671)^{(3-1)}$ or 0.3216. If the chance of not surviving $\$$ is only about one in three, then the probability of surviving S is only about $\frac{2}{3}$—not a very heartening piece of information for our searcher.

We can express the probability of not surviving in a different way by substituting $(1 - p)^{d^v}$ for the value of P_0 in the expression:

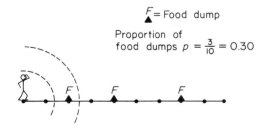

$\underset{\blacktriangle}{F} =$ Food dump

Proportion of
food dumps $p = \frac{3}{10} = 0.30$

FIGURE 13–8

Searching for food dumps in one dimension.

$$\mathcal{S} = (P_0)^{(H-v)}$$

This is the same as:

$$\mathcal{S} = [(1-p)^{d^v}]^{(H-v)}$$

Combining the exponents, the probability of not surviving is:

$$\mathcal{S} = (1-p)^{(H-v)d^v}$$

In our particular example, this is:

$$\mathcal{S} = (1-0.2031)^{(3-1)2.5^1}$$
$$= 0.3216$$

which is the same as before.

The final expression relates the probability of the searcher not surviving to all the elements in the problem—two of them describing the environment and two describing attributes of the searcher himself. Given the values we have chosen in our numerical example, our searcher does not stand much of a chance of surviving in such a poor environment with such limited abilities. But probability models are sometimes deceptive. Except for very simple situations we are not used to thinking in probabilistic ways, and our intuition can let us down very badly. To see how the situation can rapidly change, let us alter each of the variables in the problem one by one while holding the others constant.

We shall first double the storage capacity of the searcher H from three to six. Now our expression to obtain the probability of not surviving becomes:

$$\mathcal{S} = (1-0.2031)^{(6-1)2.5^1}$$
$$= 0.0585$$

By doubling the storage capacity, which now allows the searcher to move $(6-1)$ or 5 moves before starving, his chances of not surviving have been considerably reduced. Indeed, a searcher with such an ability to keep going has over five times the chance of surviving another man who has half the storage capacity. If we graph the probability of not surviving against the storage capacity (Fig. 13–9), we can see how rapidly these probabilities decline. With a storage capacity of just a single move, it is certain that the searcher will not survive. But by increasing the storage capacity to eight, there is only a very small chance that he will

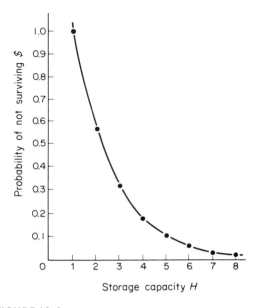

FIGURE 13–9

The probability of not surviving in the Man-Environment system $\mathcal{S}$ as a function of the storage capacity of the searcher H.

not survive even though the environment is still rather harsh and his ability to see ahead is limited.

In the same way, we can double the richness of the environment (p) while holding the three other variables constant. If there are 26 instead of 13 food piles in the space, p becomes 0.4062 and our expression for the probability of not surviving is:

$$\mathcal{S} = (1-0.4062)^{(3-1)2.5^1}$$
$$= 0.0738$$

Doubling the richness of the environment reduces the chances of not surviving over four times. Once again, we may graph the relationship between the probability of not surviving and the proportion of points with food dumps (Fig. 13–10). Obviously, when there is no food in the space the probability of not surviving is quite certain or 1.0. As the environment becomes richer and richer the probability declines rapidly so that if 0.6 or 60 percent of the points have food there is only about one chance in 100 that our searcher will not survive.

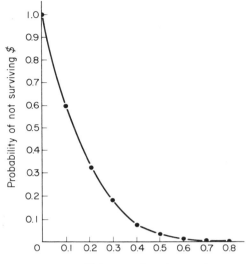

FIGURE 13–10

The probability of not surviving in the Man-Environment system $\mathcal{S}$ as a function of the richness of the environment (p).

The other variable set by the environment is the average number of branches from each point d equivalent to the number of alternative paths the searcher can take at any one time. If we double the average number from 2.5 to 5 our expression of the probability of survival becomes:

$$\mathcal{S} = (1 - 0.2031)^{(3-1)5.0^{1}}$$
$$= 0.1034$$

Thus doubling the number of alternative paths reduces the probability of not surviving about three times.

Finally, we can triple the ability of the searcher to see ahead (v), allowing him to see three steps down each of the paths from his location, and increase the storage capacity (H) from 3 to 4. Our probability of survival expression becomes:

$$\mathcal{S} = (1 - 0.2031)^{(4-3)2.5^{3}}$$
$$= 0.0288$$

With the two abilities of the searcher increasing together there is a dramatic drop in the probability of not surviving. Under these conditions a small increase in the storage capacity reduces the probability of not surviving in the system to about three chances in a hundred. This result is not intuitively obvious, and it makes the important point that our ability to think about probabilistic matters is very limited without the aid of a formal model. The outcomes of deterministic models are sometimes more obvious than are those of a probabilistic nature. This is especially true when many of the variables can change at the same time. Even in this little probability model it is impossible for us to work through in our heads the consequences of changes in the four variables describing the simple Man-Environment system. Thus the model stands as an extremely simplified and abstract statement between us and the real world, but by looking through the model our understanding of the overwhelmingly complex world around us is heightened and clarified.

Looking at the World through the Model of Search

If we look through a model of search at the world beyond, a number of apparently disparate things appear to have many basic elements in common. Our notion of environment need not be confined to a physical environment with paths and food dumps, but can be expanded to include the total social and cultural environment in which people find themselves. Many seem to be dropping out of the "system" these days, and it is worth considering, even in oversimplified terms, the reasons for this great waste of human resources measured in dissatisfied human lives.

Within our large metropolitan areas are deep pockets of chronic unemployment that are extremely difficult to deal with. Many thousands of men and women are out of work in our cities, and huge sums of money are spent every year in welfare subsidies to allow them, quite literally, to survive in the urban system. Although our model of search is very simple, we can use it to describe the unemployment problem in terms of the same four variables. To increase the survival rate of the unemployed the values of some of the variables in the system must be increased. Very little can be done about the storage capacity H, but the ability to see ahead v is clearly one variable that may be increased, presumably by expanding vocational education and increasing the flow of information through the urban employment system. There may also be ways of increasing the number of alternatives d available to the unemployed, although the *perception* of the alternatives may itself depend upon educational levels. Finally, we may increase the richness of the environment p, although this can be done in two, quite distinct ways. In the short run, and assuming moral values in the society that forbid people starving to death, we can parachute emergency food piles to locations just before a person collapses. Although such values may be assumed to exist today, in earlier times starvation in the streets was a common sight. Even today many children in the United States are chronically undernourished, and we have commented before upon malnutrition with such tragic and far-reaching consequences. In the long run, however, increasing job opportunities may be the easiest way of increasing p. Providing everyone with the dignity of a job may have many other beneficial effects besides simple survival. Ideally, of course, all variables in the problem should be increased over mere survival level—for we are talking about the lives of men, women, and children, not electronic automata recharging their batteries by plugging into an electric lattice.

If we increase our scale of observation to the regional level, we can look at depressed areas in Appalachia through the same model. Many jobless in Appalachia, for example miners and their families in areas where the coal seams have run out, may be well aware that their storage capacities are very limited as they think about searching for new employment opportunities. With no savings, the ability of a poverty-stricken man to keep going in a new urban setting while he searches for a job is very limited. He is far more likely to stay at home than to risk the uncertainty of a strange city. Moreover, his information about opportunities may be so limited that he is like our searcher who can only see a few moves ahead. For the poorly educated

person who has grown up in an area of deep depression, the average number of alternatives d also may appear to be extremely limited. Thus even if the environment is very rich in opportunities, his low storage and poor perception may ensure that he has little chance of surviving the system without parachute drops.

We can also think about locational problems for individual firms within a probabilistic framework of search. The problem of finding a new location presents far fewer dangerous possibilities for a large industry than for a smaller firm. A large firm trying to locate a branch plant has a very large storage capacity or supply of capital behind it. It may keep looking for a long time, and its chances of surviving in the system are extremely high. With research and development divisions attached to it, it can see many moves ahead, and the number of alternatives facing it at each point in the economic system is very large. Even if the environment is not very rich, the large firm will usually find a good location enabling it to use the meager opportunities available. In contrast, the small firm does not have a large amount of capital behind it, and we may think of its storage capacity as relatively small. Its information may also be much less than that available to its larger counterpart, and it may not see as many ramifications stemming from a locational decision. In very rich environments small firms can survive quite easily, but as commercial environments become more marginal the smaller and poorly managed firms are sifted out. We only have to think of the number of abandoned gas stations and closed restaurants around to see how often locations are tested by these small operations in an economic environment which has a generally high but spatially uneven demand for gasoline and prepared meals. In poor environments survival rates of small firms are even lower, so that such firms are often pushed out of the system by the more able searchers. Corner grocery stores are barely surviving, and many fail in the face of larger competitors and changing conditions. In the same way that the decay of a radioactive material can be measured in terms of a half-life, so economic functions decay in a probabilistic

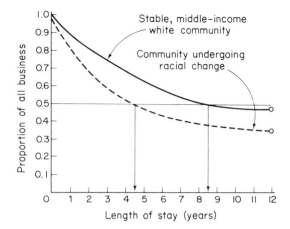

FIGURE 13–11

Decay curves for small grocery stores in two Chicago communities (adapted from Berry, Parsons, and Platt, Fig. 3.17, p. 123).

the grain." In terms of our model of search, such directional bias gives us two values of v. The variable takes on a high value whenever the searcher looks toward the center of the city or turns around and looks toward the suburbs. In between, to his right and left as he faces the city center, the value of v becomes smaller. This means that good housing opportunities could be available in many parts of the city, but the searcher would not be able to take advantage of them because of his perceptual bias. The question of the perception and evaluation of opportunities in geographic space is an important one to which we shall return.

fashion (Fig. 13–11). Unlike radioactive materials, however, the decay rates and half-lives of economic functions are influenced by the surrounding environment. Small grocery stores in one of Chicago's stable, middle-income areas had a half-life of about 8.5 years. Those in a neighborhood undergoing racial change had a half-life of only 4.5 years. Failure in the first two years was particularly likely, as small stores with few resources to sustain them over the long haul foundered and sank.

Finally, we can look through the "probabilistic lens" of our model and think of the problem of searching for housing in an urban environment. When people dissatisfied with their present housing at a particular location look for alternative dwellings, they do not search in a random fashion. When we plot many housing changes within a city, we can see a strong directional bias to the movements. People tend to search for new housing opportunities along an axis joining their present location to the center of the city (Fig. 13–12). If we record many housing moves in a city, and measure the angles to the downtown area, a clear directional bias toward or away from the center appears. Thus the probability of making a move along such an axis is much greater than of moving "across

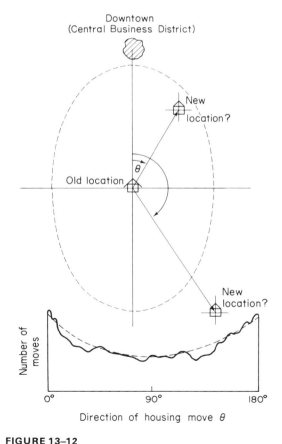

FIGURE 13–12

Directional bias in the search for urban housing along the downtown axis (adapted from Adams).

Other Ways of Searching Space

Search patterns over geographic space may assume different forms. Where systematic patterns of opportunities do not exist or have not been perceived, random paths of search may be effective. At other times it may be possible to sample individual points in the space in a random fashion. Random patterns of search at points form an unbiased sampling of the particular environment in which the searcher is interested. But there are other space-searching occasions under more difficult conditions, when the sampling of random points may be either impossible or ineffective. It is not always possible to sample space at points randomly chosen, although helicopters now increase the number of situations in which such patterns of information-gathering can be used.

Roads and similar channels of movement form some of the most common patterns we see on a map of the human landscape. In many underdeveloped countries they represent the more successful search paths from major ports on the coast to interior areas of economic potential. Most underdeveloped areas rely heavily upon the export of a few agricultural and mineral products. Like rivers draining their own catchment areas, road networks in an underdeveloped country indicate the way in which the export produce is drained to the ports with their distributary lines to the rest of the world. We shall term the process leading to such "space-draining" search as *labyrinthian*, for like Theseus unwinding the thread through the labyrinth of the Minotaur, roads must always connect commodities and people back to the starting points.

A very simplified description of transportation development in an underdeveloped economy can be cast within a framework of labyrinthian search. Imagine a country (Fig. 13–13), with a single port along its coastline. In precolonial times trading is at a low level, and except for a few, very highly valued commodities, most trading takes place over short distances. No trade takes place with a larger world economy. A colonial power wishes to search the space and tries to find areas that have economic potential for the metropolitan

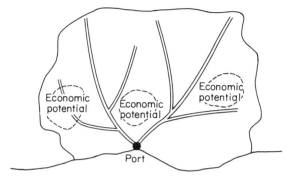

FIGURE 13–13

Road penetration lines searching space in the early years of development.

economy. For political or military reasons the first roads are usually built as lines of direct penetration into the interior. Railways also serve as major penetration lines, but these facilities always involve much larger sums of money and lack the flexibility of the roads. In the absence of much information about the economic potential of the areas, mistakes in the railway building business can prove to be very costly in later years. In West Africa, for example, some of the original lines built in the nineteenth century have been termed "the seven pillars of unequal wisdom."

Roads, on the other hand, are much more flexible instruments of continuous labyrinthian search. The capital investment for a road is usually much less than that for a railway, and roads have the great advantage that they can be built up in a number of quality stages. When a road is first put through an area, where the economic potential is difficult to judge, it may be built at dirt or laterite standards. Later on additional capital can be used to upgrade the road to gravel, and finally to tarred standards. Thus we can think of roads as the bloodhounds of the space economy, sniffing out, in a very flexible fashion at minimum capital cost, the economic potential of an area.

Typically road densities will not be evenly distributed throughout a country. Rather, the penetration lines and the links forming bridges over barren areas lying between productive

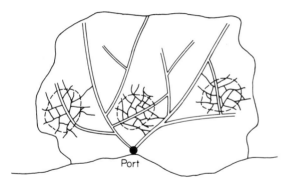

FIGURE 13–14

Feeder roads exploding to drain areas of economic potential discovered by labyrinthian search at an earlier time.

regions will form the main trunk routes, and superimposed upon these will be very dense areas of small feeder roads. These often develop in an explosive fashion when an area of economic potential has been tapped (Fig. 13–14).

Labyrinthian search, producing local space-organizing and national space-draining patterns, is obviously a result of a sequence of decisions by individual administrators. Implicit in such decisions, and in the patterns that result from them, is a process of learning that is shaped by strong flows of information that are fed back from the searching roads themselves. In the early years a road may be built at very low standards through a part of the country with very little economic potential. As a result

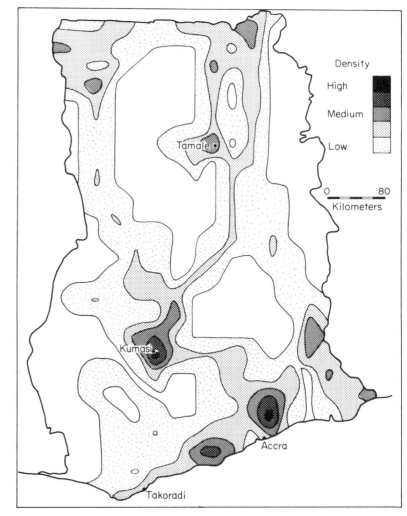

FIGURE 13–15

Road densities in Ghana in 1922: first stages in the development of the "latent image" (adapted from Taaffe et al., Fig. 1).

very little traffic develops, and such negative information or feedback will lead an administrator to decide upon the abandonment of the route, or its continued maintenance at very low standards if it is only required to give access for political reasons. But another road, also built initially at low standards, may pass close to an area of economic potential. Slowly, a modest flow of traffic develops. The traffic represents positive information and feedback so that the decision-makers learn about the potentially favorable locations and change their decisions accordingly. From the initial line of search feeder roads are driven through the area of potential, and these in turn sprout smaller paths and tracks to drain and organize the area. As a result, traffic swells on the trunk routes, and eventually these are raised by incremental steps to gravel and tarred standards. Original patterns of settlement tend to be warped into conformity with the changing surface of accessibility as people decide to move to the main routes. New central places start to grow. As flows of information increase with the generation of commodity movements and human traffic, the patterns of one time are reinforced

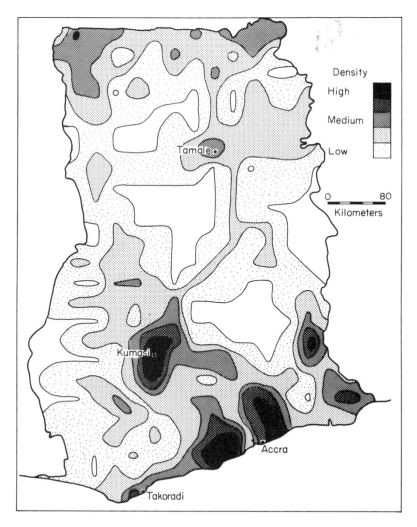

FIGURE 13–16

Road densities in Ghana in 1937: the space economy "develops" from the 1922 image (adapted from Taaffe et al., Fig. 1).

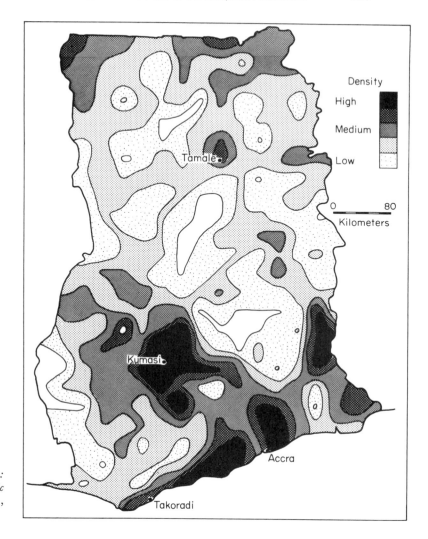

FIGURE 13–17

Road densities in Ghana in 1958: further development of the "geographic plate" (adapted from Taaffe et al., Fig. 1).

and accentuated to produce the patterns of the next. Thus many economies appear to develop their spatial structures rather like a developing photographic plate as the process of search and learning takes place through time.

The constant reinforcement of spatial success experiences can be clearly seen in many sequences of historical maps in both the developed and underdeveloped worlds. In Ghana, for example, road densities were generally low in 1922, although a few lines of penetration had already tapped areas of economic potential in which feeder roads exploded to produce nodes

of high density (Fig. 13–15). Fifteen years later, the process of labyrinthian search has been extended, and by 1937 many of the inaccessible areas have been reduced, while the original nodes of high density have expanded and coalesced (Fig. 13–16). Two decades later, in 1958, only a few pockets of complete inaccessibility remain (Fig. 13–17), while the original nodes of high density have expanded greatly as road building decisions reinforced earlier information that areas of high economic potential had been hit. Like the development of a photographic plate or a fade-in on a television program, we can think of the latent image of Ghana's space economy in 1958 existing in the simpler patterns nearly four decades previously. The process of search and learning has accentuated and developed the 1922 "plate" to produce the more intense patterns of road density four decades later.

SEARCHING AND LEARNING

In most processes of spatial search some learning takes place. Information constantly impinges upon the decision-maker, reinforcing some patterns of behavior at the expense of others As a result, spatial decisions and movements that appear to have large random components in the initial periods of trial and error often appear to settle down into more predictable and stereotyped patterns later on.

"Welcome Wagon" geography is an example of individual spatial decisions being strongly affected by learning processes. When a family moves to a new town they know very little about the shopping opportunities in the area. As a result, many local stores in the town will contribute to a Welcome Wagon in the hope that free information, reinforced by free gifts and samples, will produce patterns of desired shopping behavior. Consider the case of Mrs. Newarrival who had just moved into her new home and is faced with the problem of buying groceries from one of three supermarkets, Backme, Shopnow, and Cumon (Fig. 13–18).

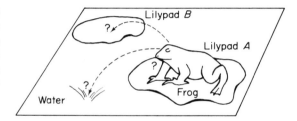

FIGURE 13–19

Frog in the lilypad-water system.

Mrs. Newarrival shopping between three supermarkets is like a frog jumping from one lilypad to another (Fig. 13–19). Each pad is a state in the lilypad-water system, and we can specify the position of the hopping frog at any one time with only a certain degree of probability. The probabilities of the frog hopping from one leaf to another, or from one state in the system to another, can be described in the form of a probability matrix (Fig. 13–20). If the frog is sitting on Lilypad A at time t_0, then at the next time period t_1 there is a 0.50 or 50 percent chance that he will still be on A, a 30 percent chance that he will have hopped to B, and a 20 percent chance that he will be in the water. In the same way, we can describe the probability of his going from the other states in the system, Lilypad B or the water, to others depending upon the state he is currently in.

Suppose at time t_0 he is cooling himself in the water. We describe his position with a *state vector:*

$$\begin{array}{ccc} \text{Lilypad } A & \text{Lilypad } B & \text{Water} \\ (\quad 0 & 0 & 1 \quad) \end{array}$$

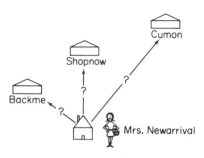

FIGURE 13–18

Mrs. Newarrival trying to decide between Backme, Shopnow, and Cumon.

		States at the next time period		
		Lilypad A	Lilypad B	Water
States at the first time period	Lilypad A	0.50	0.30	0.20
	Lilypad B	0.40	0.40	0.20
	Water	0.30	0.30	0.40

FIGURE 13–20

Transition matrix for the frog in the lilypad-water system.

where the 1 indicates that he is in the water state at time t_0. We now have all the information we require for a very simple probability model. The probability or *transition* matrix is like a small engine driving our frog with a certain probability from one state of the lilypad system to another. For example, if we multiply the state vector once by the probability matrix, we can find the probability of his being in any of the three states at time t_1:

$$(0 \quad \overset{t_0}{0} \quad 1) \cdot \begin{pmatrix} .5 & .3 & .2 \\ .4 & .4 & .2 \\ .3 & .3 & .4 \end{pmatrix}$$

$$= (.30 \quad \overset{t_1}{.30} \quad .40) \cdot \begin{pmatrix} .5 & .3 & .2 \\ .4 & .4 & .2 \\ .3 & .3 & .4 \end{pmatrix}$$

$$= (.39 \quad \overset{t_2}{.33} \quad .28)$$

The new state vector in this case will simply be the last row of the probability matrix. At time t_1 there is a 40 percent chance he will still be in the water, while there are chances of 30 percent that he may have climbed onto either of the two lilypads. If we continue the multiplication process, the probabilities within the state vector shift. Even by time t_2 there is the greatest chance that the frog will be sunning himself upon Lilypad A. Continuing the multiplication process for a few more steps, we can see how the probabilities eventually converge upon this final state vector:

$$\begin{matrix} \text{Pad } A & \text{Pad } B & \text{Water} \\ (\ 0.42 & 0.33 & 0.25 \ ) \end{matrix}$$

Thus, after some time has passed, we may expect to find the frog in the water a quarter of the time, on his favorite pad A 42 percent of the time, while he takes a change of scenery on pad B one-third of the time.

The shopping behavior of Mrs. Newarrival can be described in the same probabilistic terms as we used for the location of the frog, although she may not like the thought of being driven around town on shopping trips by a transition matrix. The first shopping trip of Mrs. Newarrival is to the Shopnow Supermarket, so the state vector at time t_0 is:

$$\begin{matrix} \text{Backme} & \text{Shopnow} & \text{Cumon} \\ (\quad 0 & 1 & 0 \quad) \end{matrix}$$

Suppose her transition from one shopping center to another is described by a probability or transition matrix:

$$\begin{matrix} & & \text{Transition} \\ \text{State Vector } t_0 & & \text{Matrix} \\ (0 \quad 1 \quad 0) & \cdot & \begin{pmatrix} .72 & .20 & .08 \\ .49 & .35 & .16 \\ .47 & .29 & .24 \end{pmatrix} \end{matrix}$$

$$\begin{matrix} \text{State Vector } t_1 \\ = (.49 \quad .35 \quad .16) \end{matrix}$$

Then at time t_1 there is roughly a fifty-fifty chance she will buy her groceries at Backme, a 35 percent chance of still patronizing the Shopnow, but less than one chance in five she will buy at the Cumon store. We might expect such a state vector during an early stage of her search for good shopping opportunities, for while she has some knowledge about the Shopnow products, the Backme store is closer to her home. For the same reason, the probability of her shopping at the Cumon store is quite low because it is the farthest away. If we continue to multiply Mrs. Newarrival's state vector with a transition matrix, it will eventually converge upon the values:

$$\begin{matrix} \text{Backme} & \text{Shopnow} & \text{Cumon} \\ (\ .633 & .248 & .119 \ ) \end{matrix}$$

Our very simple model indicates that Mrs. Newarrival will eventually learn to choose the Backme store about two-thirds of the time for her grocery purchases, using Shopnow about a quarter of the time and the Cumon store for the remaining 12 percent. If she kept a daily record of her shopping trips, and made a list of her visits after she had been in the area for a month or so, she would see how her behavior eventually settled down into a quite stereotyped and reasonably predictable pattern.

Searching Behavior	Learning Behavior	Stereotyped Behavior

SSBBSCBCBSSBCCBSSBBSBCBSSBBSBBBBSCBBBSBBBSB

The model is extremely simple as it stands because it cannot take into account the fact that the transition probabilities may themselves change through time. When we examine the real world of shopping behavior, it is obvious that the probabilities are not stable but are changing with the experiences of success and failure on the part of a learning and searching person. If Mrs. Newarrival had an unfortunate experience at Shopnow, then the probability of going back there would decline while the chances of patronizing Backme and Cumon would increase correspondingly. After a while the negative experience she had at Shopnow might fade, so that the probability of returning to the store would increase. At the same time, of course, the managers of the Backme and Cumon stores would be doing their best to retain her patronage, perhaps reinforcing her spatial behavior and growing her brand loyalty with economic fertilizer like stamps and other "give-away" programs whose real cost is hidden in the price of the food. Such learning and reinforcement processes would probably make her shopping behavior far more stereotyped and consistent than that produced by our simple model. Other theoretical models can be developed to take into account changing transition probabilities over time, but these are considerably more complex.

Learning under Conditions of Reinforcement

Searching and learning go hand in hand, and the learning process is often reinforced by new information and experience gained from past behavior. We can think of the decisions taken during a period of search as trials—which may or may not be successful. To keep the problem simple, we shall only consider situations where two alternatives are available at any one time. We can represent such choices in the form of a T-shaped maze (Fig. 13–21), where a turn to the right means failure and no reward. We assume that a reward will reinforce the decision to turn left, making it more likely that it will be repeated the next time. We might imagine, for example, a farmer thinking about two markets for the sale of his produce (Fig.

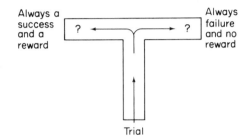

FIGURE 13–21

The T-maze with behavior rewarded for turning to the left and punished for turning right.

13–22). In the beginning he may have hardly any knowledge about the merits of the two markets and the prices they will give him for his goods. At first he will search between the two, and if one market consistently gives him higher prices we would expect his behavior to become reinforced and stereotyped very quickly:

Searching	Learning	Stereotyped

BAABBBAAABABABABAAAAAAAAAAAAAAAA

Conversely, in a situation where the prices fluctuate the markets will only partially reinforce his behavior at any particular trial. Notice how close the decision problem comes to a game with two strategies. Faced with two mar-

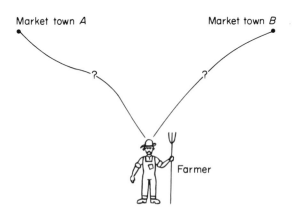

FIGURE 13–22

The farmer faced with two market towns in which he can sell goods.

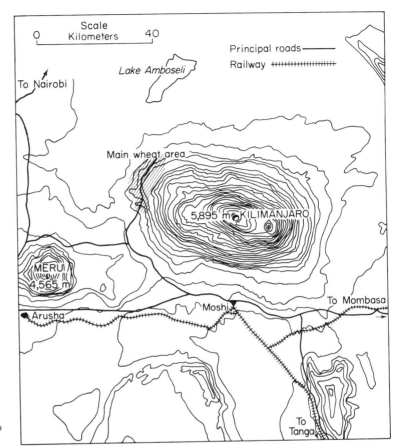

FIGURE 13–23

Main wheat growing areas on Kilimanjaro (adapted from Gould, 1965, p. 157).

kets, the farmer has the task of organizing and evaluating his past experience of market prices, trying to assess the chance of greatest reward, using the probability value to weight his payoff. Under these conditions his behavior is only reinforced a part of the time:

Searching	Learning	Stereotyped

BBABAAABBABAABAAABAAAAABAAAA

Learning models with only partial reinforcement may also describe the way in which men make decisions about adopting new innovations. For example, European wheat farmers on the northwest slopes of Mt. Kilimanjaro started wheat farming during the Second World War (Fig. 13–23). Unfortunately, new mutations of wheat rust appeared from time to time and com-

pletely ruined whole fields of wheat just before they were to be harvested. A plant breeding program in Kenya constantly developed new varieties of wheat that were resistant to the rust mutations, but in the early years the farmers were very conservative and would not accept the newly developed varieties.

In terms of our simple learning model, farmers had the choice at the beginning of each season to accept or reject the advice of the plant breeders (Fig. 13–24). In some years new rust mutations did not appear, and those who had rejected the earlier advice of the plant breeding program happened to be lucky. In other years, however, farmers who rejected the advice paid heavily for their decisions, particularly when their yields were compared to others

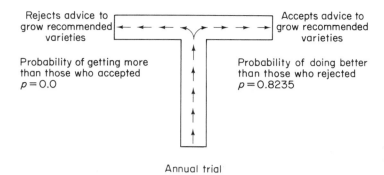

FIGURE 13-24

Simple learning model with partial reinforcement: the wheat farmer on Kilimanjaro and the plant breeder (adapted from Gould, 1965, p. 161).

who had accepted the advice to grow the new rust resistant varieties. Thus, over the years farmers who accepted the advice had their behavior reinforced in a positive fashion, while those who were stubborn and conservative gradually had to admit that they were making the wrong decision. This is not an uncommon situation in many walks of life.

Formal statements of learning models with partial reinforcement are difficult, but their graphical expression is quite straightforward (Fig. 13-25). On Mt. Kilimanjaro only a very small proportion of the wheat grown in 1943 was in recommended varieties. By 1947 a mutation known as K8 was identified on the wheat, and the "learning curve" began to rise

quickly over the next six or seven years as other wheat rusts attacked the standing grain. Ten years later, in 1957, nearly all the wheat fields were grown in recommended varieties, and today no farmer would dream of rejecting the advice of the plant breeding station. The behavior of the farmers after twenty years, or twenty trials of partial reinforcement, is now completely predictable and stereotyped. It is also worth noting the more than passing similarity of the learning curve to the previous curves describing innovation diffusion rates. Searching among alternatives and weighing decisions in the light of past experience and new information are very common aspects of human behavior whose outcomes can frequently be seen in patterns of land use and the movement of goods and people.

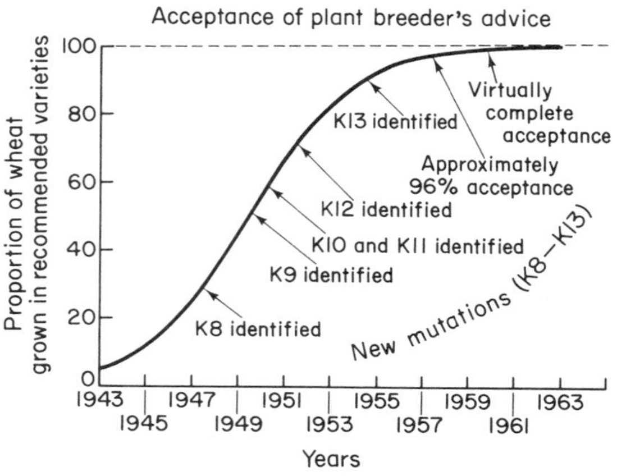

Learning to Cooperate:
Dogs in Spatial Mangers

Sometimes the outcome of a decision affects only the person who made it. Its consequences neither reach beyond the person nor is it influenced by the decisions of others. In reality such isolated decisions are comparatively rare, for usually the outcome of one man's decision is greatly affected by others made by people around him. Conversely, his own decisions may have far-reaching effects upon the outcomes of others. People can work together and often achieve more by cooperation than by working individually in competition and at cross purposes. The notion is still around that extreme economic competition maximizes the individual's gain, and so maximizes the returns to the

FIGURE 13-25

The learning curve from the model of partial reinforcement (adapted from Gould, 1965, p. 162).

society of which the individual is a part—an archaic nineteenth-century legacy, whose disastrous consequences we can see all around us on the landscape.

That is, if we can see the landscape at all. Belching smokestacks putrify the atmosphere so that many children today grow up in an atmospheric sewer. Death rates in cities correlate highly with smog, though the smog-makers are never prosecuted for homicide. The factory owner who decides to pour untreated sewage into a river may minimize his own costs, but the next man downstream shoulders the financial burden, treats the water, and in turn spills his filth back into the stream. Society—not an inanimate *thing*, but men, women, and children—pays the ultimate price in the sight and stench of garbage swills that used to be called rivers. Oil slicks make vast stretches of our coasts uninhabitable, and individuals seeking to maximize profits destroy their own grazing lands and then pressure the representatives of society to allow soil-destroying stock densities on federal lands. Any highway littered with trash—each piece representing a decision by an individual—any common scenic heritage obscured by the individual, profit-maximizing billboard, testifies that we are a nation of decision-making slobs.

That cooperation can lead to larger gains than tooth-and-claw competition may have to be learnt. Bargaining and compromise can intervene in games that men play against one another, for men are not inanimate and impersonal forces of nature, but beings with the power of evaluation and rational decision-making. Capable of seeing beyond their own selfish noses, and prepared to think about the overall consequences of cooperative decisions, men are able to join together for the common good—and when they do not do so, society is increasingly willing to force the cooperative decision upon them. Automobile safety standards long overdue; still inadequate atmospheric, sewage, and heat pollution control; tightening standards of atomic waste disposal and "black lung" compensation paid to coal miners—all testify to changing attitudes and the growing willingness of society to control the selfish decisions of some that affect others.

Slowly, with painful slowness, we are learning to cooperate. To help us think about making cooperative decisions we are going to model a simple game which allows two players to compete against or cooperate with each other. In the same way that our model of search highlighted many similarities between apparently disparate problems, so our learning-to-cooperate game will be a lens through which we can view a number of spatial problems.

Suppose we have two trucking firms, Acme and Bolt, who are both faced with two choices of routes for delivering their goods (Fig. 13–26). Each firm can use a very winding route that takes 100 minutes to cover. An alternative, which they both share, is a short and much quicker route. Unfortunately it is only one lane wide so that trucks cannot pass one another. One firm must wait until the other is through before he can drive over the narrow link to his own destination. Time is money: for every minute spent on driving, profits decline by one dollar. The value of a truckload at the start is $200. If a firm uses the one-way road, it only takes 50 minutes to get to the destination, and 20 of these are spent on the one-way portion. Thus the profits are (200 − 50) or $150 per trip.

If a firm uses the one-way road it means the

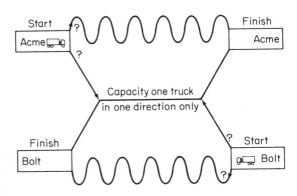

FIGURE 13–26

The alternative routes available to Acme and Bolt (adapted from M. Deutsch and R. M. Krauss, "The Effect of Threat upon Interpersonal Bargaining," Journal of Abnormal and Social Psychology, LCI: 2 [1960], Fig. 1, p. 183. Copyright © 1960 by the American Psychological Association and reproduced by permission).

other firm will have to wait patiently and let him pass. Thus waiting for 20 minutes while the other truck comes through costs him $20. Under conditions of cooperation *total* profit for the two firms will be:

Firm using one-way
  road first $200 − $50 = $150
Firm waiting for
  the other to pass $200 − $50 − $20 = $130
  Total $280

The most rational behavior would be for the two firms to cooperate, making use of the one-way road in turn so that they alternately get $150 and $130 for their deliveries. Unfortunately, men are not always rational. They may be quite aggressive entrepreneurs who see the other man only as a rival to be done down. Their personalities may be such that waiting for the other person represents a loss of face, and communication between them may be so poor or nonexistent that any discussion and bargaining that could lead to cooperative decisions may be impossible. They may even enjoy confrontations, no matter what the price.

Let us see how the game might be played. At the start, both trucks try to minimize their over-the-road time and make a dash for the one lane road. Meeting in the middle, they sit and glare at each other while precious minutes and profits melt away. Perhaps the least aggressive, say Acme, finally decides to back up to let Bolt through. Bolt has wasted time in trying to stare down Acme and only makes (200 − 100) or $100, while Acme makes $20 less, or $80. Acme has lost not only face but additional money as well.

The next time the game is played they again meet in the middle, but this time Acme does not care. He may be losing money but at least Bolt will lose just as much! Bolt thinks he will make Acme back up again, but the minutes tick away and all of Bolt's threats and bad language have no effect. They sit it out, and their profits finally melt away to nothing. In fact, they may only wake up to this state of affairs too late, and actually get negative profits as they find their way back to the starting point.

The third time Acme may have had enough

and he takes the long way around, while Bolt roars through the one-way road unimpeded. "But why," thinks Acme, "should that guy get all of it his own way?" The fourth time Acme roars into the one-way lane and jams on his brakes just in time to avoid Bolt coming the other way. This time Bolt decides that he will back up, but Acme also has second thoughts as he remembers the disastrous time when he made negative profits and he too begins to back up. Seeing each other backing away, aggression becomes the better part of valor, and both start forward again. The minutes tick away, the ledger turns from black to red. If Bolt/Acme is going to play it that way—OK! No matter what the cost, Acme/Bolt is not going to budge. Costs soar and profits are even lower than before as they forget the initial purpose of their trips in the all-consuming rivalry to control the crucial link between their origins and destinations.

On the fifth trip Bolt begins to come to his senses. Meeting Acme in the middle, he shoves and pushes for a while, but finally backs out and lets Acme through. The time taken in these aggressive gestures is still quite large, say 120 minutes, so profits are only (200 − 120) or $80 for Acme, while they are $60 for Bolt. Both realize they could do better. On the next trip Acme makes a gesture to Bolt, waits for him to come through, and then continues on his own journey. Bolt thinks Acme may not be such a bad guy after all, and lets him through first the next time. Gradually, with a few aggressive attempts once in a while, which are always "punished" the next time, the two competitors settle down on the most rational, cooperative course.

Locational Decisions as Compromise Resolutions to Conflicts

Conflicts over locations are very common. Many of the locations chosen for shops, industries, and roads are the result of compromises arising from conflicting views. Highway boards know that no matter where a new road is put it will arouse some opposition. Often the final routes are compromise solutions between the least cost routes in engineering terms and the high cost routes that would wind around every object in the way. We can consider such locational compromises in game theoretic terms, but unlike the zero-sum games we used before, our model now will be a special game that has come to be known as Prisoner's Dilemma.

Imagine two men, Tom and Mike, who have been arrested for a serious crime they did, in fact, commit. Unfortunately, there is not enough evidence to convict them for the crime, though they can be convicted on a lesser charge. They sit awaiting their trial in separate cells with no way of communicating with each other. If they cooperate with one another and do not confess, they will serve three months in jail each:

		Mike			
		Confess		Not Confess	
		Tom	Mike	Tom	Mike
Tom	Confess	18 months	18 months	Free	36 months
	Not Confess	Tom	Mike	Tom	Mike
		36 months	Free	3 months	3 months

If both confess they will serve much longer sentences of 18 months. A confession from one who turns state's evidence will let him go free, but the other will pay a very severe penalty of 36 months.

Given the structure of the payoff matrix, what should the prisoners do? Each one must be absolutely sure that the other will cooperate. But can they be sure? If Tom confesses he goes free or serves 18 months. If he does not he may only serve three, but he could end up being the sucker with 36 months while Mike goes free. It really might be better not to trust that guy and confess. After all, maybe *he* will go free. The only trouble is that Mike, sitting alone in his cell and thinking about Tom's reliability, comes to the same conclusion—and both end up with 18 months and much worse off than if they had cooperated with one another.

In general terms, the Prisoner's Dilemma game can be structured in the following way:

		Player 2	
		Cooperate	Defect
Player 1	Cooperate	R, R	S, T
	Defect	T, S	P, P

If both players cooperate they are both rewarded R, while double defection results in punishment P. If one cooperates while the other defects then the cooperative player is the sucker S. The defector in this case gets the temptation reward T. If the game is played again and again, a period of stable cooperation usually follows initial periods of great stress and distrust caused by occasional defections and counterdefections. Cooperation follows conflict.

Let us use the game as a model in which many locational decisions can be viewed. When a new highway is built, particularly through a densely populated area in a city undergoing urban renewal, many groups come into conflict over its location. This was not always the case. At one time highway engineers ran roughshod over the views of other groups, with the result that some cities now seem to consist of cloverleaves and overpasses. Cars, after all, are more important than people. Some groups in the city were martyred as they were forced to "cooperate," receiving the sucker's payoff S, while other groups got the large temptation reward T. Today groups who were formerly martyred are less likely to have the cooperative strategy forced upon them, and many initial locational choices are now challenged in the courts. As groups hurt by new highways or so-called urban renewal projects fight back and choose to defect, they gain the P reward which is better than the former sucker payoff S. Only cooperation can lead to the best of all worlds R and away from the defector trap.

As an example, consider the city of Southernville consisting of Ourtown on this side and Theirtown on the other side of the abandoned railway tracks that run through the center of town. The track could be easily purchased as a straight and level right of way for a new interstate highway (Fig. 13–27). In Ourtown, on the "right" side of the tracks, the central business district (CBD) is compact and neat. In Theirtown the CBD is strung out along a main road, and in the fastidious eyes of a number of power groups it forms an untidy strip. Thus it is easy to kill two problems with one highway. The power groups choose the defect strategy, for the other players have no power and must use their cooperative play. Let them take the sucker's

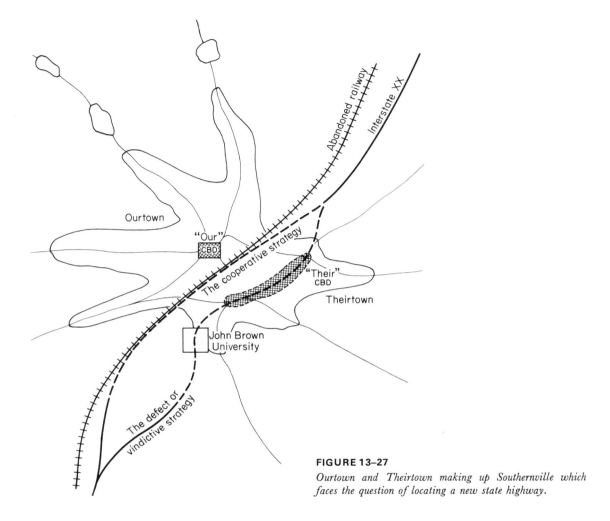

FIGURE 13–27

Ourtown and Theirtown making up Southernville which faces the question of locating a new state highway.

payoff *S* of miserably low compensation and the total destruction of their CBD. Moreover, because the uppity faculty at John Brown U thought about choosing the defect strategy, and even delayed the construction of the highway in the courts, slice the campus in half with an interstate highway and teach them a lesson. The works of man may, indeed, be seen upon the landscape.

As a final model of locational conflict, let us expand the Prisoner's Dilemma game to allow each of the players a new play called the *sanctuary* strategy. It lies between the defect and cooperate strategies—an appropriate position, for it

forms a halfway house between the attitudes of martyrdom and cooperative trust. The sanctuary position means that the weak player in a game may not be able to defect on the other player, but he can throw a wrench in the works and force a stalemate. For this reason the row and column payoffs are zero:

		Player 2		
		Cooperate	*Sanctuary*	*Defect*
	Cooperate	R, R	0	S, T
Player 1	*Sanctuary*	0	0	0
	Defect	T, S	0	P, P

With the sanctuary position available, a weak player can always stalemate a conflict when he is fed up with playing the role of the martyr, even though he may not be strong enough to hit back with the defect strategy. Communication and bargaining are needed to get him out of the sanctuary, which results in a zero payoff to all the players.

Many cities face the problem of finishing the federal highway program by driving the last remaining links through their centers. But unlike the earlier years, the last stages are arousing great opposition as more and more people question the value of such noisy, air-polluting, and people-removing projects. Expressways often seem to bring more problems than they solve for the people of the inner city, although truckers and suburbanites benefit. Some cities have files of abandoned plans for crosstown highways as new groups with new interests have challenged the older interests of traditional power groups and special lobbies. Locational choices today are made increasingly from compromise and cooperation between conflicting groups.

Information, Perception, and the Spatial Decision

From our discussion of searching, learning, and conflict resolution, it is clear that decisions and their spatial expressions will be influenced strongly by the perceptions of the environment and the information available to the men who make them. The environment in which a decision is made is not derived simply from elements of the physical world, but includes the works of man as well as the cultural, social, and political settings which are often the strongest environmental influences of all. The effects of the different parts of the environment upon the decisions of men cannot always be neatly partitioned into distinct pieces. Many decisions are influenced by both physical and cultural environmental factors, as well as by distinct personality traits and other particular characteristics of the decision-maker.

Whether considered in a normative or descriptive framework, the attitudes of men toward risk influence their decision-making and spatial behavior in games both against nature and against other men. Their very perception of risk and their evaluation of hazards in the environment influence the decisions they make. Because men are subject to varying information flows, because their ability to use information varies, and because their perception of risk changes, we must again think in probabilistic rather than deterministic terms. We can only assign certain degrees of probability to an individual's decision, although in the aggregate, when we combine the effects of many decisions, we may be able to find considerable regularity and order.

The Flood as an Environmental Hazard

One of the most dramatic, dangerous, and costly hazards in the physical environment is the flood that sweeps away the patient work of many years, carrying with it not only human lives but many human hopes and aspirations. The control of floods has always been a most difficult task, and in areas of great hazard men have had to adjust their locations and lives to such virtually uncontrollable forces.

With the advent of large-scale fluvial engineering in the United States during the twentieth century, by means of which whole river basins were dammed and controlled, man has tried to modify the hazards presented by nature. Yet after decades of extremely costly flood control programs, we face a paradox that was quite unforeseen half a century ago. Despite the billions of dollars devoted to flood control, the annual losses from floods in the United States seem to show little decline. This is not because the dams are totally ineffective, or because man in trying to control nature has been overpresumptuous. In one sense flood control is an established fact, and the vast regional networks of dams such as those under the TVA show that we have at least perceived the systemic interdependence of many parts of an area. But flood losses are still high because the very modifications we make in natural systems encourage people to use land that was formerly seen as much too hazardous for the location of homes and factories.

A dam may be built to control and regulate

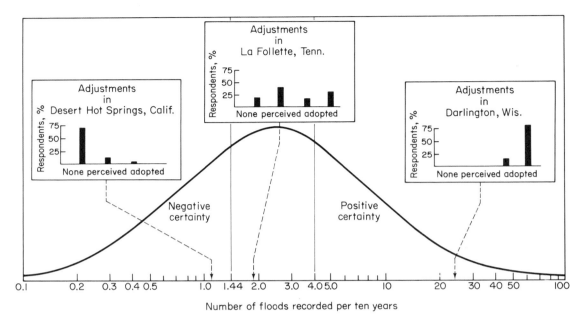

Number of floods recorded per ten years

FIGURE 13–28

Perception and adjustment to flood hazards varying with the frequency of flood arrivals: Desert Hot Springs, California; La Follette, Tennessee; and Darlington, Wisconsin (courtesy Kates, Burton, and White).

the flow of a river, and its very presence may encourage the expansion of residential and industrial areas onto formerly marginal land. Unfortunately, flood control projects are often built to deal with average conditions, with only a small allowance for extreme conditions of either too much or too little water. Seldom are they built to deal with the "hundred year flood," perhaps because men, including engineers, are often not very good at perceiving such rare and mentally remote events, or because insufficient information is available about the regime of a river at the time it is dammed. Unfortunately, nature, drawing from a probability distribution, comes up with the hundred year flood in just about the proportion we would expect from such a random procedure! The result is that America's flood losses are still extremely high, simply because men's perception and evaluation of the environmental hazard are so poor. Decisions and judgments leading to the use of land in flood plains are frequently erroneous.

In areas subject to flood a number of adjustments can be made to minimize the effect of flooding. Some are cooperative ventures such as the construction of levees and flood channels, the implementation of emergency action programs, and zoning controls established by political action. Others are individual adjustments involving structural changes on buildings and the payment of insurance premiums for flood damage. But whether the people even perceive, let alone adopt, such adjustment strategies is quite another matter. Perception and adoption of adjustments seem to depend upon the frequency with which floods are experienced (Fig. 13–28). In Desert Hot Springs, California, a flood is experienced on the average roughly once a decade. Few people know about the adjustments that can be made to alleviate flooding, and of the few who perceive the possibility of making some adjustments none have bothered to adopt them. On the other hand, in La Follette, Tennessee, where the rate of flooding on the average is two every ten years,

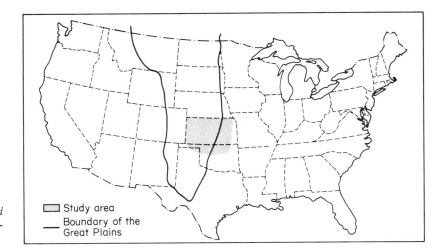

FIGURE 13–29

The Great Plains of the United States; a zone of transition (adapted from Saarinen, Fig. 5).

▨ Study area
▬ Boundary of the Great Plains

only about 20 percent of the people are totally ignorant about adjustments to floods, and over a quarter have actually adopted some form of floodproofing to minimize losses in the face of the environmental hazard. Finally, in Darlington, Wisconsin, where flood frequency is about twenty every ten years, everyone knows about flood proofing innovations, and most of the people in the flood plain have adopted them. Thus ignorance, perception of alternatives and adjustments, and the adoption of floodproofing methods against the hazard vary in a very regular way with the frequency of the hazard itself. Where hazardous events are very rare people's memories appear to be short and little is done to alleviate the problem. Where events are quite frequent and the hazard is constantly brought to the attention of the people, adjustments are nearly always adopted.

The same holds true for many other environmental hazards. In areas of frequent earthquakes, like Japan, buildings are constructed to withstand strong ground vibrations. Memories of earthquakes are constantly reinforced and adjustments are adopted whenever possible. Unfortunately, where earthquakes are rare and the memories of disasters fade with time, adjustments may not be made. Hardly any building in San Francisco is earthquake proof, for example, and in a probabilistic framework we can be almost "sure" that terrible disaster will strike again, perhaps soon.

The Perception of Drought

While some men live in places where there is sometimes too much water, others try to wrest a livelihood from areas where there is not enough. In many marginal areas of the world, where humid regions fade off into drier ones, wheat farming is a major human enterprise. Australia, Argentina, the Soviet Union, Canada, and the United States all face problems of aridity in marginal areas of grain production.

In the United States (Fig. 13–29), the Great Plains form a transitional zone between humid areas of the east and the drier areas of the west. As one moves west across the plains from eastern counties such as Adams and Barber (Fig. 13–30), to Kiowa and Cimarron counties on the western margins, the average precipitation declines by about ten inches. For men engaged in wheat farming, the environmental stress increases considerably across the region. However, the perception of drought by the farmers also changes radically with their locations in this marginal zone. The farmers in Adams and Barber counties greatly underestimate the proportion of drought years in their counties (Table 13–3). The number of years with actual drought do not differ greatly from east to west, but on the eastern margins there is a gross underestimation of this environmental hazard by the farmers. Their estimations rise through

TABLE 13–3

Actual and Estimated Droughts on the Great Plains

	Counties					
	Adams	Barber	Frontier	Finney	Cimarron	Kiowa
Farmers' estimates, % years	17	16	19.9	28.6	34.8	34.9
Actual drought, % years	42.4	46.9	41.6	47.2	48.7	47.2

Adapted from Saarinen.

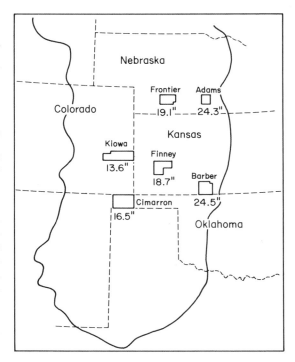

FIGURE 13–30

Six counties in the Great Plains: average rainfall figures indicate that environmental stress increases east to west (adapted from Saarinen, Fig. 2).

Frontier and Finney counties, and are even larger in Cimarron and Kiowa, where the stress is greatest. But all the farmers, no matter where their location, underestimate the actual proportion of drought years they have experienced. On the average, they are very optimistic, discounting the bad times and remembering the good ones with hope for the future.

The perception of the drought hazard does not depend only upon the degree of environmental stress. In all the counties, older farmers with considerable experience perceive and estimate the risk better than those who have spent shorter periods farming in the area. They learn from the environment and organize their experience to deal with it. But drought perception also varies with the personalities of the farmers themselves. Farmers who indicate high achievement levels in a number of psychological tests are also the most realistic, perceiving the real risks more accurately than their content-to-jog-along fellows. They are also the ones who are more open to new innovations for alleviating, and in some degree controlling, the problem of aridity and wind erosion. Thus the perception of environmental hazard varies with aridity, the amount of drought experience, and personality characteristics of the farmers. Greater aridity also narrows the range of farming choices perceived by the farmers. In the moister areas, farmers are prepared to consider changing their farm operations by bringing in other crops and various livestock combinations. In the drier, western counties very few alternatives to wheat farming are perceived as possible alternatives.

Much of the history of wheat farming in the Great Plains and other marginal areas of the world settled by Europeans can be written in terms of overoptimism and degrees of risk-taking that approach wild-eyed speculation. Some pioneer farmers appear willing to probe to the outermost limits of a marginal environment, perceiving the opportunities through rose-tinted spectacles until an arid year sends them scurrying back to safer areas or forces them to hang on grimly waiting for better times to come. In the wheat areas of Canada and southern Australia the wheat lands have "pulsed" not only in response to transportation development and fluctuations of world prices, but also according to the perception of men whose memory traces of environmental hazard fade with time only to be reinforced by a crippling year of drought and dust storms.

Mental Maps and Residential Utility Surfaces

The decision to migrate, to change one's permanent residence, is usually a major one for the individual or family. Yet it is a decision made by one-fifth of America's people each year, for Americans are the most mobile population the world has ever seen. Each move by a person or family, particularly young families, represents a decision with clear spatial implications. Since the middle of the nineteenth century, the rural population, following the major westward sweep of America's pioneers, has slowly drained into the urban areas. But not all migration in America is of this sort. Another strand in the complex pattern is the major shift of population to regions that are perceived as generally desirable.

The states of America are perceived in a wide variety of ways for residential purposes, and their images in the minds of men and women influence the decisions that are made to move from place to place. Nor do such varying perceptions and evaluations of geographic space influence only the migratory movements of individuals. Many industries today are termed "footloose," implying that they are no longer tied to the traditional factors of raw materials and the market. For example, transporting the raw materials for transistors forms only a minute portion of their total cost and value. Similarly, the transistors themselves are of such high value in relation to their weight that transporting them to the major markets does not eat into profits to any considerable degree. Thus many electronic factories, along with many other new industries and research and development firms, are choosing new locations on criteria very different from those considered critical fifty years ago. Recreational facilities, good climate, a forward-looking social milieu, and good educational facilities are now much more important factors entering into the locational decision. Areas of congestion and high urban stress in choking, smog-ridden locations, and places geographically and culturally remote do not hold the same attractions.

Much of the migratory movement in America is composed of young adults in their twenties. With the growth of higher education the mobility of these people has been greatly accentuated, for college graduates are among the most mobile of all groups in the country. With their high levels of educational and professional skills, they have a freedom of movement far exceeding that of many other groups in the country.

The perception of residential desirability by college students depends in part on the point of perception. If the residential desirability of each state is measured on a scale from zero to one hundred, and the values are recorded at the population centers of each state, we can use them to construct residential perception surfaces (Figs. 13–31 through 13–34). Like a contour map whose lines connect points of a certain height above sea level, so the perception surface shows the hills and valleys of residential desirability for a particular group of people. Where the "iso-percepts" are far apart, the perception of residential desirability changes slowly over an area. Where the spacing between the lines is small, steep perceptual gradients indicate rapid change in spatial preference.

The view from California (Fig. 13–31) is fairly typical of the outlook of western and northern students. A high ridge of residential desirability runs down the West Coast, with a steep drop to a perceptual basin located over Utah and Nevada. The mountain states of Colorado and Wyoming stand out above their neighbors, but generally the trend of the perception surface is down from the West Coast to the Great Plains states. A perceptual sinkhole is centered over South Dakota. At approximately the one hundredth meridian, the surface changes its orientation by ninety degrees, with the lowest trough on the surface in Alabama and Mississippi. From the core of the South the surface rises to Florida, as well as northward to the Midwestern and Eastern Seaboard states. Well-publicized Appalachia swings the iso-percepts northward to include the states of Virginia, Kentucky, and West Virginia. New York and the New England states are regarded favorably, and for California's students only the West Coast states score higher.

When the viewpoint is shifted to Minnesota (Fig. 13–32), much the same surface is seen

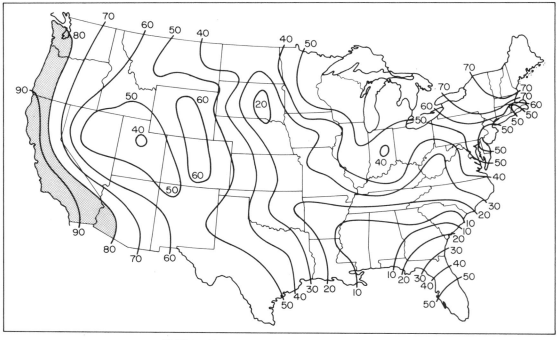

FIGURE 13–31

The residential desirability surface from California.

FIGURE 13–32

The residential desirability surface from Minnesota.

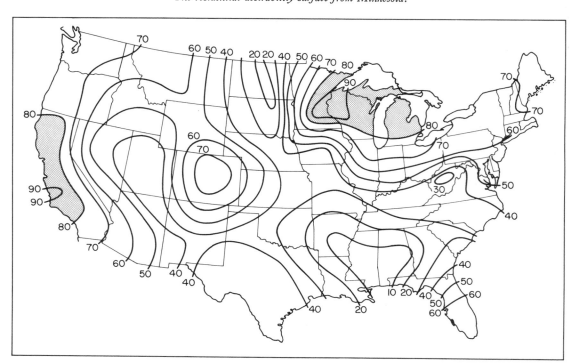

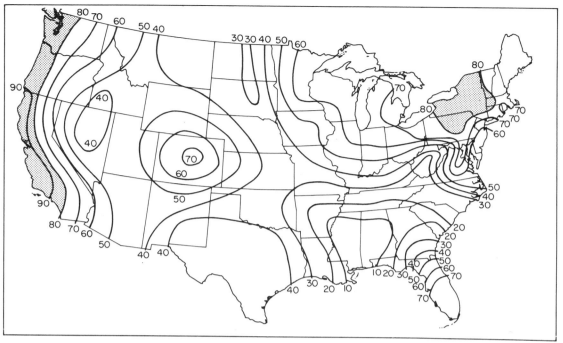

FIGURE 13–33

The residential desirability surface from Pennsylvania.

FIGURE 13–34

The residential desirability surface from Alabama.

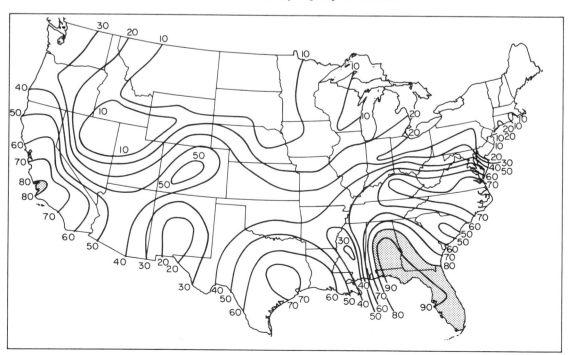

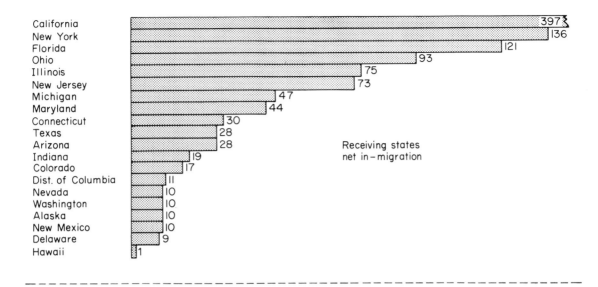

Receiving states
net in-migration

California 397
New York 136
Florida 121
Ohio 93
Illinois 75
New Jersey 73
Michigan 47
Maryland 44
Connecticut 30
Texas 28
Arizona 28
Indiana 19
Colorado 17
Dist. of Columbia 11
Nevada 10
Washington 10
Alaska 10
New Mexico 10
Delaware 9
Hawaii 1

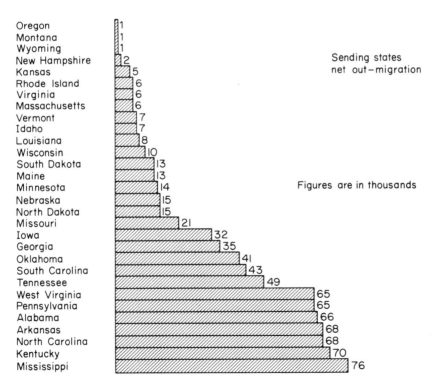

Sending states
net out-migration

Oregon 1
Montana 1
Wyoming 1
New Hampshire 2
Kansas 5
Rhode Island 6
Virginia 6
Massachusetts 6
Vermont 7
Idaho 7
Louisiana 8
Wisconsin 10
South Dakota 13
Maine 13
Minnesota 14
Nebraska 15
North Dakota 15
Missouri 21
Iowa 32
Georgia 35
Oklahoma 41
South Carolina 43
Tennessee 49
West Virginia 65
Pennsylvania 65
Alabama 66
Arkansas 68
North Carolina 68
Kentucky 70
Mississippi 76

Figures are in thousands

FIGURE 13–35

Net migration in the United States for the 25–29 year age group during the period 1950–1960. Results are shown for 1960, by states.

except for a high peak of desirability centered on Minnesota itself. The same West Coast Ridge, Colorado High, Dakota Sinkhole, and Southern Trough appear, indicating that with the exception of some local differences the students in Minnesota share much the same views as their Californian contemporaries.

A very similar perception surface also describes the residential desirability for Pennsylvanian students (Fig. 13–33), and like Californians and Minnesotans they perceive the West Coast and Colorado as being very desirable for residential purposes.

When the perception point shifts to Alabama, however, a different residential desirability surface appears (Fig. 13–34). White Alabaman students regard the southern states as most desirable, and whereas their northern counterparts "homogenize" the South with low perceptual scores, the Alabaman students discriminate very carefully between certain southern states. Alabama, Georgia, Texas, North Carolina, Kentucky, and Virginia are all perceived very favorably, while Mississippi and South Carolina form distinct sinkholes in the otherwise high surface. The North is completely blanketed by low values, and only in the West does the perception surface begin to conform to those of students elsewhere. California is still greatly desired, and once again the surface falls steeply to a perceptual basin over Nevada and Utah before a smaller Colorado peak is reached.

The three northern and western perception surfaces appear very similar, highlighting the distinctness of the southern viewpoint. Nor should this surprise us when we think about the broad shifts of population in the United States today. Highly desirable California is now our most populous state, and nothing seems to stem the tide of migrants. In the decade of 1950–60 it had a net migration more than three times that of New York, its closest rival (Fig. 13–35). With the exception of Pennsylvania, many of the Northeastern and Midwestern states were net receivers of America's migration flows. Similarly, the West Coast states, Colorado, Texas, and Florida were lodestones for America's people. In marked contrast, most of the Great Plains and Southern states sent many more

people out than they received. Thus the perception of residential desirability is translated into action by the makers of individual migration decisions.

The Formation of Perception Surfaces

The formation of perception surfaces raises numerous questions about flows of information to which people are subjected. Information flows from many media impinge upon the

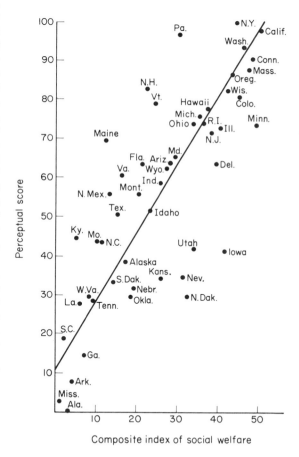

FIGURE 13–36

Residential desirability scores as a function of general levels of social welfare (adapted from Gould, 1969, Fig. 6, p. 40).

individual, and these are ordered and evaluated into images of varying desirability. The way a state is perceived by northern and western students is closely related to the general level of social welfare in those states (Fig. 13–36). California and New York, two states with very high levels of social welfare, are also perceived as very desirable areas for residence. South Carolina and Mississippi, with two of the lowest levels of social welfare in the nation, are perceived as extremely undesirable places to live in. The assessment of residential desirability by young men and women matches fairly closely a composite index of social welfare taking into account ten major social goals for Americans outlined by a Presidential Commission initiated by President Dwight D. Eisenhower.

But the perception of residential desirability also takes into account other factors. When the view is from Pennsylvania, some states like Vermont, New Hampshire, and Maine are perceived as more desirable than their levels of social welfare would indicate. Others, such as Minnesota, Iowa, Utah, Nevada, and North Dakota are greatly underestimated. The latter, with high levels of social welfare, are pulled down by their location in the poorly perceived regions, while the former gain in mental brightness because of the New England location and the recreational images associated with it. Since the perception of residential desirability is closely linked to the individual decision to migrate, the perception surfaces have direct implications for regional planning. Whether we consider older, developed countries like the United States and the United Kingdom, or a newly developing country like Ghana, residential perception surfaces derived from groups of mobile young adults have direct policy and planning implications.

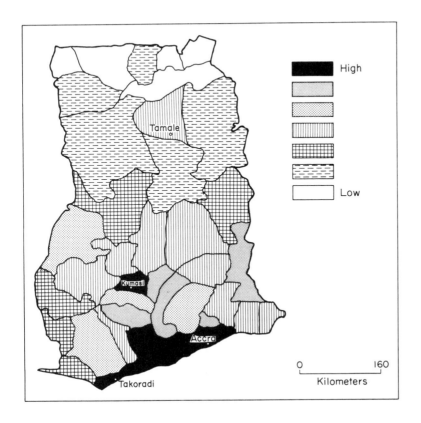

FIGURE 13–37

Values of university students' perceptions of residential desirability for districts in Ghana.

The Spatial Assignment of an Elite

In Ghana, as in many countries of Africa and the underdeveloped world, there is a major problem of getting well-educated men and ...men away from the bright lights of the principal towns and into the rural areas where they are so desperately needed. All over Asia, Africa, and Latin America there is a strong tendency for university graduates to stay in the towns where the neon lights shine so brightly in the minds of men. In Ghana (Fig. 13–37) the perception surface for university students mea-

sures the great value placed upon location in the southern core of the country, particularly along the pleasant coast and inland at the important town of Kumasi. The perception scores form such a regular pattern that a simple trend surface (Fig. 13–38) generalizes the steady decline in desirability away from the major core area of the south toward the northern part and periphery of the country. Most governments in Africa face the problem of assigning government personnel, administrative and agricultural officers, nurses, and teachers to locations in the rural areas to serve the people and hasten the development process. Today salaries can be related inversely to the hills and valleys of the perception surfaces. In this way people who want to stay in the major towns, where the supposed action is, will subsidize with their lower salaries the higher ones received by other men and women willing to serve in the rural areas. In this way the salary differentials "smooth out" the perceived desirability surface so that individual decisions more closely match national development goals.

Perception Surfaces: National and Local Parts

Problems produced by the desirability surfaces of young mobile people are not unique to underdeveloped countries. The United Kingdom also faces major planning problems involving the movement of young adults. "The drift to the south" has become a well-known phrase, describing the steady and seemingly inexorable flow to the southern counties, particularly those in the southeast immediately around the metropolitan center of London. Residential perception surfaces for pupils about to leave school and look for their first jobs have many characteristics in common though the perception points may be far apart. The view from the south (Fig. 13–39) shows the way in which the whole southern coast is blanketed with high scores. Central London is in a perceptual sinkhole, but from a Southern Plateau of Desire two prongs extend northward along the east coast and through the pleasant rural counties along the border of England and

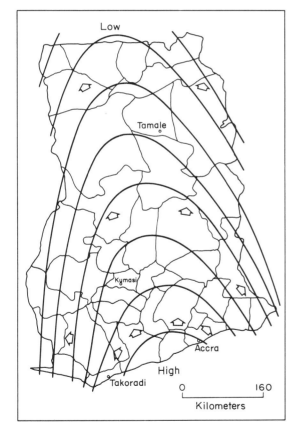

FIGURE 13–38

The general perception surface for university students in Ghana who are considering assignments as government servants.

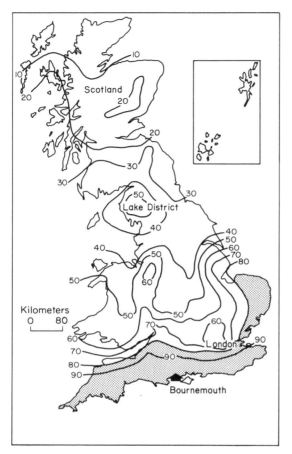

FIGURE 13-39

Residential desirability in Britain as seen by English pupils on the south coast who are about to leave school (adapted from Gould and White, Fig. 3).

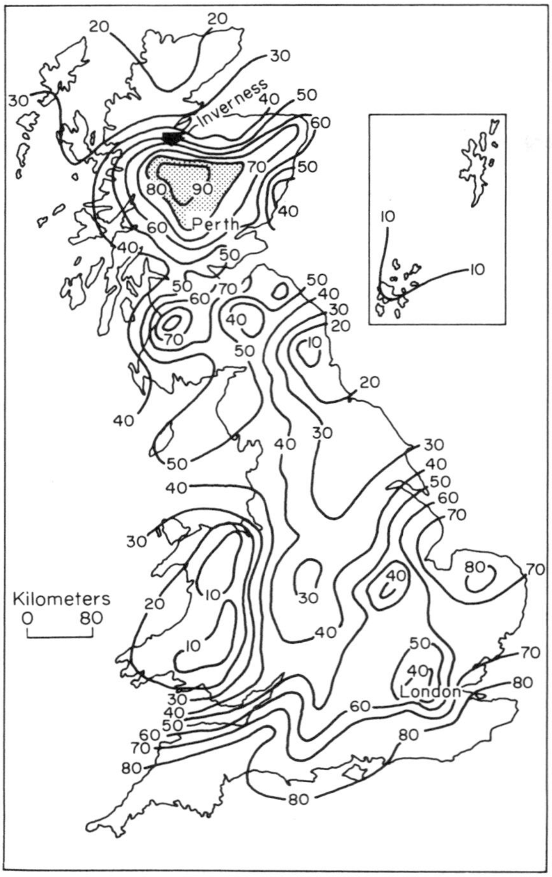

FIGURE 13-40

Residential desirability in Britain as seen by Scottish pupils about to leave school (adapted from Gould and White, Fig. 9).

Wales. Generally the surface slopes downward to Scotland, with the exception of a dome in the Lake District, one of Britain's most beautiful and well-publicized vacation areas. The view from Scotland, however, is somewhat different (Fig. 13–40), for immediately around the perception point there is a very high dome of local desirability. Even so, the perception surface over England still resembles the view from the south. The southern counties are seen as desirable even by young Scots, the same Metropolitan Sinkhole appears, and there is a

deteriorated ridge of desirability in the east, while the midlands are still generally disliked.

The perception surfaces of young, mobile people in the British Isles can be decomposed into two pieces. Using a simple model that decomposes a perception surface for a group of people at a particular location into a *general* or *national* surface, together with a *local dome* of desirability, we can imagine the general surface shared by all groups at all locations to be molded in rubber, with the local dome as a tennis ball pushing up the surface at a particular perception

Local peak emerging from general surface with northward shift in point
of perception

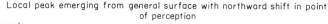

FIGURE 13–41

Decomposition of perception surfaces into a general, or national, surface shared by all, together with a local dome of desirability whose prominence varies with location (adapted from Gould and White, Fig. 20).

point. The prominence of the local dome depends upon the location of the perception point (Fig. 13–41). In the south, the local dome is hidden under the general or national surface, for here young adults share the general viewpoint that a southern location is highly desirable. But in northern England, at the towns of Liverpool and Newcastle, the effect of the local

dome begins to emerge. Finally, when the viewpoint is from Inverness in Scotland and Kirkwall in the Orkney Islands, the local dome of desirability is almost completely exposed. Thus people's perception of residential desirability is composed of two parts: a national viewpoint shared by most people in the country, together with a local dome of preference indicat-

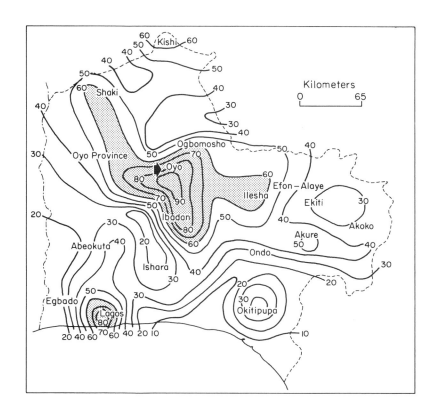

FIGURE 13–42

Residential desirability surface for 13-year-old students in Oyo, Western Region of Nigeria (adapted from Ola).

ing that many people prefer the local area that they know best and in which they feel most comfortable.

In some parts of the world the desirability of the local area declines with age. Very young people evaluate their familiar local areas much more highly than do those in their early twenties. It is little wonder that men and women in the twenty-year age group are the most mobile of all, for their mental and perceptual ties to the local area are much weaker. In the Western Region of Nigeria, for example, thirteen-year-olds at the town of Oyo have a distinct dome of local desirability immediately around the local area (Fig. 13–42). Fifteen-year-olds, however, have weaker feelings for the local area (Fig. 13–43), and the fading of the local image is even more pronounced for eighteen-year-olds (Fig. 13–44). Finally, by the time pupils are twenty-three years old (Fig. 13–45), residential desirability in a number of other places is higher than for the local town. The

local dome has faded away, and their residential preference surface matches closely the main crescent-shaped area of the region containing the major towns, railway lines, and good roads "where the action is."

We have come full circle in our consideration of spatial decisions in a descriptive framework. Men search space and learn about opportunities within it, yet their patterns of search, their abilities to learn, and their very survival in physical and social environments are affected by their perception of the world around them. Their perception, which changes with age and with circumstances, also controls the information they seek, and it is upon this information that they base their decisions. But decisions with spatial and locational implications do not always make the best or fullest use of the information available. We have seen that men seldom act rationally upon the information available, but often come into conflict with each other. Learning is not confined to dealing with

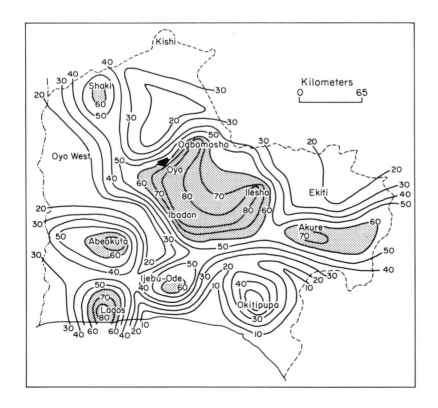

FIGURE 13–43

Residential desirability surface for 15-year-old students in Oyo, Western Region of Nigeria (adapted from Ola).

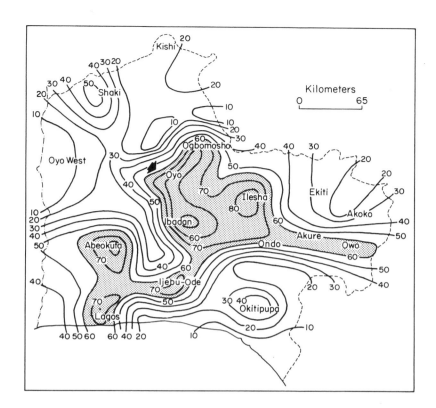

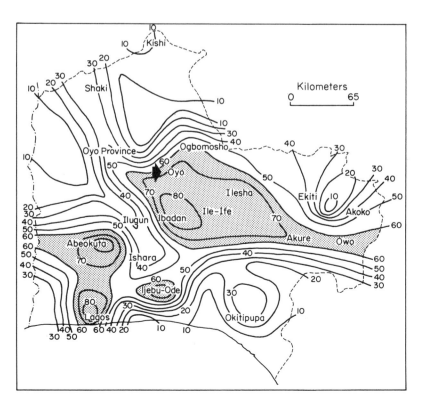

the player nature, but also involves considering other groups of men. Many of the decisions are compromises that indicate cooperation between men to resolve conflicting goals and views. Decisions made under conflicting circumstances and goals are very common, and are often at the heart of some of the most difficult spatial questions that geographers deal with. It is time to consider one of these in detail.

Suggestions for Further Reading

ADAMS, JOHN S. "Directional Bias in Intra-Urban Migration," *Economic Geography*, XLV (October 1969), 302–23.

BURTON, IAN. *Types of Agricultural Occupance of Flood Plains in the United States*. Chicago: Department of Geography, The University of Chicago, 1962. Research Paper No. 75.

CURRY, LESLIE. "Seasonal Programming and Baysian Assessment," in W. R. Derrick Sewell, ed., *Human Dimensions of Weather Modification*. Chicago: The University of Chicago Department of Geography Research Paper No. 105, 1966, pp. 127–38.

GOLLEDGE, R. G., and L. A. BROWN. "Search, Learning, and the Market Decision Process," *Geografiska Annaler*, XLIX, B: 2 (1967), 116–24.

GOULD, PETER. *On Mental Maps*. Ann Arbor: Department of Geography, University of Michigan, 1966. Michigan Inter-University Community of Mathematical Geographers Discussion Paper No. 9, mimeo.

KATES, ROBERT W. *Hazard and Choice Perception in Flood Plain Management*. Chicago: Department of Geography, The University of Chicago, 1962. Research Paper No. 78.

KEMENY, JOHN G., J. L. SNELL, and G. L. THOMPSON. *Introduction to Finite Mathematics*. Englewood Cliffs, N.J.: Prentice-Hall, Inc., 1956.

PRED, ALLAN. *Behavior and Location*, Parts I and II. Lund: Gleerup, 1967 and 1969. Lund Studies in Geography, Series B, No. 27.

RAPOPORT, ANATOL. *Fights, Games and Debates*. Ann Arbor: University of Michigan Press, 1960.

TÖRNQVIST, GUNNAR. *Contact Systems and Regional Development*. Lund: C. W. K. Gleerup, 1970.

WHITE, GILBERT F. *Choice of Adjustments to Floods*. Chicago: Department of Geography, The University of Chicago, 1964. Research Paper No. 93.

Works Cited or Mentioned

ADAMS, JOHN S. "Directional Bias in Intra-Urban Migration," *Economic Geography*, XLV (October 1969), 302–23.

BERRY, B. J. L., S. J. PARSONS, and R. H. PLATT. *The Impact of Urban Renewal on Small Businesses*. Chicago: Center for Urban Studies, The University of Chicago, 1968.

DEUTSCH, MORTON, and ROBERT M. KRAUSS. "The Effect of Threat Upon Interpersonal Bargaining," *Journal of Abnormal and Social Psychology*, LXI: 2 (1960), 181–89.

FRAENKEL, GOTTFRIED S., and DONALD L. GUNN. *The Orientation of Animals*. Oxford: The Clarendon Press, 1940.

GOULD, PETER. "Problems of Space Preference Measures and Relationships," *Geographical Analysis*, I (1969), 31–44.

———. "Wheat on Kilimanjaro: The Perception of Choice Within Game and Learning Model Frameworks," *General Systems*, X (1965), 157–66.

———, and RODNEY WHITE. "The Mental Maps of British School Leavers," *Regional Studies*, November, 1968. Reprinted in *General Systems Yearbook*, XV (1969).

KATES, ROBERT, IAN BURTON, and GILBERT WHITE. "The Human Ecology of Extreme Geophysical Events," unpublished mimeographed paper, Natural Hazard Research Working Paper No. 1.

OLA, DANIEL. "Perception of Geographic Space in Lagos and the Western States of Nigeria," unpublished master's thesis, Department of Geography, The Pennsylvania State University, 1968.

SAARINEN, THOMAS F. *Perception of Drought Hazard on the Great Plains*. Chicago: Department of Geography, The University of Chicago, 1966. Department of Geography Research Paper No. 106.

SIMON, HERBERT. *Models of Man*. New York: John Wiley & Sons, Inc., 1957.

TAAFFE, E. J., R. L. MORRILL, and P. R. GOULD. "Transport Expansion in Underdeveloped Countries: A Comparative Analysis," *Geographical Review*, LIII (1963), 503–29.

WILDE, DOUGLASS. *Optimum Seeking Methods*. Englewood Cliffs, N. J.: Prentice-Hall, Inc., 1964.

CHAPTER 14

THE QUESTION OF THE BEST LOCATION: THE GEOGRAPHER'S UNSOLVED PROBLEM

It is better to know some of the questions than all of the answers.
—James Thurber

THE LOCATION-ALLOCATION PROBLEM

Some books come to you like well wrapped packages. As you pass the halfway mark the chapters begin to close the subject off, and by the time you have reached this point all that remains is the job of tying everything up with a big knot. We think that books like this are essentially misleading, for there are few areas of human inquiry in which the knots can be tied with any firmness. A certain amount of simplification can be lived with, and is usually essential in a beginning work. But books that try to wrap up a subject are worse than misleading: they are dull and humorless, and this is difficult to tolerate under any circumstances. If a book gives you the impression that all the problems have been solved, you are perfectly justified in asking why you read the work in the first place.

Geographers face many unsolved problems. We have selected from all the open questions one with which some progress has been made, al-

though the general solution remains unknown. It looks quite easy as you begin to examine it, but a little thinking shows how complex it really is. Further thought generates that glorious sense of intellectual frustration about a spatial problem that makes a geographer's life worth living. Quite apart from its intellectual challenge, it is a very important problem from the practical point of view, for there are many obvious and immediate applications with very strong humanitarian overtones.

The problem is generally called the Location-Allocation Problem. In more prosaic language it asks: how shall we allocate one set of facilities to serve a second set of people (Fig. 14–1)? Examples quickly come to mind, for this geographical problem is a familiar one to all of us. Hospitals must be located in geographic space to serve the people with competent medical care, and we must build schools close to the children who have to learn. Fire stations must be located to give rapid access to potential conflagrations, and voting booths must be placed

Things to be
allocated:

Hospitals
Schools
Birth control clinics
Fire stations
Voting booths
Administrative centers
Branch campuses
Playgrounds
Etc.

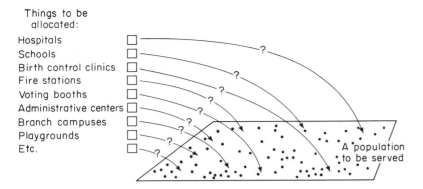

A population
to be served

FIGURE 14–1

A set of facilities to be assigned to serve a population.

so that people can cast their ballots without expending unreasonable amounts of time, effort, or money to reach the polling stations. Many of our states face the problem of locating branch campuses to serve a burgeoning and increasingly well educated population. In the cities we must create and locate playgrounds for the children. Many overpopulated countries must assign birth control clinics to reach the people with contraceptive and family planning information. In each of these cases we face the problem of locating multiple facilities to serve people with goods, services, or information.

Optimizing under Conflicting Goals

The problem in its most general form has never been solved. Yet our first impression is that it cannot be terribly difficult. After all, the basic problem seems easy to state. We want to locate service facilities in such a way that the total cost or effort of the people moving to the facilities is minimized. Efficient school locations will save children time, and will also save their parents taxes allocated to school bus programs. Properly located hospitals, maximizing the accessibility of the people to them, can save human lives in emergencies. Highly accessible voting booths have positive consequences for the democratic process. As we begin to think about the problem in more detail, however, we suddenly find ourselves in intellectual thickets through which it is difficult to see a clear and open path.

The trouble is that many things can vary in a problem of this sort. First, we can alter the

number of facilities to be located. At one extreme we can have a single hospital or school to serve all the people in the region. We are already familiar with this special case. We can build one school with enough places for all the children in the area and locate it using the principle of Weber's weight triangle. At the other extreme every person in the region can have his own facility. A complete surgical team with all the necessary nursing care can be assigned to each person, and every child can get a team of expert teachers. In most cases, actual solutions will lie somewhere between these two extremes. The inefficiency that would result from establishing only one facility may cost too much in human terms according to the set of values held by the people (Fig. 14–2). If there is only one hospital, most of the people may not reach it in time when emergencies arise. Similarly,

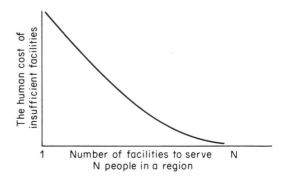

FIGURE 14–2

Relationship between the human cost of insufficient facilities and the number available to serve the people in a region.

a single school may be so far from many of the children that they are too tired to work effectively when they reach it. As the number of facilities increases, the human cost declines rapidly, until it is zero when everyone has his own, private service. On the other hand (Fig. 14–3), the monetary cost of providing many facilities will be extremely high. It is not difficult to think of situations where the cost curve soars beyond the financial limit of the people in the region. This is a familiar story to most people in the underdeveloped world, who would like many more schools and hospitals if only they had the money to pay for them.

Even as we consider one piece of the general multiple location problem, we raise the question of conflicting goals, a question of more than just passing interest. Many optimization problems of this type (including the Weberian weight triangle) are called *extremum* problems. Extremum problems that try to maximize or minimize some quantity under conflicting goals are very common in human affairs and nature. For example, we may want a structural beam of maximum rigidity and minimum weight. Unfortunately, a beam of infinite stiffness would also be a beam of infinite weight, so some compromise solution must be found. A chicken's egg should be a sphere, for this shape maximizes the volume with the least surface area so that cooling of the chick embryo is minimized. However, there is obviously some cost to be paid for spherical eggs in discomfort units by

hens, and nature compromises with the more familiar shape. Similarly, in order for corn plants to maximize exposure of their ears to the sun they should grow all of them at the top. But this would produce a very unstable plant, and to maximize stability the ears should all be placed at the bottom, where there is little sun. As a result, the corn plant becomes an extremum mechanism, producing a compromise solution under conflicting goals. In the same way, we must often find for our spatial location problems a compromise solution between a single facility and private facilities for every person in the area.

Unfortunately, the problem is even more complex than this. Assuming we could decide upon the number of facilities to allocate to the people, we would still be faced with the question of their size or capacity. We could decide upon schools and hospitals of equal size, but there is no assurance that equal sizes will maximize the accessibility of the people to them. In fact, if the people themselves are not evenly distributed across the landscape, facilities of unequal size will always be better. Finally, the locations of the facilities themselves may vary. A region has an infinite number of points from which we may choose a set of particular locations. A mathematician concerned with the sheer elegance of a solution may wish to consider the problem in this way, although for most practical purposes we usually try to choose a set of final locations from a larger but still finite number of prespecified places. Most public services need water, sewerage, electricity, and lines of transportation, so it may not make much practical sense to consider the infinite number of points available in an area.

We have a geographical problem, then, with three things that can vary: (1) the number of things to be assigned; (2) the size or capacity of the facilities; and (3) the locations to which the facilities must be allocated. Not only has the problem in its most general form never been solved, it has never even been stated in sufficiently precise terms to be approached mathematically. A mathematician would tell us that it has yet to be "well defined." Stafford Beer has described the problem in the following way. Imagine a number of men searching on a pitch black night or in a dense fog for the lowest

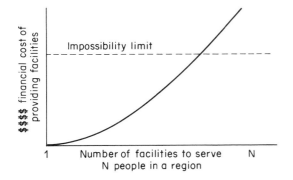

FIGURE 14–3

Relationship between the financial cost of providing facilities and the number available to serve the people in a region.

points on a surface. They grope around on their hands and knees, trying always to move toward the low ground. Yet as they move the surface starts to undulate according to their patterns of search. Usually there will be one or two who always feel they can improve their location, but their movements may set the others searching again as the surface shifts and sways. And even if all the searchers feel they cannot improve their locations by further movement, there is no guarantee that they have found the very best set of locations.

Global and Local Solutions

Beer's analogy raises two further difficulties in this most difficult of geographical problems. First is the question of the uniqueness of a solution. How can we tell, in any particular problem, whether there is a single, best solution to it, or whether there are many solutions all with the same optimal values? Perhaps three schools can be located in a region in several different ways, yet each set of locations might still minimize the total effort of the children to exactly the same degree. Secondly, if we think we have found a solution, how do we know it is the very best solution of all?

The ultimate solution, the one that truly maximizes the accessibility of the people to the facilities, is termed a *global* solution. On the way to searching for a global solution to a problem we may hit a number of *local* solutions. These will raise new difficulties as we try to decide whether they are simply local or whether they are the true and final global answer. Consider, for example, two men searching for the highest point within fenced areas on the side of a mountain at night (Fig. 14–4). By groping his way around the fence with his hands, Mr. Jones will keep moving around as long as he feels himself going uphill. As soon as he gropes around a corner and feels himself going down he will backtrack to the highest corner point A. In this particular case, the point is the global solution, and if he reaches it he has solved the problem. On the other hand, Mr. Smith faces a more difficult task. If he searches clockwise around the fence he will reach a point B, which he will assume is the highest point he can reach on the

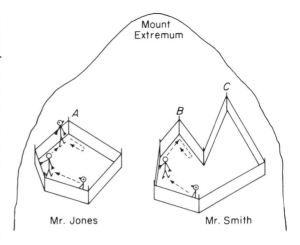

FIGURE 14–4

Searching for optimal solutions. Mr. Jones reaches the global solution to his problem (A), while Mr. Smith only reaches the local solution (B) (adapted from Wilde and Beightler, Fig. 3–1).

mountain. After B his path moves downhill and he quickly goes back to the point which happens to be only a local maximum in his problem. If he were lucky enough to search counter-clockwise, however, he would reach C, which is the global solution.

In a somewhat analogous fashion, we may find a set of places that seems to be a solution to a location problem, so that a shift of one facility raises the overall cost of movement. We may think we have the very best set of locations, when in reality we may have only a local, rather than a global solution. This means that it might still be possible to find a completely different set of locations which lowered the cost of servicing the population still further.

SOME APPROACHES TO THE PROBLEM

Difficult problems of this sort have to be simplified before we can even approach them. One way to cut into a problem is by holding constant some of the things that can vary. Suppose we choose beforehand both the number and the locations of the facilities serving the people in the region, and allow only the capacities to

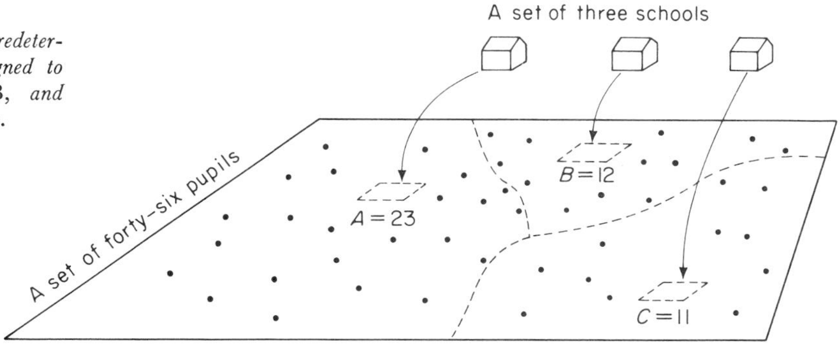

FIGURE 14–5

The proximal solution: a predetermined number of schools assigned to predetermined locations A, B, *and* C, *with capacities that can vary.*

A set of three schools

A set of forty-six pupils

A = 23 B = 12 C = 11

vary (Fig. 14–5). In this case we can simply assign the children to the school nearest to them so that *A* has 23, *B* has 12 and *C* has 11. Because the children move to the nearest facility, we shall call this a *proximal* solution. It is very simple, but we cannot lower the overall cost of movement any further. As a real problem it is not very interesting or practical, because by initially setting the number of facilities and their locations, and allowing only the capacities to vary, we have assumed away nearly all the difficulties. But let us see what other variations we can play on this basic theme of holding things constant.

Special Medical Facilities in Sweden

Instead of completely deciding upon the locations of the facilities, we can let these vary over a larger set of prespecified places. The sizes of the facilities will also vary, but the number of facilities to be assigned will be set beforehand. We are going to take a real example from Sweden, where geographers are frequently consulted when questions of optimal location for public facilities arise. In the early 1960s the Swedish government wanted to expand very specialized hospital facilities. These included iron lungs and artifical kidney machines, as well as special surgical facilities for children and cancer patients. Some of these facilities already existed in six of Sweden's main hospitals associated with university medical schools at Umeå, Uppsala, Göteborg, and Lund, while Stockholm had two hospitals of this sort (Fig. 14–6). New facilities could not be assigned to any little village or town, and available sites

were limited to five including Sundsvall, Karlstad, Örebro, Linköping, and Jönköping. The problem posed by the government to the geographers was to select two locations out of the set of five possibilities.

We can think of assigning the first hospital to any one of five locations, and the second one to any of the remaining four. Thus, there are (5 × 4) or 20 possible ways of assigning the new facilities. In trying to decide which pair of locations was the best, the geographer Gotlund used the criterion of total travel time to the hospitals. Ideally he wanted to get all the people of Sweden within four hours of these specialized medical facilities, so that a person could attend a hospital and return to his home without staying overnight. Unfortunately, with Sweden's scattered population in the far north, and with the possibility of building only two new hospitals, this goal could not be completely met.

The problem was approached by drawing isochronic maps for each pair of possible locations. That is, lines of equal travel time were drawn around each of the seven locations—the five associated with the medical schools, plus the two new ones to be assigned. The isochronic maps were then laid upon very carefully constructed maps of the predicted population distribution for 1975. The numbers of people in each time zone were then counted, and the total number of travel hours used as a criterion for the efficiency of a particular solution.

If the specialized hospital facilities are assigned to the five existing centers (Fig. 14–7), only 76 percent of the people are within four hours of the locations. If every man, woman,

FIGURE 14–6

Locations of existing specialized hospital facilities in Sweden, with five possible locations, for the assignment of new ones (adapted from Godlund).

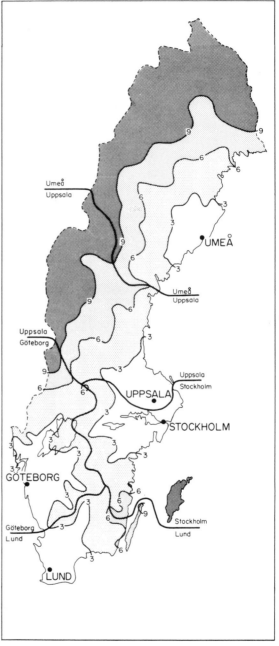

FIGURE 14–7

The patient hinterlands of the five existing locations for specialized hospital facilities (adapted from Godlund).

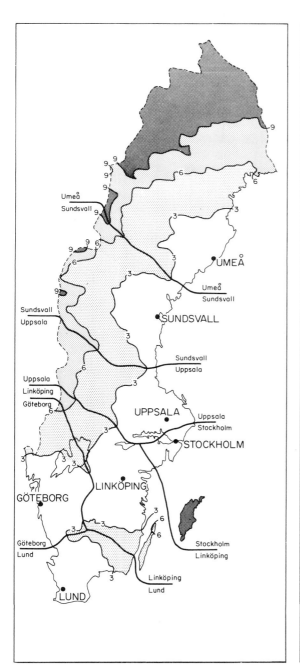

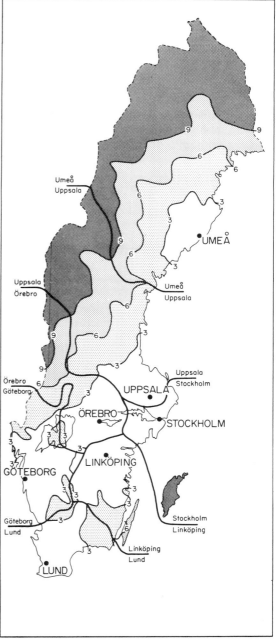

FIGURE 14–8

The patient hinterlands if two new hospitals are assigned to the towns of Sundsvall and Linköping (adapted from Godlund).

FIGURE 14–9

The patient hinterlands if the two new hospitals are assigned to the towns of Örebro and Linköping (adapted from Godlund).

and child moves to the nearest hospital, they use up a total of 3.93 million man hours. An addition of two new facilities will obviously reduce the travel time and increase the accessibility of the people to these facilities. The question is: *which* two? If the new hospitals are assigned to Linköping and Sundsvall (Fig. 14–8), 82 percent of Sweden's population comes within a four-hour isochrone. Also, the total number of hours needed to get all the people to the facilities is reduced to 3.57 million man hours, a saving of 360,000 man hours over the solution with only five locations.

Sweden's population is concentrated very heavily in the south, and there are vast areas of the north with hardly any people at all. Thus,

to maximize the accessibility of the people to the hospitals the new one was taken from Sundsvall and placed in Örebro instead (Fig. 14–9). While the proportion of people in the four-hour zone increases only slightly to 83 percent the number of man hours required for traveling drops to 3.39 million. This solution was the one chosen by the government as they planned for the future. As Gotlund noted, the plans called for expenditures of forty million dollars, while the cost of the geographical investigation was only two thousand. We shall never know the exact size of the savings that accrue to Sweden's people year after year—savings that are measured in terms of both money and human lives saved.

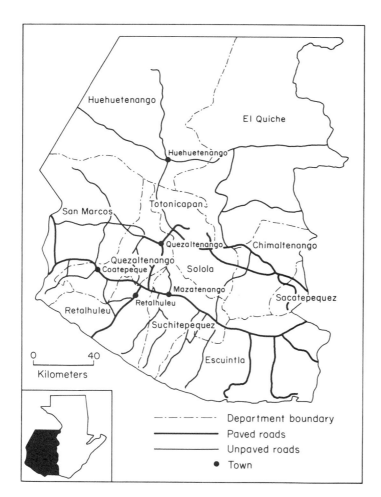

FIGURE 14–10

The road network and main towns of western Guatemala (courtesy of Gould and Leinbach and T.E.S.G.).

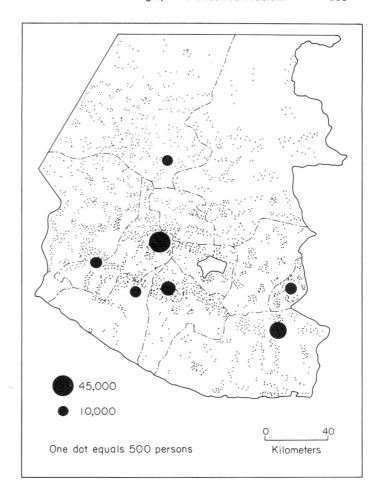

45,000

10,000

One dot equals 500 persons

0   40
Kilometers

FIGURE 14–11

The distribution of population in western Guatemala (courtesy of Gould and Leinbach and T.E.S.G.).

Hospitals in Western Guatemala

Careful spatial planning of hospital facilities need not be confined to the developed portion of the world. Many underdeveloped countries have plans to expand their educational and medical facilities, and with so little fat on their economic frames, planning for the future is even more critical than in developed nations. Solutions that minimize inputs and maximize outputs in a normative fashion make good sense in countries where every penny must be stretched to do the work of two. In western Guatemala (Fig. 14–10), the population is distributed in a very uneven fashion (Fig. 14–11). The people live in small towns and villages, but most of these have so few urban facilities, such as water, sewerage, and electricity, that they cannot form potential sites for regional hospitals. Suppose we wish to assign three hospitals to five potential sites such as Huehuetenango, Quezaltenango, Coatepeque, Retalhuleu, and Mazatenango. There are ten ways we can choose a set of three hospital sites from five potential locations. Once again, the question is: which set of three minimizes the cost of the people traveling to the new medical facilities?

We can simplify the problem by deciding to make the capacities of the three hospitals equal—at least to begin with. By making such an assumption, we can structure the problem as a transportation problem in linear programming.

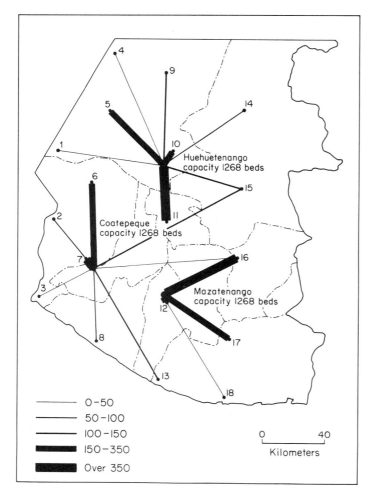

FIGURE 14–12

The cost matrix for the assignment of eighteen population cells to three out of five hospitals of equal capacity in western Guatemala (courtesy of Gould and Leinbach and T.E.S.G.).

Hospital locations to be filled
(3 to be chosen out of 5)
1 . . . . *j* 5

Population cells
to be assigned

1    $C_{1.1}$ $C_{1.2}$.

2   .

3   .

4   .

.   .

.   .

.   .

i   . . . . .C_{ij}

.

.

18                          $C_{18.5}$

FIGURE 14–13

The initial assignment of patients to hospitals with equal capacity constraints (courtesy of Gould and Leinbach and T.E.S.G.).

Huehuetenango
capacity 1268 beds

Coatepeque
capacity 1268 beds

Mazatenango
capacity 1268 beds

0 – 50
50 – 100
100 – 150
150 – 350
Over 350

0   40
Kilometers

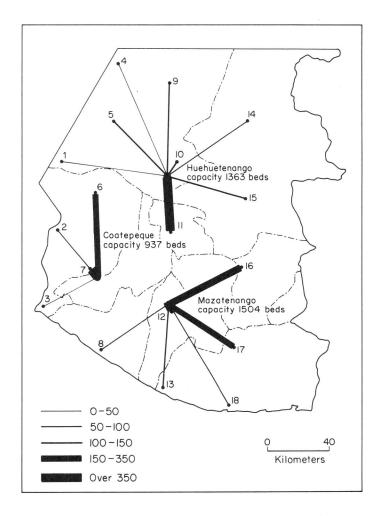

Huehuetenango
capacity 1363 beds

Coatepeque
capacity 937 beds

Mazatenango
capacity 1504 beds

	0-50
	50-100
	100-150
	150-350
	Over 350

0   40
Kilometers

FIGURE 14-14

Final solution after changing the hospital capacities (courtesy of Gould and Leinbach and T.E.S.G.).

Instead of assigning the surpluses from one set of regions to fill the deficits of another set, we now assign "surplus" patients to "deficit" hospitals. Of course, we cannot consider each person in western Guatemala separately, or we would have millions of equations to solve. Instead, we lay a hexagonal grid of eighteen cells over the population map, and concentrate all the people in a cell at its center point. By measuring the road distances between each of the population points and the five potential sites we can fill in the basic cost matrix we require (Fig. 14–12). These costs can be weighted by the quality of the roads, for in relative cost space towns and villages on rough

dirt roads are "farther away" than others on paved and tarred links.

By running the problem on the computer ten times, allowing every combination of three sites to enter in turn, we get ten different solutions assigning the people to all possible sets of three locations drawn from the five potential sites. It is now a simple matter to choose the lowest one (Fig. 14–13). This is the best solution we can achieve by linear programming methods. But notice how awkward the solution looks on the map. Many of the flows look quite sensible, but there are a number which immediately strike us as puzzling, and even irrational. Many of the people in population node 15 are assigned to the hospital at Coatepeque, when obviously Huehuetenango is closer. Similarly, we would expect people in population cell 13 to go to Mazatenango instead of Coatepeque. The reason for such seemingly irrational assignments of patients is that we have imposed equal capacities on all the hospitals. There is no question that we have the optimal solution to the problem given the equal capacity constraint. However, by increasing the capacity of some hospitals, and reducing the size of others, we can obtain even lower cost solutions.

In a series of steps, the hospitals at Huehuetenango and Mazatenango were increased to 1368 and 1504, while Coatepeque's was reduced

to 937. By moving back and forth from the computer to the map, a better assignment of hospital patients was achieved by allowing both the locations and the capacities to vary (Fig. (14–14). Whether the final solution is a local or global one we do not know. Perhaps by choosing another set of three locations, and then letting the hospital capacities vary, we may be able to find an even lower cost solution to the problem.

A particularly pertinent attribute of this locational model for regional planning is the ability to view the impact of changes in the transportation network. If, for example, we postulate a road extension in western Guatemala

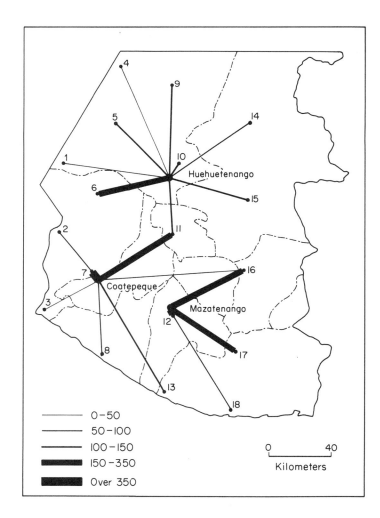

FIGURE 14–15

Changes in the patient flows resulting from upgrading portions of the road network and building a new road (courtesy of Gould and Leinbach and T.E.S.G.).

from node 6 to the hospital at Huehuetenango, and alter an unpaved road from the northern border of Huehuetenango to Quezaltenango to a tarred surface, the costs of movement change. The relative cost space warps and shrinks parts of the region. As a result, the flow patterns are also altered (Fig. 14–15). Although the optimal grouping of hospitals remains identical to that of the previous assignment, the total cost is much less as the patient flow from node 6 is now directed to the hospital at Huehuetenango instead of to the Coatepeque hospital.

The awkward cross flows from nodes 16, 13, and 11 to Coatepeque can be eliminated by successive alterations of the hospital capacities. In this way, a national government planning additions to its transportation infrastructure can simulate network development and analyze the effects on travel patterns while revising decisions with regard to hospital locations. This fact emphasizes the extent to which many planning decisions are interrelated, for in this particular example the problems of health, transport cost, network structure, and the locations and capacities of hospitals are all tightly interwoven.

Administrative Centers in Ontario

In many developed countries the problem of administering very large and complicated government services is becoming increasingly difficult. Often a single, highly centralized office is quite impracticable, and many government programs have decentralized their services. Once again, the question of multiple locations to serve the people arises. This time, however, we shall fix the number of centers and their capacities beforehand, but allow their loca-

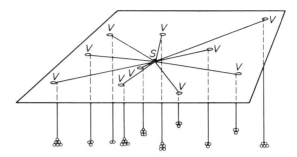

FIGURE 14–16

Weber's weight triangle solution for the location of a coffee inspection station (S) *minimizing the sum of all the distances from the producing villages* (V).

tions to vary over an infinite number of points.

When we allow locations to vary over the infinite number of points in a region, we have to tie two simpler locational models together. These are the Weberian weight triangle and the transportation problem from linear programming. Let us briefly review the two models which we are going to use to forge a new and powerful analytical tool.

Weber's weight triangle method allows us to locate a single facility to serve many other points (Fig. 14–16). Suppose we have ten villages in a coffee producing area, and we wish to set up a coffee inspection station. We can weight each village by its coffee production and find the location of the station minimizing the sum of all the distances from the villages. Whenever we face the problem of locating a single facility Weber's weight triangle can be used.

The transportation problem in linear programming assigns the surpluses from one set of areas to fulfill the deficits of others (Fig.

FIGURE 14–17

The assignment of surplus regions A, B, . . . S, *to the deficit regions* X, Y, *and* Z *using the transportation problem from linear programming.*

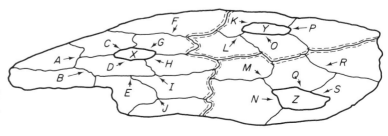

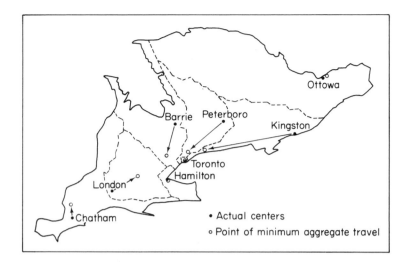

FIGURE 14–18

Initial locations, points of minimum aggregate travel, and administrative regions after the first iteration (adapted from Goodchild and Massam, Fig. 1).

		Administrative centers of equal capacity		
Townships in southern Ontario		1 . . . . . . *j* 8		
pop_1	1		•	
pop_2	2		•	
pop_3	3		•	
•			•	
•			•	
•			•	
•			•	
pop_i	*i*	. . . .	C_{ij}	
•			•	
•			•	
•			•	
•			•	
pop_{504}	504			

FIGURE 14–19

The cost matrix for the assignment of 504 townships in southern Ontario to 8 administrative centers of equal capacity by the transportation problem in linear programming (adapted from Goodchild and Massam).

14–17). Here the surplus areas $A, B \ldots S$ have been allocated to the three deficit areas $X, Y,$ and Z so that the total cost of movement is minimized. Notice how the assignment of surpluses to deficits groups the surplus areas into three regions. By tying such a regionalizing method to the Weberian weight triangle we can tackle the final multiple location problem. As an example, we shall take a problem solved by the mathematician Goodchild and the geographer Massam in southern Ontario, Canada.

Many government activities in southern Ontario are administered from eight centers. This number was chosen as the number of centers to be located in the space with its very uneven population distribution. The centers were initially located at eight main towns in the province (Fig. 14–18). These are shown as solid dots on the map. Southern Ontario is divided into 504 townships, and the populations of each were considered to be located at their geometrical centers. Thus, the problem was structured first as a transportation problem (Fig. 14–19), with the 504 surplus populations to be assigned to the eight administrative centers. The matrix recorded the costs of movement of all the township centers to the administrative points. The solution that minimized the overall costs of movement of the people to the

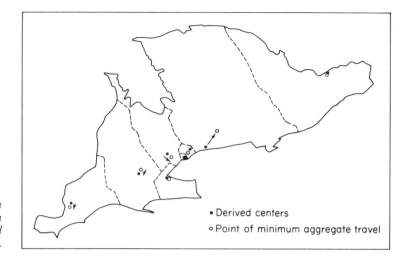

FIGURE 14–20

Locations of the administrative centers in southern Ontario after the second iteration with equal capacity constraints (adapted from Goodchild and Massam, Fig. 2).

• Derived centers

○ Point of minimum aggregate travel

administrative centers produced eight regions, whose boundaries are shown by the dotted lines.

Each of these smaller eight regions was taken one at a time, and Weber's weight triangle used to find the points of minimum aggregate travel within them (open circles). If the administrative centers are now moved from their initial locations to the points of minimum aggregate travel, the original cost matrix changes quite radically. The Toronto, Hamilton, and Ottawa centers do not shift far, but the points of minimum aggregate travel in the Kingston, Peterborough, Barrie, and London regions are a long way from the initial set of locations.

With a new cost matrix defined by the distances between each of the 504 townships and the new points of minimum aggregate travel, the transportation problem may be run again (Fig. 14–20). A new set of boundaries now defines eight new administrative regions. Once again, the Weberian weight triangle solution can be used to find the points of minimum aggregate travel within each of them. This time the minimum travel points are not quite so far from the previous set of locations. Even so, by shifting the administrative centers to the new set another cost matrix is defined between the townships and the centers. Again the transportation problem reassigns the townships to the new adminis-

trative centers so that still another set of administrative regions is obtained (Fig. 14–21). The boundaries differ only slightly from those of the previous run, and the weight triangle solutions within each of the regions show that the points of minimum aggregate travel are not far from the previous set of locations. We have here a case of spatial convergence as the locations finally settle down, or converge to a stable set of locations.

Given the fact that we required each of the administrative centers to serve an equal number of people, this is the best solution we can obtain. We do not know if it is a global solution, for it might be possible to lower the cost for the whole system by starting the initial points at quite different locations than the towns chosen in this analysis. If the points always seem to converge on the same final locations no matter where they start, we can be reasonably sure that our solution is a global one. Notice, however, that we can only be *reasonably* sure, for we do not have a tight mathematical proof that this method will find the global solution. It might instead converge on a local solution every time, leaving the global solution undiscovered and chuckling up its sleeve.

We have considered four problems in the area of spatial allocation. In the first we set the

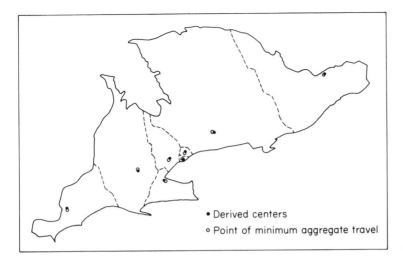

• Derived centers

○ Point of minimum aggregate travel

FIGURE 14–21

Locations of the administrative centers in southern Ontario after the third iteration with equal capacity constraints (adapted from Goodchild and Massam, Fig. 3).

number of facilities and their locations, but allowed the capacities to vary. In the second we held the number of facilities constant, confined the possible locations to a set of five Swedish towns, and let the capacities vary. In the third, we held the number of facilities constant, but allowed the capacities to vary after an initial set of locations had been discovered from a set of potential sites. Finally, we fixed the number and capacity of the facilities, but allowed the locations to vary over an infinite number of points in the region. The problem may be approached in one further way. We can allow the number and capacities to change while letting the locations also vary over a very large number of points on a grid covering the region.

THE SOLUTION OF TÖRNQVIST

Our approach changes rather drastically when we allow all three factors in the location problem to change. For our example we shall turn once again to Sweden and the studies of geographic location by Gunnar Törnqvist.

To start with, suppose we wish to locate just two plants that will make cement blocks for the building industry all over Sweden. The country is divided into 185 cells (Fig. 14–22) and in each cell we can record the weight of cement blocks usually used in a year. These figures serve as a

surrogate measure of demand for this particular product, but other values, such as retail sales, total income, or population, could be substituted for other problems. We initially locate plants A and B in the furthest cells of the space, and assign all the other cells to them on the basis of their cost-distance (indicated by small arrows). That is, we assume each cell in the country is served by the plant nearest to it— "nearness" being measured by rail cost, rather than straight line distance. This initial assignment divides the country into two regions, the northern portion being served by A, while the southern part is served by B. It is important to note that the initial locations may be anywhere in the space, for the final solution will always be the same. When the plants have been located, the demand figures for cement blocks in each cell are multiplied by the transport costs from the supplying plant. These are added together, to give us the total cost of transportation using these initial locations.

At this point the process of search starts as the plants begin looking for better locations to bring down the costs of transportation. Let us take the northern plant A as an example (Fig. 14–23). The plant moves one square north (in this starting location impossible because it hits the boundary), then one square south, then east, and finally west. After each move the total cost of supplying cement blocks to all

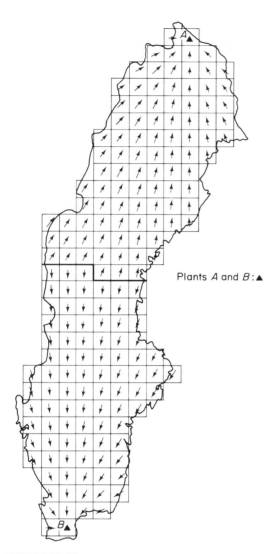

FIGURE 14–22

A grid over Sweden in whose cells we can record any suitable measure of demand. Initial locations of Plants A *and* B *are in the furthermost cells.*

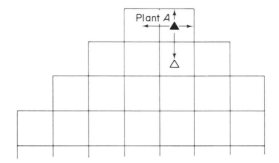

FIGURE 14–23

An enlargement of the northernmost cells in Fig. 14–22 showing the pattern of search and the first relocation of Plant A.

south, east, and west of its present position, and finds that by moving one cell north it can lower the cost of transporting blocks to the other cells still further. Now it is Plant *A*'s turn again, followed by *B*, and so on. At each turn, four cells are examined, the costs calculated, and the cell giving the lowest cost chosen as a new location. By hand it is a tedious process of search, but for a modern computer the task is easy. Finally, the plants end up at locations from which no further steps lead to an improvement and the pattern of search ends (Fig. 14–24*a*). In this solution Plant *A* locates at the town of Enköping and serves 61 percent of Sweden's total demand for building blocks. Plant *B* locates between the towns of Borås and Jönköping to serve the remaining 39 percent. At these locations the total costs of transportation are about 6 million dollars, and no further reduction is possible.

Obviously, with more plants, the costs of shipping cement blocks to the building sites (cells) in Sweden will come down. Exactly the same search process can be used with any number of plants. When four are allowed to search the space (Fig. 14–24*b*) the locations are Sollefteå, Örebro, Stockholm, and Halmstad, serving roughly equal proportions of the demand at an overall cost of $5.4 million. Six plants (Fig. 14–24*c*) reduce the cost still further to $4.8 million, while eight bring the cost down to $4.6 million (Fig. 14–24*d*).

the cells in Sweden is recalculated to see if any improvement results. In this example, a move one cell south results in a lower overall cost, so the plant relocates there.

Now it is the turn of Plant *B* in the south. It too starts to search the four cells to the north,

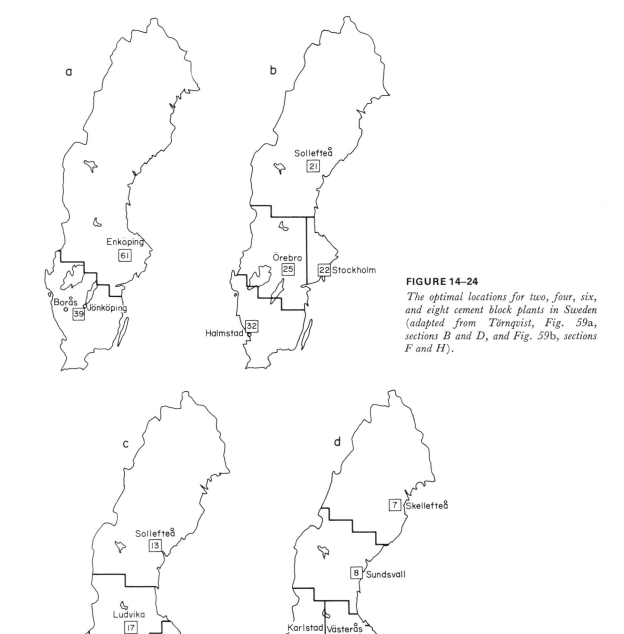

FIGURE 14–24

The optimal locations for two, four, six, and eight cement block plants in Sweden (adapted from Törnqvist, Fig. 59a, sections B and D, and Fig. 59b, sections F and H).

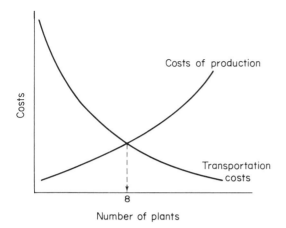

FIGURE 14–25

Balancing declining transport costs against rising production costs due to losing economies of scale. The optimum number of plants is 8.

Why not go on building cement block plants? Clearly, we could go on until we reached the ridiculous situation of each of the 185 cells with its own plant. But at this point all the advantages we gained in transport savings would have long been wiped out by the costs of building all the small and therefore inefficient plants. We would lose all the economies of scale that come from a few large and highly efficient industries. We shall have to balance transport savings against economies of scale. As the costs of transportation come down with an increasing number of plants (Fig. 14–25), the costs of production rise. The two curves cut at eight plants—our final solution.

This approach is the closest that geographers have got to solving the location-allocation problem. For all *practical* purposes the problem has been solved by Törnqvist's ingenious process of search. We also know that the solutions are global ones, providing the transport cost curves are convex downward—that is, if every mile traveled away from a plant location costs a little less than the previous one. This is the freight *taper* we have met before describing nearly all the cases we are likely to meet in the real world. Perhaps one day a complete and very elegant mathematical solution will be found in

which all three factors in the problem can vary simultaneously. In the meantime, Törnqvist's breakthrough can solve most of our multiple location problems as we plan our spatial environment to an ever greater extent.

LOCATIONS, ECONOMICS, AND POLITICS

These approaches to one of the classical problems of geography raise, once again, the question of normative solutions in spatial planning. Time and effort are precious commodities, not to be wasted in overcoming distance unnecessarily. An efficient set of locations can minimize human effort and monetary resources that can be devoted to many other things.

Unfortunately, locational decisions are often made on political grounds. In the United States, many pork-barrel projects are located in a hopelessly inefficient way by congressmen scratching each other's backs. The price of such political bargaining is paid annually by all the people in inefficient movements costing time and money. Similarly, many decisions are taken on political grounds in the underdeveloped world. Western Nigeria, for example, has an extremely poor pattern of industrial location because many decisions of spatial allocation were made almost entirely upon political grounds. The people of western Nigeria pay every year for these costly mistakes.

This is not to say that political decisions are always inappropriate. Many factors must enter into a locational decision, for there may be good grounds for locating facilities using criteria other than simple cost minimization. It may be desirable, for example, to place hospitals in sparsely populated areas because society places a high humanitarian value upon the provision of medical facilities to all the people. Similarly, in areas of high density, overcrowding, and high environmental stress, it may be desirable to decentralize many facilities—even though monetary costs rise. The important thing is that people making decisions on essentially political grounds should be aware of the costs and benefits involved. Thus, in both the developed and underdeveloped world, normative solutions can play an immensely valuable role in the spatial decision process. Faced with a normative solution, a politician or a legislative body is at least aware of the cost of deviating from such an optimal pattern of facilities. With an optimal solution in front of him, he may be far less willing to deviate for selfish reasons—especially if a free press publicizes the normative solutions to the people who elect the decision-maker.

Suggestions for Further Reading

BEER, STAFFORD. *Management Science.* Garden City, N.Y.: Doubleday & Company, Inc., 1968.

GODLUND, SVEN. *Population, Regional Hospitals, Transport Facilities and Regions: Planning the Location of Regional Hospitals in Sweden.* Lund: Gleerup, 1961. Lund Studies in Geography, Series B, Human Geography, No. 21.

GOODCHILD, M. F., and B. MASSAM. "Some Least-Cost Models of Spatial Administrative Systems in Southern Ontario," *Geografiska Annaler,* LII, B:2 (1969), 86–94.

GOULD, PETER, and THOMAS R. LEINBACH. "An Approach to the Geographic Assignment of Hospital Services," *Tijdschrift Voor Economische en Sociale Geografie,* LVII (1966), 203–6.

MORRILL. R., R. J. EARICKSON, and P. REES. "Factors Influencing Distances Traveled to Hospitals," *Economic Geography,* XLVI (1970), 161–71.

Works Cited or Mentioned

BEER, STAFFORD. *Management Science.* Garden City, N.Y.: Doubleday & Company, Inc., 1968.

GODLUND, SVEN. *Population, Regional Hospitals, Transport Facilities and Regions: Planning the Location of Regional Hospitals in Sweden.* Lund: Gleerup, 1961. Lund Studies in Geography, Series B, Human Geography, No. 21.

GOODCHILD, M. F., and B. MASSAM. "Some Least-Cost Models of Spatial Administrative Systems in Southern Ontario," *Geografiska Annaler,* LII, B:2 (1969), 86–94.

GOULD, PETER, and THOMAS R. LEINBACH. "An Approach to the Geographic Assignment of Hospital Services," *Tijdschrift Voor Economische en Sociale Geografie,* LVII (1966), 203–6.

TÖRNQVIST, GUNNAR. *Studier i Industrilokalisering.* Stockholm: Geografiska Institutionen vid Stockholms Universitet, 1963.

WILDE, DOUGLASS J., and CHARLES S. BEIGHTLER. *Foundations of Optimization.* Englewood Cliffs, N.J.: Prentice-Hall, Inc., 1967.

CHAPTER 15

THE GEOGRAPHY OF THE FUTURE
AND THE FUTURE OF GEOGRAPHY

My interest is in the future, because I'm going to spend the rest of my life there.
—Charles F. Kettering

The future is unavoidable because it has already begun. Just as the past exists as present memories, the future exists as present expectations. We cannot escape the future because we are creating part of it today, whether we do so consciously or unwittingly. The future is not an exogenous variable in human affairs. Present spatial structure is a major determinant of ensuing spatial process, and the locational decisions we make today will constrain future actions in the same way that we find our activities constrained by locational decisions made decades ago. The spatial organization of the future will develop from existing spatial structures, from trends that have already started, and from new developments which are difficult to forecast.

What must be recognized at the outset is that the conceptual tools we have for thinking about the future are very limited in number, and that the few we have are exceedingly crude. We do not know how to think rationally and productively about the future, for we are simply not accustomed to thinking about it. For that reason, the science of futuristics (technological forecasting and long-range planning) is still in its infancy. Nevertheless, we cannot avoid the necessity to try to think about the future no matter how clumsy or occasionally ridiculous some of the results of such attempts may be. Like it or not, man has gotten himself into a dilemma in which he must think about the future or he will not have a future to think about. As the scale of human operations and the sizes of the systems people build expand, the magnitude of serious mistakes grows also, probably at a greater rate than the size of the systems involved. Today, man has managed to organize the world into such large units, and he has constructed systems of such power, that there are some mistakes we simply dare not make. Any one of a number of possible (and in some cases highly probable) sequences of events must not be allowed to occur in the next hundred years, lest the species destroy itself.

FUTURE SPATIAL ORGANIZATION

The necessity to develop techniques for thinking about the future is incumbent upon all scholars, but especially on geographers. Spatial and locational decisions usually have very long "half-lives." Because large fixed investments usually accompany locational decisions, mistakes cannot easily be remedied. A shortsighted decision about the width of a street, the location of a manufacturing plant, the site of a city, or the location of a residence is rescinded only at great cost, if at all. More often than not, relocation costs are so great that many locational decisions which should have been rescinded at some later date have been allowed to stand. Once started, most locational and agglomerative processes operate in a circularly causal fashion which is very difficult to interrupt. We are only now becoming aware of the costs we incur by maintaining some of our traditional and contemporary settlement patterns, and an overriding concern of the next several decades will be to decide what kind of spatial organization we want for the future. For that reason, the most important goal to which geographers can and must devote their spatial expertise is an attempt to forecast the gross organization of people and their activities in terrestrial space. In this discussion, we shall talk largely about extrapolations of past and existing trends, although we shall touch upon the more difficult matter of forecasting the evolution of new trends and patterns.

The Trend toward Local Dispersion

The dominant trend over the last 250 to 300 years has been agglomeration, although we must be careful to specify what this term means. Urbanization, a positive change in the *proportion* of a nation's population residing in urban areas, has increased greatly in the last three centuries. But not all urbanization is attributable to movements of people from rural areas to the city. Urbanization is more a change produced by natural growth within cities themselves. Small numbers of migrants transfer their potential growth from rural to urban areas when they move, so that natural increase in the cities

rather than migration itself is mostly responsible for urbanization. We emphasize this point because urbanization is sometimes referred to as an agglomeration process, and one conjures up an image of hordes of people migrating from rural areas to the cities. In fact, our ideas about the degree to which agglomeration, in the sense of migration into already dense areas, has been a dominant process are somewhat exaggerated.

Nevertheless, a greater proportion of the world's population is living on a smaller fraction of the world's land today than was the case 250 years ago, and in this sense, agglomeration has occurred. If we look at the ways people have occupied that smaller fraction, however, it is clear that *within* urban areas agglomeration has decreased over the last hundred years. With each census, cities in the United States and other areas of the world show decreasing population densities. We saw in Chapter 10 that successive increments to cities in the last century have been characterized by progressively lower settlement densities. Contemporary additions to metropolitan areas in the United States are typically at residential densities of 3,700 people per square kilometer (9,500 per square mile), whereas the density of population in portions of large cities settled in 1860 could be as high as 38,500 per square kilometer (100,000 per square mile).

The relation between the rate of increase of a metropolitan population and the rate of increase in metropolitan area indicates that dispersion is occurring within a general context of agglomeration. If urban population were growing at the same rate as city extent in absolute space, cities would not be dispersing. But this has not happened. In economically developed nations, cities often expand areally three times as fast as their population growth, although an areal growth rate of twice the population growth rate is more nearly average. Thus cities are dispersing at a rate faster than new people are coming into them by natural increase and migration.

Transportation systems have produced the time- and cost-space convergence which makes such local dispersion possible. By permitting people to move over distances more rapidly and at lower cost, systems like trolley networks and automobiles have shrunk urban space. In relative space, cities are as compact as they have ever been, if not more so. But in absolute space, cities and their internal activities have dispersed over a far wider area than was possible in the middle of the nineteenth century, when people had to walk or ride horses to their jobs.

Because transportation is more efficient and people are more affluent, decisions about the location of residences, the largest land use in cities, can now be made much more on the basis of amenities for comfortable living than was ever possible before. Whereas the residence was at one time tied closely to the location of the job, absolute distance from the job is no longer very important. Increasingly, residential locations are chosen to maximize access to available amenities and to fulfill the desire for personal and family space. In the United States, families with children prefer the private space which is provided by single family dwellings on individual lots in suburban areas. In part, such preferences are culturally determined. Whether urban people with non-Western cultures will choose more dispersed residential patterns as the economic constraints which now prevent them from doing so are relaxed remains to be seen.

A second amenity which is promoting dispersion in absolute space is recreation. An increasing number of families in developed nations owns a second, recreational dwelling. On weekends, holidays, and vacations many residents of American metropolitan areas resort to cottages within three hundred kilometers of their weekday locations. Summer weekends are spent fishing, boating, or loafing in a more bucolic environment than is available immediately adjacent to metropolitan areas. Winter weekends are devoted to hunting, skiing, and snowmobiling. As highways are continually upgraded and as the amount of leisure time increases, a larger number of the more affluent members of many societies will spend more of their time in the second, more dispersed amenity residences.

Over the last century, but especially in the last several decades, an increasing number of people have been able to indulge their desire for pleasant surroundings in making locational decisions. Whereas at one time people were forced to live at very high densities in order to have access to employment, today the automobile makes it possible for people to give primary consideration to amenities and to consider the location of the place of work as only a secondary factor. Amenity, rather than location close to work, has become the foremost consideration in residential location for the more affluent members of the family of man.

Secondary and tertiary activities have also dispersed from earlier locations in or near the central portions of cities. Those of you who have lived in American metropolitan areas are familiar with the scattered clusters of plants and shopping centers located on the margins of metropolitan areas at points where radial and circumferential highways intersect. These junctions are very accessible locations, and are attractive both to firms which must draw their labor force from other parts of the city and to entrepreneurs who must attract large numbers of customers. Because both workers and customers are wed to automobile transportation, large parking spaces must be provided, and this is more easily done at lower cost suburban locations.

Reviewing the locations of human activities, it is clear that there is a trend toward dispersal. People engaged in primary activities like farming or forestry have usually been the most dispersed of all, since primary activity normally requires people to get as close to the resource area as possible. Although in the past secondary and tertiary activities were responsible for agglomeration, since World War II they too have begun to disperse in response to the impact of automobile transportation.

Quaternary activity also was highly agglomerated in the past, but because of developments in communications technology, it is now dispersing, and is in fact the most portable of the four kinds of economic activity. The fundamental inputs and products in quaternary activity are information, and because information is more transferable than the raw materials and products of any other industry, quaternary activities will disperse even more in the decades ahead than they have in the last twenty-five years. Like commerce, quaternary functions were long confined to the central city, but recently they have joined the migration away from the CBD. In the Washington, D.C., area, many government agencies such as the Central Intelligence Agency, the National Bureau of Standards, and the National Aeronautics and Space Administration have been dispersed to locations fifteen to thirty kilometers from the Capitol complex. Even corporate headquarters, long wed to the central city, are now being located in suburban areas with increasing frequency. The existence of large colleges and universities in small towns is evidence that education thrives in locations distant from the major central places of a nation's urban hierarchy. Our prognosis is that the spatial arrangements of people and their activities will become even more scattered at local scales as more efficient space adjusting techniques enable private, business, and government decision-makers to weight amenity even more heavily in locational equations than they do now.

Centralization

The fact that we will enjoy more dispersed patterns of activity in the future does not mean that societies will be less centralized. Because of advances in communications technology in the last thirty years, it is now possible to have centralization without agglomeration. Formerly this could not be done. When the technology of movement was primitive, centralization—control vested in a small number of people or a single person—was not possible without agglomeration. In both private and government operations small units were preferable because of the problems of coordination and control encountered in large operations, especially if they were areally extensive.

At the present time it seems as though there is virtually no upper limit on the size of the system which can be centrally monitored and controlled in a nation that has at its command modern transportation and communications techniques. The gradual evolution of centralization is perhaps most easily seen in the areal expansion of political units. Government is implicitly a space-filling function. Control and services must be available over the entire area a government claims if such claims are to be defended. Other things being equal, the distances over which a government can extend its control and services determine the size of its territory. As space-adjusting techniques have become more sophisticated, man has been able to organize increasingly larger units of space into coherent, centrally controlled blocks. At one time a clan or a tribe was the largest group of people which could be kept together and coordinated, and the amount of space which could be organized was correspondingly small. At a later date, improved transportation and communications systems enabled talented men to organize larger regions in the river valleys of the East and Middle East, but these were still small by modern standards. Even the Roman Empire, perhaps the best example of the interrelationship between communications and hegemony, was a rather small operation compared to some of today's spatio-political giants.

Modern communications and transportation technologies make it possible for a powerful government to deliver its military forces and ideas to the entire world. The United States can deliver nuclear explosives and propaganda to any place on the surface of the earth, as can the U.S.S.R. Both can deliver conventional military forces to any place not strong enough to defend itself or closely allied to the opposing power. The leaders of the world's superpowers can run their far-flung domains from almost any location on earth, thanks to the power of modern communications systems. Whereas at one time agglomeration and proximity were prerequisite to control, this is no longer the case. Control can be extended over the entire world from a single point.

At the same time that modern movement systems allow us to organize ever larger areas by shrinking absolute space, increasing communications power encourages us to administer larger areas from smaller numbers of power centers. In the United States, for example, people are more concerned with the decisions made at the federal level than ever before. Federal government policy today has far more impact on daily life than it had in the past; indeed, it usually has a greater impact than state or local government. As space shrinks, control agglomerates. In the same way that time-space convergence tends to eliminate lower and middle level central places, time- and cost-space convergence saps the power and importance of lower and middle level governments.

The development of truly national societies in which differences are more a function of class and occupation than of location is evidence of the centralizing power of modern space-adjusting systems. In the United States, centralization is increasingly focused on and in the person of the President. Mass communications media have welded the nation's population into a center-focused mass. Franklin Roosevelt instituted radio "fireside chats" in the 1930s in an attempt to counter waning confidence in our economic and political systems, both of which were being sorely tried by the Great Depression. Ever since, the President of the United States has been largely a creation of the press, radio, and television, through which most of the electorate receives what information it has about him. At few previous times in history has a nation's concern been so intensely focused on a single person as was the attention of the population of the United States during the administration of President Lyndon Johnson.

By making information about people, events, and controversies universally available, communications systems inevitably focus national attention and concern. With attention thus focused, those at the power center can usually manipulate opinion and policy more readily than they could when attention was diffused.

Private or interpersonal communications media have also played a part in promoting increased centralization. Corporate management and government agencies could not possibly exist without postal and telephonic communication. Electronic communications systems and information handling equipment have been especially important in promoting the centralization of decision-making. Decisions once were made at dispersed locations because of difficulties encountered in gathering the information necessary to reach a decision at some central location. Now we can easily collect and evaluate data at central locations. One example of this change is the amount of discretion allowed diplomats in making commitments for their governments. In the eighteenth century it took someone like Ben Franklin six weeks or more to get a message from Paris to Philadelphia. Thus he necessarily made many decisions himself, decisions a contemporary ambassador, who can consult his superiors in a matter of minutes or hours, would not dream of making. A few executives in command of all relevant information can now make decisions which might formerly have required a corps of middle- and lower-level functionaries at many dispersed locations.

It seems paradoxical that better transportation and communications promote the dispersion of activities and the centralization of control and management at the same time. Yet this appears to be the case now, and we assume this trend will continue in the future. Because these systems shrink absolute space, they make it possible for people to spread themselves over a greater area while still interacting intensely. If the amount of time- and cost-space convergence produced by movement systems in a period is greater than the dispersion during the same period, the interacting places are closer to each other in relative space after dispersion than before, despite the fact that considerable absolute distances may be involved. Because time- and cost-space convergence produced by communications systems is increasing at a more rapid rate than dispersion, centralization is increasing even as dispersion continues. Increasingly we live in societies which are spatially dispersed but functionally agglomerated because they possess highly centralized control activities.

Centrifugal-Centripetal Trends and the Location of Human Activity

Constantine Doxiadis and his associates at the Athens Center for Ekistics (the science of human settlements) predict the evolution of Ecumenopolis, a universal city which will consist of the entire inhabitable area of the earth (the *ecumene*). This world city will develop over the next fifty to seventy-five years, and will be one interconnected, functionally continuous (though not spatially contiguous) urban system. Time- and cost-space convergence will continue until the space of the earth has shrunk to the extent that all people will be as accessible to each other as are contemporary residents of a metropolitan area. Improved movement systems will make it possible to construct this worldwide network of urban life.

Here again, electronic intercommunications systems will be fundamental bases of the world city. Modern cities, and especially their central business districts, are, in the final analysis, media of communication. Cities are physical artifacts which permit large numbers of people to come into contact with one another, sometimes for industrial and commercial purposes, but often in order to exchange information. Cities substitute for the transportation which would be necessary if people were more widely scattered. Nevertheless, considerable movement is still necessary in most cities, as rush hour traffic jams indicate. Many urban residents spend inordinate amounts of time commuting. By substituting for many of the movements which people now find necessary and by making the ability to diffuse information almost completely independent of distance, future electronic communications systems will produce almost complete cost- and time-space convergence. For purposes of sending and receiving information, all places in the world will be located at a single point, just as telephone subscribers within a metropolitan area are now located at a single point in communications space—that is to say, the cost and speed of reaching anyone within the city are the same. Ecumenopolis will be a communications city. We will create the universal city by producing an isotropic communications and information surface over the entire earth.

Some portions of Ecumenopolis are already resolving. Jean Gottmann identified the first Megalopolis as the coalescing system of metropolitan areas along the eastern seaboard of the United States between Boston and Washington, D.C. Although the region is not a single city, it is rapidly becoming a megalopolitan area as separate metropolitan areas and their outlying subordinates expand and reach out to one another. Similar megalopolises are becoming evident elsewhere. The triangular area formed by Dortmund, Lille, and Amsterdam, for example, is a small area containing many

coalescing, high density urban regions. In the more extensive spaces of the United States, two additional megalopolises are developing. The Chicago-Detroit-Pittsburgh axis is one, and the San Diego-Santa Barbara or San Francisco concentration is the second. By the year 2000 we may have three identifiable megalopolitan areas in the United States, Boswash (Boston-Washington), with a population of 80 million, Chipitts (Chicago-Pittsburgh), with a population of about 40 million, and Sansan (San Francisco-San Diego), with 20 million people (Fig. 15-1).

The process of megalopolitanization is further evidence that local dispersion and agglomeration are occurring simultaneously. Agglomeration will continue as an increasing fraction of the nation's population is located in megalopolitan areas. But the increments to megalopolitan areas will be characterized by lower settlement densities to the extent that people continue to exercise their options for low densities and amenity locations. Several factors could retard or halt the trend toward lower densities. A stable population might slow the dispersion of urban areas. Also, governmental decisions to halt megalopolitan spread might be necessitated if the social costs of very low density urban regions were excessive. Over the short run, however, and probably over the long run as well, we think that the impact of space-adjusting systems will be such that peripheral areas will win out in the continuing competition between traditional cores and outlying areas to attract new growth to megalopolitan systems.

As with the death of Mark Twain, reports of the demise of central business districts and central cities have been greatly exaggerated. When we speculate about the geography of the future, we must always weigh technological possibilities against social and political probabilities. There is much more to a city than economics and efficiency. Central cities are powerful social and cultural systems, and many locational decisions are based on criteria other than simple economic efficiency. Even if it becomes less expensive for society as a whole, and for entrepreneurs as individuals, to shift to much more dispersed locations, we might choose

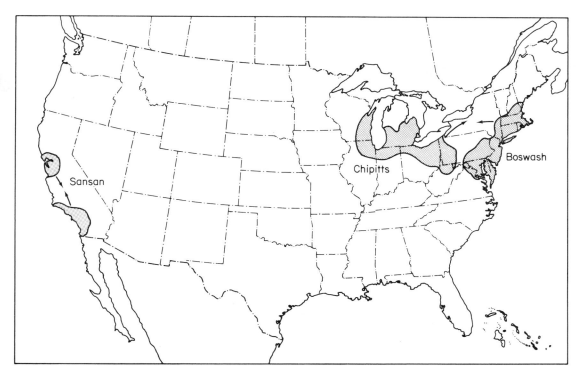

FIGURE 15–1

Megalopolitan areas in the United States in the year 2000.

not to optimize and continue to live at very high densities in central cities. Many people really like cities and the variety they offer despite their diseconomies, and as some wag once noted, it will be very hard to find a satisfactory electronic substitute for the business lunch.

Nonetheless, we feel that enough people have already voted with their locational options to indicate that in situations in which other factors are relatively equal, dispersion with amenity is preferred to increased local agglomeration. Moreover, the mere possibility of dispersion will probably provide an adequate incentive to disperse. When the automobile was first invented, many thought it would never be useful outside cities because cities were the only places with adequate roads. But the possibility of motor travel itself was a sufficient condition of the road building programs which produced

the road networks we now enjoy. Similarly, the technological capability of additional dispersal within great megalopolitan regions will promote areal extension, because it acts in conjunction with the desire for amenities which can only be provided at low settlement densities.

Lags will continue between what is technologically possible and what is accepted practice. Even when it is in a person's best economic interests to adopt an innovation, he is frequently reluctant to do so simply because the new way of doing things is different from the traditional way. It is now technologically possible for executives scattered over the nation to hold business conferences via closed-circuit television systems. Yet businessmen we have consulted feel that this medium will never be a satisfactory substitute for personal contact. Over the short run, such acceptance lags may delay dispersion of activities somewhat. But in the long run such opposition is as transitory as was the generation which refused to conduct any business on the telephone. In the same way that this generation of recalcitrants was replaced by a new generation of businessmen who had grown up with the telephone and considered it an appropriate medium for business communication, a new generation of executives, weaned on television during childhood, will make much more extensive use of electronic communications systems in the future. In doing so, they will help bring about Ecumenopolis.

PROBLEMS OF FUTURE SPATIAL ORGANIZATION

Localized dispersion within the context of further agglomeration into megalopolitan regions will produce problems whose solutions will require all our talents. We have no intention of providing a complete catalog of future problems, even if we thought we were capable of doing so. But we think it is important to mention a few that might arise. After all, people who are now students will be reaching the peaks of their careers sometime around the year 2000, and they are the people who will be called upon to help solve the problems of those decades. It is our experience that the longer any of us reflects on a problem, the more successful we are at devising solutions. Perhaps by starting to think about future problems now, you will be able to solve them more easily in the decades ahead.

Short-Term Problems (the Next Fifteen Years)

Some problems are relatively immediate, and among these is the problem of population size. Barring major catastrophes, the world may have a population of about 7 billion people by the year 2000, twice the number who were alive in 1968. Now there is nothing inherently problematic in numbers of this magnitude. So long as there is an adequate supply of natural and cultural resources to feed, clothe, educate, and otherwise care for new members of the family of man, it can expand considerably. What

concerns us so much, of course, is that today we do not possess the ability to support even half of humanity comfortably, and the chances of increasing productivity at the same rate as population growth, so as merely to maintain present levels of support, are decidedly dim. If there is a local disparity between the present ability of a region to support people and the number of people at that place, it is fairly obvious that one of three things must happen. People in such areas may not be supported, in which case they will die or be forced to live rather miserable existences, depending upon how serious the disparity is. A second alternative is for them to move to other areas where there is a positive disparity between numbers of people and resources. The third alternative is to provide such places with the resources they need.

Assuming that we reject the first alternative on humanitarian grounds, we are faced with the prospect of forcing or enticing people to move from areas of low human carrying capacity to areas which are "understocked." We must also, of course, identify areas with surplus carrying capacity and remove whatever political or cultural barriers exist to block population movements. Whether we choose to move surplus people to cultural and naturally produced resources, or to move surplus resources to people in deficit areas, we encounter again a transportation problem. We have a statistical surface of demand and one of supply, and the problem is to find the set of flows between areas of supply and demand which minimizes movement.

A second way we can help reduce man-resource imbalances is either to prevent them from occurring in the first place or to alleviate them in places where they already exist. By developing sound theory concerning the operations of spatial diffusion processes, geographers can help design effective strategies by which birth control methods can be diffused to populations in areas where human and physical resources are inadequate. The decision to use birth control measures is the acceptance of an innovation which is in some respects similar to the decision to use hybrid corn or an improved variety of wheat. Birth control often encounters more significant cultural barriers than other decisions or ideas less loaded with moral implications. Even so, geographers can develop strategies to blanket areas with information by identifying the places which are nodes of information flow through which the entire population may be reached most effectively.

A third way to solve man-resource imbalances is to increase available resources. A doubling of population in the next thirty years would place so much pressure on existing resources that geographers and other scientists will have to reevaluate the potential productivity of some parts of the earth. Increasingly, we may look at the world's seas not as barriers to travel or as convenient highways, but as sources of mineral and agricultural resources. Certain species of marine life are already being domesticated and cultivated (oysters and shrimp, for example), and this trend will continue in the future. Food production in water can be more efficient than production in soil. Similarly, many minerals are present in seawater in concentrations which may soon be economically exploitable, and some minerals are precipitated to the bottom of the sea in exploitable concentrations. We became accustomed in the 1960s to looking to outer space as the locus of opportunities for the future. In fact, the "megabucks" allocated to space exploration might have been much better spent exploring the inner space of the aquatic portions of our world. Over the short run, we will probably get far more from the oceans than we will get from space.

Knowledgeable people disagree on how productive the earth's oceans will be. Despite the current inclination to be pessimistic concerning the resources which can be produced from the oceans, we still feel that over the long run the oceans will be far more productive than we now think. Man is, after all, a land animal, and in most respects he is as out of his element in the oceans as he is in outer space. It is our suspicion that much of what exists in the seas of the world cannot now be evaluated as potentially productive largely because we are not imaginative enough at present to see how it might be used. As we devote more attention to ways for making oceanic resources productive, we will invent ways of making apparently useless materials usable, and as we do so we shall create resources where none can now be seen.

Given potential resources in oceans and seas and the pressing need for resources to support increasing population, questions of ownership of the sea and its resources will soon arise. The high seas were traditionally unclaimed largely because no government possessed the technology necessary to control and govern them. For the same reason, the sea's resources were considered to be fugitive—that is, subject to unlimited exploitation on a first-come, first-served basis. Five-kilometer (three-mile) territorial limits are spatial relics of the era when that was the range of a cannon. The more recent extension of territorial sovereignty to nineteen kilometers by many nations, and even further by some, is simply a tardy recognition of the fact that nations are now technologically capable of exercising control over large areas of water if they wish to do so.

Suppose the extension of territorial limits continues, and the sea is divided among the coastal nations of the world. What would the territorial map of the world look like after the oceans had been claimed in this fashion? Portugal and Great Britain, because of their island possessions, would dominate the North Atlantic (Fig. 15-2). For the same reason, the United States would dominate the North Pacific. Is this an optimal way of allocating this resource? Might there not be a better or more equitable way?

Portugal is a poor nation, and some would assert that such riches as Portugal might reap by owning much of the North Atlantic would help improve her economic position among the family of nations. But the United Kingdom and the United States are already wealthy and many would question the wisdom of a policy which adds to their wealth and thus increases the gap between them and most of the other nations of the world. And what of nations unfortunate enough to be without shorelines? Are they to be denied a share of the riches of the world's oceans because of their locations? A much more equitable policy would prevent any nation from claiming sovereignty over the

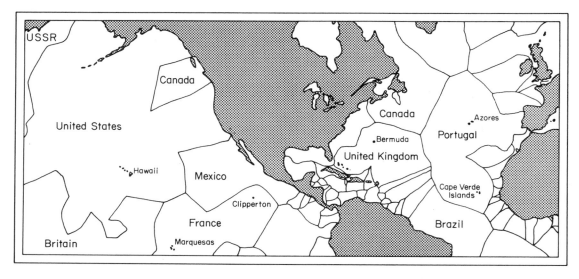

FIGURE 15–2

Division of the Atlantic and Pacific which would result if each existing nation extended its sovereignty outward at an equal rate until it met the ocean territory of adjacent and opposite states (courtesy Christy and Herfindahl, The Law of the Sea Institute, pp. 41–42).

sea. Those who wished to exploit its resources would make royalty payments to a central agency which would then disburse the proceeds to nations in proportion to their populations. Or perhaps the profits could be placed in trust for underdeveloped nations. The gap between the developed and the poor nations of the world is increasing rather than closing. Granting underdeveloped nations the resources in the world's oceans might be a way of helping them close the gap at relatively little direct cost to the already developed nations.

Getting governments to agree to such a scheme would be extremely difficult, for while many would benefit, some would certainly lose, and the ones which stand to lose the most in opportunities foregone are those which are now most powerful. Instituting such a policy would require more political integration at supranational levels than now exists. The size of political units has responded to improvements in organizational technology, and the trend toward larger units has been relatively constant. But political integration probably becomes more difficult as the sizes of the units to be amalgamated increase, simply because more people have to be convinced that integration is a wise policy. Organizations like the Common Market and the United Nations prove that some kinds of integration are possible, but the troubles they encounter also indicate how difficult supranational cooperation is. Successful melding of nation-states into larger political units is made desirable because of the economies of scale which can be realized by large units, but at the same time achieving it is made difficult because of the cultural differences and passions involved.

Human geographers must teach people the world over about differences in human beliefs and behavior, and about the barriers such differences can place in the way of effective interaction in a world in which places are converging on one another at a rapid rate. By bringing cultural differences to the level of consciousness and thus enabling people to recognize them for what they are, geographers and other social and behavioral scientists can help all the people of the world to understand the behavior of those from other culture realms. People with European attitudes, for example, are offended by the way people from the Middle East violate the sphere of personal space each of us carries around. A frequent comment by Europeans about foreigners is "they are so *pushy*." For most Europeans, most foreigners are indeed pushy because the absolute dimensions of their personal space needs are smaller than those of Europeans. Since a northwest European or an American will usually take evasive action to prevent people from getting too close to him, both parties will be upset. A typical Arab, for example, perceives the social distance which Europeans consider correct as evidence of insincerity. If both Europeans and non-Europeans are consciously aware of their own feelings about the dimensions of personal space and those of other people with whom they interact, communication is possible without conflict and misunderstanding. A compromise distance may be reached, or one party may acquiesce and maintain the other's proper distance without becoming offended, for he recognizes that his companion behaves the way he does not out of malice or ignorance, but simply out of cultural habit.

We have chosen the example of personal distance to emphasize the importance of being consciously aware of one's own behavior patterns as well as those of different people we may encounter. But the same principle holds for other kinds of differences, whether they be differences in political, economic, religious, or other behavior patterns. If we can raise to the level of *mutual* consciousness the habits of thought and behavior which cause political, economic, and social conflict, there is a good chance that such differences can be minimized and that political integration can proceed satisfactorily. Given the potential we have now acquired of destroying our species with biological, chemical, and nuclear weapons, it is fairly obvious that supranational political cooperation and integration will have to evolve if we are to have futures worth worrying about. In the realm of conflict resolution, geographers might make some useful contribution by developing a better understanding of group space and national space. To cite two recent cases of conflict, the strategic balance of power was changed little

if at all by the presence of offensive missiles in Cuba (1962) or by the establishment of a more liberal government in Czechoslovakia (1968). Yet the United States and the Soviet Union both took what in retrospect were very predictable actions to remove threats which they *perceived* to be extremely serious, largely because of their proximity to national space. We badly need information on the spatial structure of perceived threat values so that neither side in the Cold War will make a fatal miscalculation of its opponent's sensitivity to some proposed action.

People cause problems in ways other than by putting pressure on resources or by their political behavior. Sheer numbers of people *may* have adverse effects on man, although evidence is not altogether clear. It is possible that each person has a certain psychological resource which is partially consumed by each person he sees or meets in a day, week, or month. Very high population densities in central cities appear to engender stress which, in turn, aggravates existing social disorders and may create its own. Animals allowed to increase until very high population densities are attained exhibit social and physical deterioration. Under such conditions rats, for example, become homosexual and violent toward others of their own species. If people react in somewhat similar ways to crowding, we will have to redesign our cities so that high densities or the feeling of high densities are avoided and people encounter one another in much smaller groups. This is a topic about which we know little. Evidence from both the animal and human realms is confusing, conflicting, or nonexistent.

Understanding spatial behavior under various physical and cultural conditions is critical for the world of the future. Making and keeping Ecumenopolis and its constituent megalopolises psychologically suitable for human habitation will tax the spatial design talents of all of us. All the behavioral sciences must work together to design urban modules which are of viable size for the future but which at the same time are small enough to maintain pleasant human scales. Ways of interconnecting and locating such modules in space so as to maximize resident convenience and minimize system costs will be difficult to devise, yet this must be done if

Ecumenopolis is to succeed. As we saw in Chapter 14, we sometimes have spatial problems that involve conflicting goals. Designing systems which produce a spaceless world, and at the same time keeping the costs of such systems within reasonable bounds will provide plenty of problems for the science of spatial organization to solve. There are always at least two viewpoints on the operation of any spatial system, the customer's and the system manager's. Customers can afford to take the view that the world is really timeless and spaceless if systems make it so, but systems managers know better. The space- and time-smoothing systems of the future will present managers with problems of spatial organization which will dwarf those we encounter today in complexity and difficulty.

In addition to making Ecumenopolis psychologically viable, we must also ensure that it is physically habitable. The enormous consumptive and productive power of existing megalopolitan regions and the equally enormous problems of pollution which production and consumption now create lead us to predict that such problems will grow increasingly acute if population doubles in the next three decades and if the productive and consumptive power of both developed and developing nations increases. Nations like the United States have already used the air and surface waters for sewers for so long that there are serious questions as to whether some heavily polluted resources, like Lake Erie, can ever be restored to cleanliness and life. Aside from obvious pollutants like smog, sewage, and garbage, we assault ourselves with noises which set our nerves on edge and produce further stress. We are just beginning to realize how serious noise pollution is and how much it upsets our lives. Here again, the geographer can help in the task of building into Ecumenopolis a spatial structure which minimizes noise, or in which noises which cannot be avoided are confined to locations where they do not bother people.

We are only now beginning to realize what an incredibly intricate life support system the physical earth is, and how delicately it is balanced. We are also beginning to appreciate how easily it can be thrown into disequilibrium by human activity, with potentially lethal consequences for all of us. DDT concentrations, for example, are already so high that many nations and states have banned its use. Even in the oceans, levels of DDT pollution are such that some bird species which feed only in the open sea will be extinct by 1980. The food chain from plankton to fish to bird concentrates DDT into doses which can and have become lethal at some steps in the chain. Using DDT and other insecticides in the interiors of continents cannot be divorced from events thousands of kilometers away, for such relatively indestructable compounds eventually spread through the entire ecosystem.

The challenge of the future in this regard is to design and build a spatial settlement system which does not overload physical systems, which are themselves inherently spatial (e.g., wind and air mass systems, river systems, ground water, etc.). In time we may be able to manipulate physical spatial processes to make them accord with our ideas about optimal urban spatial structure. But over the short run we shall have to design urban spatial structures with a healthy respect for the spatial structures of physical systems, or we shall suffer the unpleasant and expensive consequences. Rather than dumping our refuse into physical systems in the fond hope that it will go away if we do not worry about it, we will have to recycle "waste" products by using them as raw materials. If it is true that a gram of prevention is worth a kilogram of cure, we shall find it cheaper in the long run to treat our physical environment with the respect it deserves.

Intermediate-Range Problems (1985–2000)

The problems we have mentioned thus far are those of an immediate and practical nature which grow out of man's propensity to reproduce at rapid rates, to take a rather cavalier attitude toward the slaughter of others of his own species, and to abuse the natural systems which surround him. These are vitally important problems which must be solved soon, even at the expense of inadequate attention to long-term problems. If we do not solve the interrelated problems of population growth, resource balances, political integration, and environmental pollution, we shall not be around to worry about intermediate- and long-range problems. Nevertheless, we anticipate several intermediate-range problems to which we should devote some attention.

The idea of planning or zoning land use is relatively recent even in urban areas, and we have scarcely begun to plan land use in rural areas. But cities are our spatial laboratories. Many techniques of organization developed first in the urban environment are subsequently extended to much larger areas. The technology of telephonic communications developed to serve cities in the 1920s and 1930s, for example, is now used on a continental basis in North America. Similarly, techniques of land use zoning and planning which we are now perfecting in our high density urban laboratories will, of necessity, eventually be applied to large regions, to continents, and even to the whole earth. An urban society cannot afford to let incompatible land uses and congestion stifle its progress by generating large social costs. Worldwide land use zoning will be necessary to prevent serious and irreparable damage to environmental systems, as well as to keep man's spatial artifacts in good working order. Air pollution, to take one problem, is usually thought of as a problem which can afflict individual metropolitan areas. It is increasingly evident, however, that air pollution in the United States and northwest Europe is a regional problem. In order to control air pollution in the northeastern United States, regional action must be taken over areas as large as the entire Boston-Washington urban complex. National land use zoning and planning are already practiced in some of the smaller and more densely settled nations of the earth like The Netherlands, and we can expect this kind of comprehensive planning to be extended to larger nations and to the entire earth in the future.

The location of industrial and transportation activities, for example, might have to be regulated on a worldwide basis if we continue to use fossil fuels, which release carbon dioxide (CO_2) when burned. The proportion of CO_2 in the atmosphere is critical because CO_2 acts as a heat trap. It allows solar energy to enter the earth system but prevents it from leaving, in much the same way that glass traps heat in a closed automobile on a sunny day. Plass estimated that man's use of carboniferous fuels has raised the CO_2 content of the atmosphere by about 13 percent over the last hundred years, and so far as we can tell, this increase has raised the average temperature of the earth by 0.56°C (1°F). We know nothing about possible threshold levels that might trigger serious changes in the environment. Should change in atmospheric composition and the resulting warming continue, serious environmental consequences may result.

There are several ways of avoiding increased warming. One of them is to stop releasing CO_2 by not using carboniferous fuels—and this may be the way we choose to solve this problem, for we have alternative sources such as atomic power, although that option presents its own problems. Another solution is to trap CO_2 as it is produced. Vegetation absorbs CO_2 from the atmosphere in the process of photosynthesis. One conceivable way to keep CO_2 under control is to revegetate areas to absorb more CO_2. Since the activities which produce carbon dioxide are highly agglomerated and plant growth is by nature extensive, industrial activities would have to be dispersed to prevent local concentrations of CO_2 pollution. In smaller amounts, plant life might more easily be able to absorb what is produced. Alternatively, after careful study we might feel that there would be no adverse effects of a warmer earth, that most of the effects would be positive. The earth, after all, has been tropical for most of its history and we might decide that the present temperate status is not optimal. In this case, geographers could provide advice on the location of CO_2 generating plants so as to distribute the gas as evenly as possible throughout the atmosphere in the same way that they could provide locational advice on how to minimize pollution if we decide we do not want the earth any warmer than it is. In either case, zoning of permissible land uses on a worldwide basis might be required.

We have already taken preliminary, faltering steps toward zoning certain portions of the earth's surface. As soon as artificial satellites were launched in the late 1950s, suggestions were made to restrict the use of outer space to peaceful purposes. These suggestions have not yet been formally adopted, but the basic idea of prohibiting certain activities in certain portions of the earth system is evident. Similar proposals were made concerning the sea bottom in the late 1960s. The technological capability of exploring and economically exploiting the sea bottom implies the possibility of using it for military purposes. Once again, proposals have been made to outlaw military uses of the sea bottom by international agreement, and here some progress has been made.

Some nations (particularly the United States) have got into the habit of using the sea bottom as a dump for very dangerous substances, on the theory that the oceans are pretty big and that it is unlikely that substances deposited there will ever harm anyone. Lethal military gases have been sunk in the ocean, as have atomic wastes. If the sea is to be useful to man, such arrogant practices must be stopped soon.

As the consumptive power of humanity grows both through increasing numbers and increasing affluence, we shall have to establish regional and worldwide quotas on certain kinds of production and exploitation. We have already been forced to do this for a few species. Whales, for example, are now somewhat protected by international agreements which place an upper limit on the number which can be captured each year. They are still treated as a fugitive resource, to be exploited by whoever can catch them first. As of yet we have not set up national quotas for whales. Somewhat similar agreements exist for some kinds of fish. International conventions also govern land use and ownership in Antarctica. In the future these kinds of restrictions will become increasingly common. Wherever productive or consumptive activity places strains on regional, national, or world abilities to produce resources or to absorb some of the by-products of productive and consumptive processes, some sort of regulation will be required.

A second general problem we could face over the next several decades is that of coordinating human activity over continents or over the entire earth. Mass and personal communications systems are knitting the people of the earth into a more tightly organized whole. A major problem which Ecumenopolis will have to face is that of different times. The fact that nations like the United States are divided into time zones is more than a minor inconvenience. There is now only a six-hour overlap in the normal business day (8:00 A.M. to 5:00 P.M.) on the east and west coasts of the conterminous United States. From this we must subtract the lunch hour at each place, since they do not coincide. This means that people wishing to conduct business between the east and west coasts have four effective hours to do so in the normal business day. The lack of overlap is even worse when we consider Hawaii and Alaska. There are only two overlapping business hours between Honolulu and Fairbanks, on the one hand, and New York and other east coast cities, on the other.

Obviously these differences are unacceptable if the world is to be organized into a single city. At some point in the future we shall have to adopt world standard time. A tremendous amount of impassioned oratory and visceral emotions have already been aroused by daylight saving time in many nations. It was only in 1968 that the federal government in the United States succeeded in forcing states to accept standard adoption dates for daylight time. Proposals to establish world standard time will arouse immensely greater concern than daylight time, and will raise some very serious problems as well. Most people prefer to sleep at night and work during the day. Indeed, man may be physiologically adapted to a daytime existence. This being the case, how will we decide who is to be upset by being forced to stay up all night if he wishes to interact with the rest of the family of man? Creating a single, unified mankind will have many advantages, but doing so will also require a great deal of coordination. Some changes in the very fundamental notions we have about how our lives should be structured in time and space will be one of the prices we shall have to pay for these advantages.

Long-Term Problems
(beyond the Year 2000)

It is difficult to specify the long-run problems we are likely to encounter in the next few centuries. It is legitimate to extrapolate present trends over the next fifteen years and even beyond 1985. Such extrapolations cannot be linear, of course, and they must be used carefully and with full awareness of their limitations, which are considerable. But as we try to foresee what is likely to happen more than three decades from now, we must increasingly enter the realms of forecasting and prediction. Here, the conceptual tools we could present are very limited in number, and the development of new techniques is proceeding at a very rapid pace so that any ideas we could present would soon be outdated.

A danger which we must always recognize in our attempts at forecasting and long-range planning is the possibility that we may overplan. By structuring the future too rigidly with long-term plans now, we can preclude the adoption of attractive but now unforeseeable options which might become available in the future. In the same way that we provide rockets with corrective steering mechanisms, we must build flexibility into our plans rather than shooting just once at distant goals. Our inability to describe the specific problems we might encounter does not prevent us from thinking about what the general nature of the overriding problems of the future will be.

The major problem in the decades after 2000 A.D., if it is not already the major problem of the present, will be accommodating change. No matter what realm of life we consider, the amount of change which occurs in a time period seems to be increasing at an increasing rate, and seems to have been doing so for a number of decades. Even if *rates* of change slow down somewhat, as it seems they must, changes will continue at a rapid rate. The agents which are producing change—science and the knowledge and information explosions—seem to be inherently accelerating activities. For our own well-being we must adopt the notion that change is a natural state of affairs and that static conditions are abnormal, but to do so will involve a radical restructuring of human thought.

We mentioned earlier the need to provide adequate resources for the world's people. Problems of inadequate resources are closely related to people's ideas about stability and traditional patterns of thought. What a person considers to be a resource depends on his culture and thus on the traditional values his society holds. Many areas of the world such as The Netherlands, Japan, and most metropolitan areas have little to offer in the way of naturally produced resources, yet they are wealthy and can thus purchase the natural resources they need. A major problem we now face is that of helping all areas of the world develop their cultural resources (economic, political, educational, social institutions, etc.) to the point that they can become wealthy enough to purchase or exploit the natural resources they cannot now use. In many cases, developing cultural resources on the scales required to achieve adequate or high levels of consumption will require wholesale revisions of traditional values. Cultural change is never an easy process, but all of us, developed nations as well as underdeveloped areas, must lay the foundations for more flexible attitudes toward cherished values in many realms of life if we wish to continue to enjoy high levels of living or to attain them if we have not yet achieved them.

Traditionally man has sought stability in both environmental conditions and his own behavior for very good reasons. In an unchanging or slowly changing world, conservatism is an optimal policy. Traditional practices yield the highest rewards because they have stood the test of time before being accepted. In a rapidly changing world, tradition is usually a liability because it is often not adapted to the new conditions in which it is invoked. Because the world has been relatively stable until very recently, man acted rationally when he clung to the policies and behavior of the past and used them to guide the decisions he had to make.

In a world which is changing as rapidly as ours, past behavior is often the worst guide to adopt. The spatial structures men establish when they are using past or even contemporary conditions as guides frequently become serious obstacles as conditions change and the assumptions which underlie them no longer hold. Most spatial systems man creates are based on assumptions about people's spatial behavioral patterns and the nature of space. These sooner or later become outmoded under the impact of new developments in movement technology. When we delimit the boundaries of a city, for example, we make some implicit assumptions about how large the community is likely to become in the future. Most of these assumptions have proved to be woefully erroneous in the past, with the result that many metropolitan areas in the world are underbounded. The actual settled area and functional community far exceed the boundaries of the central city. Because metropolitan areas are now composed of scores of small municipalities in addition to the central city, area-wide problems cannot be effectively attacked. Each political unit keeps its own interests foremost in its collective consciousness, to the detriment of area-wide considerations and to the long-range detriment of the entire metropolitan community.

Moreover, rather than making it easy for central cities to remedy such problems by annexing adjacent areas as they are settled, governmental structures make annexation exceedingly difficult if not wholly impossible. It is bad enough that many of our spatial, organizational decisions are made on the basis of inaccurate and shortsighted assumptions about future mobility and ranges of interaction. What is worse, however, is that many of our spatial systems preclude changes to accommodate new conditions. Hardly ever is the capacity for change inherent in a system or structure. This is best seen in government, where areal units are particularly sacred once they have been established, even though they may later become serious obstacles to the solutions of contemporary problems. But temporal lags are also evident in other areas of life. In the United States during the 1950s and 1960s the family farm—which was often inefficiently small and undercapitalized—was a sacred cow which could not be publicly criticized without serious repercussions.

The challenge which confronts us now and which will continue to challenge us is to design spatial structures which can adapt to changes because they are *designed to accommodate* change. It is true that our government structure contains provisions for change, but such provisions are ineffective. Changing the spatial structures of existing government systems always threatens the jobs of some of the politicians who must

make the final decisions about change. This being the case, politicians have a vested interest in not changing the system so long as they wish to hold their jobs. Consequently changes in the spatial structure of government are rare. A series of Supreme Court decisions was required to force most state legislatures in the United States to abide by the legal requirements for legislative redistricting contained in state constitutions. Minnesota, for example, had not redistricted since 1910, and did so in the 1960s only under Federal Court order, despite the fact that the state constitution required the legislature to be redistricted every ten years.

The dangers of maintaining obsolete spatial structures are evident in the present plight of metropolitan areas. Because state legislatures refused to redistrict after 1900, or even earlier in some cases, central city populations were never adequately represented in most state legislatures. Large-scale agglomeration of population into cities between 1880 and 1960 left rural legislative districts progressively overrepresented and urban areas increasingly underrepresented in state legislatures. By 1965 it was not at all uncommon for a legislator from an urban district to represent twenty times the number of people that elected a legislator from a depopulated rural district. Yet both legislators had an equal vote and voice in deciding state policy. As a consequence of such imbalances, the problems of central cities received very little attention at the state level over the last six decades.

To compound the tragedy, the redistricting carried out in the mid 1960s does not promise to alleviate urban problems to the degree one might think. Adhering closely to the one man-one vote principle, redistricting plans correctly give most metropolitan legislative votes to suburban areas, where the bulk of the populations of many metropolitan areas now reside. While this is morally and politically commendable, it does not help alleviate the plight of central cities. The suburban legislator's first loyalty is to his suburban electorate, and not to the central city where the problems of a metropolitan area are most critical. Thus the central cities of metropolitan areas have never been effectively represented in the state legislatures of the United States. Inflexible spatial structures and vested interests operated in concert virtually to disenfranchise central cities, and present philosophy makes it unlikely that they will acquire greater political muscle in the future. How much this lack of representation has contributed to the existence and persistence of present urban problems is difficult to say, but we think it probably has made a significant contribution. Had central cities been adequately represented during the last six to eight decades when present urban problems were in embryonic stages of development, it is possible that many of them would not be as serious as they are now.

The problems produced or aggravated by obsolete spatial structures are going to become immensely more serious as we move from metropolitan to megalopolitan scales of urban organization. Devising solutions to the problems of any metropolitan area unfortunate enough to lie astride a state boundary has usually been much more difficult than the problem of dealing with a metropolis located in a single state. Yet another independent unit of government became involved, and yet another recalcitrant legislature had to be convinced that urban problems were worth worrying about. As we move toward megalopolitan urban organization on the scales of Boswash and Chipitts, state boundaries—which are much more sacred than legislative boundaries were before 1965—may become ever more serious obstacles to the solution of megalopolitan problems. Our constitution gives states the power to subdivide, and new states can be formed from portions of two or more existing states. But such changes require the consent of all the state legislatures involved, and given the vested interests in the existing spatial structure of state governments, change along these lines is virtually out of the question.

What we need in advanced, mobile nations like the United States is an entirely new philosophy of governmental spatial organization which has a built-in capacity for changes in areal units—an inherently adaptive spatial structure. One way we might organize the political space of the nation in the future is suggested by the map of third order nodal regions (Fig. 15-3). We could abolish all existing states and set up twenty-three new govern-

mental units in accordance with the twenty-three existing functional regions. To most of you this will sound preposterous, but let us consider some of the advantages of organizing political systems on the basis of actual spatial behavior. An underlying assumption in establishing areal governmental units is that a community of interest exists or will exist among the people within the unit, and that this common interest is stronger than attachments to places outside the region for most of its inhabitants. While such a description may have been applicable to existing states at one time, this is certainly no longer the case. Maps of nodal regions reflect existing communities of interest, not only in the spatio-economic, but often in the cultural sphere of life as well. Obviously, these communities of dominant spatial behavior cut across state lines in many cases and include several states in others.

Pennsylvania, for example, is split into four major regions by the four metropolitan areas whose hinterlands extend into the state. Western and Central Pennsylvania focus on Pittsburgh. The north is tributary to Buffalo with the exception of the Erie corner, which could just as well belong to Ohio so far as most Pennsylvanians (Erie residents included) are concerned. The east focuses on Philadelphia. By contrast, several entire states and parts of states focus on Denver. Peoples' foci for goods, services, and information strongly reflect their primary loyalties. In most parts of the nation, nodal trade regions form far more coherent areal units and communities of shared interests than do states, which retain what force they have as cohesive units of interest largely through inertia and historical precedent.

The principle that governmental regions should be based on actual, changing communities of interest could be applied at lower political levels also. We could replace existing county governments in many areas where counties are small with areal units focused on powerful trade centers. Often these regions would be composed of several existing counties. Given modern transportation and communications technology, there is no reason why this enlargement should cause any particular difficulties. Law enforcement agencies, library services,

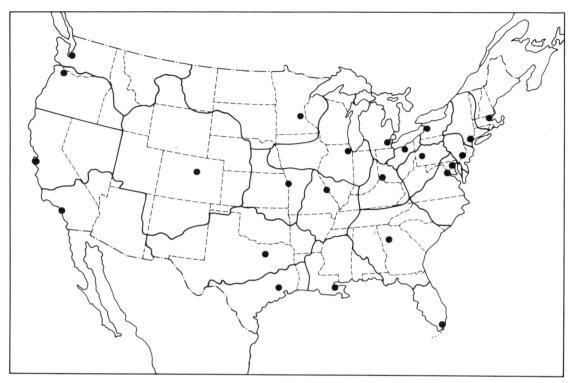

FIGURE 15–3

Third-order nodal regions in the United States in 1960 (after unpublished work by Borchert).

and citizens themselves are all immensely more mobile now than they were in 1900 or 1850. Yet in many states changes in county boundaries since then have been negligible. Unlike systems which operate in the private sphere and must rationalize their structures to accord with changing technology or accept smaller than maximum profits, there is no compelling incentive for governmental units to take advantage of improvements in space-adjusting techniques and time-space convergence. Most areal units of government at any scale we wish to examine reflect assumptions about movement technologies and the sizes of communities of interest which were valid from fifty to over a thousand years ago.

If adequate government is to be provided, provisions for change must be incorporated into the basic design of the system. Censuses will be

technologically obsolete in a couple of decades, since we will soon be recording census data on a continuing basis. We will probably continue to have a decennial "census" because the requirement that we do so is written into our Constitution. But with or without censuses, every ten years would be a reasonable interval between adjustments of the boundaries of local, regional, and "state" governments to reflect whatever changes in communities of interest have occurred in the previous decade. If boundaries of functional regions had remained stable or nearly so during the period, no changes would be required. If, on the other hand, one growing center had been expanding its trade area and influence at the expense of a declining center, governmental boundaries would be adjusted to reflect the fact that some locations had shifted their allegience to a new center. Such a system should be designed to be as flexible as possible. When Boswash, Chipitts, and Sansan reach true megalopolitan stature, for example, our governmental structure should contain provisions which will allow them to erect megalopolitan governmental units if they are required. Changing boundaries might create some problems, but with more standardized governmental units these should not be serious. The populations of states are not constant in any case. Each state has a high "turnover" of population because of the mobility which is characteristic of American life.

Some of the ways such problems of delimiting boundaries and choosing centers could be handled were reviewed in Chapter 14 when we discussed the spatial structure of administrative functions in southern Ontario. There are certainly significant methodological and theoretical problems which geographers would have to solve if we were to adopt such a self-organizing system of government. But the major problems are not technical; they are philosophical and educational. Most people are not comfortable with the pace at which the world is changing and with change itself. Their constant referent in evaluating present policy and its implications for the future is the past. So long as people continue to concentrate on a past which has been filtered through their memories, inherently adaptive spatial structures cannot be established. Spatial structures with the capacity for change and evolution built into them are antithetical to traditional definitions of "stability," with its emphasis on wholesale carryovers of ways of life and patterns of behavior from one generation to the next.

In the specific realm of government, there are other possibilities which can be considered for the long-term future. Somewhere between the attainment of megalopolitan levels of organization and Ecumenopolis, nations like the United States will reach the nationopolis stage, the era of the national functional city. Will new intermediate levels of government be appropriate for this period of technological development? If time- and cost-space convergence succeed in creating a national city, why not run the nation as a city? It will be possible, for example, to run the nation-city of the future as a pure democracy if we wish to do so. When decisions are to be made, citizens could vote electronically on each important issue after sufficient time has been allotted for experts to present the advantages and disadvantages of adopting or rejecting some proposed policy or course of action. Frequent referenda on major issues and a pure democratic form of government would create considerable problems of their own. But might not such a form of national government be appropriate to the middle of the twenty-first century when the average citizen will probably have an education equivalent to today's Ph.D.? Our experience with department staff meetings leads us to think that actually agreeing on any course of action in such a system would be difficult, if not impossible. But such alternatives, unrealistic as they may appear to us, are worth thinking about and conceivably worth advocating. To the extent that we fail to think imaginatively now about possible and probable future consequences of today's decisions, our children and grandchildren will be cursed with intractable, inflexible spatial and social structures similar to those which now plague us. Perhaps whenever we make group decisions about matters with long-term implications, we should include an individual with voting power to represent the generations of the future who are not yet here to advocate their own interests.

"Stability" will have to be redefined in our minds so that it is equated with changing conditions. In the foreseeable future, change in all realms of life will be the normal state of affairs, and stagnation will be abnormal. This being the case, traditional notions must be reversed and we shall have to accommodate ourselves to the idea that change is natural and desirable. Once our society has crossed this intellectual threshold, we will have surmounted the major obstacle to designing inherently evolutionary spatial structures.

It seems fairly obvious to us that time-, cost-, and N-space convergence are going to produce an "everywhere" world sometime within the next century. Whether people choose to follow the dominant trend and agglomerate in the megalopolitan areas of Ecumenopolis, or whether they choose to utilize the technological capabilities of tomorrow's communications facilities to their maximum extent and enjoy the amenities of low density isolation in absolute space will not make too much difference. In relative space, everyone who wants to participate in the Ecumenopolitan system will have access to do so through the communications and transportation systems we shall have at our command. In a few decades, intercontinental weekend jaunts will become as common as weekend trips to adjacent metropolitan areas or weekend homes are today, and worldwide electronic communications behavior will become identical to contemporary intrametropolitan intercommunication.

In our attempts to anticipate what the spatial structure of the future will be like, it is important that we pay close attention to existing metropolitan areas and the behavior of people within them. Today's cities are spatial models of tomorrow's world, spatial laboratories wherein we will perfect the space-adjusting techniques which will be applied to more extensive areas to produce the Ecumenic city. Cities and urban areas are the frontiers of spatial organization.

It is precisely because the city is the spatial model of the future and our spatial laboratory that today's urban problems are so critical and deserving of attention. In the same way that man's ability to organize space in the city is an index of his ability to organize the surface of the earth in the future, his ability to solve the social and physical problems which now afflict the world's cities is an index of his ability to solve or prevent the Ecumenic social problems of the future. There is much we can learn about building Ecumenopolis from existing metropolitan areas.

In the final analysis, we have no choice but to learn these lessons well if horrendous human dislocations are to be avoided. Urban, metropolitan, and megalopolitan areas are land use techniques. They are human artifacts produced by our technology. Whether we like to admit it or not, man is a prisoner of the technologies he has created. The metropolis and megalopolis are the most successful kinds of land use man has yet devised. They are spatial technologies which can support more people, at higher levels of living, than any other alternative. Like all species, man tends to expand to the carrying capacity of the ecological niche he occupies. Man is somewhat unique in that he can create his own niches, but the basic principle remains the same. Having created megalopolitan niches, we have expanded our numbers to take advantage of the increased carrying capacity they provide, and now there is no turning back. If, for some reason, metropolitan and megalopolitan areas suddenly became uninhabitable, there would be no alternative for almost half of humanity but to die. There is no alternative land use technology which could support the hundreds of millions of people now living in cities and urban regions. And given the prospect of a doubling of population by the year 2000, there is no way to go but forward with the ecumenopolization of the world. Even Zero Population Growth, a social goal which is rapidly becoming popular in many circles today, would not seriously retard the development of Ecumenopolis. As we noted earlier, it is our communication and transportation technologies which are taking us toward a single, functional world community. The evolution of what Marshall McLuhan has described as the "global village" depends only slightly upon the number of people which will inhabit it.

THE FUTURE OF GEOGRAPHY

Today's geography is, as we saw in Chapter 3, an outgrowth of the geographies of the past. Similarly, the geography of the future will be built on the foundations of the theory and substantive knowledge we develop today. Geographers—like all other scientists—tend to interpret experience in terms of the dominant themes of the societies in which they live. This human habit of viewing and interpreting the world through the lens of the dominant ethos of each era makes it possible to explain past and present geographies, and enables us to make an educated guess about the kind of geography which will evolve in the next several decades.

The Persistent Spatial Organization Theme

When Western man broke away from his previously sedentary existence, both in the ancient world and then later in the era of European exploration and discovery, the highest priority was given to locating spatial data on maps. As peoples' activity spaces and information fields expanded to embrace new worlds, the amount of spatial data to be organized and recorded grew rapidly. Until A.D. 1900, most geographers were primarily concerned with answering questions like "where?" and "what is where?"

That people asked such questions presupposes that they had some contact with distant places. But even though the modern mobility revolution had begun in the West in the fifteenth century, most men in most places could not actually visit distant places until the land transportation barrier was broken by the diffusion of railroads in the latter half of the nineteenth century. By the beginning of the twentieth century Western man had attained a very high degree of mobility. Even though a great deal of isolation still existed by today's standards, most of the places people in advanced nations wanted to visit could be reached with a speed and convenience unthinkable a century earlier.

In the same way that communications media fascinate us today, nineteenth-century thinkers were enthralled with the transportation systems their century created. A book like Jules Verne's *Around the World in Eighty Days* could only have been written in the late nineteenth century. Similarly, the geographical literature of the late nineteenth and early twentieth centuries devoted considerable attention to transportation and its consequences. Probably the best indication of the ways transportation provided satisfactory explanations of human spatial activity is found in Friedrich Ratzel's works.

Ratzel's greatest contribution to geography is the concept of relative location. The notion that the locations of places are constantly changed by transportation and communication media permeates his works. By becoming increasingly mobile, man greatly increased his capacity to change the centrality of places. What existed or occurred at a given place was no longer determined by site factors, but rather by the actions of people. The Suez Canal changed the eastern Mediterranean from an isolated backwater into one of the world's most critical nodes; railroads created boom conditions in places they connected and condemned those

they bypassed to isolation and stagnation. By the end of the nineteenth century, the human geography of the earth was largely under control and attempts to explain areal variation in human activities over the earth's surface took this fact into account. Geographers explained the locations of things by referring to culturally created places, and they explained the creation of places by referring to social, political, and (especially) economic principles of behavior, all of which operated through the medium of transportation.

As the world grew more closely knit during the first half of the twentieth century, such explanations continued to provide satisfactory answers to the questions geographers asked. The dominant theme in geography for the last seventy years has been the description and explanation of spatial organization. Important phenomena (often economic, but political, social, and cultural as well), were observed to be distributed differentially in space, and the availability or absence of transportation media and the effect they had were the explanatory variables cited to explain areal variation. As a single, functional world system developed in the first half of the twentieth century, more and more attention was devoted to spatial interaction—the dynamic dual of the spatial organization of apparently static distributions—but the explanatory variables remained the same. Culturally created locations and mobility processes enable us to account for the areal variety we experience.

We think geography is now on the threshold of a shift in explanations, a shift which will not alter the geographer's traditional concern with spatial structure and process, but a shift to a new set of explanatory variables more in harmony with the ethos of society and technology in the last quarter of the twentieth century. Ratzel could not formulate the notion of relative location and call attention to its importance until the transportation revolutions that created relative locations were in full swing. Similarly, contemporary geographers could not formulate a geography of spatial organization and interaction until further progress had made the world (or at least large portions of it) a single interacting system. Transportation and relative location have dominated our thinking and explanations up to now because they were indeed some of the most important elements of social, economic, and political life. An economy and society which gains its livelihood by producing goods and services must possess integrated transportation systems. Transportation is literally the lifeblood of the society engaged in secondary or tertiary activities.

But advanced nations are now reaching post-industrial stages of economic development, and *quaternary* activities and enterprises are the growth industries of today and tomorrow. Quaternary societies are organized around flows of information and ideas. Increasing numbers of people are engaged in the generation, transmission, processing, and dissemination of information and knowledge of one kind or another. And because of the ease with which information and ideas can be transmitted, relative location is now becoming less important as an explanatory variable.

Up to now, we explained the locations of things by constructing explanatory narratives which concluded that the phenomena in question had to be where they were (or at least it was better for them to be there) in order for men to have access to some natural or cultural resource or set of resources. When differences in accessibility to such resources existed, some places were good places to do particular things and others were not. Places determined the locations of activities by virtue of their locations within interaction systems. Such explanations are not confined to the economic realm of human activity. When we explain cultural variations, we tend to look at place and location as determinants of the quality and quantity of information received, and therefore of cultural, social, and political organization.

In the last quarter of the twentieth century, information will be the primary resource to which producers must have access. Data and ideas are rapidly becoming the lifeblood of economic, social, and political activities, and as our communications technologies continue to make all information actually or potentially ubiquitous, the traditional relationship between location and information is being reversed. Whereas the location of people once determined the information they received, information now

determines the places where people choose to be. People are now able to choose among a larger and larger number of places, and often their choice is constrained only by those possibilities of which they have knowledge. A person who wishes to work as a craftsman in a steel mill has a limited number of locational options. But someone who wishes to be involved in information processing can do so at an immensely larger array of places. A business executive, for example, can and does change locations and jobs much more easily than the steelworker.

The isotropic information surfaces which modern communications technologies are rapidly creating reduce the importance of relative location as an explanatory variable. Explanations of continued migration to California cast in terms of relative location are not very satisfying. People go to California because they *want* to. The fact that they can make a good living there makes migration possible, but such economic considerations are only necessary conditions of migration; they are not sufficient. Peoples' *desires* for a more pleasant physical milieu and a more hedonistic social environment are more to the point. California is, after all, rather poorly located in terms of accessibility to the rest of the national system.

By producing complete time- and cost-space convergence and thus annihilating relative location with respect to information, communications technologies will set men free to give aesthetic and emotional criteria primary consideration in making spatial and locational choices. People will be able to live where they wish for as long as they wish. Whatever the form and density of Ecumenopolis in absolute space, it will be a single place in relative, functional space, and for that reason many if not most people will enjoy complete mobility within it.

The geography of the future is the geography of human choice, for human preference for experience will create the spatial organization and spatial interaction of the future. Geographers will continue to be vitally concerned with the organization of structures and processes in space, but the explanatory variables involved will change considerably. Greater concern with amenity environments, with the qualities of locations and regions as places to live, will produce a more environmental, more regional geography than exists now.

Practicing a geography in which human preference plays a dominant role will be extremely complicated and challenging, far more difficult than today's or earlier geographies. The complexity of a world organized in response to traditional locational principles is, indeed, considerable. But the intricacy of a world in which people are freed from economic locational constraints and the tyranny of distance will make today's spatial organization appear childishly simple. Giving human preference free reign will produce a world with greater areal differentiation than has ever existed before.

The fact that choice will become a dominant theme in the geography of the future does not mean that those of us with more contemporary viewpoints and models of the world will be unemployed. A necessary condition of Ecumenopolis (or Ecumenhedonopolis) will be the existence of highly rational and efficient communication, transportation, and other life support systems. Those who prefer to investigate the spatial implications of traditional social, economic, and political principles will find plenty to keep them busy. Designing a least-cost communications or transportation network to serve a world in which existing constraints on people's locational behavior have been greatly relaxed or completely eliminated will require very sophisticated analytical, predictive, and creative skills. As consumers of goods, services, and places, the inhabitants of Ecumenopolis may live in an isotropic world. As producers, they will confront the fact that rationally organized space-adjusting systems are needed to create isotropy. Within any social science like geography, then, more diversity of philosophy, method, and problem application will exist in the coming decades than now exists.

The Geography of the Future

The diversity of geography will increase, but the fundamental core of locational questions which have given unity to the discipline throughout 2,000 years will continue to be asked in the future. The basic questions about where things

are and why things are where they are will still be at the core of geographical inquiry as we continue our third millenium as a distinct science. Questions will shift from those about where things are or where things have been to a greater concern with where things *can* be and where things *should* be, but the fundamental concern with explanations of location will persist. Geography will continue to be an explanatory, predictive, and prescriptive science, concerned with the location of things and people in space. Even complete N-space convergence will not eliminate man's passionate concern with the spatial and locational dimensions of his existence, nor will it eliminate the spatial diversity which provokes that concern.

Thus geography will grow out of its past, but this growth will be much more than an extension of its previous work. All disciplines build upon their pasts, but they also change direction and emphasis. Geography, too, must develop beyond what previous generations have done. Geographers will increasingly be expected to practice preventive spatial medicine as we move towards Ecumenic levels of spatial organization. As the single world city continues to evolve and grow, its spatial structure will require continual manipulation to alleviate problems which are arising and to prevent others from developing. Systems as complex and vital as megalopolitan regions and Ecumenopolis must constantly be monitored to detect disturbances the instant they appear or, preferably, in advance of their appearance. Just as cancer is easier to cure if it is detected almost as soon as it starts, the spatial problems of Ecumenopolis will be much easier to remedy if they are caught early.

Controlling an advanced society is in many ways analogous to driving a powerful automobile. Up to now, it seems, we have been content to drive blindly into the future at full throttle. We try to repair the consequences of the inevitable accidents only after they have occurred, which is, of course, too late. We are now coming to realize that it would be much less expensive and traumatic in both the short and long run to start driving defensively and thus avoid social, political, and economic "accidents." By specifying the alternative futures which may result from acting or not acting on current opportunities, geographers can help their societies avoid some of the more serious problems which will surely develop in the absence of such a defensive approach to the future.

Geography must become a more future-oriented science than it has been in the past. What has happened in the past is certainly important, for the past is what we can explain, and the spatial processes and structures we can explain are those about which we can make predictions. But increasingly the emphasis in geography will be on explanation of past and present experiences as a key to the future, not as an end in itself. We do not intend to criticize the present focus on explanation of past and present, for it is necessary and productive. Our point is that if the world doubles and redoubles its numbers, and if man's spatial artifacts grow immensely more complex than they are now, we shall have no choice other than to study and structure the future more than we do now.

We mentioned earlier some of the dangers of *over*structuring the future. The giant reptiles of the Pennsylvanian geological era were too rigid to adapt to rapidly changing conditions, and consequently they perished. It is for this reason that we are wary of long-term plans and planning, and even more chary of the involvement of geographers in this process. Our own predilection is to try to keep geography focused on what *can* be, on the range of possible spatial futures which can be built upon the foundations of the present. We cannot avoid thinking about what *should* be, of course, for when we have laid out an array of possible futures which are the consequences of present spatial alternatives, a choice among them must still be made. But we feel that this choice should not be made by geographers, by architects, by professional planners, or by any other special interest group. Ultimate choices among possible futures should be made by an informed electorate or their representatives, after a variety of options have been presented, criticized, and evaluated.

Building this kind of self-adjusting capacity into our political, economic, and social decision-making processes will not be easy. It will require that geographers and other social scientists

convince people that the future is worth thinking about, and to do that we must eliminate our existing bias toward the past. This, we believe, is the great challenge of the next three decades for the social sciences. Geographers must help produce spatial structures and systems which are not technology-specific, but which are inherently adaptive and flexible. Decision processes must be so designed that we can walk the thin line between spatial overplanning and spatial chaos. There is no greater challenge and no more worthy task than making spatial and locational change a natural, stress-free process.

For the science of spatial organization, the prospects over the next several decades are thus very exciting. Our ideas about spatial events in the world will assume new importance because of the rapidity of change. We usually talk about theory and reality as if they were separable, but it is increasingly apparent that theories create reality. The things which we recognize as existing—actually or potentially—are those things which fit into the theoretical frameworks we have created. Price indices and unemployment data are parts of our economic reality because we have created theoretical structures which require their existence. Before they became theoretically important, price indices and a host of other economic indicators did not exist. Because they *do* exist now, we can formulate alternate futures and specify precisely their widely different economic characteristics.

Similarly, a spatial hierarchy of central places did not exist for any practical purposes until Walter Christaller created a theoretical framework based on his observations of such a hierarchy. Because they exist within central place theory, thresholds, ranges of goods and services, and settlement lattices are now a part of the geographer's reality. The existence of such spatial concepts enables us to incorporate these elements of reality into our thinking and to formulate alternative futures with different settlement plans and systems of spatial organization.

Whatever else geography may be or become in the future, it must continue to be theoretical and geographers must produce better theory than they have to date. Cultural geographers will have to devise new and more sophisticated theories about cultural behavior in space, political geographers will have to formulate more advanced theories of political spatial behavior, and those interested in the geography of human preference will have to develop theories concerning spatial and locational regularities where few now exist. Once formulated and tested, such theories will create new dimensions of reality which will enable us to construct more comprehensive alternate futures for our societies.

For over 2,000 years before man had much of the earth and many of its processes under control, spatial theory had to be adjusted to reality in order to resolve conflicts between the two. Now, because we control much of our ecosystem, and because the part we do not yet directly control is shrinking rapidly, resolving differences between theory and reality is no longer a one-way street. The various theories of human spatial behavior we are building are as much alternate versions of the world of tomorrow as they are mirrors of the way the world is today. The increasing congruence of theory and reality will certainly be partly attributable to continued advances in theory, but it will also be attributable to a growing tendency to manipulate reality to make it more accordant with our theories and moral judgments concerning how the world should be organized. Spatial theory and spatial reality will become more nearly identical not by the adjustment of one to the other, but by a circular process of convergence. By building spatial theory today, we are helping design the spatial organization of the world of tomorrow. Our theories are our futures, and it behooves us to build them carefully, and above all, compassionately.

We suppose every generation feels that its short sojourn on earth is the most important period in human history, and we are no exception. Whether or not we are being temporocentric, we are convinced that the latter half of the twentieth century is and will continue to be a very exciting time to be alive. In our lifetimes, for the first time in several million years, man —not as an individual, a tribe, a kingdom, or a nation—but man as a *species* has acquired the power to destroy himself. Simultaneously he has acquired the potential to enrich the existence of his species to an unparalleled degree. The next thirty years will be filled with hazards, but they will also be filled with opportunities. In most games of chance, the higher the risks, the larger are the rewards of success or the penalties for failure. If we fail to build successful Ecumenic systems in the future, we will fail abysmally. If we build on the present wisely and soundly, however, we can expect future benefits far out of proportion to present costs. We shall be successful if we are guided by the knowledge that the present is the beginning of the future, not the end of the past.

Suggestions for Further Reading

BERRY, BRIAN J. L. "The Geography of the United States in the Year 2000," *Ekistics*, XXIX: 174 (May 1970), 339–51.

BOULDING, KENNETH E. *The Meaning of the Twentieth Century*. New York: Harper & Row, Publishers, 1967.

BRIGHT, JAMES R., ed. *Technological Forecasting for Industry and Government*. Englewood Cliffs, N.J.: Prentice-Hall, Inc., 1968.

CALHOUN, JOHN B. "Space and the Strategy of Life," *Ekistics*, XXIX:175 (1970), 425–37.

COCHRAN, THOMAS C. "The Social Impact of the Railway," in Bruce Mazlish, ed., *The Railroad and the Space Program: An Exploration in Historical Analogy*. Cambridge, Mass.: The M. I. T. Press, 1965.

The Futurist. Published six times a year by the World Future Society, P. O. Box 19285, Twentieth Street Station, Washington, D.C. 20036.

GABOR, DENNIS. *Inventing the Future*. Baltimore: Penguin Books, Inc., 1963.

HALL, EDWARD T. *The Hidden Dimension*. Garden City, N.Y.: Doubleday & Company, Inc., 1966.

DE JOUVENEL, BERTRAND. *The Art of Conjecture*. New York: Basic Books, Inc., 1967.

KAHN, HERMAN, and ANTHONY J. WIENER. *The Year 2000: A Framework for Speculation on the Next Thirty Years*. New York: The Macmillan Company, 1967.

McHALE, JOHN. *The Future of the Future*. New York: George Braziller, Inc., 1969.

McLUHAN, MARSHALL. *Understanding Media: The Extensions of Man*. New York: McGraw-Hill Book Company, 1964.

PLATT, JOHN R. *The Step to Man.* New York: John Wiley & Sons, Inc., 1966.

———. "What We Must Do," *Science* (November 28, 1969), 1115–21.

TÖRNQVIST, GUNNAR. *Flows of Information and the Location of Economic Activities,* Lund Studies in Geography, Series B, No. 30. Lund: Gleerup, 1968.

"Toward the Year 2000: Work in Progress," *Daedalus,* XCVI: 3 (Summer 1967).

Works Cited or Mentioned

CHRISTY, FRANCIS T., Jr., and HENRY HERFINDAHL. *A Hypothetical Division of the Sea Floor.* Map published by the Law of the Sea Institute, 1968.

DOXIADIS, CONSTANTINOS. *Ekistics: An Introduction to the Science of Human Settlements.* New York: Oxford University Press, Inc., 1968.

GOTTMANN, JEAN. *Megalopolis: The Urbanized Northeastern Seaboard of the United States.* Cambridge, Mass.: The M.I.T. Press, 1964.

PLASS, GILBERT N. "Carbon Dioxide and Climate," *Scientific American* (July 1959).

RATZEL, FRIEDRICH. *Anthropogeographie,* 2 vols. Stuttgart: Engelhorns, 1921 and 1922.

———. *Politische Geographie,* 3rd ed. Munich: Oldenbourg, 1923.

VERNE, JULES. *Around the World in Eighty Days.* New York: The Macmillan Company, 1966.

INDEX